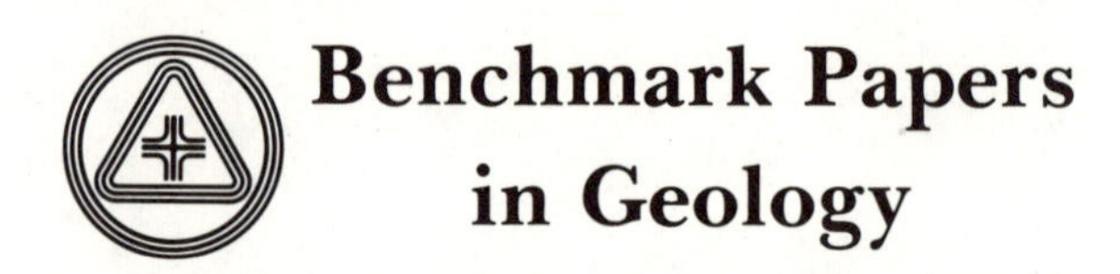

Series Editor: Rhodes W. Fairbridge
Columbia University

Published Volumes and Volumes in Preparation

ENVIRONMENTAL GEOMORPHOLOGY AND LANDSCAPE CONSERVATION, VOLUME I: Prior to 1900 / Donald R. Coates
RIVER MORPHOLOGY / Stanley A. Schumm
SPITS AND BARS / Maurice L. Schwartz
TEKTITES / Virgil E. and Mildred A. Barnes
GEOCHRONOLOGY: Radiometric Dating of Rocks and Minerals / C. T. Harper
SLOPE MORPHOLOGY / Stanley A. Schumm and M. Paul Mosley
MARINE EVAPORITES: Origin, Diagenesis, and Geochemistry / Douglas W. Kirkland and Robert Evans
ENVIRONMENTAL GEOMORPHOLOGY AND LANDSCAPE CONSERVATION, VOLUME III: Non-Urban / Donald R. Coates
GLACIAL ISOSTASY / John T. Andrews
WEATHERING AND SAPROLITE / Dan Yaalon
EARTHQUAKES: Geological Effects / Clarence R. Allen
ROCK CLEAVAGE / Dennis S. Wood
ENVIRONMENTAL GEOMORPHOLOGY AND LANDSCAPE CONSERVATION, VOLUME II: Urban / Donald R. Coates
BARRIER ISLANDS / Maurice L. Schwartz
VERTEBRATE PALEONTOLOGY / J. T. Gregory
PALEOMAGNETISM / Kenneth Creer
PALEOECOLOGY / John Chronic, Jr.
SOLID STATE GEOPHYSICS / Thomas J. Ahrens
OCEANOGRAPHY / Takashi Ichiye
PERMAFROST / J. D. Ives
PLAYAS AND CLAY-PANS / James L. Neal
GEOCHEMISTRY OF SULFUR / Isaac Kaplan
MODERN CARBONATES / Gerald Friedman
PHILOSOPHY OF GEOLOGY / Claude C. Albritton

A *BENCHMARK* ™ Books Series

SLOPE MORPHOLOGY

Edited by
STANLEY A. SCHUMM and M. PAUL MOSLEY
Colorado State University
University of Sheffield, U.K.

Library of Congress Catalog Card Number: 72–95135
ISBN: 87933–024–4

Library of Congress Cataloging in Publication Data

Schumm, Stanley Alfred, 1927- comp.
Slope morphology.

(Benchmark papers in geology)
1. Slopes (Physical geography)--Addresses, essays, lectures. I. Mosley, M. Paul, joint comp. II. Title.
GB406.S39 551.4'3 72-95135
ISBN 0-87933-024-4

Manufactured in the United States of America.

Exclusive distributor outside the United States and Canada:
John Wiley & Sons, Inc.

Acknowledgments and Permissions

ACKNOWLEDGMENTS

American Association for the Advancement of Science—*Science*
"The Convex Profile of Bad-land Divides"

American Geophysical Union—*Transactions, American Geophysical Union*
"Factors Affecting Sheet and Rill Erosion"

American Geophysical Union—*Journal of Geophysical Research*
"Seasonal Variation of Infiltration Capacity and Runoff on Hillslopes in Western Colorado"

Cambridge University Press—*Geological Magazine*
"On the Disintegration of a Chalk Cliff"

The Geological Society of America—*Geological Society of America Bulletin*
"Mathematical Models of Slope Development"
"Intravalley Variation in Slope Angles Related to Microclimate and Erosional Environment"

Geological Society of Australia—*Journal of the Geological Society of Australia*
"Some Problems of Slope Development"

Geological Society of London—*Quarterly Journal of the Geological Society*
"Note on the Movement of Scree-Material"
"Second Note on the Movement of Scree-Material"

Geological Society of South Africa—*Transactions, Geological Society of South Africa*
"Slope Form and Development in the Interior of Natal, South Africa"

Imprimerie Nationale—*Les Formes du terrain*
"Du Façonnement des versants"

Indiana Academy of Science—*Indiana Academy of Science, Proceedings*
"The Misleading Antithesis of Penckian and Davisian Concepts of Slope Retreat in Waning Development"

Royal Geographical Society, London—*Geographical Journal*
"The Nodal Position and Anomalous Character of Slope Studies in Geomorphological Research"

U.S. Government Printing Office—*Geology of the Henry Mountains* (U.S. Geographical and Geological Survey of the Rocky Mountain Region)
"Land Sculpture"

U.S. Government Printing Office—*U.S. Geological Survey Bulletin 730–B*
"Erosion and Sedimentation in the Papago Country, Arizona"

Vanderhoeck and Ruprecht, Gottingen—*Nachrichten Akademie der Wissenschaften in Gottingen, Mathematisch-Physikalische Klasse*
"Importance of Soil Erosion for the Evolution of Slopes in Poland"
"The Debris Slides at Ulvådal, Western Norway—An Example of Catastrophic Slope Processes in Scandinavia"

PERMISSIONS

The following papers have been reprinted with the permission of the authors and the copyright owners.

American Association for the Advancement of Science—*Science*
  "Rates of Surficial Rock Creep on Hillslopes in Western Colorado"

American Society of Agricultural Engineers—*Transactions of the American Society of Agricultural Engineers*
  "Effect of Slope Shape on Erosion and Runoff"

American Society of Agricultural Engineers—*Agricultural Engineering*
  "Erosion Equations Predict Land Slope Development"

Edinburgh Geological Society—*Transactions, Edinburgh Geological Society*
  "The Uniformitarian Nature of Hillslopes"

Gebruder Borntraeger—*Zeitschrift für Geomorphologie*
  "A Comparison of Theoretical Slope Models with Slopes in the Field"
  "Characteristic and Limiting Slope Angles"

Institute of British Geographers—*Institute of British Geographers Special Publication 3*
  "An Application of the Concept of Threshold Slopes to the Laramie Mountains, Wyoming"

Institute of British Geographers—*Institute of British Geographers Transactions*
  "Some Observations on Slope Development in South Wales"
  "Contrasts in the Form and Evolution of Hill-side Slopes in Central Cyprus"

Institution of Civil Engineers—*Geotechnique*
  "Stability of Steep Slopes on Hard Unweathered Rock"

Macmillan (Journals) Ltd.—*Nature*
  "Soil Movement by Denudational Processes on Slopes"

Oxford University Press—*Journal of Soil Science*
  "The Description of Relief in Field Studies of Soils"

The University of Chicago Press—*Journal of Geology*
  "Measurement and Theory of Soil Creep"

Yale University—*American Journal of Science*
  "Equilibrium Theory of Erosional Slopes Approached by Frequency Distribution Analysis"
  "Cliff Retreat in the Southwestern United States"
  The Role of Creep and Rainwash on the Retreat of Badland Slopes"

Series Editor's Preface

The philosophy behind the Benchmark Papers in Geology is one of collection, sifting, and rediffusion. Scientific literature today is so vast, so dispersed, and, in the case of old papers, so inaccessible for readers not in the immediate neighborhood of major libraries that much valuable information has become ignored—by default. It has become just too difficult or too time consuming to search out the key papers in any basic area of research, and one can hardly blame a busy man for skimping on some of his "homework."

This series of volumes has been devised, therefore, to make a practical contribution to this critical problem. The geologist, perhaps even more than any other type of scientist, often suffers from twin difficulties—isolation from central library resources and immensely diffused sources of material. New colleges and industrial libraries simply cannot afford to purchase complete runs of all the world's earth science literature. Specialists simply cannot locate reprints or copies of all their principal reference materials. So it is that we are now making a concerted effort to gather into single volumes the critical material needed to reconstruct the background of any and every major topic of our discipline.

We are interpreting "geology" in its broadest sense: the fundamental science of planet Earth, its materials, its history, and its dynamics. Because of training and experience in "earthy" materials, we also take in astrogeology, the corresponding aspect of the planetary sciences. Besides the classical core disciplines such as mineralogy, petrology, structure, geomorphology, paleontology, and stratigraphy, we embrace the newer fields of geophysics and geochemistry, applied also to oceanography, geochronology, and paleoecology. We recognize the work of the mining geologists, the petroleum geologists, the hydrologists, and the engineering and environmental geologists. Each specialist needs his working library. We are endeavoring to make his task a little easier.

Each volume in the series contains an Introduction prepared by a specialist (the volume editor)—a "state of the art" opening or a summary of the objects and content of the volume. The articles, usually some thirty to fifty reproduced either in their entirety or in significant extracts, are selected in an attempt to cover the field, from the key papers of the last century to fairly recent work. Where the original references are in foreign languages, we have endeavored to locate or commission translations. Geologists, because of their global subject, are often acutely aware of the oneness of our world. The selections cannot, therefore, be restricted to any one country, and whenever possible an attempt is made to scan the world literature.

To each article, or group of kindred articles, some sort of highlight commentary is usually supplied by the volume editor. This should serve to bring that article into historical perspective and to emphasize its particular role in the growth of the field. References, or citations, wherever possible, will be reproduced in their entirety—for by this means the observant reader can assess the background material available to that particular author, or, if he wishes, he too can doublecheck the earlier sources.

A "benchmark," in surveyor's terminology, is an established point on the ground, recorded on our maps. It is usually anything that is a vantage point, from a modest hill to a mountain peak. From the historical viewpoint, these benchmarks are the bricks of our scientific edifice.

Rhodes W. Fairbridge

Contents

III. FORM

IV. EVOLUTION

V. PROCESS

VI. APPLICATIONS

Contents by Author

Introduction

Slopes have been a favorite subject for research by geomorphologists for many years, but they have also been intensively studied by agricultural and civil engineers, and by soil conservationists. As a result an extremely diverse literature is available; papers range from the purely descriptive to highly theoretical, from the wholly academic to the rigorously practical. The editors, therefore, were faced with a vast range of many hundreds of publications dealing wholly or partly with slopes, from which 32 have been selected.

The study of hillslope form and evolution may, under ideal conditions, be a simple process, and this is the situation in rapidly evolving badlands. However, in most cases lithology, structure, and geologic history considerably complicate the interpretation of slopes. The effects of climatic change, especially during the Pelistocene and Holocene, and of tectonic movements have left their mark, but their precise influence on hillslope form is frequently unknown. More recently, man has become capable of affecting the operation of geomorphic processes by clearing wood and scrubland for agriculture and introducing grazing animals. It is extremely difficult to decide, without a thorough historical study, whether the vegetation presently found in an area is the natural climax vegetation, or whether it has been modified by man's activities. If the present vegetation cover is substantially different from that in the past, then present rates of erosion and processes of slope evolution may also be very different.

Due to these complications, it is impossible to apply uncritically the laws of physics in studies of slopes. The techniques of soil mechanics, hydraulics, and rock mechanics may be invaluable in many situations, but the complexity of the field situation precludes the development of generalizations applicable to all slopes. Hence, the rigor attained in studies of river morphology, in which many empirical relationships and laws have been discovered, has yet to be achieved in slope studies. At present, it also seems unlikely that hillslopes may be classified solely in terms of their morphology because slopes of a given form may be the result of the operation of different processes at different rates (Schumm, 1966). In fact, Dylik (1968) recently felt it necessary to define the vocabulary used by students of slopes, because of "the amazing absence of any precise definition of the slope, and the scarcity of opinions concerning the meaning of that term in geomorphology."

The approach taken in the preparation of this volume has been to select representative papers which reveal the nature of hillslope research work, and which cover some of the topics of major concern to students of slopes. Thus theoretical and empirical, academic and applied approaches are represented, as are papers on slope form and evolution and the erosional processes operating on slopes.

Although many of the selections are accepted classics in their field, our aim has been broader than to gather together into one volume the most important contributions to slope studies. We attempt to show where and in what fashion the most fruitful progress has been made in the past and may be made in the future. Although basic research should continue, there should also be an emphasis on the application of the findings of geomorphologists and geologists to environmental problems. If significant results are made more widely available, the status of hillslope studies will be enhanced, and valuable feedback from other disciplines will result.

The papers selected have been placed into six parts dealing with early work, theory and controversy, form, evolution, process, and application. Many papers could, of course, be placed in more than one of these parts, and we will not quarrel with those who would rearrange the classifications. Topics such as pediment form and evolution or mass movement per se were felt to be somewhat beyond the scope of this volume, and so our selections concentrate on the form of, evolution of, and processes operating on valley-side slopes, or hillslopes.

The recent publication of two books devoted to hillslopes (Young, 1972; Carson and Kirby, 1972) provides the student of slopes with a comprehensive bibliography of relevant publications, and the reports of the Commission pour l'Etude de l'Evolution des Versants of the International Geographical Union (1956, 1964, 1967, 1970) are another major source of information.

Editors' Comments on Paper 1

As the first selection in this collection of readings, we introduce a short paper by R. J. Chorley, which discusses the place of slope studies in geomorphology, the complexity of the problem, and the manner in which the frame of reference and attitude of the investigator may influence his interpretation of a landscape. It is thus ideally suited to serve as the final part of this introduction.

Reprinted from *Geog. J.*, **130**, 70–73 (1964)

1 II. THE NODAL POSITION AND ANOMALOUS CHARACTER OF SLOPE STUDIES IN GEOMORPHOLOGICAL RESEARCH

RICHARD J. CHORLEY

THE STUDY OF THE FORM, significance and development through time of erosional slopes has long held a nodal position in geomorphology, a study upon which have been focused most of the subject's methodological disputes. Thus, for example, the early nineteenth century 'diluvial' *versus* uniformitarian controversy, the mid-nineteenth century marine *versus* subaerial wrangle, the Davis *versus* Penck debate and the most recent conflicting landscape interpretations based either on cyclic *versus* non-cyclic notions, or on 'contemporary' *versus* past processes seem to have been contracted and intensified when viewed in the context of erosional slope studies. The importance of erosional slopes lies not only in their widespread character, but in their geometrical importance in constraining regional morphometry in such a manner as to permit characteristic 'regional landscapes' to emerge. Strahler (1950, p. 685) has pointed out that, together with drainage density, relief and the upper slope curvature, maximum valley-side slope angles represent a most important morphometric parameter whereby whole erosional landscapes may be characterized. The third, and most important, reason for the nodal position assumed by the study of valley-side slopes (i.e. their broad significance in terms of the force/resistance ratio) will form the point of return at the end of this paper.

Slope studies have always been among the most perplexing which have faced the geomorphologist for several reasons. To begin with, the detailed forms of erosional slopes have been difficult to obtain without precise field surveying. This is both because the human eye is extremely deficient in estimating slope angles and forms in the vertical plane (Vernon, 1937, p. 139), so that, for example, dominantly straight slopes usually appear concave to the visual observer, and because valley-side profiles (unlike longitudinal stream profiles) cannot be at all satisfactorily obtained from even the most accurate contour maps in common use.

The second difficulty lies in the essentially varied and complex character of slope forms and in the multivariate nature of slope processes. Individual slope profiles are mathematically complex, so that in the past attention has been variously concentrated on the upper convexity (Lawson, 1932), the lower concavity (Lake, 1928), or the middle straight segment (Strahler, 1950). To increase their difficulties, geomorphologists have been constantly concerned as to whether the actual surface or the bed-rock surface constitutes the 'real' slope profile. Similarly, this complexity of individual slope form is believed to mirror a complexity in the assumed controlling structures, processes or erosional histories. Nor does the wide range of characteristic slope angles between different geomorphological regions (Strahler, 1950, p. 680) entirely mask the local variations in slope angle which probably have so much to teach us regarding the circumstances responsible for them (Strahler, 1950, p. 812–13; Melton, 1960; Carter and Chorley, 1961). In all these respects multivariate techniques (Melton, 1957; and Krumbein, 1959) have much to offer the researcher, but they must be used rationally so that co-variation due to some undetected common cause, or occurring by chance, can be distinguished from that possibly due to cause and effect.

The third impediment facing slope studies is the doctrinaire attitude of most researchers, which is both a cause and an effect of the methodological importance which has been assumed by these studies. Little need be said here regarding the sources of partiality resulting from strongly held cyclic, poly-cyclic, posthumous climatic, climatic morphometric or, for that matter, dynamic/equilibrium notions, to which some reference will be made later.

A further impeding source of partiality lies in the assumed minor importance of studies of present processes. This attitude emanates largely from a preconceived notion of the importance of correlations between present forms and the long-past events which are wholly believed to account for them; from the recognition of the complexity of present processes; from the current lack of knowledge relating to

➔Mr. R. J. Chorley is a lecturer in geography at the University of Cambridge, and Fellow of Sidney Sussex College.

the magnitude and frequency of operation of geomorphic processes; and, finally, from the belief that slope processes are either too slow or too catastrophic to make useful measurement possible. Of these four assumptions suffice it to be said that the first assumes the existence of an answer to perhaps the most important question under investigation; the second is no reason for assuming that present process studies are irrelevant in the investigation of form; the third has been demonstrated as capable of relevant analysis (Wolman and Miller, 1960); and the work of Young (1960) and Schumm and Lusby (in Press) is showing that even the slowest mass movements are susceptible of measurements which have relevance to the associated slope forms.

The last major difficulty facing students of erosional slopes lies in the often baffling 'hen and egg' (e.g. feedback) nature of the problem. Intuition prompts the geomorphologist to believe that slope forms should be capable of understanding on the basis of the processes producing them through time, yet any detailed study of an individual slope problem shows that the magnitude and often the character of the processes affecting a given slope seem to be conditioned by the existing slope geometry. However, one of the most interesting features of contemporary work in geomorphology is a weakening in the control over slope processes assumed to be exercised by slope angle. For example, Young (1960) has shown the minor role of slope angles of less than 35° over the rate of soil creep in some upland areas in Britain, and Melton (1957) has demonstrated in a more general sense that the factors controlling maximum valley-side slope angles in parts of the western United States are those which would be largely independent of slope angle itself. This relegation of the assumed dominance of slope inclination over process parallels the earlier fluvial research by Leopold (1953).

The effect of these difficulties has been to highlight the anomalous character of slope forms, so that neither the manner of evolution of the slope nor the processes responsible can be unambiguously deduced from the slope form alone. In other words, similar slope forms (e.g. a pronounced upper convexity) can be produced under a number of widely-varying situations. This ambiguity has allowed so much elbow-room in the interpretation of slope form that no branch of geomorphology is more currently a prey to preconception than slope studies. For example, the interpretation of cliff-top convexities in the south-western United States by Ahnert (1960) on the basis of past pluvial conditions seems to have no more field evidence to support it than the doctrinaire belief that past pluvial conditions must have left some imprint on the present landscape. Much of this rounding seems to be more a function of existing rock structures and erosional processes than of any hypothetical past pluvial conditions—particularly in view of the apparently rapid retreat of many of the cliff-formers (Schumm and Chorley, in preparation). If 'Occam's razor' is to operate in slope studies it must often lead one to accept the prime importance of contemporary rock structures or weathered debris characteristics in controlling present slope forms. This is especially true in poorly-vegetated sub-humid areas and, as has been previously suggested (Bryan, 1922, p. 39–55 and Mabbutt, 1955, p. 78), the forms of most desert slope profiles can only be understood in terms of the manner of both the *primary* and *secondary* reduction achieved by rock weathering. All erosional slope forms present ambiguities if one attempts to reason simply and directly from form to process, and it is now patently apparent that no distinctive slope form is uniquely linked to a given climatic or tectonic erosional environment.

It may seem a gross omission that, even in this brief review of the problems and achievements of current research on erosional slopes, no mention has been made of the application of mathematical models (e.g. Scheidegger, 1961) to this work. Such models are often considered as the acme of modern research on slope development and, as such, they have drawn much of the criticism which was really intended for modern quantitative work in general. In fact, the lack of precision both in the description of slope forms and in the analysis of the processes possibly responsible for their transformations has meant that the application of mathematical models to slope research has progressed little in the past 100 years. There can be no doubt that ultimately such techniques will prove to be powerful research tools, but currently their employment is contributing little to the study of erosional slopes. Despite the problems attendant upon quantitative studies of erosional slopes, contemporary work is showing conclusively that slope forms and their associated processes are so multivariate in character that it is no longer possible to hold rigid views as to any unique manner of slope development, even within a given region of uniform structure, lithology and climate. Within a single climatic environment some slopes may recline whereas others retreat parallel (Schumm, 1956*a*); in the same lithological environment and 'stage' of dissection adjacent recline and parallel retreat can be deduced (Strahler, 1950, p. 804). It is possible, even within a small geographical compass, to find examples of many types of slope development through time, and occasionally to deduce steepening of valley-side slopes through part of their history (Carter and Chorley, 1961). Such deductions are emerging both from direct measurement of contemporary changes (Schumm, 1956*a* and 1956*b*) and from inferences regarding the relative positions in a time sequence occupied by different slopes at a given instant (e.g. Savigear, 1952).

Now, as always, studies of erosional slope development are reflecting the most pertinent methodological problems facing the whole science of geomorphology. The central position assumed by slope studies derives largely from the very different rates at which slope processes are currently operating, varying from the rapid adjustments such as were implied by Strahler (1950) or measured by Schumm (1956*a*), to the very slow rates of change occurring, for example, in some heavily-vegetated temperate regions. Thus slope studies cover a wide spectrum between the rapidly adjusting geomorphological forms associated with a high energy/resistance ratio and the slowly changing ones of low ratio (Chorley, 1962, p. 6). It is possible, therefore, either to view slope forms in the same manner as the light from a distant star in which what is perceived is merely a reflection of happenings in past history, or to regard them as aspects of the dynamic adjustment of present forms to reasonably contemporaneous processes. In fact, all slopes possess, in highly variable proportions, both the aspects of 'historical hangover' and 'dynamic equilibrium', and one of the most pressing needs of current slope studies is that the co-existence of these attributes should be recognized and their relative importance evaluated.

References

Ahnert, F. 1960 The influence of Pleistocene climates upon the morphology of cuesta scarps on the Colorado Plateau. *Ann. Ass. Amer. Geogr.* **50**: 139–56.

Bryan, K. 1922 Erosion and sedimentation in the Papago Country, Arizona. *U.S. Geol. Surv. Bull.* 730–B.

Carter, C. A., and R. J. Chorley 1961 Early slope development in an expanding stream system. *Geol. Mag.* **98**: 117–30.

Chorley, R. J. 1962 Geomorphology and general systems theory. *U.S. Geol. Surv. Prof. Paper* 500–B.

Krumbein, W. C. 1959 The 'sorting out' of geological variables, illustrated by regression analysis of factors controlling beach firmness. *J. Sediment. Petrol* **29**: 575–87.

Lake, P. 1928 On hill slopes. *Geol. Mag.* **65**: 108–16.

Lawson, A. C. 1932 Rain-wash erosion in humid regions. *Bull. geol. Soc. Amer.* **43**: 703–24.

Leopold, L. B. 1953 Downstream change of velocity in rivers. *Amer. J. Sci.* **251**: 606–24.

Mabbutt, J. A. 1955 Pediment land forms in Little Namaqualand, South Africa. *Geogr. J.* **121**: 77–83.

Melton, M. A. 1957 An analysis of the relation among elements of climate, surface properties and geomorphology. Office of Naval Research Project NR389-042, Tech. Rept. 11, Dept. of Geol., Columbia Univ., New York.

—— 1960 Intravalley variation in slope angles related to micro-climate and erosional environment. *Bull geol. Soc. Amer.* **71**: 133–44.

Savigear, R. A. G. 1952 Some observations on slope development in South Wales. *Trans. Inst. Brit. Geogr.* **18**: 31–51.

Scheidegger, A. E. 1961 Mathematical models of slope development. *Bull. geol. Soc. Amer.* **72**: 37–50

Schumm, S. A. 1956a The role of creep and rainwash on the retreat of badland slopes. *Amer. J. Sci.* **254**: 693–706.

—— 1956b Evolution of drainage systems and slopes in badlands at Perth Amboy, New Jersey. *Bull. geol. Soc. Amer.* **67**: 597–646.
—— and G. C. Lusby (In press) Seasonal variations of erosion and hydrology on Mancos Shale hillslopes in western Colorado.
—— and R. J. Chorley (In preparation) Rock weathering and cliff recession in the Colorado Plateau.
Strahler, A. N. 1950 Equilibrium theory of erosional slopes approached by frequency distribution analysis. *Amer. J. Sci.* **248**: 673–96 and 800–14.
Vernon, M. D. 1937 *Visual perception.* Cambridge University Press.
Young, A. 1960 Soil movement by denudational processes on slopes. *Nature* **188**: 120–2.
Wolman, M. G., and J. P. Miller 1960 Magnitude and frequency of forces in geomorphic processes. *J. Geol.* **68**: 54–74.

Early Work

I

Editors' Comments on Papers 2, 3, and 4

Description precedes explanation, and so a great deal of the early work on hillslopes and on landforms in general consists of qualitative description. Nevertheless, perhaps one third of the contributions to the study of slopes published before 1925 may be profitably read today (Young, 1972). Outstanding among these are the descriptions of slope form and process by explorers in the American West and Southwest. Their prose and often poetic description are contributions to both science and literature. The reports by Powell, Dutton, and Gilbert are the most widely known, and we here include an extract from Chapter 5 of G. K. Gilbert's "Geology of the Henry Mountains" as a representative of this work. The selection includes only those portions of Chapter 5 which deal with slopes. (Much of the remainder is reprinted in the "River Morphology" volume of this series.)

Gilbert's United States Geological Survey report is a classic, and Chapter 5, "Land Sculpture," is justly regarded as a foundation of modern geomorphology. Although his arguments are qualitative in nature, Gilbert's logic and fertility of ideas are outstanding. The careful observations of this eminent geologist, made under the difficult conditions of nineteenth century southern Utah, are to be admired and emulated by the modern student. His work is exciting because he not only described what he saw, but sought to explain it.

One problem which Gilbert considered, the cause of the convexity of badland divides, was discussed by W. M. Davis in a short note published in 1892, his only article dealing solely and specifically with hillslopes. The length of Davis's major papers preclude their inclusion in this volume, but his work is well known and his major writings are available in Johnson's (1954) collection of "Geographical Essays." We thus include Davis's note on the convexity of badland divides, more as an acknowledgment of his influence on geomorphology than as a representative of his work. This short discussion is noteworthy for its consideration of the actual processes which produce specific landforms. As Bryan (1940) remarked, "Long-continued reading of Davis's papers yields very little material on slopes. Phrases and catchwords may be found here and there, but there is no study of the processes that produce slopes . . . ," and, "Slightly bemused by long, though mild intoxication on the liquid prose of Davis's remarkable essays" the geomorphologist "wakes with a gasp to realize that in considering the important question of slope he has always substituted words for knowledge, phrases for critical observation." It is apparent that Davis was capable of considering slope-forming processes when he thought it necessary, and that Bryan's much-quoted evaluation applies more to the American geomorphologist than to Davis himself.

An acknowledged, but sadly neglected, classic is "Les Formes du Terrain," by G. De La Noe and E. De Margerie. This work was, like that of Gilbert, sponsored by a government agency. Chapter 3, "Du faconnement des versants," anticipated many of the concepts suggested by geomorphologists writing in other languages and in later years. For example, like Davis, De La Noe and De Margerie concluded that slopes decline, attributing this to the action of rainwash; like Gilbert they discovered the concept of equilibrium. They stated that slope inclination depends on "résistance de la roche, de la nature des agents qui produisent sa désagrégation, et de la durée de leur action"; that is, on structure, process, and time. However, they laid especial stress on the effects of geological inhomogeneities, using both mathematical (geometric) and scale models to demonstrate these effects upon hillslope form.

Other topics discussed by De La Noe and De Margerie include valley asymmetry, which they attributed to differences in exposure or to structural conditions, and cliff retreat (see also Melton, Paper 15, and Koons, Paper 18, of this volume). Of especial interest to the modern geomorphologist is their concern with man-induced accelerated soil erosion (see Everard, Paper 20).

But perhaps the most significant feature of "Les Formes du Terrain" is that it is "process oriented." The arguments take full account of the processes operating, and do not depend solely on inference from morphology alone, as have so many later studies. A short English summary of Paper 4 begins on page 57.

For a more complete discussion of early studies of hillslopes and of the history of geomorphology, the interested reader should refer to Chorley, Dunn, and Beckinsale (1964).

DEPARTMENT OF THE INTERIOR

U. S. GEOGRAPHICAL AND GEOLOGICAL SURVEY OF THE ROCKY MOUNTAIN REGION

J. W. POWELL IN CHARGE

2

REPORT

ON THE

GEOLOGY OF THE HENRY MOUNTAINS

BY G. K. GILBERT

SECOND EDITION

WASHINGTON
GOVERNMENT PRINTING OFFICE
1880

II. SCULPTURE.

Erosion may be regarded from several points of view. It lays bare rocks which were before covered and concealed, and is thence called *denudation.* It reduces the surfaces of mountains, plateaus, and continents, and is thence called *degradation.* It carves new forms of land from those which before existed, and is thence called *land sculpture.* In the following pages it will be considered as land sculpture, and attention will be called to certain principles of erosion which are concerned in the production of topographic forms.

Sculpture and Declivity.

We have already seen that erosion is favored by declivity. Where the declivity is great the agents of erosion are powerful; where it is small they are weak; where there is no declivity they are powerless. Moreover it has been shown that their power increases with the declivity in more than simple ratio.

It is evident that if steep slopes are worn more rapidly than gentle, the tendency is to abolish all differences of slope and produce uniformity. The law of uniform slope thus opposes diversity of topography, and if not complemented by other laws, would reduce all drainage basins to plains. But in reality it is never free to work out its full results; for it demands a uniformity of conditions which nowhere exists. Only a water sheet of uniform depth, flowing over a surface of homogeneous material, would suffice; and every inequality of water depth or of rock texture produces a corresponding inequality of slope and diversity of form. The reliefs of the landscape exemplify other laws, and the law of uniform slopes is merely the conservative element which limits their results.

Sculpture and Structure; the Law of Structure.

We have already seen that erosion is influenced by rock character. Certain rocks, of which the hard are most conspicuous, oppose a stubborn resistance to erosive agencies; certain others, of which the soft are most conspicuous, oppose a feeble resistance. Erosion is most rapid where the resistance is least, and hence as the soft rocks are worn away the hard are left prominent. The differentiation continues until an equilibrium is reached through the law of declivities. When the ratio of erosive action as dependent on declivities becomes equal to the ratio of resistances as dependent on rock character, there is equality of action. In the structure of the earth's crust hard and soft rocks are grouped with infinite diversity of arrangement. They are in masses of all forms, and dimensions, and positions; and from these forms are carved an infinite variety of topographic reliefs.

In so far as the law of structure controls sculpture, hard masses stand as eminences and soft are carved in valleys.

The Law of Divides.

We have seen that the declivity over which water flows bears an inverse relation to the quantity of water. If we follow a stream from its mouth upward and pass successively the mouths of its tributaries, we find its volume gradually less and less and its grade steeper and steeper, until finally at its head we reach the steepest grade of all. If we draw the profile of the river on paper, we produce a curve concave upward and with the greatest curvature at the upper end. The same law applies to every tributary and even to the slopes over which the freshly fallen rain flows in a sheet before it is gathered into rills. The nearer the water-shed or divide the steeper the slope; the farther away the less the slope.

It is in accordance with this law that mountains are steepest at their crests. The profile of a mountain if taken along drainage lines is concave outward as represented in the diagram; and this is purely a matter of sculpture, the uplifts from which mountains are carved rarely if ever assuming this form.

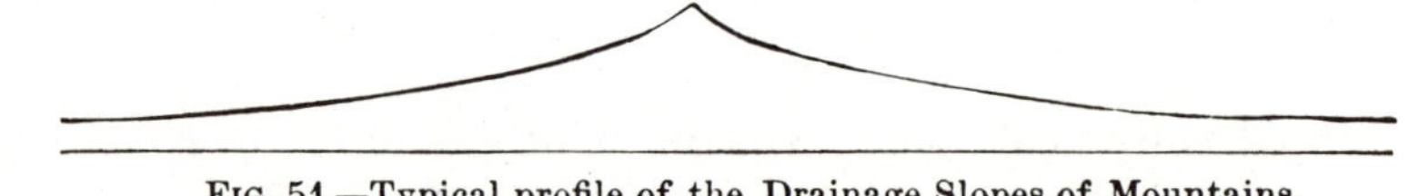

FIG. 54.—Typical profile of the Drainage Slopes of Mountains.

Under the *law of Structure* and the *law of Divides* combined, the features of the earth are carved. Declivities are steep in proportion as their material is hard; and they are steep in proportion as they are near divides.

The distribution of hard and soft rocks, or the geological structure, and the distribution of drainage lines and water-sheds, are coefficient conditions on which depends the sculpture of the land. In the sequel it will be shown that the distribution of drainage lines and water-sheds depends in part on that of hard and soft rocks.

In some places the first of the two conditions is the more important, in others the second. In the bed of a stream without tributaries the grade depends on the structure of the underlying rocks. In rock which is homogeneous and structureless all slopes depend on the distribution of divides and drainage lines.

The relative importance of the two conditions is especially affected by climate, and the influence of this factor is so great that it may claim rank as a third condition of sculpture.

Sculpture and Climate.

The Henry Mountains consist topographically of five individuals, separated by low passes, and practically independent in climate. At the same time they are all of one type of structure, being constituted by similar aggregation of hard and soft rocks. Their altitudes appear in the following table.

	Altitude above the sea.
Mount Ellen	11, 250 feet.
Mount Pennell	11, 150 feet.
Mount Hillers	10, 500 feet.
Mount Ellsworth	8, 000 feet.
Mount Holmes	7, 775 feet.

The plain on which they stand has a mean altitude of 5,500 feet, and is a desert. A large proportion of the rain which falls in the region is caught by the mountains, and especially by the higher mountains. Of this there is abundant proof in the distribution of vegetation and of springs.

The vegetation of the plain is exceedingly meager, comprising only sparsely set grasses and shrubs, and in favored spots the dwarf cedar of the West (*Juniperus occidentalis*).

Mount Ellen, which has a continuous ridge two miles long and more than 11,000 feet high, bears cedar about its base, mingled higher up with piñon (*Pinus edulis*), and succeeded above by the yellow pine (*P. ponderosa*), spruce (*Abies Douglasii*), fir (*A. Engelmanni*), and aspen (*Populus tremuloides*). The pines are scattering, but the cedars are close set, and the firs are in dense groves. The upper slopes where not timbered are matted with luxuriant grasses and herbs. The summits are naked.

Mount Pennell sends a single peak only to the height of the Ellen ridge. Its vegetation is nearly the same, but the timber extends almost to the summit.

Mount Hillers is 650 feet lower. Its timber reaches to the principal summit, but is less dense than on the higher mountains. The range of trees is the same.

Mount Ellsworth, 2,500 feet lower than Mount Hillers, bears neither fir, spruce, pine nor aspen. Cedar and piñon climb to the summit, but are not so thickly set as on the lower slopes of the larger mountains. The grasses are less rank and grow in scattered bunches.

Mount Holmes, a few feet lower, has the same flora, with the addition of a score of spruce trees, high up on the northern flank. Its summits are bare.

In a word, the luxuriance of vegetation, and the annual rainfall, of which it is the index, are proportioned to the altitude.

Consider now the forms of the mountain tops.

In Figure 55 are pictured the summit forms of Mount Ellen. The crests are rounded; the slopes are uniform and smooth. Examination has shown that the constituent rocks are of varying degrees of hardness, trachyte dikes alternating with sandstones and shales; but these variations rarely find expression in the sculptured forms.

In Figure 56 are the summit crags of Mount Holmes. They are dikes of trachyte denuded by a discriminating erosion of their encasements of sandstone, and carved in bold relief. In virtue of their superior hardness they survive the general degradation.

The other mountains are intermediate in the character of their sculpture. Mount Pennell is nearly as smooth as Mount Ellen. Mount Ellsworth is

Fig. 55.—The Crest of Mount Ellen, as seen from Ellen Peak.

Fig. 56.—The Crest of Mount Holmes.

nearly as rugged as Mount Holmes. One may ride to the crest of Mount Ellen and to the summit of Mount Pennell; he may lead his sure-footed cayuse to the top of Mount Hillers; but Mounts Ellsworth and Holmes are not to be scaled by horses. The mountaineer must climb to reach their summits, and for part of the way use hands as well as feet.

In a word, the ruggedness of the summits or the differentiation of hard and soft by sculpture, is proportioned inversely to the altitude. And rainfall, which in these mountains depends directly on altitude, is proportioned inversely to ruggedness.

The explanation of this coincidence depends on the general relations of vegetation to erosion.

We have seen that vegetation favors the disintegration of rocks and retards the transportation of the disintegrated material. Where vegetation is profuse there is always an excess of material awaiting transportation, and the limit to the rate of erosion comes to be merely the limit to the rate of transportation. And since the diversities of rock texture, such as hardness and softness, affect only the rate of disintegration (weathering and corrasion) and not the rate of transportation, these diversities do not affect the rate of erosion in regions of profuse vegetation, and do not produce corresponding diversities of form.

On the other hand, where vegetation is scant or absent, transportation and corrasion are favored, while weathering is retarded. There is no accumulation of disintegrated material. The rate of erosion is limited by the rate of weathering, and that varies with the diversity of rock texture. The soft are eaten away faster than the hard; and the structure is embodied in the topographic forms.

Thus a moist climate by stimulating vegetation produces a sculpture independent of diversities of rock texture, and a dry climate by repressing vegetation produces a sculpture dependent on those diversities. With great moisture the law of divides is supreme; with aridity, the law of structure.

Hence it is that the upper slopes of the loftier of the Henry Mountains are so carved as to conceal the structure, while the lower slopes of the same mountains and the entire forms of the less lofty mountains are so carved as to reveal the structure; and hence too it is that the arid plateaus of the Colorado Basin abound in cliffs and cañons, and offer facilities to the student of geological structure which no humid region can afford.

Here too is the answer to the question so often asked, "whether the rains and rivers which excavated the cañons and carved the cliffs were not mightier than the rains and rivers of to-day." Aridity being an essential condition of this peculiar type of sculpture, we may be sure that through long ages it has characterized the climate of the Colorado Basin. A climate of great rainfall, as Professor Powell has already pointed out in his "Exploration of the Colorado," would have produced curves and gentle slopes in place of the actual angles and cliffs.

Bad-lands.

Mountain forms in general depend more on the law of divides than on the law of structure, but their independence of structure is rarely perfect, and it is difficult to discriminate the results of the two principles. For the investigation of the workings of the law of divides it is better to select examples from regions which afford no variety of rock texture and are hence unaffected in their erosion by the law of structure. Such examples are found in *bad-lands.*

Where a homogeneous, soft rock is subjected to rapid degradation in an arid climate, its surface becomes absolutely bare of vegetation and is carved into forms of great regularity and beauty. In the neighborhood of the Henry Mountains, the Blue Gate and Tununk shales are of this character, and their exposures afford many opportunities for the study of the principles of sculpture. I was able to devote no time to them, but in riding across them my attention was attracted by some of the more striking features, and these I will venture to present, although I am conscious that they form but a small part of the whole material which the bad-lands may be made to yield.

If we examine a bad-land ridge, separating two drainage lines and forming a divide between them, we find an arrangement of secondary ridges and secondary drainage lines, similar to that represented in the diagram, (Figure 58.)

The general course of the main ridge being straight, its course in detail is found to bear a simple relation to the secondary ridges. Wherever a secondary joins, the main ridge turns, its angle being directly toward the secondary. The divide thus follows a zigzag course, being deflected to the right or left by each lateral spur.

The altitude of the main ridge is correspondingly related to the secondary ridges. At every point of union there is a maximum, and in the intervals are saddles. The maxima are not all equal, but bear some relation to the magnitudes of the corresponding secondary ridges, and are especially accented where two or more secondaries join at the same point. (See profile in Figure 59.)

I conceive that the explanation of these phenomena is as follows: The heads of the secondary drainage lines laid down in the diagram are in nature tolerably definite points. The water which during rain converges at one of these points is there abruptly concentrated in volume. Above the point it is a sheet, or at least is divided into many rills. Below it, it is a single stream with greatly increased power of transportation and corrasion. The principle of equal action gives to the concentrated stream a less declivity than to the diffused sheet, and—what is especially important—it tends to produce an equal grade in all directions upward from the point of convergence. The converging surface becomes hopper-shaped or funnel-shaped; and as the point of convergence is lowered by corrasion, the walls of the funnel are eaten back equally in all directions—except of course the direction of the stream. The influence of the stream in stimulating erosion above its head is thus extended radially and equally through an arc of 180°, of which the center is at the point of convergence.

Where two streams head near each other, the influence of each tends to pare away the divide between them, and by paring to carry it farther back. The position of the divide is determined by the two influences combined and represents the line of equilibrium between them. The influences being radial from the points of convergence, the line of equilibrium is tangential, and is consequently at right angles to a line connecting the two points. Thus, for example, if *a*, *b*, and *c* (Figure 58) are the points of convergence at the heads of three drainage lines, the divide line *ed* is at right angles to a line connecting *a* and *b*, and the divides *fd* and *gd* are similarly determined. The point *d* is simultaneously determined by the intersection of the three divide lines.

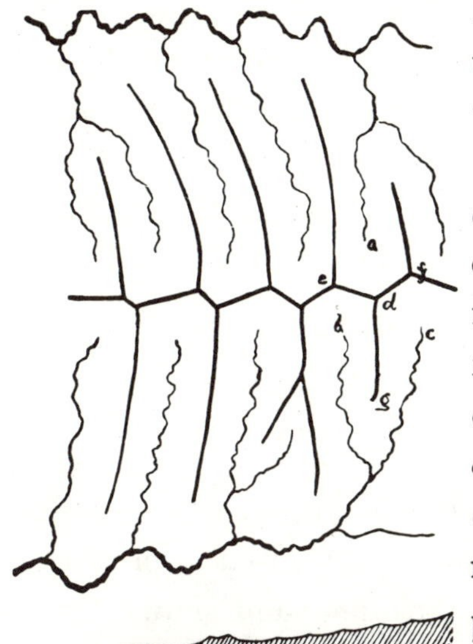

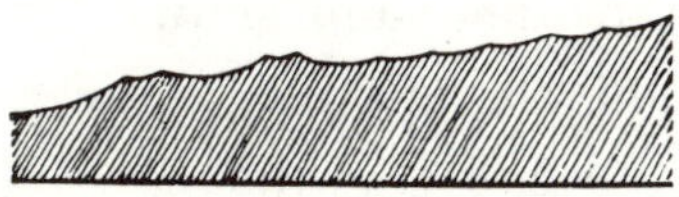

FIG. 58.—Ground-plan of a Bad-land Ridge, showing its relation to Waterways. The smooth lines represent Divides.
FIG. 59.—Profile of the same ridge.

Furthermore, since that point of the line *ed* which lies directly between *a* and *b* is nearest to those points, it is the point of the divide most subject to the erosive influences which radiate from *a* and *b*, and it is consequently degraded lower than the contiguous portions of the divide. The points *d* and *e* are less reduced; and *d*, which can be shown by similar reasoning to stand higher than the adjacent portion of either of the three ridges which there unite, is a local maximum.

There is one other peculiarity of bad-land forms which is of great significance, but which I shall nevertheless not undertake to explain. According to the law of divides, as stated in a previous paragraph, the profile of any slope in bad-lands should be concave upward, and the slope should be steepest at the divide. The union or intersection of two slopes on a divide should produce an angle. But in point of fact the slopes do not unite in an angle. They unite in a curve, and the profile of a drainage slope instead of being concave all the way to its summit, changes its curvature and becomes convex. Figure 60 represents a profile from *a* to *b* of Figure 58. From *a* to *m* and from *b* to *n* the slopes are concave, but from *m* to *n* there is a convex curvature. Where the flanking slopes are as steep as represented in the diagram, the convexity on the crest of a ridge has a breadth of only two or three yards, but where the flanking slopes are gentle, its breadth is several times as great. It is never absent.

Thus in the sculpture of the bad-lands there is revealed an exception to the law of divides,—an exception which cannot be referred to accidents of structure, and which is as persistent in its recurrence as are the features which conform to the law,—an exception which in some unexplained way is part of the law. Our analysis of the agencies and conditions of erosion, on the one hand, has led to the conclusion that (where structure does not pre-

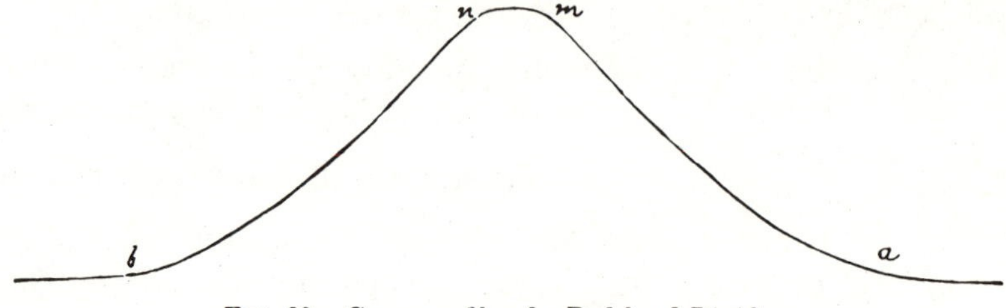

FIG. 60.—Cross-profile of a Bad-land Divide.

vent) the declivities of a continuous drainage slope increase as the quantities of water flowing over them decrease; and that they are great in proportion as they are near divides. Our observation, on the other hand, shows that the declivities increase as the quantities of water diminish, up to a certain point where the quantity is very small, and then decrease; and that declivities are great in proportion as they are near divides, unless they are *very* near divides. Evidently some factor has been overlooked in the analysis,—a factor which in the main is less important than the flow of water, but which asserts its existence at those points where the flow of water is exceedingly small, and is there supreme.

Equal Action and Interdependence.

The tendency to equality of action, or to the establishment of a dynamic equilibrium, has already been pointed out in the discussion of the principles of erosion and of sculpture, but one of its most important results has not been noticed.

Of the main conditions which determine the rate of erosion, namely, quantity of running water, vegetation, texture of rock, and declivity, only the last is reciprocally determined by rate of erosion. Declivity originates in upheaval, or in the displacements of the earth's crust by which mountains and continents are formed; but it receives its distribution in detail in accordance with the laws of erosion. Wherever by reason of change in any of the conditions the erosive agents come to have locally exceptional power, that power is steadily diminished by the reaction of rate of erosion upon declivity. Every slope is a member of a series, receiving the water and the waste of the slope above it, and discharging its own water and waste upon the slope below. If one member of the series is eroded with exceptional rapidity, two things immediately result: first, the member above has its level of discharge lowered, and its rate of erosion is thereby increased; and second, the member below, being clogged by an exceptional load of detritus, has its rate of erosion diminished. The acceleration above and the retardation below, diminish the declivity of the member in which the disturbance originated; and as the declivity is reduced the rate of erosion is likewise reduced.

But the effect does not stop here. The disturbance which has been transferred from one member of the series to the two which adjoin it, is by them transmitted to others, and does not cease until it has reached the confines of the drainage basin. For in each basin all lines of drainage unite in a main line, and a disturbance upon any line is communicated through it to the main line and thence to every tributary. And as any member of the system may influence all the others, so each member is influenced by every other. There is an interdependence throughout the system.

Reprinted from *Science*, **20**, 245 (1892)

3

THE CONVEX PROFILE OF BAD-LAND DIVIDES.

BY W. M. DAVIS, HARVARD COLLEGE, CAMBRIDGE, MASS.

In Mr. Gilbert's analysis of land sculpture, constituting chapter V. of his "Geology of the Henry Mountains," he explains why the surface of an eroded region possesses slopes that are concave upwards and steepest near the divides, and shows that it is for the reasons there stated that mountains — that is, mature and well-sculptured mountains, such as are of ordinary occurrence — are steepest at their crests (p. 116). The *arêtes* of the Alps illustrate this perfectly. Gilbert calls this generalization the "law of divides."

But in discussing the forms assumed by eroded bad-lands, or arid regions of weak structure with insignificant variety of texture, he finds an exception to the law of divides. The two lateral concave slopes of a bad-land ridge do not unite upwards at an angle, forming a sharp divide, but are joined in a curve that is convex instead of concave upwards. "Thus in the sculpture of the bad lands there is revealed an exception to the law of divides,— an exception which cannot be referred to accidents of structure, and which is as persistent in its recurrence as are the features which conform to the law,— an exception which in some unexplained way is part of the law. Our analysis of the agencies and conditions of erosion, on the one hand, has led to the conclusion that (where structure does not prevent) the declivities of a continuous drainage-slope increase as the quantities of water flowing over them decrease; and that they are great in proportion as they are near divides. Our observation, on the other hand, shows that the declivities increase as the quantities of water diminish, up to a certain point where the quantity is very small, and then decrease; and that declivities are great in proportion as they are near divides, unless they are *very* near divides. Evidently some factor has been overlooked in the analysis,— a factor which in the main is less important than the flow of water, but which asserts its existence at those points where the flow of water is exceedingly small, and is there supreme" (pp. 122, 123).

It has for some time seemed to me that the overlooked factor is the creeping of the surface soil; and, as I have not seen mention of this process as bearing on the form of the crest-lines of divides, a brief note on the subject is here offered.

The superficial parts of rock-masses are slowly reduced to rock-waste or soil by the various processes included under the term, *weathering*. Unconsolidated materials are in the same way reduced to finer texture near their surface. The loose and often fine material thus provided at the surface is carried away by various processes, of which the chief are moving water, moving air, and occasionally moving ice; but there is an additional process of importance, involving dilatation and contraction of the soil, and in consequence of which not only the loose particles on the surface are transported, but a considerable thickness of loose material is caused to creep slowly down-hill.

Dilatation is caused by increase of temperature, by increase of moisture, and by freezing. Vegetable growth may probably be added to this list. The movements are minute and slow. They are directed outwards, about normally to the surface. Contraction follows dilatation, when the soil cools or dries, or when its frost melts. The movement of the parts is then not inward at a normal to the surface, but vertically downwards, or even downwards along the slope. As the two motions do not counterbalance each other, a slow down-hill resultant remains. This is greatest near the surface, where the dilatations and contractions are greatest; but it does not cease even at a depth of several feet, perhaps of many feet. Hence the down-hill dragging of old-weathered rock, often well shown in fresh railroad cuttings in non-glaciated regions. I presume all this is familiar to most readers; although from the frequent inquiry concerning the means by which valleys are widened it is evident that the creeping process is not so generally borne in mind as that by which running water washes loose material down-hill.

The form assumed by the surface of the land depends largely on the ratio between the processes of washing and creeping. Wherever the concentration of drainage makes transportation by streams effective, the loose material is so generally carried away (except on flood-plains) that the action of creeping is relatively insignificant. But on divides, where drainage is not concentrated but dispersed, the ratio of creeping to washing is large, even though the value of creeping is still small. This is especially the case in regions of loose texture and of moderate rainfall; that is, in typical bad-lands, where the supply of loose surface-material ready to creep is large, and where the loose material is slowly taken away by washing. On the divides of such regions, the surface form is controlled by the creeping process. The sharp-edged divides, that should certainly appear if washing alone were in action, are nicely rounded off by the dilatations and contractions of the soil along the ridge-line. The result thus determined by the slow outward and downward movements of the particles might be imitated in a short time by a succession of light earthquake shocks.

Mr. Gilbert has himself given several beautiful illustrations of the close dependence of sharp or rounded divides on rainfall; structure remaining constant. If the rainfall should increase in bad-land regions, would not all their divides become sharper; and if the rainfall were continuous, so as to carry away every loose particle as soon as it is loosened, would not the divides assume the sharp ridge-line expected from Mr. Gilbert's analysis but not found in the actual arid bad-land climate? In the eastern and well-watered part of our country, I have often seen clay-banks much more sharply cut than the equally barren surface of the western bad lands; but even on clay-banks, the minute divides between the innumerable little valleys are not knife-edge sharp; they are rounded when closely looked at. Perhaps they are sharper in wet weather and duller in dry spells.

If rainfall remain constant and structure vary, then the harder the structure, the less the supply of soil for creeping and the sharper the divides; the weaker the structure, the more plentiful the supply of soil for creeping and the duller the divides. Numerous examples of this variation might be given.

Reprinted from *Les Formes du terrain*, 20–33, 39–47 (1888), Imprimerie Nationale

CHAPITRE III.

Du Façonnements des versants*

4

G. DE LA NOE and E. DE MARGERIE

DU FAÇONNEMENT DES VERSANTS.

Grâce à la couche de matériaux désagrégés qui constitue la surface du sol, celui-ci se trouve tout préparé pour l'entraînement facile de ces matériaux par les eaux de pluie, d'où résultera son modelé. *Cette préparation est un fait qu'il importe d'avoir présent à l'esprit* pour comprendre facilement le mécanisme de l'érosion : *l'eau, seule, n'est pas capable de diviser mécaniquement les roches compactes*, et la démolition des crêtes formées de matériaux résistants serait difficile à concevoir sans cette désagrégation préalable.

Nous allons maintenant voir comment l'eau agit pour opérer le transport des matériaux meubles lorsqu'elle ruisselle sur les pentes, en réservant pour un autre chapitre l'examen de son action, lorsqu'elle se trouve concentrée en ruisseaux, rivières, etc.

C'est un fait d'observation journalière que les eaux de pluie, en ruisselant à la surface du sol, entraînent les matériaux meubles qui le recouvrent : il n'y a pas d'orage sans eau trouble; les talus et tranchées artificielles sont rapidement dégradés; sur certaines pentes raides, par exemple celle du «Vignoble» aux environs de Salins, les matériaux désagrégés du sol sont constamment entraînés, si bien que le vigneron est obligé de remonter chaque année, à l'aide de sa hotte, les terres descendues à la partie inférieure de sa vigne. Cette action est favorisée par la diminution de poids que l'eau fait subir aux matériaux qu'elle transporte.

L'EFFET DE LA PLUIE SUR LES VERSANTS EST DE DIMINUER DE PLUS EN PLUS LEUR PENTE. ANALYSE DU PHÉNOMÈNE.

L'effet de la pluie sur les versants est de diminuer de plus en plus leur pente : pour un même terrain cette diminution sera d'autant plus rapide que la désagrégation de la surface sera plus active, que les matières décomposées seront en débris plus fins, que la pente sera plus forte et que le volume d'eau sera plus grand. Comme d'ailleurs la pente minimum sur

*An English summary follows.

laquelle les matériaux peuvent être entraînés dépend de leur grosseur, s'il n'y a pas de limite à la division de ces particules, *la pente du versant n'aura d'autre limite que celle nécessaire pour que les eaux chargées des particules les plus fines puissent encore s'écouler*. Comme on peut admettre que, quelle qu'ait été la grosseur des matériaux au moment où ils venaient d'être détachés de la roche en place, ils peuvent à la longue, sous l'action répétée des agents de décomposition et aussi des chocs mutuels qu'ils subissent dans un transport même peu considérable, être réduits à leur tour *en particules suffisamment fines*, on doit admettre également que tous les terrains, quelle que soit leur constitution minéralogique, sont susceptibles d'atteindre *une pente extrêmement faible, à la seule condition que l'action des pluies soit suffisamment prolongée*. En un mot, l'action de la pluie a pour effet final d'aplatir tous les versants.

Examinons maintenant de plus près comment cet effet se produit (voir pl. III, fig. 8 et 9).

Soit AB le versant d'une vallée, ouverte dans une seule et même roche, au pied B duquel passe un cours d'eau dont le niveau est supposé invariable : l'eau de pluie ruisselant sur AB va entraîner les débris de la surface dans le cours d'eau qui les transportera au loin, déblayant constamment le pied du versant. Au fur et à mesure que les produits de la désagrégation disparaîtront, il s'en formera de nouveaux, aux dépens de la roche mise à nu, qui seront à leur tour enlevés. Cependant la démolition du versant ne marchera pas par tranches parallèles telles que AA′BB′, A′A″B′B″ (fig. 9), parce que les matériaux arrivés en B″ ne pourraient être entraînés sur la pente trop faible B′B. Au contraire, la dégradation marchera de telle sorte que le profil du versant sera représenté successivement par les lignes A′B, A″B (fig. 8), qui présenteront toujours une pente continue venant aboutir constamment au point B. L'observation montre que les choses se passent ainsi dans la nature : on voit constamment des pentes douces surmonter directement des pentes raides, ou réciproquement, sans être séparées par un palier horizontal, et de telle sorte que l'arête de séparation des deux pentes est en même temps le sommet de l'une et la base de l'autre.

REPRODUCTION EXPÉRIMENTALE DU PHÉNOMÈNE.

On peut imiter ce phénomène par l'expérience suivante (pl. III, fig. 10): soit A B C D une caisse contenant du sable, du plâtre ou toute autre matière pulvérulente suffisamment tassée pour qu'elle puisse se maintenir sans s'écouler sous une inclinaison A B plus grande que celle du talus B N qui correspond à la pente maximum de ces mêmes matériaux à l'état meuble; si l'on fait arriver en A un courant de sable ou même de matière semblable à celle qui remplit la caisse, son frottement sur la surface A B aura pour effet d'y creuser une rigole dont le profil sera représenté successivement par les lignes A′S′B, A″S″B, etc., à pente continue si l'on ne tient pas compte des petites rugosités de la surface, et passant toutes par le point B, inattaquable. Le déblayement opéré au pied des versants par le cours d'eau qui les baigne est ici remplacé par la chute à l'extérieur des matériaux amenés jusqu'au point B, lesquels, par suite, ne causent pas un encombrement qui arrêterait le phénomène. Les deux cas sont donc absolument comparables.

Le profil d'un versant (pour une même roche) est une courbe à pente continue, concave vers le ciel.

Or, dans l'expérience qui précède, on remarque que les lignes A′S′B, A″S″B, etc., ne sont pas droites, mais se présentent sous la forme d'une courbe concave vers le ciel.

Cela tient à l'augmentation de vitesse du courant de sable en passant de A en B, grâce à laquelle la force d'entraînement des particules contenues dans la caisse augmente dans le même sens : par suite, dans la courbe du profil, l'angle formé par la tangente à cette courbe avec l'horizon sera d'autant plus faible que le point considéré est plus voisin de B. Il ne faut pas perdre de vue en effet que *la pente d'un élément de la surface est d'autant plus faible que la force du courant agissant sur cet élément est plus grande.*

Dans la nature, on observe la même concavité du profil, due à une cause analogue; mais ici l'effet est encore accru par cette circonstance, que le volume d'eau passant en chaque point de la surface est d'autant plus grand que le point considéré est plus près du pied du versant. Ceci n'est pas applicable aux pentes très faibles, du moins en ce qui concerne l'augmentation de vitesse, parce que les frottements dus à la résistance du sol interviennent pour diminuer celle-ci; mais ce que nous avons dit de l'augmentation de volume d'eau subsiste, et lorsque cet effet sera prépondérant, c'est encore au pied du versant que se produira la pente la plus douce, malgré la diminution de vitesse.

C'est par le pied des versants que commence la régularisation de leur profil.

Dans l'expérience de la caisse, l'inclinaison de la surface primitive et la nature des matériaux qui composent le courant restant les mêmes, le résultat du frottement de ce dernier sur le fond est l'établissement d'un profil définitif A″S″B, par exemple, à partir duquel il n'y a plus aucune matière enlevée; mais on remarque que, au début du phénomène, les profils successifs tels que A′S′B, au lieu de former une courbe continue, présentent des rugosités qui forment dans le courant comme de petites cascades; ces rugosités commencent à disparaître à partir du bord B de la caisse; l'élément voisin de ce point prend le premier la pente continue, à surface unie, qui fera partie du profil définitif A″S″B, et les éléments adjacents acquerront successivement leur forme définitive en remontant de proche en proche vers l'amont. En un mot, le modelé définitif de la surface se fait de B vers A, c'est-à-dire de l'aval à l'amont. Dans la nature, les choses doivent se passer de la même façon, et nous verrons plus tard pour les thalwegs (que l'on peut assimiler dans une certaine mesure à la ligne de plus grande pente d'un versant) que dans la période transitoire du creusement du lit, les eaux se présentent souvent en cascades dans la partie moyenne ou supérieure de leur cours, alors même qu'elles s'écoulent en aval sur une pente continue.

À UNE MÊME ROCHE NE CORRESPOND PAS UNE PENTE UNIQUE.

Puisque l'effet final des eaux de pluie est d'aplatir les versants, *on ne peut pas dire qu'à une même roche corresponde une pente unique* et lui appartenant en propre: c'est ce dont les plaines de la Champagne nous fournissent un exemple frappant. La craie, qui en constitue le sous-sol, y forme des collines et des vallées dont les pentes présentent la plus grande diversité.

Les versants des vallées, en marchant vers leur aplatissement final, passent par une série d'états transitoires : leur inclinaison est d'autant plus raide que les actions qui tendent à les aplatir s'exercent depuis moins longtemps, toutes choses égales d'ailleurs. Aussi la topographie des contrées formées par une même roche est-elle loin de présenter partout des caractères identiques : les parties granitiques des hautes montagnes (Alpes) ne ressemblent pas à celles des régions moyennes (Vosges) et encore moins à celles des pays de collines (Bretagne).

PROFILS DES VERSANTS SUR LESQUELS AFFLEURENT DES ROCHES DE NATURE DIFFÉRENTE.

Si nous considérons les versants de vallées découpées à travers des roches de nature différente et soumises pendant le même temps aux mêmes actions de désagrégation et d'entraînement, il est bien évident que la pente des versants de chacune de ces vallées dépendra uniquement de la *vitesse* avec laquelle la roche composante se désagrège : plus cette vitesse sera grande, plus la pente sera faible. Si nous considérons maintenant le cas d'un versant sur lequel affleurent des roches de différente nature, chaque partie du profil présentera une pente différente en relation inverse avec la vitesse de désagrégation de la roche à laquelle cette partie correspond.

La figure 11 (pl. VI), qui reproduit le versant de droite de la vallée de l'Ozerain vu du Mont Auxois, montre ces différences.

Un autre exemple très frappant est représenté dans la figure 12 (pl. VI), où l'on voit un versant formé de gradins successifs assez rapprochés les uns des autres. Les roches qui affleurent à la surface du sol sont composées de couches stratifiées d'une épaisseur relativement faible alternativement résistantes et faciles à désagréger; l'entraînement de ces dernières a détruit successivement la base qui supportait les couches supérieures et a forcé, par un recul progressif de ces dernières, le profil du versant à prendre la forme d'un escalier géant. La disposition en gradins de certains amphithéâtres, tels que le cirque de Gavarnie, sont, sur une échelle beaucoup plus grande, le résultat d'une opération semblable.

Nous ferons remarquer à cette occasion que la dégradation de la roche facilement décomposable, au lieu de produire une pente douce raccordant directement la crête et le pied des deux couches résistantes adjacentes, se traduit, comme dans la figure 13 (pl. III), par une excavation plus ou moins profonde du profil qui a pour effet de laisser la roche supérieure en surplomb. Lorsque la stratification est horizontale, cette excavation d'une même couche produit au milieu de l'escarpement général une rainure horizontale d'une longueur parfois notable, que certains observateurs ont attribuée à tort à l'action d'un cours d'eau qui aurait autrefois coulé à la hauteur correspondante sur le flanc de la vallée. Cette manière de voir est contredite immédiatement par ce fait d'observation que là où les couches sont contournées, comme dans les cluses du Jura par exemple, la rainure suit exactement le contournement : elle a donc été pratiquée dans une même couche et elle provient de la faible résistance de celle-ci aux agents atmosphériques.

En réalité, la loi qui régit le profil des versants n'est pas tout à fait aussi simple que nous l'avons dit à cause de l'excès de volume des eaux que reçoivent les couches les plus basses; de cet excès il résulte que si une même roche formait deux couches distinctes, l'une au sommet et l'autre vers la base du versant, la première présenterait une pente plus raide que la seconde.

De même, la couche qui forme la base du versant présente une exception qu'il importe de signaler (pl. VII, fig. 14) : ainsi que nous l'avons dit, pour le cas d'une couche unique, son profil ne peut que s'abaisser en tournant autour du pied B du versant, tandis que le profil des roches qui la surmontent recule parallèlement à lui-même, puisque le point P, qui en forme le pied, se déplace. En effet, nous avons vu que le profil s'établissait par le pied : le profil B P′ s'établira donc indépendamment de ce qui existe au-dessus : si la dégradation de N P n'a pas marché aussi vite, la roche N P se trouvera en surplomb sur une partie de P′ P variable suivant les vitesses relatives de la désagrégation, et elle s'écroulera, suivant une tranche correspondante, sur la pente inférieure d'où elle sera entraînée ultérieurement. Cette action de *sapement* se présentera toutes les fois qu'une roche résistante surmontera une roche facile à désagréger; et cette circonstance accélérera le modelé des versants en hâtant la disparition de roches dont la cohésion aurait opposé une longue résistance aux agents

4

IMPRIMERIE NATIONALE.

météoriques, si ces derniers avaient dû agir seuls pour produire leur désagrégation et les préparer pour l'entraînement.

Il est facile, avec ce que nous venons de dire, d'expliquer les formes variées qu'affecte le profil des versants composés de plusieurs roches superposées. Si elles se succèdent suivant un ordre tel que la vitesse de désagrégation de chacune d'elles aille toujours en croissant de la base au sommet, le profil correspondant sera concave vers le ciel (pl. VIII, fig. 15 *a*); il sera convexe dans le cas contraire (fig. 15 *b*); enfin il sera concavo-convexe (comme dans la fig. 15 *c*), quand il y aura superposition des couches dans un ordre qu'il est facile de prévoir; et l'on peut imaginer toutes sortes de combinaisons intermédiaires donnant aux versants les profils les plus variés. Il y aura lieu naturellement de tenir compte de ce que, toutes choses égales d'ailleurs, aux roches les plus basses correspondent les pentes les plus douces, pour la raison que nous avons indiquée plus haut; cet état de choses contribue à l'élargissement du fond des vallées, lorsque la roche placée à la base du versant présente une résistance notablement moindre que toutes les autres.

Raccord des pentes et suppression des arêtes.

Dans les exemples ci-dessus, nous avons donné aux profils une forme polygonale; mais le plus souvent, dans la nature, ce profil est au contraire une ligne à courbure continue, soit que les angles rentrants aient été comblés par les débris, soit que les arêtes aient été émoussées par les frottements. D'une façon générale, nous pouvons observer d'ailleurs que les arêtes vives ne doivent se présenter qu'à l'état d'exceptions : considérons en effet (pl. VII, fig. 16) deux versants AS, BS, se rencontrant en S suivant un angle aigu; comme nous l'avons dit plus haut, si toutes les circonstances sont les mêmes et que partout la roche soit unique, la désagrégation s'effectuera suivant deux bandes limitées par des lignes telles que A'P, B'Q, parallèles à la surface primitive AS, BS; on voit donc que la région voisine de l'arête, c'est-à-dire celle dont la section est représentée par le rectangle PMQS, subira l'effet d'une double attaque, par la surface SQ et la surface SP simultanément, et que, par conséquent, elle sera plus rapidement désorganisée et susceptible, par suite, d'être entraînée plus vite que le reste de la surface des versants.

MARCHE IRRÉGULIÈRE DU FAÇONNEMENT DES VERSANTS.

L'aplatissement des versants ne se produit pas avec la régularité qui semblerait résulter de notre analyse; mais l'effet final n'en reste pas moins le même : en réalité, c'est par une série d'à-coups que la démolition se produit : sur les pentes argileuses, le fendillement dû à la dessiccation détermine des fissures longitudinales dans lesquelles les eaux s'infiltrent; ces eaux, délayant l'argile, donnent lieu à la formation de surfaces de glissement, généralement cycloïdales[1], sur lesquelles sont entraînées les masses supérieures qui, poussées en avant par leur poids, engendrent sur la surface du versant autant de creux et de protubérances (pl. VIII, fig. 17). C'est pour cette raison qu'aux environs de Toul, par exemple, les pentes extérieures du plateau corallien, qui descendent vers la plaine de la Woëvre, présentent au-dessous du couronnement une série d'irrégularités dues à l'affleurement des argiles oxfordiennes; le même effet s'observe pour les marnes supra-liasiques qui forment le pied de la crête parallèle située immédiatement à l'Est; il en est de même pour chacune des autres ceintures du bassin parisien, en tant du moins que leur base présente des terrains argileux, et notamment pour la *falaise* tertiaire des environs de Reims et d'Épernay. Dans tous ces points, les versants ont une pente plus forte que celle qui convient au talus d'équilibre des matériaux considérés, et nous les trouvons encore, pour ainsi dire, en voie de formation.

Lorsque les terrains qui affleurent sur un versant sont composés de couches minces alternativement dures et tendres, on voit les couches dures s'incliner, dans le voisinage de la surface, vers le fond de la vallée par suite de la disparition plus rapide des matériaux meubles interposés (pl. VIII, fig. 18 *a*).

Cette préparation pour l'entraînement se manifeste encore d'une autre manière : lorsque sur un versant une couche dure surmonte une couche tendre (fig. 18 *b*), par suite des infiltrations qui se produisent à travers sa masse, les matériaux tendres qui lui servent de base peuvent être entraînés en partie, et, privée alors de support, elle se fendille et les dif-

[1] Voir A. Collin, *Recherches expérimentales sur les glissements spontanés des terrains argileux*. Paris, 1846.

4.

férents blocs ainsi découpés descendent en échelons, dont chacun disparaîtra à son tour, soit progressivement, soit par un mouvement en masse. Il n'est pas rare de voir encore, sur les pentes marneuses que dominent des escarpements calcaires, des blocs isolés qui témoignent de ce mode d'acheminement (fig. 18 *c* et 18 *d*); tandis qu'ailleurs la roche divisée en menus débris forme sur la partie inférieure des versants un manteau superficiel destiné à gagner tôt ou tard le fond de la vallée.

APPLICATION DES CONSIDÉRATIONS PRÉCÉDENTES AUX DIVERSES DISPOSITIONS QUE PRÉSENTENT LES VERSANTS DES VALLÉES.

1° *Profil des versants en terrains stratifiés horizontalement.* — Considérons une vallée ouverte à travers des roches de nature différente stratifiées horizontalement : si les agents de désagrégation et les quantités de pluie tombées sur chaque versant sont les mêmes, le profil des deux versants sera forcément le même, et la vallée offrira l'aspect représenté dans la figure 19 (pl. VII).

Si les actions sont inégales, le versant qui sera le plus exposé à la dégradation et à l'entraînement verra son profil s'aplatir plus rapidement que l'autre (pl. VII, fig. 20) et peut-être faut-il chercher dans cette cause l'explication de certains cas de dyssymétrie systématique des deux versants de vallées dont l'orientation générale est la même.

2° *Profil des versants en terrains stratifiés inclinés.* — Si les couches, au lieu d'être horizontales, sont légèrement inclinées, il en résulte forcément une dyssymétrie complète des deux versants. Supposons en effet, dans la figure 21 (pl. VII), le fond de la vallée réduit à sa ligne de thalweg : toutes choses égales d'ailleurs, le profil des deux versants, dans la roche inférieure, s'établira suivant les lignes TA et TB, formant le même angle avec la verticale; dans la roche supérieure, les talus AD et BC auront également la même inclinaison par rapport à cette même verticale; et le profil général sera CBTAD, complètement dyssymétrique, dans lequel les éléments correspondant aux mêmes couches auront la même inclinaison, mais non la même longueur : l'inspection de la figure montre immédiatement comment cet effet est lié à l'inclinaison des couches.

Il importe de signaler un cas particulier (pl. IX, fig. 22) : c'est celui où l'inclinaison des couches est assez grande pour que le talus T A de la figure 21, qu'aura pris à la longue la couche inférieure, arrive à coïncider avec la surface supérieure T M de la couche immédiatement sous-jacente; dans ce cas, il y a tendance à la disparition complète des terrains qui constituaient le versant TAD, et alors le profil de la vallée se présente comme on le voit dans la figure 22, réalisant un cas extrême qui se rencontre souvent dans la nature. Les vallées dont les versants sont formés de la sorte rentrent dans la classe des vallées *monoclinales*[1], caractérisées, comme on le voit, par ce fait que l'un des versants est formé par le dos de la couche inférieure et l'autre par la tranche des couches supérieures; cette disposition peut d'ailleurs provenir d'un mécanisme différent, comme nous le verrons plus tard, à savoir le déplacement latéral des cours d'eau sur la surface de la roche résistante T M. On peut remarquer que, surtout dans le premier cas, cette dyssymétrie géologique ne correspond pas toujours à une dyssymétrie géométrique; car il peut arriver qu'au-dessus de TM il n'y ait qu'une seule et même roche : dans ce cas le profil de la vallée se réduit à B′B T M, où il peut se faire que B T ait précisément la même inclinaison que T M.

3° *Plis et cluses.* — C'est surtout dans les régions plissées que la proportionnalité des pentes à la vitesse de désagrégation des roches conduit à des effets remarquables.

Lorsqu'un pli saillant se trouve traversé par un cours d'eau qui l'a entaillé assez profondément pour mettre à jour des couches différentes, la *cluse* ainsi formée (pl. X, fig. 23) présente les dispositions suivantes :

Sur les versants, de part et d'autre, les différentes couches se présentent sous l'aspect de demi-cylindres concentriques semblables à des voûtes en plein cintre, dont les retombées viennent rencontrer perpendiculairement le thalweg; si l'on considère un profil fait par un plan vertical passant par l'axe commun à ces voûtes (pl. X, fig. 24), il présente la même succession de pentes qu'une vallée composée de mêmes roches stratifiées horizontalement; mais ces profils se modifient si on les considère dans des plans différents, parallèles au premier; et, en particulier, dans le voisinage de

[1] L'adjectif *monoclinal* a été proposé, dès 1842, par W.-B. et H.-D. Rogers (*On the physical structure of the Appalachian Chain,* Transact. Amer. Assoc. of Geologists, p. 485).

l'entrée ou de la sortie de la cluse, où les couches successives sont à peu près verticales, les couches résistantes se présentent sous la forme de murs isolés (pl. X, fig. 25) entre lesquels les couches tendres forment le fond d'un couloir plus ou moins encaissé. Cela tient à ce que ces dernières, se désagrégeant plus vite, ont pris, à partir du thalweg, un talus beaucoup plus doux que les roches résistantes intercalées, lesquelles sont demeurées en saillie.

On pourrait analyser davantage les formes que présentent les cluses et y trouver des exemples de toutes les variétés de profils qu'on peut observer dans la nature, parce que des roches différentes s'y rencontrent entremêlées et y présentent toutes les inclinaisons possibles.

Sur la coupe de la figure 24, nous avons projeté la partie du terrain située derrière le plan de profil. La ligne AMB représente en particulier la projection de l'intersection de la surface extérieure du pli avec la surface qui correspond au versant de la cluse : il est remarquable que, en aucun point, cette ligne ne présente une pente plus forte que celle qui correspond au talus naturel des matériaux meubles. Cela prouve que la surface extérieure du pli a subi des ablations assez considérables pour que la limite, à partir de laquelle les matériaux désagrégés ne peuvent être entraînés que grâce à l'intervention de l'eau, ait été dépassée. C'est le cas général de tous les flancs des voûtes calcaires qui constituent les plis du Jura, par exemple.

Sur ces flancs, en particulier, le profil s'établit suivant une disposition qu'il est utile de signaler : soient A et C deux couches calcaires concentriques (pl. IX, fig. 26) séparées par une couche B marneuse : le profil MNPQ sera déprimé de N en P de façon à former une sorte de gradin dont la surface supérieure correspond à NP; l'explication de cette dépression est la conséquence naturelle du principe déjà développé : la pente de P à N est plus douce que dans le reste du profil, parce que c'est entre ces deux points qu'affleure sur le versant la couche la moins résistante.

REPRODUCTION EXPÉRIMENTALE.

Il est facile de reproduire expérimentalement et d'une façon très simple quelques-uns des aspects présentés par les versants sur lesquels affleurent des roches de nature différente.

Sur une petite planchette ABCD (pl. IX, fig. 27), on répand, à l'aide d'un tamis, en couches minces alternantes, du grès pulvérisé et du plâtre fin, en ayant soin que le tas ainsi formé déborde la planchette de toutes parts. Si alors on soulève la planchette en la maintenant horizontale, on emporte avec elle une partie de la matière entassée, dont les surfaces latérales prennent, à partir des bords, le talus d'équilibre qui leur convient en chaque point; c'est-à-dire que, dans les parties où affleure le grès pulvérisé, la pente est douce; elle reste raide au contraire dans celles qui correspondent au plâtre; et l'ensemble présente une succession de pentes douces et de pentes raides qui rappellent exactement celles de certains versants dans la nature.

Si, au contraire, on plaçait d'abord sur la planchette le plâtre, par exemple, de façon à lui donner la forme d'une partie de cylindre convexe (fig. 28), et que, par-dessus, on disposât des couches concentriques minces alternées de grès et de plâtre, en faisant en sorte que le tas ainsi formé dépassât, au moins de l'un des côtés, le rebord de la planchette qui est perpendiculaire aux génératrices du cylindre, en soulevant la planchette, on verrait se former sur le rebord correspondant une surface d'équilibre qui, par l'alternance des pentes raides et douces et la disposition générale en plan des couches correspondantes, reproduirait exactement sur la face AD l'aspect caractéristique des versants des cluses.

Dans les expériences précédentes, si, tenant toujours la planchette horizontale, on lui donne de légers chocs, on produit l'aplatissement progressif des surfaces latérales : leur pente générale diminue par une rotation autour des rebords de la boîte, mais elles gardent constamment leur disposition en gradins.

Le mode de démolition des couches à parois verticales formées par le plâtre est particulièrement intéressant à observer. La couche de grès, en s'établissant suivant la pente douce qui lui est propre, laisse en surplomb des bandes de plâtre qui bientôt s'écroulent et disparaissent entraînées sur les pentes inférieures.

C'est, comme on le voit, la reproduction exacte des phénomènes naturels, tels que nous les avons décrits.

EN GÉNÉRAL, LE TALUS D'UNE ROCHE S'ÉTABLIT INDÉPENDAMMENT DE LA STRATIFICATION.

C'est le moment de faire remarquer qu'en s'établissant sous la forme d'un talus plus ou moins incliné, la surface des versants, du moins quand les couches sont à peu près horizontales, vient couper les plans de stratification suivant des angles d'autant plus aigus que la pente de ces versants est plus faible, ce qui produit la disposition en biseau sur laquelle nous avons attiré l'attention dès le début. Les eaux de pluie, agissant sur les matériaux désagrégés de la surface des roches, façonnent les versants aux dépens de la roche vive, sans être notablement influencées par les changements d'allures de la stratification (épaisseur et inclinaison), absolument comme un rabot aplanit une pièce de bois, sans que les changements d'inclinaison des couches concentriques apportent une modification quelconque au résultat. D'une façon générale on peut donc dire que l'*inclinaison des versants est indépendante de la stratification.* Un exemple très frappant de ce fait se rencontre dans la vallée du Doubs entre la Rasse et Maison-Monsieur (pl. IX, fig. 29). Le passage du Doubs en ce point paraît avoir été déterminé par un pli synclinal dont l'un des flancs (situé sur la rive droite) est légèrement renversé. Dans l'état primitif, ce profil transversal de la dépression structurale devait être $m\,n\,p\,q\,r$, comme l'indique l'inclinaison des couches à la surface des versants actuels. Dans cet *état,* la roche se présentait, en certains points, en surplomb, et là elle a dû disparaître promptement sous l'action de la pesanteur. Mais partout ailleurs, et même au-dessous des parties arrachées, sont intervenues les actions désagrégeantes et entraînantes, qui ont donné, en fin de compte, à la vallée son profil actuel, à pentes modelées et relativement douces, rencontrées sous les angles les plus divers par les plans de stratification. L'effet des agents de désagrégation et de transport a donc été de raboter successivement la surface primitive jusqu'à ce qu'ils aient donné aux versants de la vallée la pente continue qui convient à la roche correspondante, dont les contournements auraient pu être différents sans que la forme actuelle cessât d'être sensiblement la même.

Sans être aussi importants dans les résultats, les effets de cette indépendance relative de la surface du sol et de la stratification se traduisent aux environs de Morez d'une façon assez singulière pour que nous en donnions ici la description (voir pl. IX, fig. 30). La rive droite de la vallée de la Bienne, en amont de Morez, entre cette ville et Les Rivières, est formée par la surface des couches calcaires qui plongent vers le thalweg; cependant la surface du versant n'est pas constituée par la surface d'une seule et même couche : celle-ci a subi des désagrégations et des ablations qui ont eu pour effet final de donner au versant son modelé actuel : la surface primitive s'est trouvée remplacée de la sorte par une surface très peu différente, qui coupe la première sous un angle très faible. Cet angle varie d'ailleurs suivant les sinuosités longitudinales de la surface du versant, qui n'est pas un plan. Ce rabotage a eu pour effet de mettre au jour les plans de stratification des couches sous-jacentes qui, grâce à l'obliquité des intersections, dessinent des rebroussements complets; de telle sorte que l'observateur, qui instinctivement se figure avoir sous les yeux une coupe pratiquée à travers des couches dont les génératrices seraient horizontales et s'appuieraient sur les lignes correspondant aux surfaces séparatives des couches, en conclurait que la roche qu'il a sous les yeux a été l'objet de contournements violents poussés jusqu'au renversement.

Il est facile de réaliser ces apparences de la façon suivante (pl. IX, fig. 31) : si autour d'un cylindre en bois, on enroule une feuille de papier, et que, en râpant la surface supérieure, on produise des facettes peu profondes, atteignant la quatrième ou cinquième enveloppe par exemple, suivant les directions relatives que l'on donnera au coup de râpe, on produira à volonté des lignes qui présenteront des points de rebroussement; le bord des feuilles ainsi découpées représentera les joints de stratification de l'exemple précédent, tandis que les feuilles figureront les couches elles-mêmes. Il nous a été facile en opérant de la sorte de reproduire toutes les apparences du versant droit de la vallée de la Bienne. Citons encore un cas très curieux sur les bords de l'Ain, où la surface des couches, présentant localement une disposition hémisphérique, son intersection par une surface hémisphérique de plus grand rayon, à la suite du façonnement du versant, a produit l'apparence paradoxale d'une stratification cylindrique dont toute la section serait visible[1].

[1] Le même fait peut être observé en grand dans la région du Mont-Perdu (*Annuaire du Club alpin français*, 13e année, 1886, p. 613, 616).

[A discussion of outcrop patterns and contour patterns in plan (pp. 34–37), and a statement regarding slopes and sinuous valleys (p. 38) were deleted.]

DÉFINITION GÉOMÉTRIQUE DE LA SURFACE DES VERSANTS.

Si l'on veut donner de la forme des versants une définition géométrique qui, il est vrai, ne sera jamais rigoureusement applicable dans la nature, mais qui a l'avantage de préciser les considérations précédentes en les systématisant, on peut dire que *la surface des versants est engendrée par une génératrice qui aurait en chaque point la pente maximum correspondant à la roche affleurant au point considéré, cette génératrice coïncidant toujours avec la ligne de plus grande pente de la surface et s'appuyant constamment sur la ligne qui forme le pied du versant comme directrice.*

Rappelons que par *valeur maximum* de la pente nous entendons la pente la plus grande qu'ait encore atteinte la roche dans l'*état donné* de désagrégation et d'entraînement où elle se trouve.

Il importe de rappeler en effet, que les reliefs du sol sont, pour ainsi dire, toujours en voie de modification; nous les voyons dans un état *transitoire*, qui est plus ou moins avancé suivant le degré de résistance des roches, l'activité des causes modifiantes et la durée de leur action. Telle roche, dans un climat sec, aura subi, au bout du même temps, des dégradations beaucoup moins considérables que dans un climat humide; dans le premier cas, elle se présentera avec une raideur de talus et une vivacité d'arêtes qui seront remplacées dans le second par l'arrondissement des angles et l'aplatissement des reliefs. C'est par des considérations de ce genre que, sous un même climat, on pourrait fixer l'âge relatif des différents accidents de la surface du sol.

MISE EN SAILLIE DES PARTIES DURES.

Une conséquence importante de l'entraînement plus rapide des roches peu résistantes est la *mise en saillie des roches dures*[1]. C'est une règle générale dans la nature que toutes les saillies du sol, toutes les arêtes qui se présentent dans le profil des versants, les crêtes de ces versants et la surface des plateaux contiguës sont formées par les roches les plus résistantes de la série locale; tandis que les roches les moins résistantes forment les

[1] L'importance du principe de la résistance des parties dures a été bien mise en lumière dans diverses publications de MM. Lesley, Gilbert et Dutton, en Amérique; Ramsay et Geikie, en Angleterre; Heim, Penck et Supan, en Allemagne; Brögger, en Scandinavie.

dépressions. La coupe du Mont Terri (pl. XII, fig. 38) montre, dans un petit espace, *cinq* alternances de roches tendres et de roches dures, auxquelles répondent des saillies et des dépressions qui satisfont à la règle que nous venons d'indiquer.

On pourrait multiplier ces exemples à l'infini, car c'est là un des traits les plus généraux de la surface des continents. Nous nous contenterons de citer encore à cause de son effet pittoresque, particulièrement saisissant, le cas des dykes si fréquents dans toutes les régions volcaniques : dans la planche XIII, fig. 39, on peut voir une sorte de muraille naturelle haute de plus de 10 mètres et dont l'isolement est dû à la grande résistance du trachyte qui la constitue, tandis que les argiles au milieu desquelles cette roche s'est épanchée ont été entraînées depuis longtemps, grâce à leur faible cohésion. Des exemples beaucoup plus grandioses encore d'obélisques de basalte, isolés par l'enlèvement des masses au milieu desquelles ils avaient surgi, et arrivant à dépasser leur base de plus de 600 mètres, sont décrits par Dutton [1].

La mise en saillie des parties dures se produit également sur les affleurements d'une même roche, quand celle-ci présente des inégalités de résistance dans ses différentes parties, et, comme ces inégalités peuvent être distribuées d'une façon très irrégulière, il en résulte parfois des effets singuliers. C'est évidemment à cette cause qu'il faut attribuer par exemple les roches de Saint-Mihiel (voir pl. XIII, fig. 40).

En aval de cette localité, sur la rive droite de la Meuse, immédiatement au pied du versant en pente douce de la vallée et à moitié enterrées dans son talus, s'alignent sept colonnes massives composées de roches calcaires stratifiées horizontalement, dont les couches prolongées se raccorderaient exactement avec celles de même composition générale qui affleurent sur le versant, où les actions pluviales les ont tranchées obliquement. Sous l'influence d'une cause qu'il resterait à déterminer, les parties de la roche aujourd'hui en saillie avaient atteint un degré de cohésion beaucoup plus grand que les parties voisines, lesquelles ont été entraînées par les agents qui ont modelé le versant, tandis que les premières restaient debout,

[1] *Mount Taylor and the Zuñi Plateau* (*6th Ann. Rep. U. S. geol. Survey*, 1884-1885, p. 164-182, nombreuses figures).

comme pour témoigner du mode d'action des eaux de pluie dans le façonnement des versants.

C'est à la même *sélection* qu'est due la présence, dans certaines régions, à la surface des roches granitiques, de blocs dans lesquels quelques observateurs ont cru voir des débris erratiques, alors qu'ils ne représentaient que les parties les plus dures de la roche en place, seules respectées par la désagrégation [1].

Certaines surfaces raboteuses doivent leur rugosité à la même cause. Un exemple remarquable nous est fourni par les *Lapiaz* ou *Karrenfelder* des Alpes; d'après les observateurs les plus autorisés, les roches calcaires, qui présentent à leur surface cette disposition, par suite de légères différences locales dans leur composition, ont été l'objet de dissolutions irrégulières, qui ont eu pour effet d'y créer une foule de petites inégalités, donnant à l'ensemble un aspect hérissé, les dépressions ainsi produites ayant été ensuite élargies et réunies en canaux continus par le travail mécanique des eaux de pluie [2].

MODIFICATIONS APPORTÉES PAR L'HOMME À LA FORME DES VERSANTS.

D'une façon générale, on peut dire qu'il n'existe aucune roche qui, même considérée sous un faible volume, présente dans tous ses points une composition et une texture absolument identiques. L'inégalité de désagrégation qui résulte de ce défaut d'homogénéité doit donc produire des irrégularités dans la forme de la surface.

Il suffit d'observer les régions en friche pour s'en convaincre : c'est l'homme qui, en préparant le sol pour la culture à l'aide de la pioche et plus tard de la charrue, donne au paysage cet aspect arrondi et poli, ce fini qui le caractérise, son modelé en un mot. Dans certaines régions, on peut saisir sur le fait cette transformation : au milieu de champs cultivés surgissent çà et là des gibbosités rocheuses dont la pioche du cultivateur arrache chaque année un lambeau et qui disparaîtront un jour, sous ce travail répété, pour faire place à des pentes uniformes.

[1] Voir notamment J. Crevaux, *Faux blocs erratiques de la Plata, prétendue période glaciaire d'Agassiz dans l'Amérique du Sud* (*Bull. Soc. Géol. de Fr.* [3] IV, 1876, pl. VII, p. 304-308).

[2] Voir A. Heim, *Einiges über die Verwitterungsformen der Berge*, Zürich, 1874, pl. I, fig. 1, p. 9-12.

6

IMPRIMERIE NATIONALE.

L'action de l'homme produit encore d'autres modifications : ainsi en poussant sa charrue à travers certains thalwegs, à pente faible et à versants peu inclinés, suivant leurs horizontales, le laboureur efface, pour ainsi dire, la forme angulaire primitive de leur surface et lui substitue une forme cylindroïdale dont les horizontales dessinent des demi-circonférences. Telles sont les dépressions qui descendent de toute part du Mont Bernon près d'Épernay.

Parfois, au contraire, l'homme détruit la régularité de la surface pour les besoins de la culture; ainsi dans les terrains à pente raide et dont la roche se décompose en parcelles très menues, que les eaux de pluie entraîneraient trop facilement, il dispose le sol en terrasses[1], soit à l'aide de murs de soutènement, soit en entaillant directement la roche, lorsqu'elle est susceptible de se tenir sous un talus assez fort, afin de décomposer le profil en une série de plates-formes superposées. La première disposition est très fréquente dans les terrains secs et rocheux du midi de la France; la seconde s'observe dans des roches assez diverses : coteaux de grès rouge au nord de Belfort; marnes jurassiques aux environs de Langres, Nancy, etc.; craie des environs d'Épernay, où cette disposition artificielle prend un tel développement que les habitants de la contrée appellent *les Escaliers* la localité correspondante (pl. XV, fig. 41). Ces gradins ne sont pas, comme certains observateurs l'ont cru, produits par des failles en escalier. Les fouilles que nous avons fait pratiquer nous ont montré qu'il n'y avait aucune fissure dans la roche; ce ne sont pas davantage les berges successives d'un cours d'eau, dont le niveau se serait abaissé progressivement, parce que les lignes suivant lesquelles elles sont dirigées, outre qu'elles ne sont pas toujours horizontales, sont discontinues et, au lieu de se montrer toujours parallèles à la direction des versants, sont quelquefois perpendiculaires au thalweg à la traversée des ravins.

Certains terrains où l'on n'a pas pris la précaution de retenir les terres végétales à l'aide de terrasses sont devenus aujourd'hui impropres à la culture; c'est ce qui est arrivé par exemple pour la partie supérieure des coteaux situés en aval de Tonnerre : ils sont aujourd'hui en friche et les débris de la surface meuble primitive sont accumulés à leur pied sous forme d'éboulis.

[1] Appelées *restenques* en Provence et *traversières* dans les Cévennes.

DES ÉBOULIS.

Ainsi que nous l'avons dit, la surface du sol est formée en général par l'affleurement de la roche vive; il est des cas cependant où celle-ci est recouverte localement, sur une grande épaisseur, de matériaux désagrégés et amoncelés.

Lorsqu'une roche résistante, susceptible de former un escarpement, surmonte une roche à surface peu inclinée, les débris de sa décomposition, sous l'influence de la gelée principalement, tombent par l'effet de la pesanteur, et, si cette désagrégation est plus rapide que l'entraînement des débris par les eaux, ils s'accumulent au pied de l'escarpement sous une inclinaison supérieure, en général, à celle des versants où affleurent des roches tendres et déjà façonnés par les eaux de pluie. Cette inclinaison, qui ne dépend que de la pesanteur, est celle que prennent tous les matériaux meubles disposés en tas; elle est à peu près indépendante de la grosseur et de la nature des matériaux et égale à 36 ou 37 degrés environ, valeur maximum qu'elle puisse atteindre. Aussi, en dehors des escarpements, qui constituent une exception, ne trouve-t-on dans la nature aucune pente supérieure à 37 degrés, bien qu'une illusion d'optique leur fasse souvent attribuer une plus grande valeur. La raideur des talus d'éboulis se maintient d'ailleurs grâce à cette circonstance que les eaux de pluie, qui tombent à leur surface, y pénètrent immédiatement sans ruisseler par dessus.

FAUX ÉBOULIS.

Souvent les talus raides d'une roche surmontée par des escarpements peuvent être pris pour des éboulis formés des fragments qui en seraient tombés, parce qu'ils ont à peu près la même pente et qu'ils sont recouverts d'un manteau formé par les débris de la roche supérieure; mais il suffit d'examiner les ravins découpés par les torrents à travers ces pentes pour constater que la roche dure arrive très près de la surface, sur laquelle les débris descendent d'une façon continue, au fur et à mesure qu'il s'en forme de nouveaux. De pareils exemples se présentent fréquemment : nous citerons ceux du versant droit du lac de Nantua et ceux du versant oriental de la chaîne de la Grande Chartreuse en amont de Grenoble

6.

(pl. XV, fig. 42). Ces *faux éboulis* sont plus fréquents qu'on pourrait le croire.

* * * * * * *

CLASSEMENT DES ROCHES SUIVANT L'INCLINAISON DE LEUR TALUS.

On ne peut pas établir une classification des roches, basée sur la valeur relative de l'inclinaison de leur talus, qui soit vraie dans tous les cas imaginables. Car cette inclinaison dépend *de la résistance de la roche, de la nature des agents qui produisent sa désagrégation et de la durée de leur action.* Or dans certaines régions tel agent prédomine, tandis que tel autre fait complètement défaut: si nous considérons en particulier les effets produits par la gelée, il est évident que telle roche, très gélive, pourra, sous une latitude élevée, occuper dans la série un rang très différent de celui qu'elle occupera sous l'équateur, où il gèle rarement. Nous ne pouvons donc faire qu'une classification très grossière, en considérant d'ailleurs ce qui se passe dans une région à climat moyen comme la France.

Les roches qui s'établissent en fin de compte sous l'inclinaison la plus faible sont celles susceptibles de former avec l'eau une boue fluente capable par conséquent de s'écouler sur des pentes très faibles. Telles sont les argiles, lorsqu'elles sont disposées de façon à pouvoir s'imbiber d'eau, ce qui arrive toujours à la longue, à la suite des alternatives de sécheresse et d'humidité qui les fissurent et permettent à l'eau de s'y infiltrer. Les marnes, qui sont un mélange d'argile et de calcaire, présentent des pentes encore douces, mais cependant plus fortes que les roches précédentes; elles tiennent en quelque sorte le milieu entre les argiles et les calcaires. Les calcaires compacts, en effet, peuvent se maintenir parfois suivant des escarpements verticaux; la dissolution de leurs éléments par les eaux est trop lente, en général, pour n'être pas devancée, sous nos climats, soit par l'action de la gelée, soit par l'action de sapement que détermine l'enlèvement des roches inférieures moins résistantes; et dans l'un et dans l'autre cas, la paroi de l'escarpement recule en restant verticale. Ce sont en général les calcaires qui présentent dans la nature les escarpements les plus remarquables et les formes les plus hardies et les plus accentuées. Quelquefois cependant les grès présentent des aspects analogues, mais en général avec une structure plus massive; toutefois certains d'entre eux se tiennent sous des talus à inclinaison moyenne, supérieure le plus souvent à celle des marnes et des calcaires marneux. Enfin, au point de vue de la pente, les granites peuvent être rangés à peu près au même rang que les grès dont nous venons de parler en dernier lieu. Cependant les cimes des hautes montagnes granitiques présentent des arêtes aiguës et dentelées qui semblent en contradiction avec ce que nous venons de dire; c'est parce que les parties décomposées sont enlevées, au fur et à mesure qu'elles se désagrègent, avec une rapidité telle que la surface est, pour ainsi dire, constamment avivée. C'est ainsi que des argiles peuvent se tenir sous une inclinaison très grande si elles sont sapées par leur pied et entraînées avant que les parties contiguës aient eu le temps de s'imprégner d'humidité : telles sont les falaises d'une partie de la côte du Calvados, constituées par les argiles oxfordiennes.

Dans la figure 43 (pl. XVI), nous avons représenté le versant gauche de la vallée de Bellifontaine à la sortie de Morez (Jura). Les ravins à arêtes si aiguës qu'on y remarque sont creusés dans une accumulation de terrains de transport, qui n'a pas moins de 100 mètres d'épaisseur. Les matériaux roulés sont cimentés fortement par un limon, dont l'origine paraît être glaciaire, ce qui leur permet de se tenir sous une forte inclinaison, la surface du versant étant d'ailleurs sans cesse avivée par l'action des eaux de pluie.

Nous avions donc raison de dire, en commençant le présent paragraphe, qu'il est impossible de faire un classement général des roches suivant l'inclinaison de leur talus, puisque ces derniers varient suivant les altitudes, les latitudes et même le mode de destruction que ces roches subissent. On peut dire toutefois qu'en général les pentes sont plus fortes dans les climats secs et plus douces dans les climats humides, fait qui se comprend aisément si l'on se reporte aux considérations développées ci-dessus, sur le rôle des eaux pluviales dans l'adoucissement du profil des versants; c'est ainsi que les alluvions quaternaires du Sahara, affleurant dans les *goûr* ou buttes isolées, et sur les flancs des *ouadis* desséchés, se montrent suivant des pentes rarement atteintes par les roches correspondantes dans nos climats.

RÔLE DE LA VÉGÉTATION.

On sait que, à la suite du déboisement d'un versant, les eaux de pluie entraînent immédiatement les matériaux meubles de sa surface et produisent des désastres dont on n'avait point à se préoccuper auparavant. La végétation intervient donc pour maintenir la surface du sol sous une certaine inclinaison; sans elle la désagrégation, et surtout l'entraînement par les eaux sauvages des matériaux désagrégés, continueraient à se produire jusqu'à l'aplatissement à peu près complet du relief des continents. Mais, pour que la végétation ait pu s'établir, il a fallu que la rapidité de son développement fût plus grande que la facilité d'entraînement des matériaux meubles. Cet état particulier d'équilibre s'est trouvé atteint plus ou moins vite suivant la nature du sous-sol et suivant l'ensemble des circonstances extérieures plus ou moins favorables à la végétation. Ainsi, dans les climats chauds et humides, la végétation luttera en quelque sorte avec l'abondance des pluies de façon à maintenir, malgré la tendance de ces dernières, les talus sous une pente notable. D'une façon générale, la végétation a donc donné aux roches une pente d'équilibre et assuré la stabilité ou la fixation des formes du sol, qui sans elle auraient continué à se dégrader et à s'aplatir indéfiniment. *La notion de l'équilibre des versants est donc inséparable de l'idée d'une végétation qui le détermine.*

Les preuves du rôle conservateur du tapis végétal se présentent de toutes parts: lorsque l'ingénieur veut maintenir les talus des tranchées artificielles et des remblais, il les plante d'essences susceptibles de croître rapidement et de fixer le sol, grâce à leur manteau protecteur. Malgré cette précaution, si l'on a donné au talus une pente trop forte et si les conditions atmosphériques ont été défavorables, des dégradations se produisent. Une preuve frappante de cette action de la végétation nous est fournie par l'exemple bien connu des torrents des Alpes. Dans les régions qu'on a déboisées la dégradation spontanée du sol a recommencé et l'on sait quels ravages en ont été la conséquence. D'après Prony, les défrichements exécutés dans le bassin du Pô depuis la fin du moyen âge auraient produit un effet analogue, qui se serait traduit par un accroissement notable du delta de ce fleuve.

Ainsi la végétation assure la conservation des reliefs; mais cela n'arrive qu'à partir du moment où ceux-ci ont acquis une certaine pente, variable pour une même roche avec le climat et l'ensemble des circonstances extérieures, pente en quelque sorte *limite*, correspondant à un moment où la dégradation devient si lente qu'elle ne peut plus empêcher la végétation de s'établir.

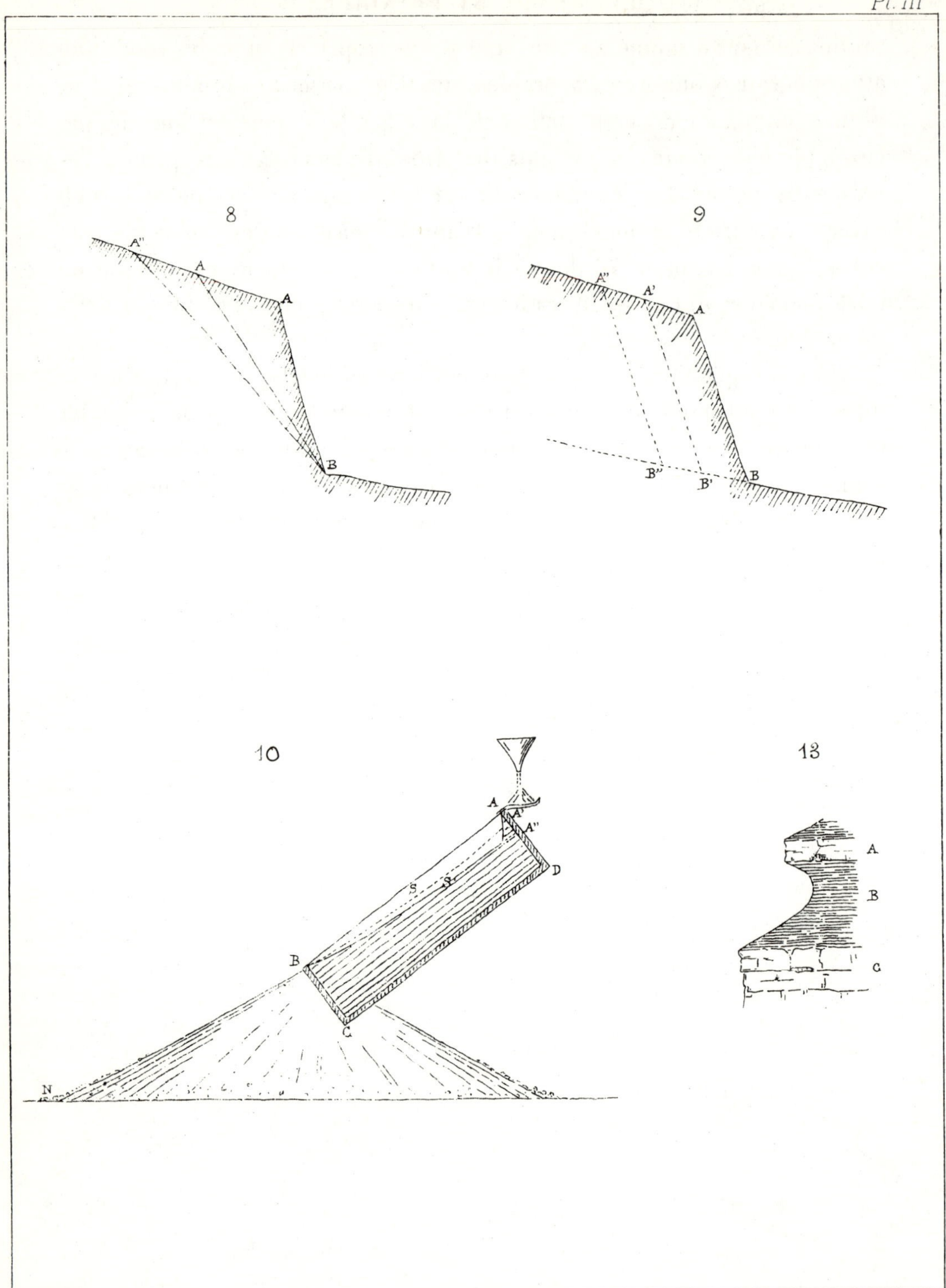
8
A''
A
A
B
9
A''
A'
A
B''
B'
B
10
A
A'
A''
S
S'
D
B
C
N
13
A
B
C

12
La Vallée de la Bienne (Jura)

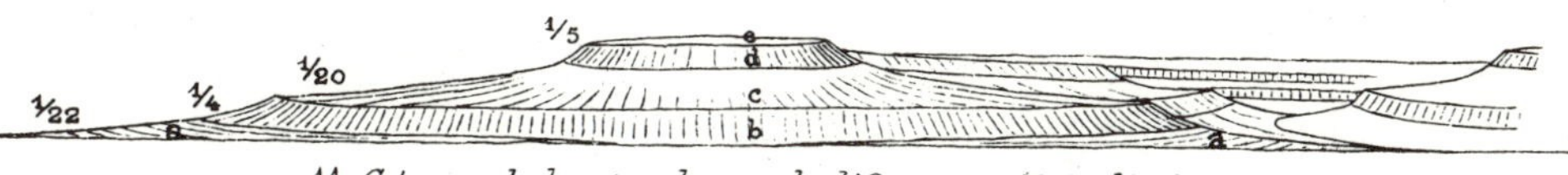

11. *Coteaux de la rive droite de l'Oserain (Côte d'or)*
(1:20000)

a-Marnes infres du Lias
b-Calc. à gryphées géantes
c-Marnes supres du Lias
d-Calcaire à entroques
e-Calc. marneux blanc jaunâtre

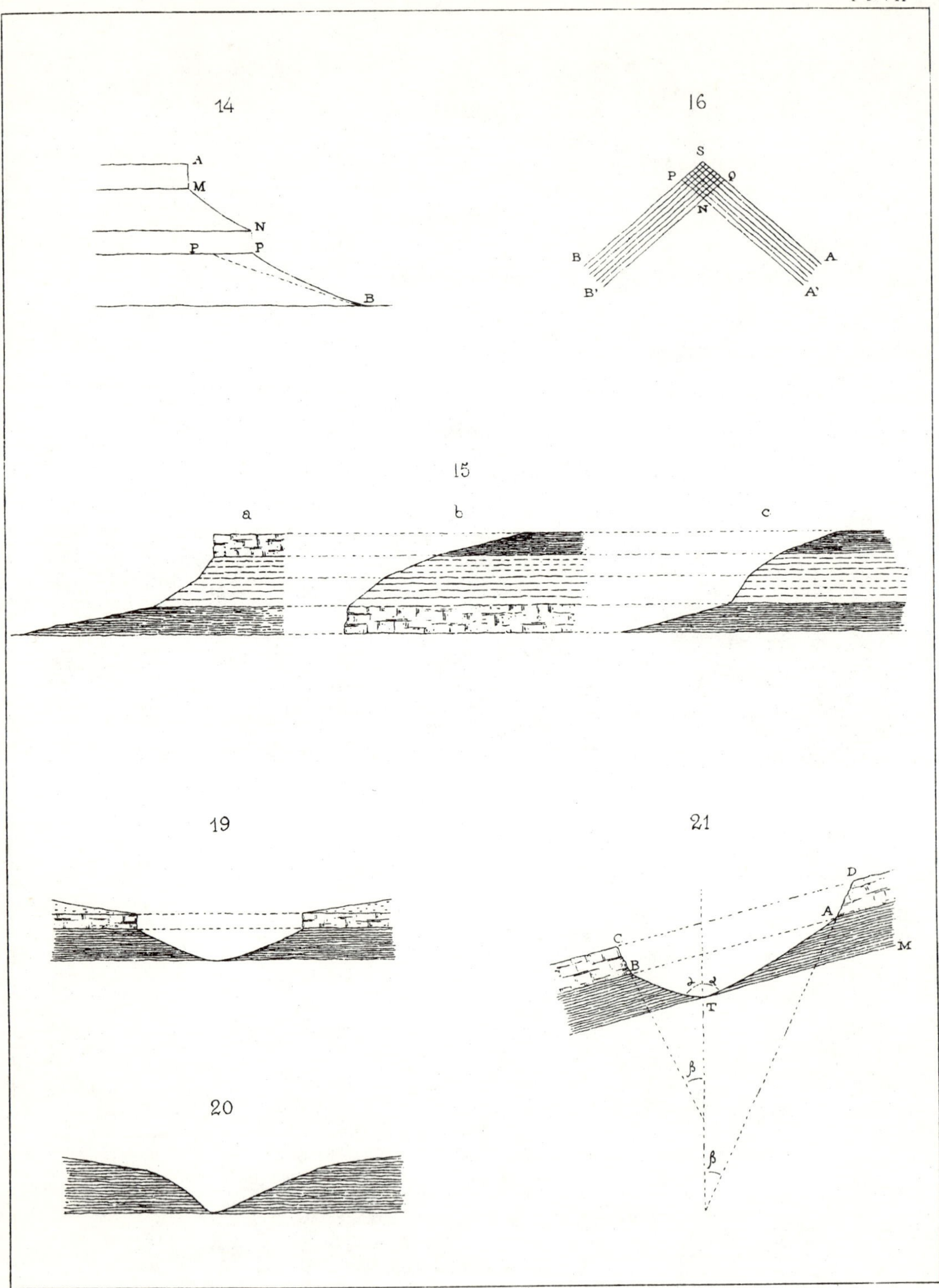

14
A
M
N
P
P
B
16
S
P
Q
N
B
B'
A
A'
15
a
b
c
19
21
D
A
C
M
B
T
α
α
β
β
20

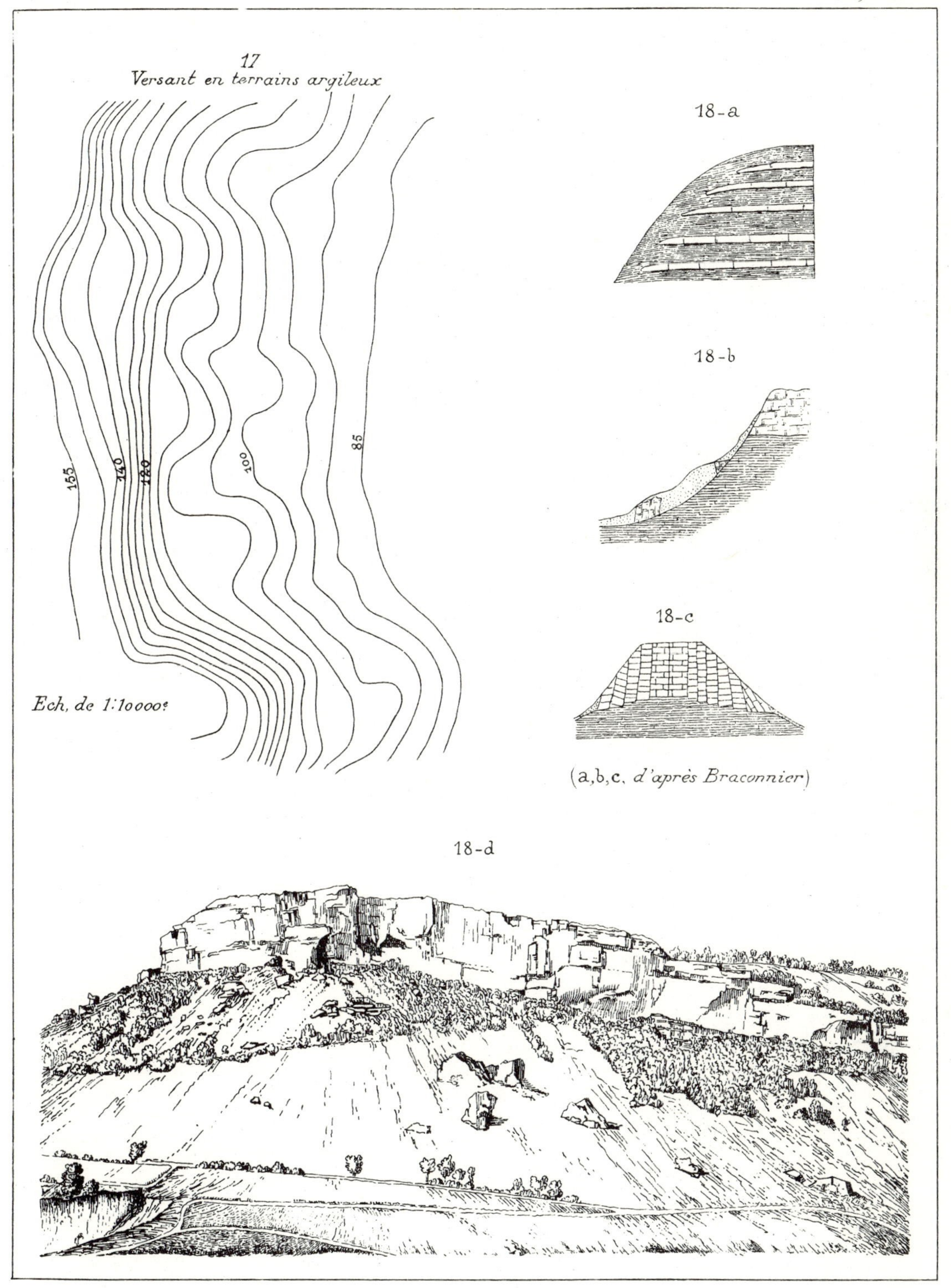
17
Versant en terrains argileux
155
140
120
100
85
Ech, de 1:10000e
18-a
18-b
18-c
(a,b,c, d'après Braconnier)
18-d

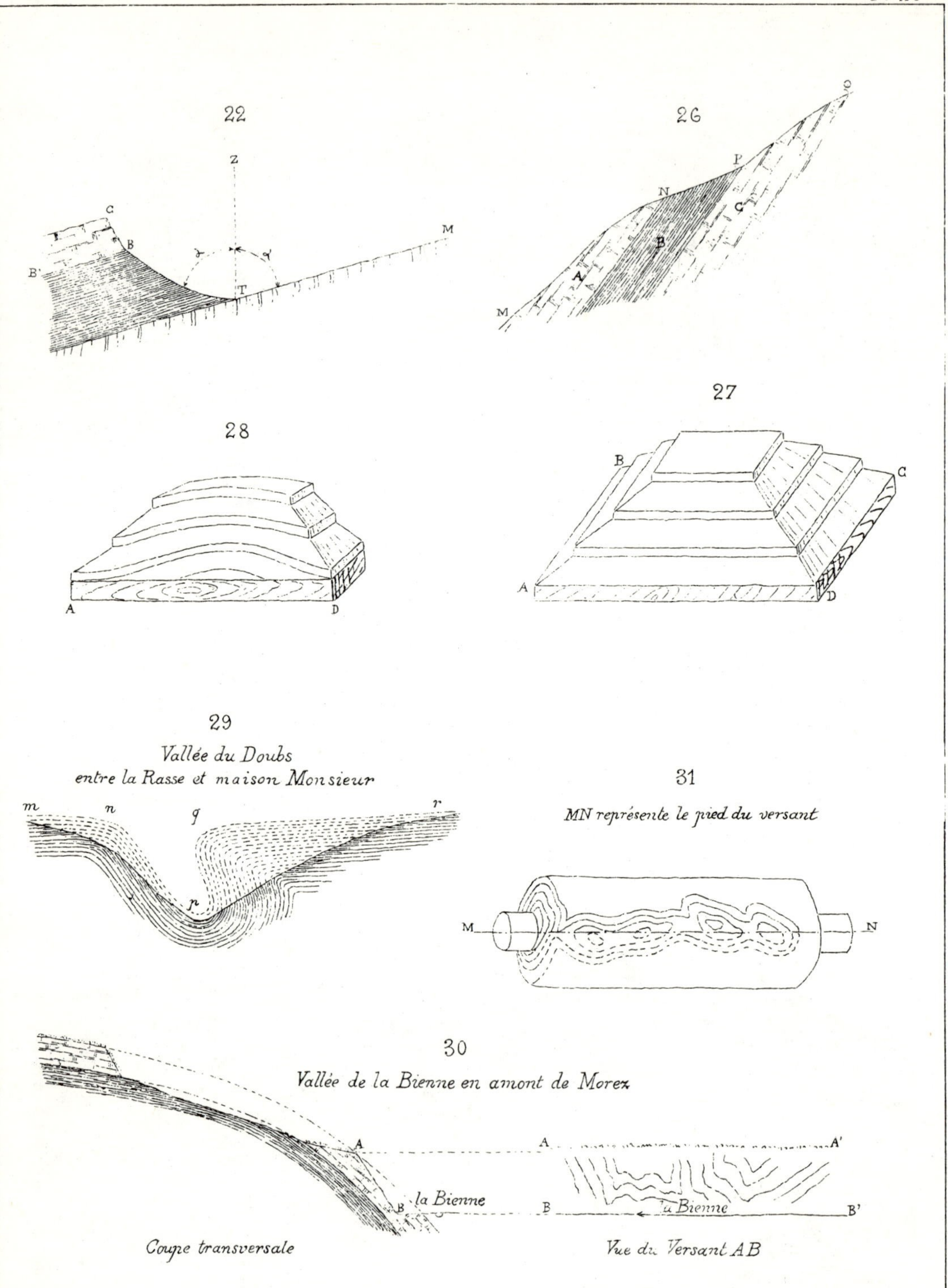
22
Z
C
B
B'
M
T
26
O
P
N
C
B
A
M
28
A
D
27
B
C
A
D
29
Vallée du Doubs
entre la Rasse et maison Monsieur
m
n
q
r
p
31
MN représente le pied du versant
M
N
30
Vallée de la Bienne en amont de Morez
A
B
la Bienne
A
A'
B
B'
la Bienne
Coupe transversale
Vue du Versant AB

Pl. X

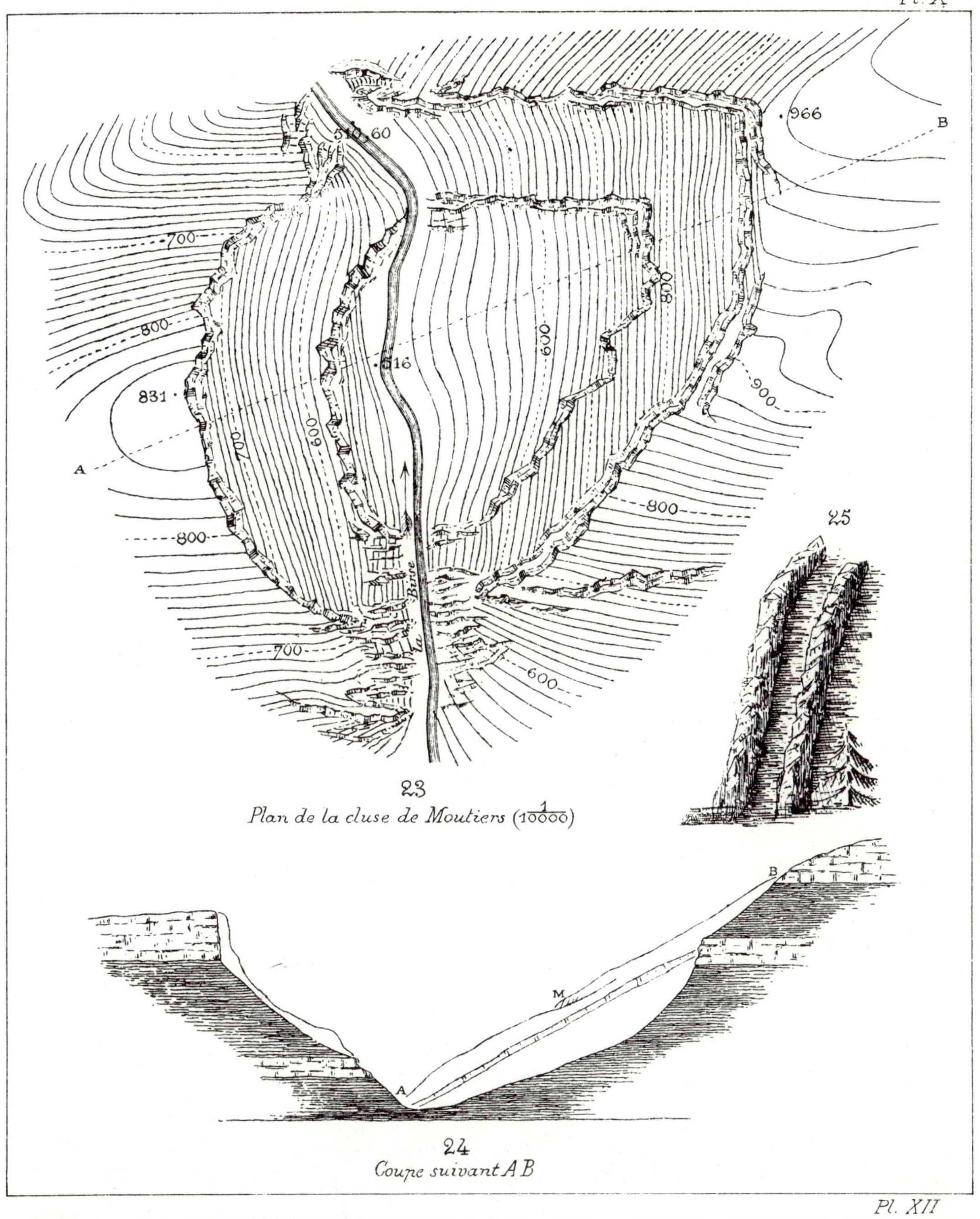

23
Plan de la cluse de Moutiers ($\frac{1}{10000}$)

24
Coupe suivant A B

Pl. XII

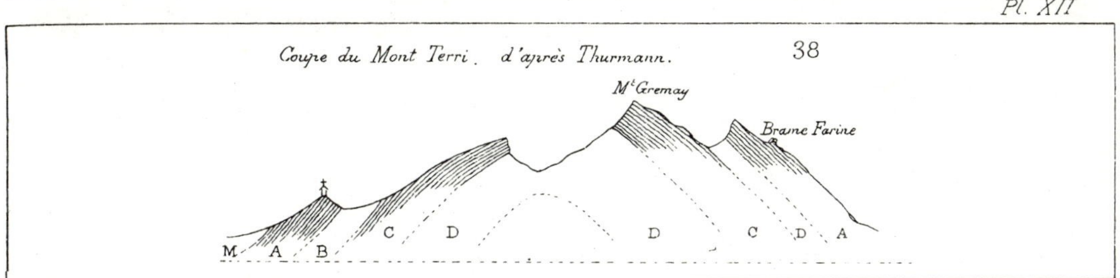

Coupe du Mont Terri, d'après Thurmann. 38

39
Dyke sur le Navajo Creek (Arizona) d'après W.H. Holmes

40
Les Roches de St-Mihiel

41

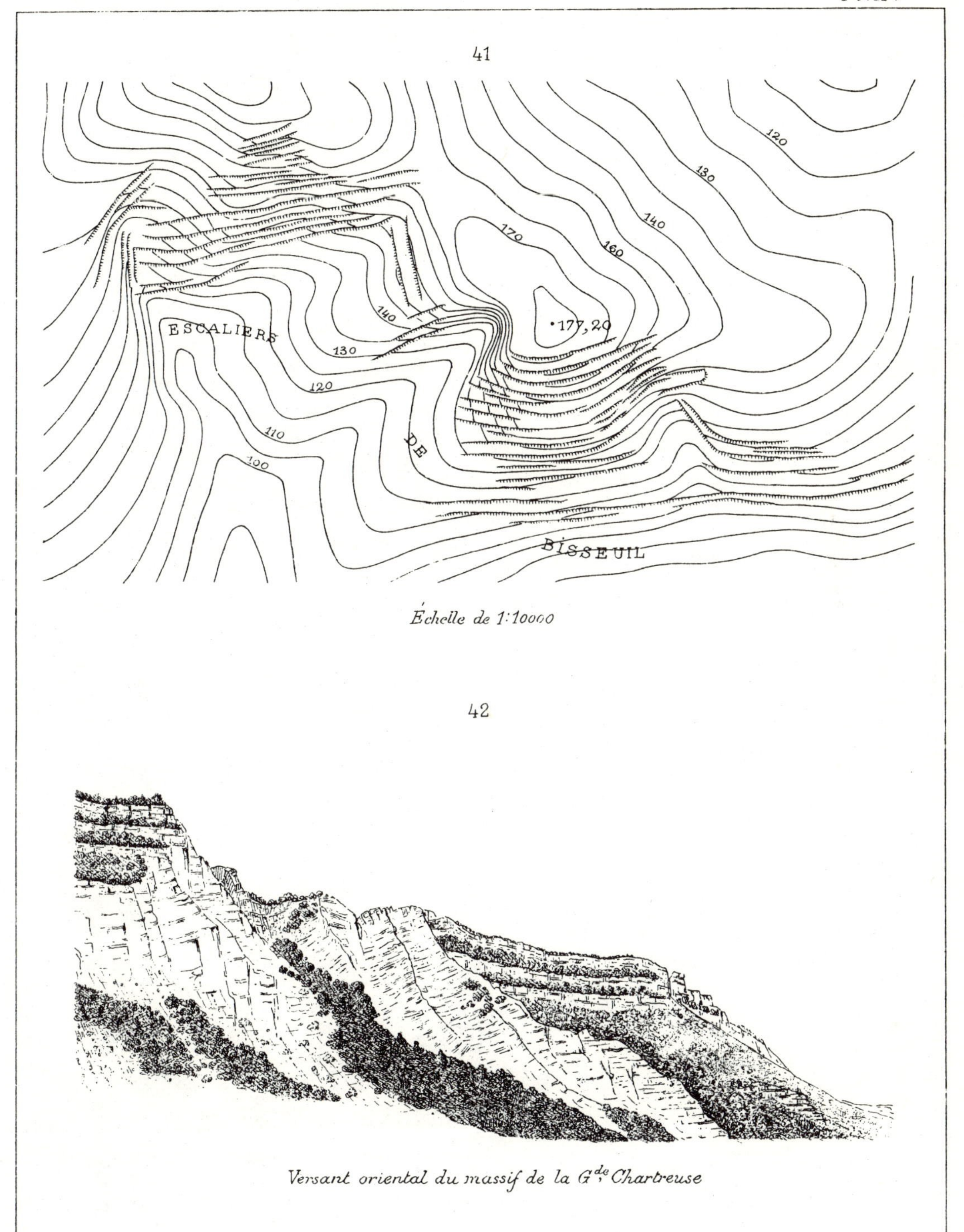

Échelle de 1:10000

42

Versant oriental du massif de la Gde Chartreuse

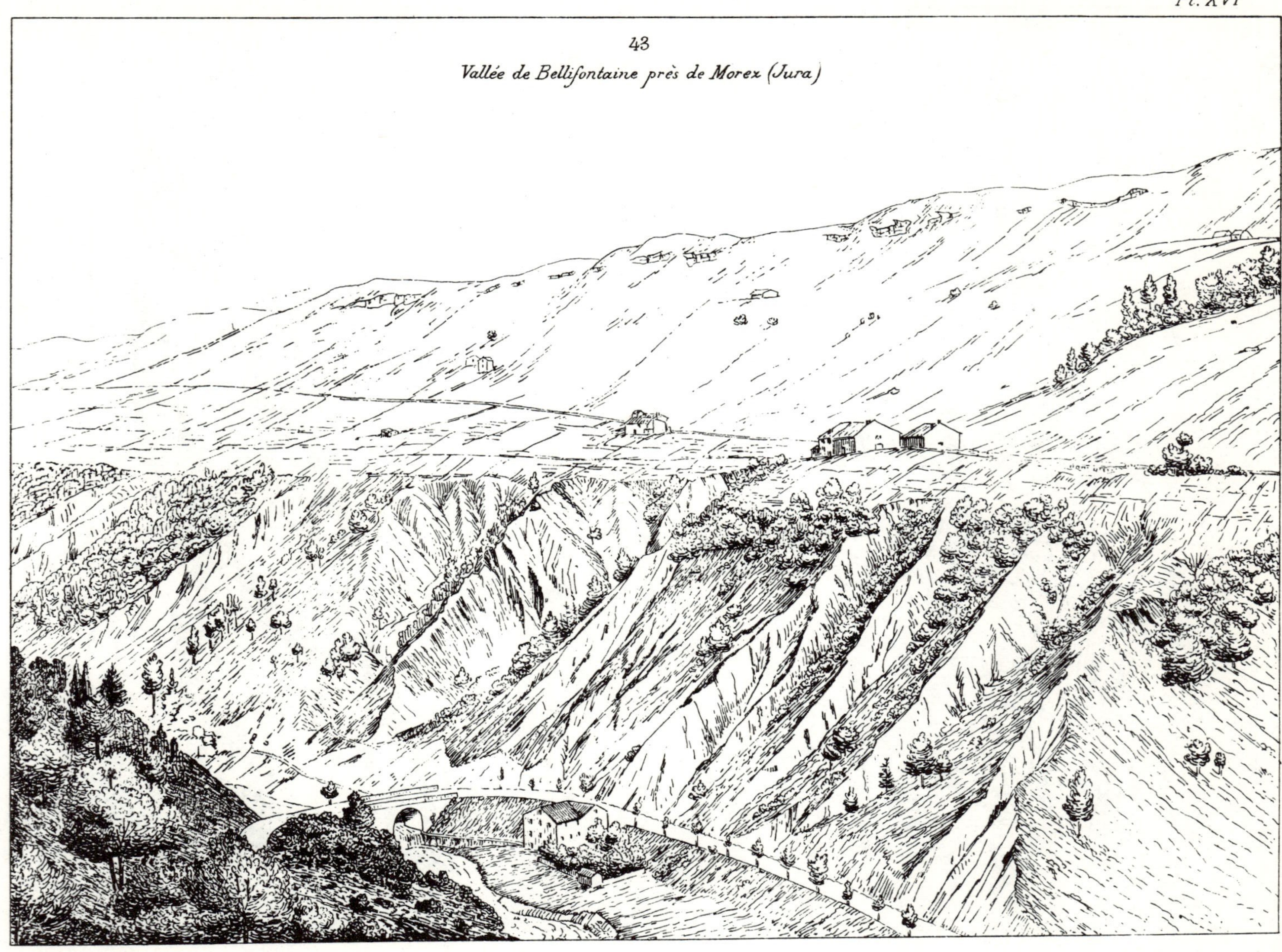

43

Vallée de Bellifontaine près de Morez (Jura)

The Formation of Slopes*

Weathering sets the stage for erosion. Water alone is not capable of mechanical disintegration of rock. In this chapter the movement of water over hillslopes and its role in transportation of movable material is examined.

The Effect of Rain on Slopes is to Progressively Diminish Slope Angles

Analysis of the Phenomena

The reduction in slope angle with time is more rapid as the effect of weathering is greater, as slopes steepen, and as the quantity of water increases. The lowest slope possible is that required for water with its sediment load to move over and away from it. Hence, as particles are reduced in size by weathering and abrasion, all landscapes whatever their geology will eventually attain a very gentle slope. The action of rain has as a final result the reduction of slopes.

To illustrate this, AB of Figs. 8 and 9 (Plate III) is a valley side of homogenous material. The erosion of the slope does not occur as in Fig. 9 because the material reaching point B cannot be transported across the gentle slope B–B′. On the contrary, erosion progresses as in Fig. 8 with the slope base always at B. The authors attempt to illustrate this process by describing an experiment with sand in a box (Fig. 10, Plate III). In this case B is fixed, and a series of slopes of declining angle form in the box as the material moves past point B. They point out that these slopes are concave upward (A′, S′, B, Fig. 10) because slope angles decrease as the erosive power of the flow acting on a slope increases downslope.

As the final effect of rainwash is to reduce slope, there is not a unique slope for each rock type. For example, in the Champagne area the hills and valleys are

*This is an English summary of Paper 4.

underlain by chalk, which presents a great diversity of slopes; the topography of regions of the same lithology do not show the same characteristics.

Effect of Lithology on Slope Profiles

The inclination of slopes on different rocks depends uniquely on the rate at which the materials disintegrate; the greater the speed the gentler the slope. Therefore, a slope composed of different rocks will have segments the steepnesses of which are inversely related to the rate of weathering (Figs. 11 and 12, Plate VI). If a weak bed outcrops between two resistant beds and all are horizontal, a concavity will occur in the scarp face (Fig. 13, Plate III).

The laws governing slope evolution are complex, and many exceptions exist. For example (Fig. 14, Plate VII), a single bed forming the base of a slope can only flatten by rotating around B at the foot of the slope, whereas the profile of the rocks above this bed recedes parallel to itself, because its base P is displaced. If bed PN is very resistant, then sapping beneath it will form a concavity and eventually failure will occur.

If a series of rocks are arranged with increasingly weak beds occurring upward, then the profile will be concave (Fig. 15a, Plate VII), but the profile will be convex in the reverse case (Fig. 15b, Plate VII). One may imagine other types of profiles resulting from different arrangements of weak and resistant beds (Fig. 15c, Plate VII).

Junction of Slopes and Reduction of Crests

In the above examples slopes are portrayed with polygonal profiles, whereas in nature they are more often a smooth curve.

In Fig. 16, Plate VII, it can be seen that the crest of a slope is reduced by a greater amount, SN, than the retreat of straight intersecting slopes B–B′, A–A′ would indicate.

Irregular Progress of Slope Development

The rate of slope flattening may not proceed in a regular fashion. For example, on clay slopes sliding may occur, which will produce a number of concavities and convexities along the slope face (Fig. 17, Plate VIII). When this occurs the slopes have gradients steeper than the equilibrium slope.

When the rocks outcropping are a series of horizontal, thin, but alternately hard and soft beds, the hard beds will dip toward the bottom of the valley, as the weak material is eroded and carried downslope (Fig. 18a, Plate VIII).

When a hard bed lies over a weak bed (Fig. 18b), the removal of the weak material leads to collapse of the resistant bed and blocks of this material descend the slope in turn. Frequently one sees isolated blocks, which are evidence of this process (Fig. 18d).

Application of the Preceding to the Diverse Forms of Valley Slopes

Slope Profiles on Horizontally Stratified Formations

Profiles will be similar on both sides of the valley (Fig. 19, Plate VII) if the agents of weathering and precipitation are the same on each slope.

If the actions are unequal, the slope most exposed to weathering will flatten more rapidly (Fig. 20, Plate VII). This may explain certain types of valley assymetry.

Slope Profiles on Dipping Stratified Formations

If the formations are slightly inclined, the two slopes will be dissimilar, as in Fig. 21, Plate VII. As erosion progresses the updip side of the valley may be completely stripped off an underlying stratum (Fig. 22, Plate IX).

Folds and Valleys

When a fold is cut by a river (Fig. 23, Plate X), the different beds form semicircular outcrops, with their limbs perpendicular to the river, but a profile across the axis (A–B) resembles that across a succession of horizontally bedded rocks (Fig. 24, Plate X). These profiles are modified as their location changes. For example, at the mouth or outlet of the valley the vertically dipping resistant beds form isolated walls (Fig. 25, Plate X).

Experimental Reproduction

It is easy to reproduce experimentally slopes formed on rocks of different resistance. For example, on a small board ABCD (Fig. 27, Plate IX), alternate layers of fine sand (grit) and plaster of Paris are spread with a sieve. If this material extends beyond the board, when the board is lifted, the material remaining takes on the character of a valley with alternate weak and resistant beds.

On the other hand, if the material is spread over a half-cylinder of plaster on the board it is possible to reproduce the conditions of a folded series of strata (Fig. 28, Plate IX).

If the board is tapped in the preceding experiments, a progressive flattening of the side slopes takes place.

In general, talus accumulations form independently of stratification. In addition, slope inclination is independent of stratification. For example, the slopes developed on an assymetrical syncline show that the final valley profile ignored the dip of the formation (Fig. 29, Plate IX). This fact produces some interesting outcrop patterns on the right side of the Bienne Valley near Morez (Fig. 30, Plate IX). The erosion of the valley side exposes the dipping rock in an irregular manner, which gives the appearance of severe deformation. Actually one can reproduce this pattern by wrapping several sheets of paper around a cylinder of wood and then scraping the surface. The edges of the sheets of paper crop out in the cuts as the strata do in the valley at Bienne (Fig. 31, Plate IX).

Geometric Definition of Slopes

The surface of the earth is always undergoing modification. We see it in a transitory state, which is more or less advanced depending on rock resistance and the duration and intensity of modifying agents. Modification is less in dry regions. Steep slopes and sharp ridges of dry regions are replaced in humid regions by rounding of angles and flattening of relief. It is by considerations of this type that one may determine the relative age of landforms under the same climate.

Projection of Hard Parts

It is a general rule of nature that all projections of the earth, all the ridges, crests, and high surface, are formed of the most resistant rocks, while the least resistant form depressions (Fig. 38, Plate XII). Dikes form high natural walls in volcanic regions (Fig. 39, Plate XIII), and sometimes variations in the resistance of one rock type or formation produces similar effects (Fig. 40, Plate XIII).

Man's Modification of Hillslope Form

Due to variations of rock resistance most surfaces will be irregular, but man has in many cases eradicated these irregularities and produced smoothly curved and straight slopes by his agricultural activities. Sometimes, however, slope regularity is destroyed, especially on steep slopes where terracing may produce a stepped profile (Fig. 41, Plate XV), which in the chalk area near Epernay is termed the Staircase.

Talus Slopes

When a resistant rock weathers its debris accumulates as talus at slopes of about 36–37° maximum inclination.

False Screes

Frequently what appears to be talus deposits at the base of a slope is actually bedrock, and this can be detected in ravines incised into the weak bedrock (Fig. 42, Plate XV).

Classification of Rocks

One cannot classify rocks by slope inclination because inclination depends on rock resistance, process, and time. In certain regions one process may be dominant, in others a completely different one is most important.

Rocks that produce the gentlest inclination are clays or shales. The marls, which are a mixture of clay and limestone, also produce gentle slopes. The massive limestones form escarpments, as do sandstones. Granites form slopes as steep as the sandstones.

In Fig. 43 is shown the left slope of the Valley of Bellefontaine in the Jura Mountains. The sharp ridges are cut in alluvium; hence, it is impossible to make a general classification of rocks based on slope angles. However, slopes are steeper in arid climates and gentler in wet.

The Role of Vegetation

Vegetation is obviously important in retarding erosion, and the notion of slope equilibrium is inseparable from vegetational effects. Vegetation assures the conservation of relief, but that will not occur until degradation becomes so slow that it cannot prevent the establishment of vegetation.

Theory and Controversy

II

Editors' Comments on Papers 5 Through 10

From the very first, students of hillslopes have relied heavily on the construction of deductive, conceptual models of slope development. The reasons for this are suggested by Chorley in Paper 1 of this volume. In most sciences, theoretical and experimental studies are closely related, since standard scientific method is to develop a hypothesis to explain a phenomenon, and then to test it against observed facts. The continuing controversy surrounding hillslope evolution has been largely a result of the development of theoretical models, but without their subsequent testing. Before 1950, virtually no one had bothered to objectively measure slope angles in the field, and without such data it was impossible to resolve the Davis–Penck debate as to whether slopes decline or retreat parallel to themselves. This perhaps explains why the symposium on Walther Penck's contributions, including slope retreat, (Von Engeln, 1940) had inconclusive results, and why the controversy dominated the hillslope literature until the mid-1950s, by which time field measurement had been applied to the problem.

As noted by Bryan (1940), it is difficult to find a concise presentation of Davis's ideas on slope development in any of his numerous publications, and the obscurity of Penck's (1953) arguments is well known. Some geomorphologists ignored the controversy and went about more productive work, and except for the inclusion of Tuan's brief discussion, this volume does the same. Later sections demonstrate that both parallel retreat and slope decline are possible, and that mode of slope evolution is influenced by many factors—factors which were often ignored during the first half of this century. Controversies of this type usually are beneficial in that they stimulate research, but the Davis–Penck debate resulted in an increase, not in field investigation, but in unsubstantiated theorization. This is especially unfortunate since, as Tuan remarks, Penck's concepts differed less from the views expressed by Davis than is sometimes supposed (Simons, 1962).

A third main model of slope development has been forcefully proposed by L. C.

King in "The Uniformitarian Nature of Hillslopes." He suggests that hillslopes evolve in the same manner, by parallel retreat and pedimentation, throughout the world (excluding certain exceptional areas). In effect, King states that climatic control of landforms is unimportant since hydraulic relationships apply the world over. The attempt to generalize and to provide a firm basis for hillslope studies in the laws of physics is welcome but unfortunately, as noted earlier and in C. R. Twidale's reply, the problem is too complex for this approach to have any wide success at the present time.

Many papers in this volume provide sufficient evidence to show that King's claims are excessive. Like those of Davis and Penck, his model is probably valid under certain conditions, invalid under others. For example, he ignores the effects of erosional environment, shown by Strahler and Melton (Papers 12 and 15) to be of prime importance. There is ample evidence that hillslope forms are influenced by erosion processes, such as creep and rainwash, which are in turn influenced by climate. Furthermore, Twidale's reply stresses the importance of climatic change, and points to the very different rates of erosion and sediment yield from different climatic regions (Stoddard, 1969) as suggesting the widely varying rates at which hillslope processes may operate.

Mathematical models have also been used for many years to investigate slope evolution (Hirano, 1968; Scheidegger, 1970). Such models have rarely been tested in the field and so their value has been seriously limited. A very early example of a mathematical model of slope evolution is that of Fisher, which has since been extensively refined (Scheidegger, 1970, pp. 120–127). Scheidegger has provided a concise and useful summary of several models in the paper reprinted here, and he shows how hillslopes might be expected to develop, given certain assumptions. However, the fact that theoretically derived profiles may resemble natural slope profiles means little, since the models may be manipulated to generate any profile, and because they provide no insight into process or sequence of development. This mathematical approach is very similar to the conceptual models of Davis and Penck. They are hypotheses and in themselves explain nothing. However, they provide a direction for the next logical step—the testing of the hypothesis. Ahnert's attempt to compare theoretically derived slope profiles with those found in nature is an encouraging sign that mathematical models may in the future contribute more to our knowledge and understanding of slope form. Without this type of comparison, however, the construction of such models will remain solely a mental exercise.

Reprinted from *Indiana Acad. Sci., Proc.*, **67**, 212–214 (1958)

5

The Misleading Antithesis of Penckian and Davisian Concepts of Slope Retreat in Waning Development

YI-FU TUAN, Indiana University

In the last thirty years or so, geomorphologists have shown intermittent but lively interest in the problem of slope development. A segment of this large and difficult problem has taken the form of a controversy, sometimes phrased rather loosely as follows: under the condition of a stable base of erosion, does the valley slope retreat parallel to itself or does its inclination decline with age? The former position is attributed to W. Penck and the latter to W. M. Davis. The presumed contrast in slope forms resulting from these concepts is illustrated by Davis in a simple diagram in his paper on "Piedmont Benchlands and the Primärrümpfe" (1). This diagram (Fig. 1) has

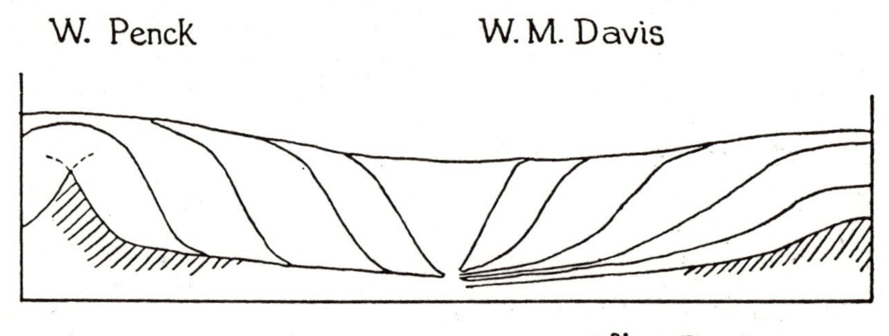

Figure 1.

been reproduced, and served as a basis for discussion, at least three times. It appears, for example, in the symposium on W. Penck held by the Association of American Geographers in 1940 (2), in C. A. Cotton's popular textbook "Landscape" (3), and, more recently, in L. C. King's paper on the "Canons of Landscape Evolution" (4).

Davis's diagram served to emphasize the opposition between Davisian and Penckian concepts of slope retreat. His simplified presentation tends to mislead because it neglects an important implication in Penck's analysis; namely, during waning development (absteigende Entwicklung), only the cliff or rock wall (Steilwand or Felswand) of the valley slope retreats parallel to itself. The valley slope as a whole becomes concave and its gradient declines in a way similar to Davis's own conception (5). The cliff (A, B, or A_5, C_5 in Fig. 2) is assumed at the beginning of Penck's deductive treatment in order to allow graphic presentation. If one starts the analysis with a valley slope of gentler inclination than the cliff, as, for instance, the slope B, D_n, C_n in Fig. 2, then the subsequent stages of slope flattening become almost identical

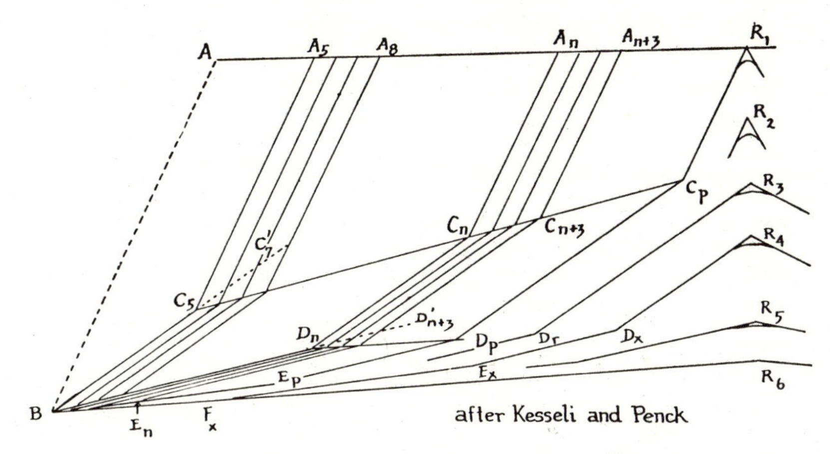

Figure 2.

with those conceived by Davis. The slope B, D_n, C_n is a smooth, concave curve. The break of slope in the valley side occurs only at the junction between the cliff (Steilwand) and the uppermost segment of the denudational slope (Haldenhang), along the line C_5, C_p where there is a change in transportational process from free fall to creep, sliding and rainwash. The denudational slope below the cliff is a smooth curve, since a. the inclination of each segment of the denudational slope is determined by the mobile size of the debris, which shows a smooth gradation; b. as soon as the Haldenhang appears beneath the receding cliff, processes of denudation would operate to produce a gentler subjacent slope segment. In Fig. 2, on the other hand, a slope segment of gentler inclination than the Haldenhang is assumed to appear only after five units of time have elapsed. The Haldenhang is thus drawn as a straight line that meets lower and gentler slope segments, also represented as straight lines, at distinct breaks. This departure from logical rigor is inherent in the graphic method of analyzing continuous and simultaneous processes.

Davis's diagram further stresses the contrast between the Penckian and Davisian concepts of slope retreat by his presentation of their views on the development of the crests of the interfluves. According to Davis, the crests of the interfluves become broadly convex in the late stages of the cycle (Fig. 1). Penck, on the other hand, argued for the prevalence of concave slopes in waning development. But he did not conclude that the residual hill should be a sharp peak as Davis depicted it in his diagram. Penck recognized and attempted to explain the rounding of hill-tops independent of his premise on rates of uplift (6). Hence the difference in viewpoint between Penck and Davis regarding the curvature of interfluves in waning development narrows down to a difference in the length of the convex arcs when the valley slopes are seen in profile. Recent observations by King and Sharp suggest that Davis has exaggerated the length of the convex arc in a region of waning development (7).

The convolute prose of Penck's treatise, "Morphological Analysis of Land Forms," is well known. His arguments are not always complete and are invalid in places. Some of Penck's statements regarding slope development have been expanded and lucidly re-stated by J. E. Kesseli (8). A careful reading of Penck's works shows that his concept of slope retreat in waning development differs less from the view expressed by Davis than is sometimes supposed. Thus in a paper on "The Piedmont Benches of the Southern Black Forest," published posthumously in 1925, Penck (9) has a simple sketch (Fig. 3) that

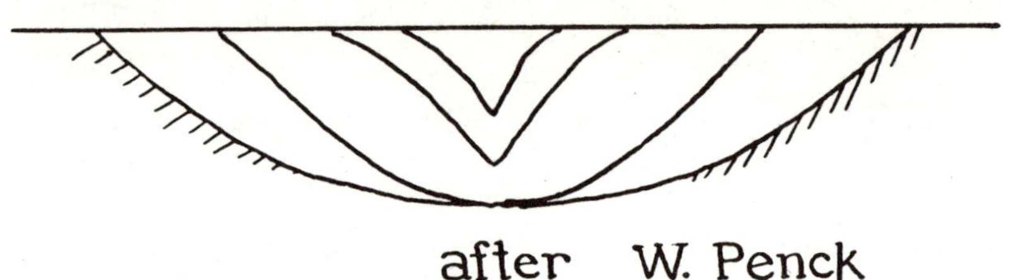

after W. Penck

Fig. 3

illustrates the gradual flattening of a valley slope in the true Davisian manner. Davis, on his part, has modified his stand on slope flattening as a universal principle. He realized that the boulder-clad cliff or Felswand is a common slope element in arid regions, and admitted that such a cliff may undergo parallel retreat, in distinction to the basal slope which flattens (10).

Literature Cited

1. DAVIS, W. M. Piedmont benchlands and the Primärrümpfe. 1932. Geol. Soc. Amer. Bull. **43**: p. 409.
2. VON ENGELN, O. D. 1940. Symposium: Walther Penck's contribution to geomorphology. Annals Assoc. Amer. Geog. **30**: pp. 222, 256-258.
3. COTTON, C. A. 1948. Landscape, as developed by processes of normal erosion, 2nd edition. pp. 230-233.
4. KING, L. C. 1953. Canons of landscape evolution. Bull. Geol. Soc. Amer. Bull. **64**: pp. 723, 729, 748.
5. PENCK, W. 1953. Morphological Analysis of Land Forms. Macmillan and Co., pp. 139-141. (Translated from the German by Hella Czech and K. C. Boswell).
6. PENCK, *op. cit.*, pp. 141-143.
7. KING, *op. cit.*, pp. 725, 735-736; see also R. P. Sharp. 1957. Cima dome, Mohave desert, California. Geol. Soc. Amer. Bull. **68**: p. 275.
8. KESSELI, J. E. 1940. The Development of Slopes. Berkley. (Mimeographed)
9. PENCK. W. 1925. Die Piedmontflächen des südlichen Schwarzwaldes. Zeit. Gesellsch. Erdk. Berlin, p. 89.
10. DAVIS, W. M. 1940. Rock floors in arid and in humid climates. Journ. Geol. **38**: 149.

Reprinted from
INDIANA ACADEMY OF SCIENCE
Volume 67—1958

Reprinted from TRANS. EDIN. GEOL. SOC.,
Vol. 17, Part 1, pp. 81-102,
1957

The Uniformitarian Nature of Hillslopes

LESTER KING

(MS received 14th May 1957)

ABSTRACT

The nature of hillslopes is examined, and their importance in landscape is assessed. The denudational processes operating upon hillslopes are examined and found to be manifold; producing ideally four specific elements in hillslopes, each of which has a distinct mode of evolution. The evolution of hillslopes under the action of denudational processes is evaluated in terms of physical science—involving surface flow of water and mass-movement of soil and rock, and is found to be dependent upon intrinsic strength of the bedrock and available relief. It is almost independent of climate, *per se*, and similar hillforms may be found under like conditions of bedrock and relief in all climatic environments short of glaciation or wind-controlled, sandy deserts.

CONTENTS

I. INTRODUCTION

Our concern is with the study of the earth's surface and its development under the normal agencies of denudation; and particularly with the evolution of hillslopes. Upon this topic, one of the most important advances was made by James Hutton in whose *Theory of the Earth* appears the first clearly-stated, all-embracing vision of the dynamic evolution of landscape. Therein, under the concept of Uniformitarianism, hillslopes are regarded as modified under denudational agencies operating now as they have operated in the geological past, and as they will undoubtedly continue to operate in the future. From this dynamic approach in the study of landscape modern researches begin.

They continue with the rise of geomorphology, or physiography as it was then called in America, towards the end of the last century when much attention was directed to two agencies of landscape evolution (*a*) erosion by rivers, and (*b*) weathering of interfluves, the emphasis in each case being, and remaining, empirical. The rivers were deemed to cut their beds down rapidly to a state of grade, related to base-level, and when this phase was passed, weathering in turn reduced the interfluves so that the relief was diminished to the ultimate stage of a peneplain (Davis, 1909).

Of the precise nature and evolution of the hillslopes that intervened between the river courses and the divides little account was taken though these hillslopes constitute, indeed, almost the whole of a normal landscape. One searches the literature almost in vain for any discussion of hillslope forms under evolution : even where the master geomorphologist William Morris Davis (1909, p. 734) set forth his so-called " Normal Cycle of Erosion " he found no better exposition than that the hillslopes flatten progressively and become convex in cross-section, a statement that, alas, all too frequently finds no support in natural landscapes. Only towards the end of his life (1930) did Davis begin to see and describe hillslopes truly as they are displayed in nature (see King, 1953; Cotton, 1955).

Meanwhile, the uncompleted studies of Walther Penck (1924), now available in an excellent English translation (Penck, 1953), suggested very strongly that major hillslopes do not flatten progressively during a long evolution but that, having achieved a stable form consonant with local geological controls, the hillslopes thereafter retreat parallel to themselves without further flattening. This conclusion is as basic to landscape study as Hutton's uniformitarianism. Unfortunately, Penck's brief life did not suffice for the clear exposition of what he saw truly in the landscape, and his accounts of hillform evolution are confused by the introduction of erroneous and irrelevant concepts relating hillform to rates of deformation of the earth's surface. These irrelevant concepts were readily refuted as they deserved to be; but unfortunately most geomorphologists "threw the baby out with the bath water " and rejected Penck's accurate observations of hillslope form as well.

Only one prominent exponent of " parallel scarp retreat " remained during the thirties of this century, the American Kirk Bryan (1940), who escaped the general condemnation by confining such practices solely to desert landscapes and leaving the " humid " landscapes free to flatten as prescribed by tradition. Considerable research was directed towards the " peculiar " processes by which desert topography was thought to evolve, but the precise operation of hillslope processes under humid regimes was considered too well established to require investigation.

The viewpoint that landscapes evolve very differently under arid, semi-arid and humid influences has long been promulgated, and is widely accepted at the present time. Our thesis will be that the basic physical controls on landscape remain the same in all climatic environments short of the frigid or extremely arid. Water moves over the earth's surface by linear or by laminar flow and in no other way; mass movements of soil and rock are executed under immutable physical laws. Horton (1945) expresses this concept: " The geomorphic processes we observe are, after all, basically the various forms of shear, or failure of materials which may be classified as fluid, plastic or elastic substances, responding to stresses which are mostly gravitational but may also be molecular . . . The type of failure . . . determines the geomorphic process and form." Only in their relative proportions may the agencies responsible for epigene landscape evolution vary, and then only within narrow limits. Hence, as we hope to demonstrate in the sequel, the manner of hillslope evolution is essentially uniformitarian in all climatic realms outside the frigid zones and the erg deserts.

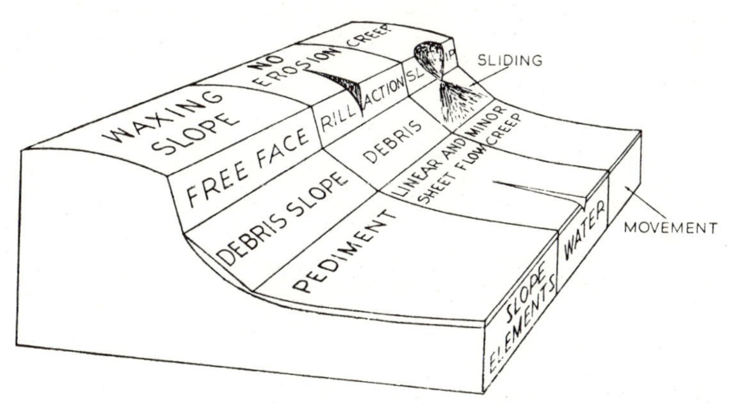

FIG. 1. Elements of fully developed hillslope.

II. Hillslope Elements

The starting point for modern hillslope study is the analysis of hillslope form by Wood (1942) who distinguished the four elements that appear in a fully developed slope. These are, from the top : the *waxing slope*, the *free face*, the *debris or talus slope* and the *pediment* (Fig. 1). Each element undergoes a semi-independent evolution though, of course, on a given slope the elements to a greater or less extent react upon each other. Any element may be absent from a given slope, which is then not fully developed.

These matters have already been discussed (King, 1953) and been made the subject of a number of Canons of Landscape Development.

Wood's contribution is sufficiently important to rank with the fundamental discoveries of Hutton and of Penck, and like them it is basically uniformitarian in character for it applies to hillslopes all over the world, in all stages of development, and indeed of all geological ages.

Let us briefly review the four elements and state the essential qualities of each. The *waxing slope* is the convex crest of a hill or scarp, usually related to the zone of weathering and measuring from an infinitesimal portion to half the total length of hillslope. It is usually longest where the depth of surficial detritus is greatest, *e.g.* in formerly periglacial regions, or where rivers and streams have recently become incised and the hillslopes have not had time to adjust to new stable conditions.

The *free face* is the outcrop of bare bedrock exposed, perhaps as a scarp face, on the upper part of the hillslope. It is the most active element in backwearing of the slope as a whole.

The *debris slope* consists of detritus slipped or fallen from the free face and resting at its angle of repose against the lower part of the scarp face. As this debris is weathered to finer detrital grades it is removed under erosion and so the debris slope retreats in essential conformity with the free face above. Clearly, if the free face retreats more rapidly the quantity of waste supplied will build up the debris slope to bury the lower scarp face, so that a balance is generally struck between these two elements.

The *pediment* is a broad concave ramp extending from the base of the other slope elements down to the bank or alluvial plain of an adjacent stream. Frequently its profile approximates to a hydraulic curve and it is unquestionably fashioned under the action of running water, though the cutrock surface may often be mantled with surficial deposit.

These four slope elements are to be found in hillslopes all over the world and in all climatic environments; and though locally one or more of the elements may be suppressed, such departures afford no contradiction of the normality of full development.

III. The Agents of Denudation

Hillslope evolution, with production of the several elements, is clearly a function of denudation, involving two phases (*a*) production of land waste, and (*b*) removal of land waste. Accordingly as denudation produces, and removes, land waste so the several elements may appear in virtually any order and in any combination. To understand the development of hillslopes, therefore, we must enquire into the working of the agents of denudation that operate upon the landscape.

The principal agents responsible for denudation are (*a*) running water and (*b*) mass movement. Both are expressions of gravity control and involve overall movement of waste in the downward sense. Wind and glacier ice, which on occasion may carry land waste to higher levels, are of local and restricted dominance. The " flow " of land waste over normal landscapes is downward and lateral, and this is fundamental to the development of such landscapes.

At this juncture a study of the physical laws under which water flows and debris moves over the skin of the earth becomes imperative. Empirical study of existing hillslopes, which has dominated geomorphological thinking hitherto, fails to give the quantities that we need to know, and fails to afford a dynamic concept of the operation of the forces at the surface of the earth.

IV. The Flow of Water over Land Surfaces

The flow of water over a land surface is a matter of hydraulics, and to hydraulics geomorphologists must now turn for enlightenment, following the lead so admirably given by Horton (1945). But although the quantities have mathematical expression to be found in text books we shall eschew the exactitude of symbols and formulae for more general statements in conformity with the constantly fluctuating conditions present upon natural land surfaces across which water is passing.

For water flowing away from the crest of a divide, there is a critical distance over which no erosion takes place. The reason may be stated simply: flowing water cannot erode a slope until erosive power exceeds the resistance of the soil to erosion, or in other words until the turbulence of the water can overcome the cohesion of the soil. At the top of each hill therefore appears naturally a zone of no erosion under water flow, with a non-concave profile corresponding morphologically to the waxing slope of Wood's analysis. Clearly, the greater the proportion of infiltration into the soil, the wider this zone is likely to be, so that on porous soils and bedrock the waxing slope attains maximum width. As examples we may quote the broad convexity of hilltops upon the boulder clay of midland Britain and the chalk country of southern Britain and France, and contrast them with the minimal summit convexity expressed on hard granite and quartzite terrains in Africa where the proportion of runoff is high.

As distance from the water-parting increases, so also does the volume and erosive power of water flowing in rills over the surface of the ground. Even upon a slope of originally uniform gradient, once the critical distance from the water-parting is passed, erosion by rill-wash supervenes and steepens the profile, improving still further the erosive power of the rills. This portion of the hillslope rapidly becomes steeply concave and is relatively sharply marked off from the waxing slope above.[1] Frequently, the closely spaced rills cut through superficial deposits to the bedrock which may be exposed as " free face " either locally or in continuous exposure as a scarp. Even where a " free face " is not developed the soil is thinnest and poorest and is removed most rapidly in this zone.

Under water wash the middle slopes of major hills inevitably become the steepest, and most actively eroded, element, and hence the retreat of hillslopes and scarps is equally inevitable once the critical limit from the water parting is passed.

Corrasion implies a corollary of waste removal and so the land waste that is actively produced in the upper part of a steep slope is necessarily washed downwards on to the lower part where also the velocity of the surface wash is checked as the controlling base-level of a neighbouring stream is approached. Corrasion diminishes, or even ceases, and the greater part or even the whole energy of the flowing water is devoted to transporting the debris to the local stream. Its efficiency in this task is measured by the extent of the debris zone (the third element in the hill-slope profile) below the free face. Only under very favourable conditions of hard bedrock yielding little waste does the debris zone vanish entirely, leaving a sharp angle carved in bedrock at the foot of a steep hillside. " Inselbergs " afford perhaps the best and most numerous examples of this (Plate VII, A, B).

Following the cutting of a steep hillside by rill and gully action, and perhaps also retreat of the scarp so formed, surface water requires to be discharged across a relatively flat terrain to an adjacent stream channel. This relatively smooth, flat zone is the pediment. The amount of water to be discharged across the pediment is a maximum including all the rainfall of the upper slopes, minus infiltration, together with precipitation upon the pediment itself. In cross section, therefore, mature pediments exhibit the smooth concave profile of a hydraulic curve on which the gradients range commonly between 7° and ½°. Longitudinally, pediment surfaces are often surprisingly smooth with a rise and fall along the inner edge at the foot of the scarp face, the *low* points corresponding to the emergence of gullies through the scarp. Throughout, the pediment, a product patently of water flow, is so adjusted as to impart the maximum efficiency to water flow and in times of heavy rainfall it may often disperse the water flow in sheets. The operation of sheet flow upon

[1] The rapidly increasing convexity often observable towards the edge of the waxing slope may be interpreted as a transitional sub-zone, dependent upon fluctuation in rainfall and runoff, where sometimes the tendency will be to form a convex surface and sometimes a concave surface accordingly as runoff is lighter or heavier from hour to hour or minute to minute. It is sometimes well developed in mist belts.

pediments has been observed in nature (King, 1953) to conform with hydraulic principles and especially the operation of the quantity called Reynold's Number. Under favourable conditions true laminar flow (without turbulence) has indeed been observed towards the inner edge of pediments (King, 1953). But such a system cannot be perfect in nature over more than a small area and within a limited time. Hence the zone of laminar flow, dependent upon a favourable conjunction of precipitation rate and surface smoothness, is often elided, and rill-cutting and gullying appear extensively upon many pediments. The remarkable thing, bearing in mind the general irregularity of ground surfaces, is that the zone of laminar flow should appear in nature at all.

As the sheet of flowing water thickens beyond the inner zone of the pediment, laminar flow breaks down and turbulent sheet flood supervenes. This is the agent primarily responsible for the hydraulic profile developed by pediments.

Under diminished incidence of rainfall, water coming on to the pediment from upper slopes, and the water falling upon the pediment itself, may be insufficient to form sheets and then passes across the pediment in rills only. Where this occurs frequently, pediments are, indeed, scored by rill channels.[2]

Land waste transported across pediments is usually in a relatively fine state, and when deposited in hollows often helps to smooth out any irregularities in the rock floor that have not been eliminated by erosion.

Pediments are best exemplified (King, 1953) on hard rock in semi-arid environments, where the transport of land waste across the terrain is most efficiently conducted. Any departure towards arid or humid extremes is attended by the accumulation of waste in the landscape, and cutrock pediments may to some extent become masked or buried. Nevertheless, pediments are fundamental elements of hillslopes formed necessarily, as we have seen, under the physical action of running water, and hence they appear with greater or less prominence in all epigene landscapes. They are by no means restricted, as American geologists would have us believe, to semi-arid and arid environments but may be admirably studied even in the Appalachians as Davis's paper of 1930 showed. English landscape, too, exhibits some clear pediments in the non-glaciated southern counties, *e.g.* Wiltshire Downs.

So we find that *the four basic elements of a hillslope, as analysed by Wood, all form naturally under the normal flow of meteoric waters across the landscape*: they reflect in form and

[2] In these studies, effects attributable to climatic fluctuation during the Quaternary are important, especially when related to the *incidence* of rainfall, which factor is more important than annual or seasonal total.

evolution the natural results of the application of hydraulic forces.

As, the manner of water flow over land surfaces is prescribed by physical laws that are invariable over the globe, the four-element hillslope is the basic landform that develops in all regions and under all climates wherein water flow is a prominent agent of denudation. This includes all regions except those wherein the dominant agents are glacial ice or wind.

The question now arises, *must* all four elements be developed or may certain elements disappear under appropriate circumstances? The standard example that we have considered involves certain assumptions, the first of which is that, after stream incision, the relief available is sufficient for all four elements to appear one above another. Briefly, in regions of low relief insufficient height is available for all elements to be developed, and the first to disappear is usually the free face, followed by the debris slope. Also, where streams are closely spaced, the critical distance from the water parting over which erosion by water flow is ineffective may occupy a large proportion of the hillslope profile and the interfluves become in consequence almost wholly convex. Wide stream-spacing, on the other hand, favours the development of broad pediments and a landscape may result wherein convexity is almost absent from interfluves and more than nine-tenths of the landscape consists of pediments (*e.g.* parts of Rhodesia).

Further, a strong bedrock affording a clear free face tends to provide a fully developed slope with all four elements clearly displayed; but a weak bedrock readily breaking down under weathering, tends to eliminate the free face and produce a degenerate, smooth convexo-concave profile in which the waxing slope from above meets the pediment from below in a smoothly reflexed curve. (Such slopes are common in parts of northern Europe and are achieved principally by prolongation downwards of the waxing slope from above. In Africa, however, the pediment slope is usually prolonged upward. These differences may arise from lighter and heavier incidence of rainfall respectively, or from the prevalence of solifluction in northern latitudes following the Quaternary ice age.)

Considerable range of hillslope form is therefore possible under differences of relief, bedrock, stream spacing, and other variables, but none of the resultant landforms depart from the basic elements; and the fully developed, four-element slope remains the most vigorously retreating, active type of all. Other slope forms, involving a reduction from the maximum of four elements, are decadent and may even atrophy (as shown by a failure of active retreat; unless the migration of Baulig's inflexion point represents retreat).

The optimum hillform, in physical terms, is clearly to be

sought in regions of hard bedrock and strong youthful relief. It is exemplified by the bornhardt or " inselberg " in which summit convexity is small, most of the hillslope consists of bare, rocky free face, the debris slope is minimal, and the main hill-form rises abruptly from an extensive pediment (King, 1948). While many bornhardts owe their form in part to decomposition of the rock or to splitting off of slabs at curved joints, in the hardest gneisses such disintegration is suppressed and the surface of the bornhardt is channelled and fluted by rill action, *e.g.* in Rio de Janeiro and Espirito Santo.

On hard rocks which disintegrate to give only very fine debris, readily washed away, inselberg landscapes may appear in miniature even though the relief is very small (Plate VIII, A).

The margins of inselbergs, it should be made clear, bear no necessary relation to local geological structures. The same rock types (and the strike) pass without distinction from the upper slope into the pediment (Plate VIII, B).

V. Mass Movement of Rock Waste over Land Surfaces

In addition to transport of land waste by flowing surface water much rock detritus moves from higher to lower levels under the influence of gravity: by creep, slip, slide and flow (Sharpe, 1938). Such movements are controlled by the physical laws governing the behaviour of earth materials under stress.

Illuminating data upon the behaviour of ideal materials are available in text books of physics, and describe the behaviour of perfect-elastic, perfectly viscous, viscous-elastic and firmo-elastic bodies, also the nature of friction—static and sliding. These are topics with which the geomorphologist must be familiar if his researches are to have real meaning, but which find scant reference in a majority of geomorphic texts. The first report of the Commission for the Study of Slopes (Rio de Janeiro, 1956), for instance, makes little mention of such vital topics and most of the contributions are in the old empirical tradition of geomorphology.

But hard and fast classifications of the phenomena are difficult to maintain because of the frequently heterogeneous nature of the materials involved. Most geological materials are of the viscous-elastic type (the Maxwell body of physicists which deforms according to the equation:

$$\tau = \mu \, \gamma^{\cdot} = G \; T \, \mathrm{rel} \, \gamma^{\cdot})$$

(Stress = viscosity × strain = Young's Modulus × Time of relaxation × strain)

This means that they deform in either or both of two ways according to the manner of application of stress. If the stress is moderate and is applied suddenly the material deforms elastically and if the stress is released the material returns to its original undeformed state. Should the stress be very great that material may be stressed beyond its yield limit (as defined by the cohesion of the molecules), when the material fractures, so that the stress is suddenly released. On release of stress there is no return to the original state.

Alternatively, however, small stress may be maintained over a prolonged time interval, when release of strain may be achieved gradually by plastic or viscous deformation within the body. The resulting deformation is permanent, and is a function not only of stress but also of time.

All hillslope studies must take cognisance of this dual mode of failure of earth materials—by fracture and by flow—especially as the superficial types of rock matter (sands, silts, clays, weathered and unweathered sandstones, shales and limestones) are generally susceptible of either mode of deformation within the lapse of time covered by a fraction of a cycle of erosion. Illuminating discussions on these topics are by Sharpe (1938), Ward (1945) and Horton (1945).

We may now apply these concepts to the study of mass movement of land waste down a typical hillslope. Towards the crest or divide the amount of waste produced under weathering and requiring to be transported is small; in consequence the natural slope is probably gentle, and soil relatively stable. The only lateral mass movement likely is by creep (a result of simultaneous action of both gravitational and molecular stresses), and even this may be minimised where slopes are very gentle, as is shown by the development of lateritic duricrusts (a product of advanced chemical weathering) upon very flat, ancient divides.

With increasing distance from the crest, the quantity of material requiring to be transported increases in direct proportion if the bedrock weathers at a uniform rate. Hence for disposal of the waste by creep the slope must steepen progressively, *i.e.* the slope must become convex and the waning slope of Wood's analysis develops. Creep is a function not only of stress but also of time so that low divides may ultimately develop very broad convexity.

As the steepness of a convex slope increases with distance from the divide so also the stresses increase until accommodation by creep becomes inadequate. On a slope of considerable relief stress may thus be increased beyond the elastic strength of the material (*i.e.* the yield limit is passed) and rupture ensues, accompanied by slipping and sliding of material. (Motion follows because the stresses built up prior to fracture are much greater than the stresses required to move the mass once the bonds across the surface of failure are overcome.)

Slipping occurs characteristically upon arcuate (nearly circular) sections and hence such slipping leaves behind *a steepened arcuate section* in the hillside. In this way, free movement of material tends to steepen middle or upper hillslopes; in other words, it produces a free face, which will continue to retreat by calving off further slices consecutively behind the slip scar. This phenomenon is all too familiar to engineers and geologists concerned with stability of hillslopes and road cuts.

On suitably steep situations, material may not slip out upon curved fractures but may fall freely, roll or bounce from higher to lower levels. Static friction is greater than sliding friction and movements once triggered may continue for a distance and time limited only by the relief and the nature of the country rock.

The detritus from slips and slides then accumulates as a talus below, adding the third element, " the debris slope," to the hillslope profile. Resting in place a talus weathers and the finer products of weathering pass downward through and over the mass, the finer being transported the farthest, so that the toe of the talus flattens into a concave profile, a pseudo-pediment or perhaps a bahada, thus adding the fourth slope element. At this stage the action of running water can no longer be ignored, even in deserts, and the final transport of rock waste over the flat, lower hillslopes is seldom, if ever, by gravity alone.

Like water flow, the natural action of mass movement tends, therefore, to produce a fully developed, four-element hillslope, each component of which is capable of semi-independent evolution. This both agencies can do even from an originally uniform slope. For this reason we must mistrust all analyses which treat hillslopes as though they were simple landforms from top to bottom. Even mathematical treatments on such a basis possess no real validity. Thus the purely mathematical treatment of ideal homogeneous media by Bakker and le Heux (1ier Rapport, Comm. de l'Etude des Versants) does not really aid geomorphologists, who derive much more assistance from the mathematical studies by engineers who have learnt to deal practically with earth materials (see later).

We return to the former question asked under water flow, *must* all four hillslope elements arise under mass movement, or may elements be suppressed under certain circumstances? The answer given is the same. Full development of all four elements is dependent upon local factors, chief of which are adequate relief and strong bedrock. Failing these, the free face tends first to disappear, followed of necessity by the debris slope. A decadent convexo-concave hillslope results. As relief is necessarily low over most of the country in the later stages of a cycle of erosion, a broad convexity often appears spreading laterally from the divides in the penultimate stage of interfluves already reduced by dominant pediplanation (scarp retreat and pedimentation). This *semblance* of peneplanation by downwearing, acting during the penultimate stages of the erosion cycle, has been mistakenly thought to have been operative *throughout the cycle*. All stages may be observed consecutively in spurs reaching forward from a major scarp. Near the scarp they are steep-sided and dominantly bi-concave in transverse section. At the distal extremity they are low and become convexly rounded. Baulig (1956, fig. 1) illustrates by a migrating inflexion point the evolution of such decadent slopes and spurs.

Pallister (1956) found in Uganda the normal four elements of hillslope profile, and also that parallel retreat was typical until the final stages of degradation when flattening occurs. This is in a humid tropical region of deep weathering and soil, and again it shows the independence of slope development from climate. The relief is only 400 feet and quite half of this is lost relatively quickly during the stage of scarp retreat so that the remnants of laterite pertaining to the initial " mid-Tertiary " summit surface are now scattered and small in area. Most of the terrain thus now falls under the class wherein relief is so small that slope flattening now dominates by both water-wash (which maintains concave pediment forms) and mass movement (which tends to cover and obscure them and introduces convexity on the lower slopes). Where this latter is prominent (Dixey 1955) " we arrive at a very gently undulating surface that is very close to the general conception of a peneplain." But, of course, this is only in the senile landscape; the full vigour of change from the " mid-Tertiary " to the modern aspect of Uganda was accomplished (as Dixey and Pallister point out) by scarp retreat and pedimentation—there was no peneplanation in the Davisian sense in the production of the later cyclic landscape from the first.

After landscape has been reduced to low relief by scarp retreat, *two* kinds of old-age surface may appear (*a*) where water-wash is dominant, and upon hard rocks (such as may be widely seen in Africa), the land surface is multi-concave upwards, (*b*) where mass movement is dominant, and upon weak rocks, the surface may show both concavities and convexities perhaps with the latter more obvious in the landscape, as in parts of northern Europe.

Our manner of treatment has overemphasised the independent roles of water flow and mass movement, idealising two separate systems as it were. But in nature these seldom act independently, rather they aid and abet each other, and the phenomena appropriate to one system pass by gradation into those of the other. Thus dry materials commonly exhibit high cohesion and tend to deform elastically and slowly until the yield limit is passed, when rupture takes place and sudden motion ensues along localised planes of maximum shear. But the same materials in the wet state may lose much of their elastic strength and deform plastically by flow. Any clots of elastically strong materials are carried along upon the flow, as blocks of hard rock may travel " bobbing like corks " upon the surface of a mudflow.

A plastic flow will stop on a slope when the yield limit is raised or the stress falls below the yield limit. Hence most mudflows come to rest upon lower hillslopes where the gradient flattens though some may continue until they block stream channels in the axis of a major valley. Critical at this stage is the amount of water present in the mass (and consequent fluidity). When the proportion of water is very high, fluid rather than plastic flow ensues and flow does not stop until the end of the slope is reached. Thus the phenomena pass over into the water dominant realm of hydraulics, the solid particles being carried discretely in a medium of low viscosity, that is, in suspension in water. The phenomena of denudation under mass movement and under water flow are thus amenable in the ultimate to a single, unified physical treatment. In consequence, also, much of the supposed distinction between " peneplanation and pediplanation " (Cotton, 1955; Holmes, 1955; Baulig, 1956) disappears from the consciousness of the field observer who finds all types of gradational processes operating upon hillslopes in every quarter of the globe —and every type of resultant hillslope.

The general physical relations of hillslope elements are now clear. In nature, of course, many complexities may be introduced by erosional factors such as renewed or continuous stream incision and lateral stream corrasion at the foot of slopes. Nonhomogeneous bedrock, of course, may produce strongly scarped free faces on resistant rocks above, while the ends of minor spurs on weak rock below are convexly rounded, a combination which illustrates clearly the fallacy of peneplanation throughout the cycle. Dixey, Pallister and Fair, all of whom subscribe to the doctrine of pediplanation dominant in the cycle, severally discuss the function of a hard caprock in slope form. Such structures do not involve any departure from the general principles of slope development. Infinite variation exists to delight the geomorphologist in the field, but the basic plan is surely clear and simple.

Yet there is one case that we must consider further, bearing in mind the dependence of hillform, and particularly the development of an actively retreating free face, upon the factor of *relief*. Beyond a critical measure of relief, upon any type of bedrock, a free face *must* develop under the physical agencies discussed above.

Where landscapes have been warped or tilted, therefore, by an amount exceeding this critical measure a scarp inevitably develops upon the displaced surface and separates higher and lower portions of what were once the same landscape now appearing at different levels. Where the cumulative uplift is sufficiently great such effects may be multiple, giving rise to a stepped landscape whether the uplift was intermittent or continuous.

Thus the piedmont-treppen concept of Penck, demonstrated in

the field in many lands, is proved valid and inevitable under the natural physical laws governing the development of tectonically active terrains. In the light of modern knowledge, the Davisian analysis (1932) is quite unsound.

VI. Some Typical Earth Materials and their Behaviour under Stress

Laboratory study of earth materials as carried out by engineers concerned with the stability of works yields data of importance to students of hillslopes. We note, therefore, some of the more useful contributions from this source.[3]

Sand. Even so simple and inert a substance as quartz sand exhibits quite contrasted properties when stressed, accordingly as its state is dense or loose. A densely compacted sand increases in strength when stressed for deformation implies an increase in volume. Loosely compacted sands, however, decrease in strength when stressed as the grains fit more closely together and the volume decreases. Loose sands cannot stand more steeply than the angle of internal friction (angle of repose), and this is the same whether the sand be wholly immersed in water or in air, but compacted sands may stand in vertical or even overhanging faces when capillary water or salts give intergranular cohesion.

Clays. Not only are clay minerals more complex in constitution, but the grains are commonly flattened so that, fitted together, they leave a much larger proportion of pore space than do normal sands. Clays as a mineral group are therefore relatively porous. Furthermore, the particles bear a negative electric charge which becomes large when compared with gravity or the mass of a particle. The charge is sufficiently strong to ionise adjacent water, and hence a layer of water about two molecules thick is held rigidly in contact with the grains. Because of this, clays do not compact like sand, but retain an open void structure between the grains and remain porous.

Only when sufficient stress is applied to overcome the rigidity of the bound layers of water and destroy the open structure of clay mass do clays compact. Such overstressed clays are no longer porous and lose many of the properties normally associated with clays. These properties are not regained with time, as may be observed in clays of significant geological age that have been subjected to stratigraphic loading. Destruction of the original structure leaves a consolidated clay that is widely traversed by minute fissures. Relief of load, as by denudation of overlying strata, permits the fissures to open slightly and admit water extensively through the mass. Such clays, of which the London Clay is an example, are very weak. Many details of surface

[3] In the preparation of this section I have been greatly aided by Mr. N. Hobbs, Lecturer on Soil Mechanics at the University of Natal, South Africa, to whom my sincere thanks are here expressed.

weathering in stiff Tertiary clays, *e.g.* crumbling, may be traced to this factor.

The shear strength of clays varies hyperbolically with water content; with some qualification upon the type of test or loading conditions in the field. A slowly applied load squeezes out the water and the strength increases. With rapid application of load the clay is subjected to conditions comparable with an undrained test, when with constant water content the strength remains constant.

Valid differences exist between sodium or marine clays and calcium clays. The former have a strong adsorbed layer of water and are more resistant to stress. But leaching of salt may reduce the strength of the clay considerably. Loss of strength from this cause may be cured by pumping in salt water. Most clays in water-laid sedimentary sequences are of this type.

The calcium clays, on the other hand, have only a thin adsorbed layer and comparatively large pore space. Hence such clays disintegrate more readily and the particles are readily dispersed by wind to accumulate elsewhere as loesses, parnas and calcium-rich aeolian soils.

Application of properties such as these can aid geomorphologists materially in understanding the behaviour of earth materials upon hillslopes; but even more useful are the analyses of stress relations in hillslopes and engineering works. What, for instance, are the conditions of stability pertaining to a vertical wall cut into sand or clay?

The treatment is standard: such a cut can retain its form until the height becomes so great that the shear stress overcomes the cohesion between the particles. The stress conditions are illustrated by Fig. 2, the corresponding equation for which is

$$\text{Horizontal Pressure } P_A = \frac{1-\sin\phi}{1+\sin\phi}\, 4\,\gamma\,\frac{H}{2}$$

For a specific material the expression $\frac{1-\sin\phi}{1+\sin\phi}$ will be a constant (K_A). *The horizontal pressure increases with depth, and there will be a critical height beyond which any given material cannot stand unsupported.*

For clean sand, which has no cohesion *per se* the height of such an excavation is clearly zero. But if the material has cohesion (c), then the critical height may be derived:

$$H_{crit} = \frac{4c}{\gamma\sqrt{K_A}}$$

if ϕ equals 0, then $H_{crit} = \frac{4c}{\gamma}$

and as an example we may quote a dense clay in which c equals 1,000 lb./sq. ft. when γ equals 200 lb./sq. ft. The critical height is 20 ft.

Or, in a rock with a cohesion of 3,000 lb./sq. in. and a K_A of ·25, subjected to a stress of 288 lb./sq. ft.:

$$H_{crit} = \frac{4 \times 3{,}000 \times 144}{288 \times \cdot 5} = 12{,}000 \text{ ft.}$$

which approaches the height of some of the highest mountain faces in the Himalaya.

These examples show the control exercised by rock cohesion upon (a) *the development of a free face,* (b) *the maintenance of a free face.* Also, the concept of critical height explains clearly the necessity for adequate relief in a landscape if a fully developed, four-component slope is to be generated, and also the inevitable appearance of piedmont treppen in tectonically rising regions.

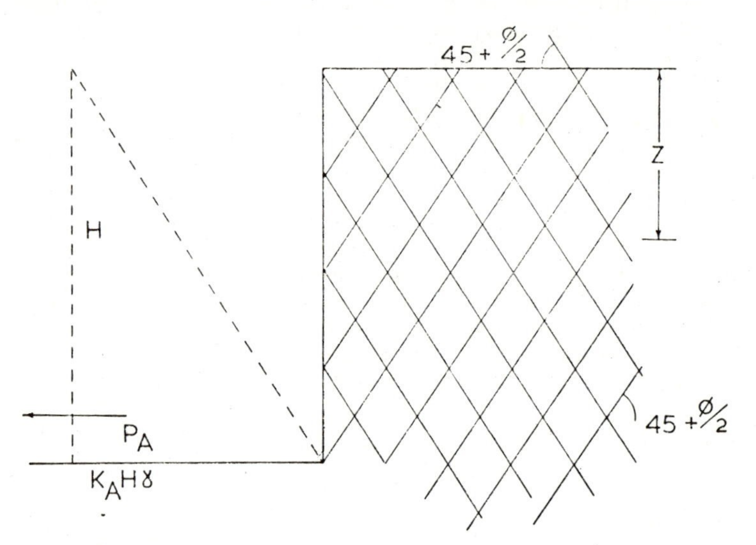

FIG. 2. Vertical wall stresses.

In nature few rocks are massive and devoid of minor planes of weakness. Where such planes exist, failure takes place much earlier and occurs not upon the theoretical planes of mechanical analysis but upon the weak planes themselves. This is especially so if these planes dip towards the face, when the problem of stability involving static friction along the plane may be solved by inclined-plane mechanics.

Virtually the only rocks in which few weak planes appear are those of plutonic origin: the granite-gabbro suite with regional gneisses and charnockites. It is amid these massive rocks therefore that the greatest smooth rock faces are displayed giving rise to the famous bornhardt or inselberg type of landscape wherein the free face attains maximum development among the tetrarchy of slope elements (*e.g.* Africa, Rio de Janeiro). As I remarked in an earlier work (King, 1948) these characteristic landforms are the combined result of (*a*) rock type, (*b*) scarp retreat following (*c*) incision in a new cycle of erosion. Also, such rock types may be expected to develop free faces in a minimal relief, as indeed they do (Plate VIII, A). These landforms owe their existence not to any peculiarity of climate or weathering (indeed they occur in all major climatic environments) but to the massive quality of the rock between joint planes.

It is to be hoped that geomorphologists will be encouraged to employ the observations and methods of engineers more freely in their future studies of hillslopes.

VII. Some Secondary Factors

The transport of waste over land surfaces does not take place without physical and chemical changes in the materials that may result in the production of soil. Raw soil, as developed in deserts, may consist only of disintegrated fragments of rock and crystals. Such primitive material has no coherence and is readily washed and blown away.

Under weathering, the feldspars decay and produce clay. Coherence is an essential property of clays, meaning in terms of geomorphology that the surface deposits, bound by fresh molecular forces, now exert a definite activity in the development of hillforms that is in essence opposed to that of the agents of denudation. To clay is added the binding effect of roots as plant growth is established, and ultimately the addition of humus, a fresh product with bonding properties.

The soil thus passes through a series of changes in which its physical constants, as they affect hillslopes, are profoundly modified. As coherence is in general increased with advancing soil maturity the effectiveness of denudational forces is diminished—both water flow and mass movement—and the sharp distinctions between the four elements of hillslopes and other asperities tend to blur until, finally, smoothly degenerate, convexo-concave slopes tend to be produced.

These changes may be linked with climatic and vegetational factors, which has led to the erroneous view of control of landforms by climate, but the basic controls are still the physical constants of specific earth materials.

Soils

The development of soils implies a local sedentariness in the mantle of waste; but widespread removal of soils is no less real, and apart from the duricrusts no ancient soils seem to have survived from much before the Quaternary Era. Mature soil profiles are only a few inches or feet in thickness, and a thin planing or skinning of the surface, almost negligible to the geomorphologist, may nevertheless result in the total loss of a soil profile acquired only after 100,000 years, or more, of surficial stability. Familiar though I was with the poverty and precariousness of African soils, I was surprised on a recent study tour of Australia[4] to find in some districts how widely and how often, even in that continent of extensive duricrust, the soil mantle had been stripped in geologically recent time. The stability of the waste mantle represented in the production of soils seems therefore to be, in geomorphic essence, transient.

Vegetation

Even though soils themselves may be geologically transient they exert an influence out of all proportion upon the stability of hillslopes through the nourishment which they provide to a vegetal cover. How many hillslopes, stable beneath a close mat of grass with its binding roots, would remain so if the grass were destroyed? And is not the soil conservationist's primary objective to make grass grow? The presence of vegetation, particularly grass with its close binding roots, alters the physical cohesion of soil aggregates and so minimises the development of a free face, either by gullying or stripping and encourages the development of convexo-concave slopes.

Climate

Perhaps it is significant that most of the supporters of climatic interpretation of denudational cycles have come from the ranks of geographers, not geologists. Most of the landscape differences that they quote arise not from climate at all, but partly through the nature of vegetation. Thus the influence of carpet grasses, as contrasted with tufted grasses, upon the physical properties of soils and their erodability under water wash and mass movement may be profound, and productive of significant differences in the elemental proportions of a hillslope profile. Basically these are still controlled by the physical properties of bedrock and waste mantle. Climate is not the fundamental controlling factor at all.

Bornhardts (inselbergs), for instance, were first described from East Africa, and from studies in Southern Africa Passarge referred to them as a special product of *arid* denudation. The multitudinous bornhardts in the maritime states of Brazil were later attributed by de Martonne to a special kind of *humid tropical* denudation. Similar hillforms occur occasionally in the *sub-*

[4] Under the guidance of Mr. Bruce E. Butler of the Soils Division, Commonwealth Scientific and Industrial Research Organisation, Canberra. Australia.

tropical coast belt of Natal; and numerous others, of which Stone Mountain, Georgia, is an example, appear in regions quoted as of "*normal humid*" climate. Comparable rock faces are known in the south-west of New Zealand which is *wet, cold temperate;* and even from the Antarctic in Queen Maud Land. Bornhardt hillforms are thus recorded from every climatic environment on earth. What are the real common factors in all these, and other, occurrences? Hard, massive rock types, and plenty of relief! The physical properties of earth materials govern the development of bornhardts, as indeed of all hillslopes, wherever they are.

Currently, many slope profiles are being surveyed in attempts to probe the manner of slope development; but how seldom are these profiles related to adequate studies of the physical elastic constants of the rock and soil variants below? As matters of observation iron falls faster than cork, and cork faster than feathers—but the physicist has long since appreciated the basic principle of gravitation. So must the geomorphologist appreciate what he is measuring, in physical terms, on a hillslope before proceeding to theorise upon the evolution of such slopes.

Reverting to the use made of data upon earth-materials by engineers concerned with the stability of works, may we remind ourselves that though these practitioners measure carefully the physical properties of the materials with which they deal, they seldom consult the local climatologist!

And, lastly, during the past six years I have travelled over 100,000 miles in five continents, patiently observing all types of terrain and endeavouring humbly to let it teach me, and the abiding impression I have received—despite admittedly minor differences due to local structural, tectonic and climatic factors—is of the essential and basic uniformity of landforms and their evolution in all environments. *As a matter of direct observation all forms of hillslope occur in all geographic and climatic environments.*

The basic homology of all landscapes needs to be appreciated before the differences which, in literature, have been unduly magnified. Within the basic homology, as stated formerly (King, 1953), the standard evolution of landscape is at its optimum under (*a*) "semi-arid" regimes, when waste is most efficiently removed from the surface and retreat of scarps is perhaps at a maximum. Deviation from the semi-arid mean decreases the effectiveness particularly of water wash, and hence waste tends to accumulate at points within the landscape and clog the physical system.

At extremes of aridity wind becomes a potent agency and may dominate over running water as in the *erg* or desert sand sea, while at the opposite extreme the action of glacial ice supersedes that of freely flowing water. These are extremes when the action of the usual physical forces are partly opposed or suspended.

But before either extreme is reached there exists a field wherein abnormal developments occur. Periglacial conditions, including frost action, have exercised widespread effects upon soil and superficial deposits (cryopedology). Of the real extent of such modifications geomorphologists have only recently become aware. All of North America beyond a southern fringe has felt the periglacial touch of the Pleistocene ice age; and according to a statement by Edelman, in Europe only the *terra rossa* soils of the Mediterranean are free of periglacial influence. In New Zealand such effects are now recognised near Wellington (Te Punga).

How then can the landforms of such regions be quoted as " normal " or standard for comparative purposes either now or for the long corridors of the geologic past? They can no more be accepted as " normal " or standard among geomorphologists than a contaminated substance can be accepted as a standard of purity among chemists.

Studies upon the hillslopes of these regions, lacking uniform control during their history, need further analysis in terms of multiple controls, and full of interest though they may be, fall for the time being beyond our present scope.

VIII. Conclusion

As the physical laws of water flow and mass movement are invariable over the land surface of the globe, and as the rock materials of which the upper crust is composed are sensibly the same in all lands, we reach the simple conclusion of a basic homology between the landforms developed under these physical agencies the world over. This has been borne out by field observation and laboratory study. So we conceive an *uniformitarianism of hillslopes* dominating the evolution of epigene landscapes that would surely bring a slow nod of approval from the shade of James Hutton, gentleman and geologist, late of Edinburgh.

IX. References

BAULIG, H., 1956. Peneplanes and pediplanes. *Soc. Belg. d'Etud. Geog.*, **25**, 25-58.

BRYAN, K., 1922. Erosion and sedimentation in the Papago Country, Arizona. *Bull. U.S. Geol. Surv.*, **730**, 19-90.

—— 1935. The formation of pediments. *XVIth Internat. Geol. Cong. Compt. Rend.*, **2**, 765-775.

—— 1936. Processes of formation of pediments at Granite Gap, New Mexico. *Zeit. für Geomorphologie*, **9**, 125-135.

—— 1940. The retreat of slopes. *Ann. Ass. Amer. Geog.*, **30**, 254-268.

COTTON, C. A., 1955. Peneplanation and Pediplanation. *Bull. Geol. Soc. Amer.*, **88**, 1213-1214.

COTTON, C. A., and TE PUNGA, M. T., 1954. Solifluxion and periglacially modified landforms at Wellington, New Zealand. *Trans Roy. Soc. N.Z.*, **82**, 1001-1031.

DAVIS, W. M., 1909. *Geographical Essays*, ed. by D W. Johnson. New York.

DAVIS, W. M., 1930. Rock floors in arid and humid regions. *Journ. Geol.*, **38**, 1-27.
—— 1932. Piedmont benchlands and Primärrümpfe. *Bull. Geol. Soc. Amer.*, **43**, 399-400.
DIXEY, F., 1955. Erosion surfaces in Africa: some considerations of age and origin. *Trans. Geol. Soc. S. Africa*, **58**, 265-280.
—— 1956. Some aspects of geomorphology of Central and Southern Africa. *Trans. Geol. Soc. S. Africa*, Annexure: du Toit Memorial Lecture.
FAIR, T. J. D., 1947. Slope form and development in the interior of Natal, South Africa. *Trans. Geol. Soc. S. Africa*, **50**, 105-120.
HOLMES, C. D., 1955. Geomorphic development in humid and arid regions: a synthesis. *Amer. Journ. Sci.*, **253**, 377-391.
HORTON, R. E., 1945. Erosional development of streams and their drainage basins; hydrophysical approach to quantitative morphology. *Bull. Geol. Soc. Amer.*, **56**, 275-370.
KING, L. C., 1948. A theory of bornhardts. *Geog. Journ.*, **112**, 83-87.
—— 1950. The study of the world's plainlands : a new approach in geomorphology. *Quart. Journ. Geol. Soc.*, **106**, 101-131.
—— 1953. Canons of landscape evolution. *Bull. Geol. Soc. Amer.*, **64**, 721-753.
PALLISTER, J. W., 1956. Slope development in Buganda. *Geog. Journ.*, **122**, 80-87.
PENCK, A., 1924. *Die Morphologische Analyse.* Stuttgart.
—— 1953. *Morphological Analysis of Land Forms.* English translation.
SHARPE, C. F. S., 1938. *Landslides and related phenomena: a study of mass movements of soil and rock.* New York.
WARD, W. H., 1945. The stability of natural slopes. *Geog. Journ.*, **105**, 170-197.
WOOD, A., 1942. The development of hillslide slopes. *Proc. Geol. Ass.*, **53**, 128-138.

University of Natal,
Durban,
South Africa.

Explanation of Plate VII

A. The absence of a " debris slope." The observer places his boot in the angle between the base of a gneissose " free face " and an equally solid apron of the same rock. Planes of gneissosity pass directly from the slope to the apron. Nearly vertical joint planes contain enough soil for grass to grow under a rainfall of 25 in. per annum. Hluhluwe Game Reserve, Zululand.

B. A junction between inselberg " free face " and " pediment " with very little debris. Namaqualand.

Photo. J. A. Mabbutt.

Explanation of Plate VIII

A. " Free faces " in a minimal relief. Midget inselbergs in hard rock yielding little waste. The foreground is thinly littered with broken crystals but, in the absence of chemical weathering, there is no true soil. Namib Desert near Swakop Canyon.

Photo. J. A. Mabbutt.

B. Transition in hard gneiss from " free face " to pediment. The geologist reclines in the angle between the two slope elements, which is also bridged by the motor-van. All debris has been removed by water-wash to the depression in the distance where the trees grow.

Photo. J. A. Mabbutt.

A

B

A

B

Reprinted from *J. Geol. Soc. Australia*, **6**, 131–147 (1960)

SOME PROBLEMS OF SLOPE DEVELOPMENT

By C. R. TWIDALE

7

(WITH 2 PLATES AND 2 TEXT FIGURES)

(*Read before South Australian Division*, 22*nd October*, 1959; *received* 27*th November*, 1959.)

ABSTRACT

Slopes are the fundamental unit of the physiographic landscape. Yet the reasons for variations in their morphology and inferred mode of development are not well understood. The problem is complex, but the slope budget, which results primarily from the interplay of climatically induced factors and structure is significant: where the erosion of the upper slope elements surpasses that on the lower, parallel retreat prevails, but when the reverse is the case, slope decline is dominant. Complications are introduced by such factors as climatic change and fluctuations of baselevel, both local and general. But structural factors can and do override all others.

A knowledge of geomorphic history, and of processes and their effectiveness under various circumstances, in advance of that usually at our disposal at present is necessary before slopes can satisfactorily be analysed. It appears unlikely that there is a universal law of slope development.

I. INTRODUCTION

Over the past fifty years a considerable number of papers concerned with various aspects of hillslope development have appeared in the literature. Many have, either directly or indirectly, discussed the behaviour or developmental trend of slopes: whether they tend progressively to decline or to maintain a constant inclination and retreat parallel to themselves. It appeared that a reasonable measure of agreement, to the effect that the overall behaviour of slopes is conditioned largely by climate, had been attained. In the moist temperate lands, for example, it was generally believed that slope decline prevails, as initially postulated by Davis (1909, pp. 249-278), whereas in the humid tropics (Freise, 1935, 1938) and in hot, semi-arid lands (Bryan, 1922, 1940; Fair, 1947; King, 1949; Pallister, 1956) parallel retreat is dominant. Both types of development, it is claimed, have been statistically demonstrated in regions of contrasted climates (Strahler, 1950, 1954; Tanner, 1956). Occasionally behaviour contrary to this climatically induced pattern has been advocated, but on the whole, the scheme has found support, as evidenced by a sequence of papers from Davis (1930) to Birot (1949, pp. 34-77) and Galon (1954, *vide* Tricart 1957).

Recently, however, the whole of this well-established concept has been called to question by Lester King (1957) in an article entitled "The uniformitarian nature of hillslopes", wherein it is forcefully argued that the same hillslope elements are found "all over the world and in all climatic environments" (*Ibid.*, p. 84); that climate has but little influence on slope form and development; and, by inference, that parallel retreat is everywhere operative. This claim is subsequently qualified with the statement that the same four hillslope

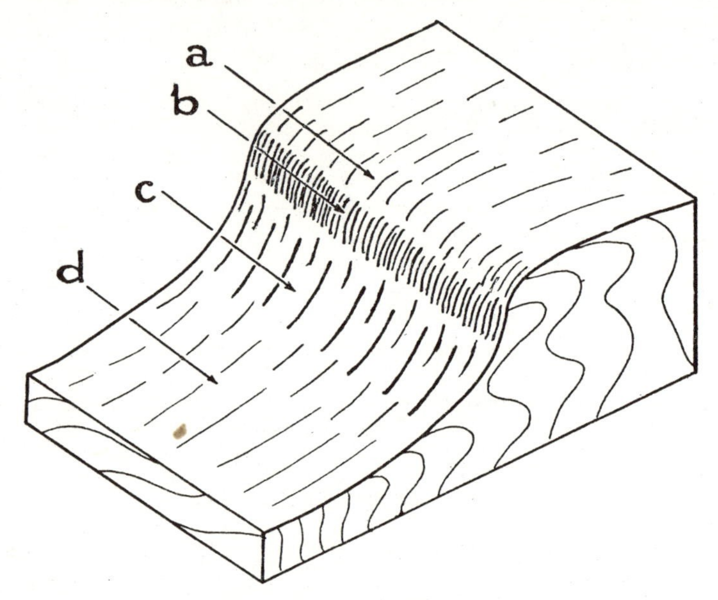

Fig. 1. Hillslope Elements.
For key see Pl. 1, Fig. 1.

elements form "the basic landform that develops in all regions and under all climates wherein water flow is a prominent agent of denudation. This includes all regions except those wherein the dominant agents are glacial ice or wind" (*Ibid.*, p. 88). Still later (*Ibid.*, p. 100), periglacial regions are cited as displaying "abnormal developments".

The scope of King's claims should be made clear, for the exceptional circumstances he quotes prevail over sizeable areas. There are various ideas as to what precisely constitutes a desert and thus what the boundaries and extent of such regions are, but both de Martonne (1925, p. 941) and Meigs (1951) indicate about 30 p.c. of the earth's land surface as being desert; however, only 20 p.c. or so could be considered to be "wind-dominated" (Meigs' "arid" and "extremely arid" categories). As for regions dominated by glacial ice, Flint (1957, p. 51) states that 10 p.c. of the present land surface has an ice cover.

However, during the Pleistocene a further 20 p.c. (*Ibid.*, p. 53) was glaciated, an event of which some of these areas bear the unmistakable imprint. To these must be added those periglacial or cryergic areas of modern times not included in any of the above categories, and also (for former cryergic phases have in some areas left distinct and important traces) some ancient cryergic landscapes. The full measure of these is not known, but may amount to as much as another 10 p.c. of the earth's land area over and above those parts included in the glaciated areas; it includes many areas in the high latitudes too arid to develop ice sheets either during the Pleistocene or at present, for example, a large extent of northern and eastern U.S.S.R., as well as several mountainous districts. Thus the land surface removed from King's generalisations may constitute as much as 60 p.c. of the whole and is in any case a significant proportion.

Nevertheless, the claims that have been made remain far-reaching, for they are said to apply to regions with moist temperate, Mediterranean, savanna, semi-arid and humid tropical climates, to name only some of the more important and extensive varieties.

It is true, as King states, that essentially similar slope forms occur in widely contrasted climatic regions. For example, faceted forms[1] comprising the four elements of Wood (1942), shown in Fig. 1 (see also Plates I and II), and considered by King to be the fully developed fundamental slope, are found in the arctic, sub-arctic, temperate, Mediterranean, hot semi-arid, hot desert, humid tropical lands and elsewhere also. Judged on surface morphology alone, these slopes, found in widely varying conditions of temperature and rainfall, are indistinguishable. Similarly, smoothly curved slopes occur in different climatic zones. If parallel slope retreat prevails, slopes in any given area should be of the same morphology; and if climate has no influence on slope development, all else being equal, slopes everywhere should be similar. From observations in many areas this clearly is not so, and King endeavours to account for such variations in terms of the non-development of some of the basic slope elements, especially the bluff and the debris slope, as a result of differences in relief amplitude and drainage texture.

It is true that both these factors are significant, but others are also: relief amplitude and drainage texture are but two of the several factors which affect the morphology and incidence of slopes. Although mountainous regions display

---

[1] Slopes are of two broad morphological types, faceted and smoothly curved (Savigear, 1956). The nomenclature used for the various slope elements is unsatisfactory. Originally proposed by Wood (1942) and extensively utilised by King, valid objections have been made by Cotton (1948, pp. 226-227) and Baulig (1956; personal communication), and others readily come to mind. These adverse criticisms arise mainly because the terms carry connotations that many would argue are not of universal application. But as is often the case, it is easier to be critical than to make constructive suggestions. In view of past and present controversies, terms descriptive of either the appearance or the location of the slope element are needed, not terms which either inherently or by virtue of past usage carry genetic implications. A set of descriptive terms has been proposed elsewhere: upper slope, (rock) bluff, debris slope and planate slope (Twidale, 1959).

far more faceted slopes than do lowlands, it cannot be said that for practical purposes there is a critical limit of relief amplitude below which the bluff and/or the debris slope do not develop, simply because factors other than relief amplitude are operative. In the Adelaide Hills, for example, faceted slopes occupy valley side walls ranging from a few inches to several hundreds of feet in height. In the same region there is apparent support for the theory that relief amplitude influences slope form in the systematic variation of slopes down valley for from head to mouth there is commonly a sequence of smoothly curved then, as relief increases, a change to faceted followed, with the diminution of relief, by a reversal to the smoothly curved type; but to present such changes simply in terms of the one factor is misleading. Relief amplitude is itself a result of variations in several factors. Some bluffs are developed, and persist, on account of the presence of a rock stratum particularly resistant to the local complex of weathering. The bluff and debris slope develop through the slope being consumed from below by incising streams, and this occurs not only downstream in the same valley, but also as a result of a widespread recent rejuvenation (causing bluff and debris slope to re-form on once smooth slopes) and of undercutting by laterally corrading streams (see Plate II, Fig. 3). Further, the absence of a bluff on some slopes, far from its never having developed, is attributable to its having been eliminated by the extension through weathering of a usually convex upper slope. The present writer would stress not relief amplitude itself, but structure; weathering and erosional processes, which vary with climate, lithology (structure) and baselevel; and the sequential development of landforms in time, a concept which surely includes consideration of relief amplitude.

Similarly, drainage texture varies according to the stage attained by the landscape, increasing in intensity at first, but later decreasing as integration of the network proceeds. It may also fairly be stated that here again an important factor governing drainage texture is climate.[1]

Thus the postulated influences of relief amplitude and drainage texture are susceptible to alternative interpretations; they themselves are effects rather than causes; they alone do not explain the variations of form and distribution of slopes recorded in the literature and in the landscape. No statistical analyses can be cited to support these contentions (the absence of accurate and detailed topo-

[1] Here it may be stressed that it is the whole complex of climate that is important. By way of illustration, Cotton (1958) has recently drawn attention to the contrasts between the intensity and spacing of stream dissection in parts of central Europe and in New Zealand. Both regions bear the imprint of periglacial or cryergic processes, but the latter is far more dissected despite having approximately the same rainfall amount. To account for this apparent anomally it has been suggested (Mortensen, 1959) that the effectiveness of rainfall, which depends in large measure on the time of occurrence, is at the root of the contrast; in central Europe the main rainfalls are in the summer when temperatures are high and evapotranspiration losses also high; in New Zealand, on the other hand, the main rains come in winter when evapotranspirational losses are lower. Mortensen cites another similar case in Chile to support his argument.

graphic maps at present precludes such a possibility), but numerous observers have commented upon the remarkably constant slope inclinations and the prevalence of faceted types in semi-arid lands; on the mixture of types found in temperature regions, though smoothly curved types appear to predominate once the landscape has attained maturity; on the gradual change from plain to upland in temperate regions contrasted with the abrupt transition frequently encountered in hot arid and semi-arid areas; on the prominence of concave forms and the great breadth of valleys in the tropics; and on the sharpness of divides in the humid tropics.

To take but two recent examples of observations which refute King's suggestions, Frye (1959) has described contrasted slopes developed on similar bedrocks in Ohio, Kansas and Texas, and attributes the differences to the climatic factor working both directly and indirectly through its influence on soils and vegetation and hence on type and rate of weathering, amount and incidence of rainfall and run-off. And Campana (1958, p. 26), discussing the Mt. Lofty-Olary arc in South Australia, states:

> "Another important factor to be considered in studying the physiographic aspects of the Mt. Lofty-Olary area is the diversity of climate under which the landscape development took place. Mediterranean in the south with seasonal rainfall of 20-30 inches yearly, the climate becomes gradually arid in the north where the annual episodic rainfall rarely reaches 8 inches. The agents of denudation vary accordingly, so that the landforms of the Mt. Lofty Ranges first merge in a relief of steppes, but give way further north to typical subdesert forms."

The factors cited by King are significant, but to consider only these two is to overlook several other very important influences, and, moreover, leaves much unexplained. In view of the opinion and observations recorded in the literature which refute King's hypothesis, full consideration should be accorded these other factors in order to evaluate their effectiveness and so better assess the contrasted viewpoints.

II. CLIMATE, STRUCTURE AND THE BUDGET

One of the important factors governing slope development is the degree of (bedrock) exposure to, or alternatively protection from, epigene processes. Almost a century ago it was realised that the accumulation of debris on the lower parts of a slope would produce a convex sub-debris (bedrock) slope (Fisher, 1866; the notion has been elaborated by Lehmann (1933)); and as early as 1872 it was appreciated that a rock bluff could retreat at a constant inclination as long as it remained exposed to atmospheric weathering (Fisher, 1872). More recently, Savigear (1952) has stressed the significance of exposure and protection, concluding that whereas the unimpeded evacuation of debris causes parallel retreat, hindered transportation leads to slope decline. These findings are based on field work in South Wales, but some support is forthcoming from investigations in Labrador, where talus accumulates on lower slopes and where hillslopes, considered as entities, are declining; but where the bluff maintains a constant angle when exposed and unprotected (Twidale, 1959). However, it is gradually eliminated by the development of the upper slope above. In northern and central Australia, on the other hand, in spite of the periodic or episodic nature, and in many areas the small amounts, of rainfall, most hillslopes are free of all but an ephemeral veneer of weathered material, indicating that evacuation is efficient; parallel retreat also occurs in these regions. A similar approach is evidenced in a recent contribution (Tricart, 1957) in which the importance of the "budget" or balance (*bilan*) of the slope, that is the rate of production of weathered material compared with the rate of its evacuation, is emphasised.

The budget of a slope, however, depends on several factors, the most important of which is climate, working in the devious ways indicated by Frye (1959). If rock disintegration is fast, as is the case in cryergic or periglacial areas (Corbel, 1959), but the absolute evacuation of the weathered debris slow, detrital accumulation and slope decline occur. In central Labrador, for example, temperature oscillations around the critical 32° F. mark are frequent; the vegetation cover is sparse; there is ample soil moisture; congelifraction and gelifluxion active, particularly on fissile and argillaceous rocks; debris is readily moved down-slope by various forms of mass movement; but apart from the major waterways the drainage network is sparse and thus there is no way in which detrital material can be transported from lower slopes. Debris fans tend to develop and lower slopes are augmented. Meanwhile the exposed upper slopes and bluffs continue to be reduced under the attack of frost and thaw, and therefore there is a decrease in slope amplitude. The slopes suffer an overall decline.

The budget concept can also be applied to tropical arid slopes, such as those of northern and central Australia. According to Fair (1947), Pallister (1956) and others, the essential features of slope development in hot semi-arid regions are:

1. Parallel retreat of steep slope elements as long as the initial surface, duricrusted or otherwise rendered resistant to weathering and erosion, remains intact.
2. Slope flattening once the initial surface has been eliminated. Slopes of varied inclinations occur in this final stage.

Here, as King stresses, drainage texture is important, for the closer the individual elements of the drainage net, the sooner will the upper slope elements be eliminated. However, as has also been indicated, the texture of the network depends in large measure on climate; here is an important factor contributing to the long life of these elements in arid and semi-arid lands.

The preservation of the upper slope elements is manifestly of critical importance; indeed, because of the critical changes occasioned by its elimination it may be thought that a hard capping of some kind is essential for a constant inclination to be maintained. There can be no doubt whatsoever that such a capping greatly assists the retention of the upper slope and bluff and many slopes displaying the elements shown in Fig. 1 in fact have such a structure. But factors other than structure are involved. King (1956) has reaffirmed that

Fig. 2. Hillslope elements in granite country to S. of Cloncurry, N.W. Queensland.

the great Drakensberg scarp in southern Africa traverses strata of varied lithology, and in northern Australia there are faceted slopes formed of rocks which either show no vertical differentiation (sub-vertically disposed quartzites, slate with near vertical cleavage, porphyries, each scarcely affected by weathering) or are apparently weaker at higher than at lower levels. As an example of the latter, there are to the south and south-west of Cloncurry granites which, though sound at the foot of slopes, have a deeply weathered upper zone of rock which can be crumbled in the hand; yet this rotten rock displays the four slope elements flanking mesas and buttes (see Fig. 2).

In semi-arid lands the incidence of heavy, if infrequent, storm rains does not permit appreciable accumulation of debris for any length of time: the weathering processes here — hydration, granular disintegration, exfoliation — are active all over the slope, though hydration is probably more effective on middle and lower slope elements on account of their closer proximity to the water-table. But whereas run-off increases in volume downslope, none of the agents of transportation is very effective on bluff and upper slope. Thus there is no renewal of weathering; on the contrary a hard skin or arid crust commonly develops. The bluff and upper slope, scarcely touched by weathering, remain intact and retreat primarily through being undercut during the erosion of the debris slope below. Hence the massive joint-blocks frequently found scattered over the upper part of the debris slope immediately below the bluff. It is of interest that Jack, unbeknown to the present writer, long ago advocated such slumping following undercutting to account for hillslope features he observed in the north-west of South Australia. He wrote thus:

> "Erosion proceeds by the undermining of these scarps and the falling away of blocks of the crust until isolated table-top hills ('tent hills') are formed and these in turn lose their capping and assume a rounded outline, and finally the hills are reduced to the base level of the drainage system" (Jack, 1915, p. 12).

Such undercutting is possible only where the lower slope elements are susceptible to erosion and manifestly would be out of the question where debris tends to accumulate on lower slopes, as is the case in central Labrador; there large blocks found on debris slopes result from the work of freeze and thaw. On the other hand, where river rejuvenation and landscape revival are taking place, as mentioned previously, the erosion of slopes from below leads to the development of rock bluffs, regardless of climate.

On the whole, in semi-arid lands the decline of slopes is negligible and parallel retreat dominant, but it may be opportune to suggest here that even during parallel retreat there is nevertheless some slight lowering of the initial surface, as evidenced, for example, by the truncation of laterite profiles on some plateaux, and as hinted diagrammatically elsewhere (Davis, 1932, p. 409).

The presence of a hard capping serves only to emphasise the contrast in budget between the higher slope elements on the one hand and the lower ones on the other; however, in some regions where slope decline would otherwise occur, the presence of a hard capping may prove critical and induce parallel retreat, overriding all other factors and tendencies.

In the above discussion the retreat of the slope is assumed to have taken place on a slope in adjustment with local and seasonal climatic conditions, but it will be appreciated that the inclination of the debris strewn lower slope is most likely not the initial gradient, but one attained only after a number of adjustments have taken place. In one case the run-off on the debris slope is more than equal to the task of transporting the available load of detrital material, in which case erosion occurs both vertically and headwards,[1] frequently following heavy storms (Bryan, 1940a), exposing a greater surface to the atmosphere, thus leading to the production of an increased volume of weathered debris. In the other, the run-off may lack both capacity and competence to move the available debris. Detrital aggregates produce a smooth slope and protect the underlying bedrock from weathering and the higher part of the debris slope is steepened by accumulation or erosion and steepening may occur at the foot of the slope where the run-off attains a greater volume than on higher parts of the slope.

In these ways a critical angle of debris slope can be achieved and maintained in response to climatic factors, the gradient of the slope being visualised as fluctuating slightly from one season to the next. The debris slope is the crucial element, and the precise inclination of the bluff depends in large measure on the structure of the country rock. In both cases cited, the retreat of the constant angle slope carries it into bedrock which, on the debris slope, is usually masked by a discontinuous veneer of debris. The gradient of steep slope elements that can be maintained depends on the relative rates of supply and removal of debris which, in turn, depends on climate and lithology.

Eventually the retreat of slopes from all sides causes the bluff and upper slope to be eliminated, an event which, as others have stated, occasions a drastic change in the mode of slope development (see Pl. II, Fig. 4). A major source of debris, the bluff, no longer exists and erosion of the debris slope tends to increase on this account; but the gentler gradients consequent upon the inevitably more marked lowering of the hillcrest cause a reduction in the whole tempo of denudation. However, wash and creep gradually expose bedrock at the hillcrest, this is weathered and thus reduction slowly proceeds. At this stage gentle concavities—either pediments or valley side slopes—extending headwards, merge in a gentle hilltop convexity. This stage has been attained over extensive areas in northern Australia, but in the Barkly Basin and the Carpentaria Plains gentle convexities predominate, a feature no doubt related to the high incidence of permeable or porous soils and rocks, and to the preponderance of creep over concentrated surface run-off (Baulig, 1940). On unconsolidated material permeability is clearly important in determining the morphology of slopes.

III. OTHER FACTORS

Several workers have invoked processes and mechanisms similar in principle to those outlined above to explain the variety of hillslopes they believe

[1] Thus the bluff is undercut and eventually slumping takes place. Some of the caverns frequently found at the debris slope-bluff junction may be caused by such headward erosion by gullies, though others apparently are not. Cavernous weathering is aided by marked groundwater weathering at this and similar junctions (similar circumstances have been cited by Clayton (1956) in relation to the development of linear depressions), but the weathered debris has to be evacuated before caverns form and here the headward extension of ephemeral streams and gullies may be significant.

to exist in various parts of the world. Common to all is a belief in the development of a sequential series of forms with the passing of time, and a conviction that climate plays a considerable part in defining the precise morphology and behaviour of slopes. If, however, it is conceded that there are relationships, direct and indirect, between climate on the one hand and slope morphology and development on the other, the problem, though clearer, becomes undeniably more complex, for there is incontrovertible evidence that the recent geological past has witnessed climatic variations of considerable magnitude. It follows, therefore, that in some areas the complex of denudational processes called the morphogenetic system must also have changed in response to these climatic variations, both in intensity and type, most probably on more than one occasion. Some areas saw several fluctuations, others centrally located in broad climatic belts probably experienced little or no significant change. The great difficulty arises from ignorance and uncertainty. The range of migration of climatic belts is not known, nor is the precise relationship between climate and process. Even if the climatic history has been unravelled, only occasionally can this be related with any confidence to the evolution of slope profiles. The susceptibility of different rock types to denudation varies with climate and can, in fact, be quite contrasted (Birot, 1949, pp. 79-116), which causes further difficulty: in this regard more studies such as those of Birot and of Derruau (1952) are necessary. Another complexity is introduced by vegetational changes induced by climatic fluctuations which, as King (1957) has stressed, influence the rate and type of physiographic processes and which therefore condition the form of the land surface. Moreover, these same climatic changes have caused the repeated growth and recession of glaciers which, in turn, has caused fluctuations of sea level, the fundamental physiographic baselevel. River rejuvenation directly influences slope form, as mentioned previously, though in this regard the work of man in clearing woodland cannot be overlooked.

Quite apart from these factors the duration or longevity of each of the morphogenetic systems must be considered. Here is involved the problem of both absolute and comparative rates of erosion. Very little is known of modern rates of erosion on slopes. This is due in part to the complexity of the problem, in part to the inherent difficulty of measuring erosion experimentally. There are in the literature numerous incidental observations to the effect that erosion (or deposition) has recently occurred, but rarely is there quantitative information which can be accurately dated.

There are no measurements so far recorded in the literature which are sufficiently detailed to meet geomorphological requirements. Sharpe (1941, p. 237) cites an area in which erosion averaged ¼ to ¾ of an inch *per annum,* with a maximum rate of 2 inches in as many months. Beare (1941), discussing erosion in cleared orchards, which are amongst the steepest of the cultivated slopes in the Mt. Lofty Ranges, writes of overall soil losses of 6-12 inches "over the greater part of the area" since the orchards were established between 30 and 50 years ago. Although such information is useful, a great deal is left unanswered, particularly from the point of view of slope studies. Similarly with studies concerned with sedimentation rates and stream loads. Does the erosion take the form of sheetwash or gullying? If the former, on which segment of the slope is it most marked, where is there "erosion" and where aggradation? What relation does the "erosion" bear to slope inclination, to meteorological conditions, to variations in vegetation on the slope? How important in a particular area is the lowering of the land surface due to losses of material in solution? (The data cited by Corbel (1959) and Leopold (1956) and others suggest it can be considerable.)

The general impression is that although gullying is the more spectacular, sheetwash is quantitatively more significant; that sudden heavy showers are more effective than steady rains; that erosion is spasmodic in incidence; and that the rate of erosion is governed by a complex of factors amongst which climate, angle of slope, soil type, vegetation cover and cultural influences (see Strahler, 1956; Leopold, 1956) may be numbered. The incidental observations of exposed tree roots, fence posts and the like by workers engaged on other projects can be of the utmost value, provided careful measurements are made and every opportunity is taken to verify the date of erosion. The experiments being pursued on selected plots both in U.S.A. and in N.S.W. may prove valuable, but in any case, need to be multiplied many times over under a variety of environmental conditions. However, there is as yet no long-continued and widespread series of observations on hillslopes patterned on the stream bank erosion experiments carried out by Wolman (1959), whose observations were made regularly (and at critical periods virtually daily) and for several complete years. Such observations have been instigated in the Adelaide Hills, initially to investigate the possibilities and practical difficulties of this approach.

As for comparative rates of erosion, Corbel (1959) has recently given an illuminating, though necessarily tentative, account. From a review of the literature it is concluded that of the regions considered, on the lowlands erosion by streams[1] is slowest in hot, dry climates (where it might be added wind achieves no erosion of major significance), next slowest in equatorial regions, and regions having "climate with cold winter", "intermediate maritime climate" and "hot moist climate" are approximately equal, though, in fact, the latter is said to display the most rapid erosion. In mountainous areas where, of course, erosion is very much faster altogether, hot moist areas are least fast, hot dry lands more rapid, Mediterranean high mountain next, and the peri-

[1] In some areas this limitation is clearly misleading and proper allowance must be made for other agencies; but some basis of comparison is presented.

glacial (or cryergic) most rapid of all. Thus on plainlands changes of climate do not have such drastic physiographic effects as in mountainous areas or areas with moderate relief. In the latter even a brief spell of extreme cryergic conditions can leave a long-lasting impression on the landscape; on the other hand, lengthy periods of arid or semi-arid climate, according to Corbel's assessment, have but little effect in comparison with cryergic conditions.

The conclusion that many hillslopes, like many other landforms, are, in fact, palimpsest surfaces bearing the imprint not only of past climates but also of past baselevels of denudation cannot be avoided. But given the general climatic and geomorphic history, and with the judicious use of data such as that presented by Corbel, it should be possible to elucidate the history of slope evolution in a reasonable, though inevitably somewhat speculative, manner. Without doubt each region and locality will present its own problems calling for individual solution: no general laws seem possible.

By way of illustration, in central Labrador, after allowance has been made for late Tertiary warping, the plateaux and ridge crests display a remarkable evenness that has been interpreted as indicating an ancient erosion surface. The development of modern hillslopes dates from the disruption of this surface, by warping and faulting, in the late Tertiary. The subsequent river rejuvenation and landscape revival were themselves followed by the Pleistocene glaciations. The ice last retreated from the area only 6-7,000 years ago (Grayson, 1956, pp. 206-207). Since then there have been several climatic fluctuations, but the few pollen analyses from the region (Grayson, 1956, pp. 200-201; Br. Sylvio, University of Montreal, personal communication) show persistent subarctic conditions, with a recent spell of even greater frigidity; in particular the climatic optimum or hypsithermal interval widely evidenced elsewhere is absent. The hillslopes of the region have been modified by streams and rainwash, by ice and by the modern cryergic system of erosion dominated by congelifraction and gelifluxion. The latter presumably also operated during part of interglacial times. The major outlines of the relief may be attributed to all three types of denudation, but there is reason to believe that the finer relief, and in particular the precise morphology of hillslopes, was developed under a cryergic system of denudation: the valley walls of certain late glacial drainage channels, of till slopes and of other postglacial features display a range of forms intrinsically similar to those of other, presumably polygenetic, slopes. The only difference of note is that these postglacial features have rather more congeliturbate debris on lower slopes. It is concluded that the slope forms of central Labrador have developed rapidly under the influence of cold-induced processes acting on glaciated or glacierized surfaces, but allowance must be made for transportation accomplished by the ice, which accounts for the comparative scarcity of frost-shattered debris. Certainly the dominance of convexities in the landscape seems comprehensible only in these or similar terms. The principal points of the hypothesis (Twidale, 1959) are:

(a) The development of an upper convexity of some magnitude by differential weathering (congelifraction).

(b) The transportation of debris downslope by mass movement of material (gelifluxion), its accumulation on lower slopes because of the inefficiency of the ill-developed stream system, and hence, following Fisher (1866), Lehmann (1933) and Wood (1942), the evolution of a convex sub-debris slope (see Fig. 4).

(c) The evacuation of debris by glaciers, and the exposure of this convexity.

(d) The parallel retreat, during its life, of the bluff, which is consumed by weathering from above and buried from below by talus. However, the overall tendency is for a decrease in relief amplitude, for slope decline.

Such an hypothesis explains the evolution and incidence of the various slope forms encountered in the region to which reference is made and is not at variance with what is known of past climates and the relative rates of work of past and present morphogenetic systems.

Where the climatic history is not well known, as in semi-arid Australia, the position is more difficult. Certain observations, some of which are and some of which may be of significance in this respect have been recorded (see, for example, Whitehouse, 1940, pp. 62-72; Hills, 1955; Browne, 1945), but no correlated sequence that can be used with confidence as a basis for other work has yet been unravelled. If the Quaternary climatic changes took the form of the compression equatorwards of approximately latitudinally disposed climatic belts, the degree of change would be less in low than in high and mid-latitudes. Northern Australia might have escaped significant climatic change. Alternatively, there may have been periods of climate even less effective physiographically than the modern regime, e.g., a more arid climate. Certainly the slopes of the region appear broadly compatible with present climatic conditions or something very similar and it may well be that the only significant effect of Quaternary climatic variations has been through the medium of baselevel (sea-level) movements.

It is well known that many modern temperate regions have suffered much colder conditions in the recent past, and it is becoming increasingly clear that some present temperate landscapes bear the impress of these former cryergic conditions. In this general connection it is of interest that ancient cryergic features are well known in New Zealand (Cotton and Te Punga, 1955) and that Ollier and Thomasson (1957) are inclined to attribute the asymmetry of certain valleys in southern England to processes active in immediate postglacial times. It has been said that the cryergic landscape is "alive" or dynamic (Tricart, 1951), whereas the temperate scene is comparatively "dead" or static, an opinion which has found recent quantitative corroboration (Corbel, 1959); and further, that the major relief features of temperate lands date from periods of past frigidity (Cailleux, 1948). The well-developed summit convexity, so typical of many temperate hills, in some areas may be a relict of such former times.

Such considerations are constantly to be borne in mind. It is not denied that structure plays an important rôle both regionally and in detail: both structure and tectonics enter deeply into this problem. It is not intended to discuss the Penckian concept of the tectonic control of relief, which is purposely omitted from discussion, but the interplay between rate of weathering (which depends to some extent on climate) and rate of stream incision (which depends on tectonics, structure and climate) has an important bearing on slope form, as witness the convex slopes so prevalent around Adelaide. Various tectonic movements have influenced the elevation of land above baselevel and thus susceptible to erosion. And structure taken to include both lithology and disposition of strata, jointing, bedding, etc., is also of great, sometimes overriding, significance. A chalk capping in some of the English scarplands, for instance, causes a precipitous bluff to be maintained for many miles, despite its being composed predominantly of soft, often poorly consolidated sediments. The chalk capping is removed partly by solution, partly by undercutting and sapping, often by springs. Similar circumstances may be cited to a greater or lesser degree, in the Jurassic scarplands. It is fair to comment that in neither case are pediments, at least as the term is commonly understood, found at the scarp foot.

IV. CONCLUSION

The burden of this discussion has been that King's main thesis, stimulating as it is, is untenable in view of slope profiles and assemblages, and past and present physiographic processes, observed or evidenced by many workers. Many of his statements are correct and their emphasis is timely; some are susceptible to other, perhaps better, interpretations; others require modification or restriction, sometimes on a drastic scale; but the thesis that slope morphology and genesis are the same wherever running water is dominant must be rejected as an oversimplification of the case. Just as debris fans, patterned ground and other features can develop in different climates and have similar superficial appearances but be genetically different, so can hillslopes attain similar morphological features but have evolved in different ways and through different agencies. The problem is complex, involving the interaction of many factors: structure, lithology, vegetation, present climate and morphogenetic system, and palaeoclimates, stage in geomorphic cycle and recent geomorphic history. It is so complex, and there is as yet so relatively little data on which to base considerations, that universal laws cannot be formulated. As Tricart

(1957) has written: "*Le forme d'un versant ne repond pas a un canon valable pour toute la surface du globe*". All that can be done at present is to suggest that the budget of slopes is of great importance, and where the erosion of lower slopes exceeds that of upper elements, causing the bluff to retreat primarily through undercutting and subsequent slumping, parallel retreat prevails, whereas if the upper elements are reduced more rapidly than the lower are eroded, slope decline is dominant; that decline and parallel retreat are not mutually exclusive, for slopes which as a whole decline contain bluffs which, while they exist, maintain a constant inclination, and during parallel retreat there is, nevertheless, a slight lowering of the initial plateau surface; and that structure is tremendously important, especially in detail and in regions of some relief amplitude, so much so that it may become of overriding importance and even the tentative generalisations enunciated above can be inapplicable. No more precise statement is as yet justifiable.

As slopes are the fundamental unit of the landform assemblage and no facile generalisations may be made concerning their origin, it follows that none can be formulated to explain the diverse landscapes of the world.

REFERENCES

BAULIG, H., 1940: Le Profil d'équilibre des Versants. *Ann. Géogr., 49*, pp. 81-97.

——, 1956: Pénéplanes et Pédiplains. *Bull. Soc. belge Étud. geogr., 25*, pp. 25-58.

BEARE, J. A., 1941: Erosion of Orchard Soils. *J. Dep. Agric. S. Aust., 44*, pp. 639-646.

BIROT, P., 1949: *Essai sur quelques problèmes de Morphologie Generale.* Lisbon, Institute para a alta cultura, Centro de Estudos Geographicos. 176 p.

BROWNE, W. R., 1945: An Attempted Post-Tertiary Chronology for Australia. *Proc. Linn. Soc. N.S.W., 70*, pp. v-xxiv.

BRYAN, K., 1922: Erosion and Sedimentation in the Papago Country. *Bull. U.S. geol. Surv., 730*, pp. 19-90.

——, 1940: The Retreat of Slopes. *Ann. Assoc. Amer. Geogr., 30*, pp. 254-268.

——, 1940a: Gully Gravure – a Method of Slope Retreat. *J. Geomorph., 3*, pp. 89-107.

CAILLEUX, A., 1948: Le Ruissellement en Pays temperé non montagneux. *Ann. Géogr., 57*, pp. 21-39.

CAMPANA, B., 1958: The Mt. Lofty-Olary Region and Kangaroo Island. Ch. I, pp. 3-27, *in* "The Geology of South Australia". (Ed. M. F. Glaessner and L. W. Parkin.) *J. geol. Soc. Aust., 5* (2).

CLAYTON, R. W., 1956: Linear Depressions (*Bergfussniederungen*) in Savannah Landscapes. *Geog. Stud., 3*, pp. 102-126.

CORBEL, J., 1959: Vitesse d'Érosion. *Ann. of Geomorph., 3*, pp. 1-28.

COTTON, C. A., 1948: Landscape. 2nd Ed. Christchurch, Whitcombe and Tombs, 509 p.

——, 1958: Alternating Pleistocene Morphogenetic Systems. *Geol. Mag., 95*, pp. 125-136.

——, and TE PUNGA, M. T., 1955: Solifluxion and Periglacially Modified Landforms at Wellington, New Zealand. *Trans. roy. Soc. N.Z., 82*, pp. 1001-1031.

DAVIS, W. M., 1909: *Geographical Essays.* Boston, Dover. 777 p.
——, 1930: Rock Floors in Arid and Humid Climates. *J. Geol., 38,* pp. 1-27, 136-158
——, 1932: Piedmont Benchlands and Primarrümpfe. *Bull. geol. Soc. Amer., 43,* pp. 399-440.

DERRUAU, M., 1952: Les Caracteres Differentiels des Roches du Socle dans l'ouest et le Sud-ouest du Massif Central Français. Paris, Soc. d'Edition, "Les Belles Lettres". 51 p.

FAIR, T. J. D., 1947: Slope Form and Development in the Interior of Natal. *Trans. geol. Soc. S. Afr., 50,* pp. 105-119.

FISHER, O., 1866: On the Disintegration of a Chalk Cliff. *Geo. Mag., 3,* pp. 254-256.
——, 1872: On Cirques and Taluses. *Geol. Mag., 9,* pp. 10-12.

FLINT, R. F., 1957: *Glacial and Pleistocene Geology.* New York, Wiley. 553 p.

FREISE, F. W., 1935: Erscheinungen des Erdfliessens im Tropenwaldes, Beobachtungen aus Brasilianischen Küstenwalden. Z. *Geomorph., 9,* pp. 88-98.
——, 1938: Inselberge und Inselberg-Landschaften im Granit und Gneissgebiete Brasiliens. Z. *Geomorph., 10,* pp. 137-168.

FRYE, J. C., 1959: Climate and Lester King's "Uniformitarian Nature of Hillslopes". *J. Geol., 67,* pp. 111-113.

GALON, R., 1954: Les Principaux paysages morphologiques du monde du point de vue des profils synthetiques que les caractérisent. *Czas. Geogr., 25,* pp. 26-37.

GRAYSON, J. F., 1956: The Postglacial History of Climate and Vegetation in the Labrador-Quebec Region as Determined by Palynology. Ph.D. Thesis, University of Michigan. 252 p.

HILLS, E. S., 1955: Die Landoberfläche Australiens. *Erde, 3-4,* pp. 195-205.

JACK, R. L., 1915: The Geology and Prospects of the Region to the South of the Musgrave Ranges and the Geology of the Western Portion of the Great Artesian Basin. *Bull. Dep. Min. S. Aust., 5,* 72 pp.

KING, L. C., 1949: On the Ages of African Land Surfaces. *Quart. J. geol. Soc. London, 104,* pp. 439-459.
——, 1956: Drakensberg Scarp of South Africa: a Clarification. *Bull. geol. Soc. Amer., 67,* pp. 121-122.
——, 1957: The Uniformitarian Nature of Hillslopes. *Trans. Edin. geol. Soc., 17,* pp. 81-102.

LEHMANN, O., 1933: Morphologische Theorie der Verwitterung von Steinschlagwänden. *Vjsch. naturf. Ges. Zürich, 78,* pp. 83-126.

LEOPOLD, L. B., 1956: Land Use and Sediment Yield. Pp. 639-647 *in Man's Rôle in Changing the Face of the Earth* (Ed. W. L. Thomas, Jr.). Chicago, Univ. Chicago Press. 1193 p.

DE MARTONNE, E., 1925: *Traité de Géographie Physique.* Tome II Paris, Armand Colin, pp. 499-1048.

MEIGS, P., 1951: World Distribution of Arid and Semi-arid Homoclimates. Pp. 203-210, *in Reviews of Research on Arid Zone Hydrology.* Paris, U.N.E.S.C.O. 212 p.

MORTENSEN, H., 1959: Warum ist die rezente Formungsintensität in Nauseeland stärker als in Europa. *Ann. of Geomorph., 3,* pp. 98-99.

OLLIER, C. D., and THOMASSON, A. J., 1957: Asymmetrical Valleys of the Chiltern Hills. *Geogr. J., 123,* pp. 71-80.

PALLISTER, J. W., 1956: Slope Development in Buganda. *Geogr. J., 122,* pp. 80-87.

Fig. 1. Hillslope elements in granite, 15 miles S. of Selwyn, N.W. Queensland. a — upper slope (the waxing slope of Wood); b – rock bluff (free face of Wood); c – debris slope (constant slope of Wood); d — planate slope (waning slope of Wood, pediment of King). Reproduced by kind permission of C.S.I.R.O.

Fig. 2. Gosse's Bluff, about 100 miles W. of Alice Springs. Dissected debris slope well seen – bedrock with only a veneer of debris. Reproduced by kind permission of C.S.I.R.O.

Fig. 3. Little Para River at Lower Hermitage, 12 miles N.E. of Adelaide. Note how the laterally shifting stream is undermining the bank and how, as a result, the whole lower slope is regrading, causing the exposure of outcrop and the development of a small but distinct bluff.

Fig. 4. Dissected hill country 40 miles W. of Cloncurry. The plateau surface is still intact at left, but in the centre and right the surface has been eroded and slope decline has occurred. Reproduced by kind permission of C.S.I.R.O.

SAVIGEAR, R. A. G., 1952: Some Observations on Slope Development in South Wales. *Trans. Inst. Brit. Geogr., 18,* pp. 31-51.

——, 1956: Technique and Terminology in the Investigation of Slope Forms. Pp. 66-75, *in l'ère Rapport de la Commission pur l'Étude des versants.* Amsterdam, 155 pière.

SHARPE, C. F. S., 1941: Geomorphic Aspects of Normal and Accelerated Erosion. *Trans. Amer. geophys. Un., 1941,* pp. 236-240.

STRAHLER, A. N., 1950: Davis' Concepts of Slope Development Viewed in the Light of Recent Quantitative Investigations. *Ann. Ass. Amer. Geogr., 40,* pp. 209-213.

——, 1954: Statistical Analysis in Geomorphic Research. *J. Geol., 62,* pp. 1-25.

——, 1956: The Nature of Induced Erosion and Aggradation. Pp. 621-638 *in Man's Rôle in Changing the Face of the Earth* (Ed. W. L. Thomas, Jr.). Chicago, Univ. Chicago Press. 1193 p.

TANNER, W. F., 1956: Parallel Slope Retreat in Humid Climate. *Trans. Amer. geophys. Un., 37,* pp. 605-607.

TRICART, J., 1951: Le Système d'Érosion Périglaciaire. *L'Information Géographique,* pp. 187-193.

———, 1957: Mise au Point: l'Évolution des Versants. *L'information Géographique,* pp. 108-115.

TWIDALE, C. R., 1959: L'évolution des versants dans la partie centrale du Labrador-Nouveau-Quebec. *Ann. Géogr., 73,* pp. 54-70.

WHITEHOUSE, F. W., 1940: Studies in the Late Geological History of Queensland. *Pap. Dep. Geol. Univ. Qd.,* (N.S.), 2, No. 1, 74 pp.

WOLMAN, M. G., 1959: Factors Influencing Erosion of a Cohesive River Bank. *Amer. J. Sci., 257,* pp. 204-216.

WOOD, A., 1942: The Development of Hillside Slopes. *Proc. Geol. Ass., Lond., 53,* pp. 28-138.

Supplement to
"Journal of the Geological Society of Australia", Vol. 8, *Pt.* 2.

ERRATUM

The following is supplied to replace the abstract on p. 131 of C. R. Twidale's paper "Some Problems of Slope Development", *J. geol. Soc. Aust., 6,* pp. 131-147, 1960.

ABSTRACT.

Slopes are the fundamental unit of the physiographic landscape. Yet the reasons for variations in their morphology and inferred mode of development are not well understood. The problem is complex, but the slope budget, which results primarily from the interplay of climatically induced factors and structure is significant: where the erosion of the upper slope elements surpasses that on the lower, slope decline prevails, but when the reverse is the case, parallel retreat is dominant. Complications are introduced by such factors as climatic change and fluctuations of baselevel, both local and general. But structural factors can and do override all others.

A knowledge of geomorphic history, and of processes and their effectiveness under various circumstances, in advance of that usually at our disposal at present is necessary before slopes can satisfactorily be analysed. It appears unlikely that there is a universal law of slope development.

Reprinted from *Geol. Mag.*, **3**, 354–356 (1866)

8

IV.—On the Disintegration of a Chalk Cliff.

By the Rev. Osmond Fisher, M.A., F.G.S.

As a slight contribution to the elucidation of questions of denudation, and at the same time an exemplification of the application of mathematics to a geological problem, I send the following:

Noticing a lofty chalk cliff, forming the face of an old quarry in the neighbourhood of Lewes, I remarked that the action of the weather upon the surface was to disintegrate it equally all over, so that the face of the quarry remained vertical, while the stuff that fell down formed a talus, whose surface was approximately a plane, inclined to the horizon at that particular angle at which such materials will stand. The question then occurred to my mind—***What will be the profile of the solid chalk behind the talus?***

The solution of this question is given in the note. The form of the solid surface is there shown to be a semi-parabola, whose vertex is at the base of the original cliff.

It is evident that an old sea cliff deserted by the sea, or a river cliff from which the stream had receded, would disintegrate after the same law. If any subsequent circumstance, such as the return of the sea or river, should remove the talus (and it may be observed that the corollary proves that an undermining action would do so completely) we should have the parabolic form disclosed. This is the shape observed in that form of cliff known as a "Nose." The undermining action of the sea upon a chalk cliff would of itself produce a vertical wall. When, therefore, we see a cliff of a parabolic form, or vertical below, and parabolic in the upper part, it seems to me that we may expect to find that we have an ancient cliff, once deserted by the waves, now attacked again, and the peculiar form due to disintegration exposed by the removal of the talus.

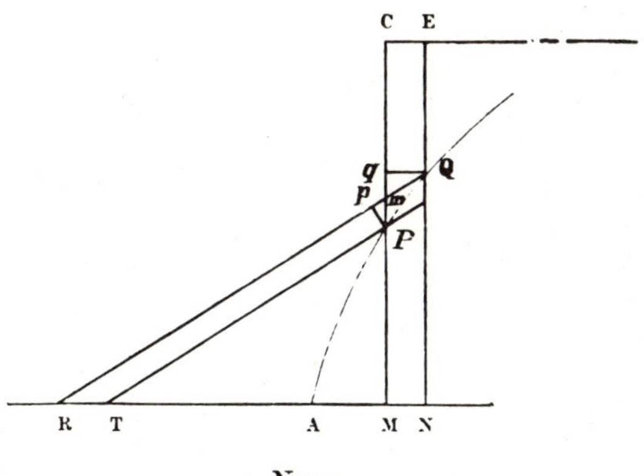

NOTE.

Suppose C P to be the face of the cliff: A the bottom of the quarry: T P the surface of the talus: and let the disintegration of the face to Q E raise the talus to R Q. Then P Q will be a portion of the curve whose form we are seeking.

P p is perpendicular to R Q, Q q to C M, and R Q meets C M in m. C M = a, and the angle P T M $= \alpha$. Then, because the material disintegrated from P C has raised the talus to R Q, we shall have in the limit—

$$C P \times Q q = T P \times P p.$$

or, if $A M = x$, $M P = y$

$$(a - y)\, d x = \frac{y}{\sin \alpha} \times P m \cos \alpha$$

$$= \frac{y}{\sin \alpha} (d y - d x \tan \alpha) \cos \alpha$$

$$= y\, dy \cot \alpha - y\, d x$$

$$\text{or } a\, d x = y\, dy \cot \alpha$$

Integrating, and observing that x and y begin together—

$$a x = y \frac{y^2}{2} \cot \alpha$$

$$\text{or } y^2 = 2\, a \tan \alpha\, x$$

which is the equation to a parabola, of which the vertex is A, and the *latus rectum* is $2\, a \tan \alpha$. The proportions of the curve will therefore be increased by an increase in the height of the cliff, or by an increased capacity in the talus to stand at a high angle.

We have $$\frac{d y}{d x} = \frac{a}{y} \tan \alpha$$

$$= \tan \alpha \text{ when } y = a$$

Hence we have the very elegant result that the curve will be continued unbroken to the very top of the cliff, at which point its inclination to the horizon will be identical with that of the talus.

A. E. SCHEIDEGGER *Imperial Oil Research Lab., 339–50 Ave. S.E., Calgary, Alberta, Canada*

Mathematical Models of Slope Development 9

Abstract: Various theoretical postulates regarding the origin of degradational slopes have been investigated. These postulates represent specific physical conditions which can be expressed in the form of nonlinear hyperbolic partial differential equations. A number of such differential equations, corresponding to various possible physical conditions, has been integrated numerically, and the results of the computations are presented. Thus one obtains a number of slope profiles upon which the observational scientists may draw. In every case, a comparison between theoretical and observed slope profiles will ascertain the physical conditions that produced the latter.

CONTENTS

1. INTRODUCTION AND ACKNOWLEDGMENTS

The explanation of the origin of land forms visible on the earth's surface has been an intriguing problem for a long time. In the development of landscapes two types of processes are involved which are termed endogenetic and exogenetic. The endogenetic processes originate in the interior of the earth and involve such phenomena as the creation of continents, the building of mountains, and the genesis of volcanoes. The exogenetic processes originate outside of the solid earth and involve such phenomena as the sloping of valleys, the planation of mountains, and the building up of flood plains.

In an endeavor to apply theoretical methods to the earth sciences, the writer has been concerned primarily with endogenetic processes. (*See* Scheidegger, 1958.) However, it is hardly justifiable to separate the two type of processes, since an interrelation must exist between them. The division between exogentic and endogenetic geodynamics is thus more for convenience than for valid physical reasons.

The theory of endogenetic processes has been studied far more intensively than the theory of exogenetic processes in spite of the fact that there is considerably more room for unfounded speculations in the former than in the latter. The data for exogenetic processes based on direct observations of landscapes, are supplied by geomorphologists. No tenuous inferences

Geological Society of America Bulletin, v. 72, p. 37-50, 16 figs., January 1961 with regard to processes and conditions in the interior of the earth are necessary, as the agents can usually be observed at work.

One of the chief problems in exogenetic geodynamics is the development of slopes. Slopes are described in many textbooks on geomorphology; much numerical work on their physiography has been done by Strahler (1952; 1956; Smith, 1958). Geomorphologists have listed various agents that may bring about slope developments; among others are erosion, accumulation, and weathering. The combined effect of these phenomena is called denudation.

A brief review of earlier mathematical work is given. Various models based on a variety of physical considerations are then set up. The consequences of these models are calculated out, and the slope profiles ensuing from them are presented. Since these slope profiles are thus tied up with specific physical conditions, the comparison of the theoretical profiles with any particular slope in nature indicates the physical conditions that produced the slope.

The present study was completed during the writer's appointment as Visiting Professor of Geophysics at the California Institute of Technology in Pasadena. The writer is greatly indebted to Robert P. Sharp and Frank Press of the California Institute of Technology for the invitation to spend an extended period of time in Pasadena, and to E. Donald Wilson and James W. Young of Imperial Oil Limited, for arranging for the leave of absence from Calgary. The writer also wishes to acknowledge the help obtained from the staff of the Computing Center at the California Institute of Technology, particularly that of Robert Nathan and Kendrick J. Hebert in debugging the program; and he wishes to thank Joseph N. Franklin for many invaluable suggestions regarding the numerical-analysis aspects of the approximation procedures involved. Robert P. Sharp has read and commented on the manuscript; his interest and valuable suggestions in connection with the present investigation have been extremely helpful. The atmosphere created by the co-operation of the various authorities at the California Institute of Technology made the writer's stay in Pasadena an extremely pleasant one.

2. REVIEW OF PREVIOUS MATHEMATICAL WORK

Several geomorphologists have set up mathematical models of slope development. The best known such model is probably that proposed by Lehmann (1933), which is applicable to the decay of Alpine cliffs: a steep slope of slope angle β is bounded by two horizontal planes which are the (vertical) distance h apart. Each individual amount weathering off the slope will pile up at the bottom, and the surface of the pile will form a scree angle α. Some of the material, however, may get lost (or the total volume may increase because of a decrease in density) during the transfer, so that one has to set

$$V_R/V_D = 1 - c, \tag{2.1}$$

where V_R is the slope material removed and V_D the corresponding volume of debris piling up at the bottom. According to Lehmann, the scree building up protects the slope under-

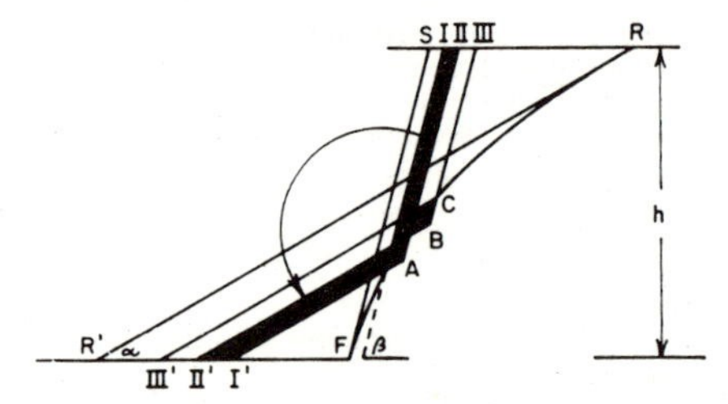

FIGURE 1.—PARALLEL WEATHERING OF A STEEP SLOPE WITH SCREES PILING UP AT THE BOTTOM
(After Lehmann, 1933)

neath from further weathering, so that for each infinitesimal amount of slope recession (Fig. 1) one obtains:

$$\frac{dx}{dy} = \frac{h \cot \beta + (\cot \alpha - c \cot \alpha - \cot \beta)y}{h - cy}, \tag{2.2}$$

where y is the vertical and x is the horizontal co-ordinate of a slope cross section. This is a differential equation which can be integrated. It then leads to the equation of the curve being formed by the cross section of the slope building up underneath the scree. Bakker and the school following him took up Lehmann's theory and greatly expanded it (Bakker and Le Heux, 1946; 1952; Bakker and Strahler, 1956; Van Dijk and Le Heux, 1952; Looman, 1956); they calculated curves for various values of the parameters represented by the solution of the differential equation (2.2); examples are shown in Figure 2. Similarly, graphical methods for the construction of developing slope profiles have been devised.

In the above theory it is assumed that during each infinitesimal time interval, the exposed part of the slope recedes by parallel slope recession. Bakker and Le Heux (1947; 1950) achieved a notable generalization of the theory by replacing this assumption by another

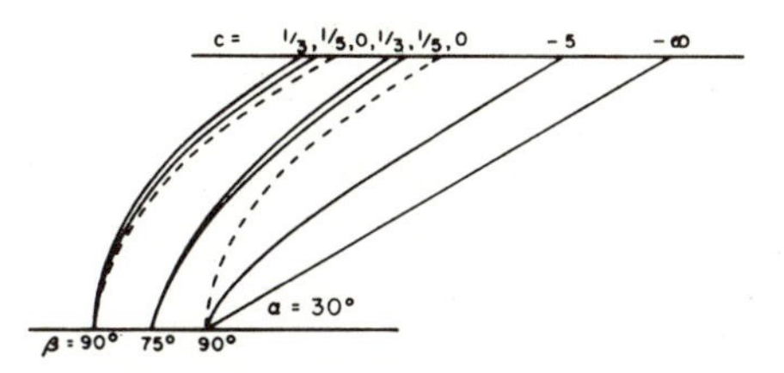

FIGURE 2.—INFLUENCE OF THE CONSTANT *c* ON THE SHAPE OF THE SLOPE UNDERNEATH THE SCREES IN PARALLEL SLOPE WEATHERING

(After Bakker and LeHeux, 1952) Note that $a = \cot \alpha$, $b = \cot \beta$; α is the screes angle and β the original slope angle.

postulating that one is faced with central rectilinear slope recession during each infinitesimal time interval (Fig. 3). This assumption represents an attempt to allow for the condition that weathering may be greater at larger heights as compared with smaller ones. The

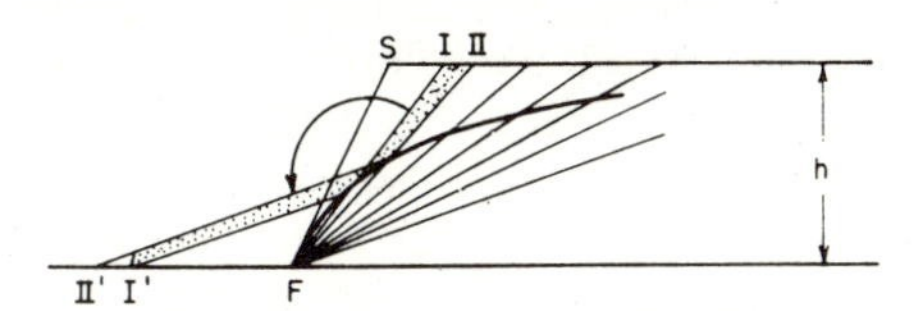

FIGURE 3.—CENTRAL RECTILINEAR SLOPE RECESSION

(After Bakker and Le Heux, 1947)

basic differential equation of slope recession (2.2) becomes now somewhat more complicated, but its solutions are of the same general type as those obtained earlier. Typical cases are shown in Figure 4.

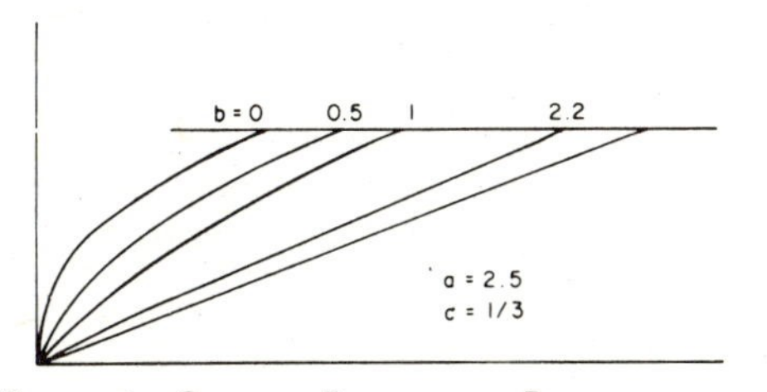

FIGURE 4.—CENTRAL RECTILINEAR RECESSION OF A PLATEAU

(After Bakker and Le Heux, 1947) Note that $a = \cot \alpha$, $b = \cot \beta$; α is the screes angle and β the original slope angle.

The above models are concerned with slope development due to weathering only. It has been assumed that the slopes are building up under a scree. This is somewhat unsatisfactory, since a slope is actually an erosion residual which evolves at the edge of the scree. Let us therefore investigate the change of a slope by erosion. A model like the above is particularly applicable if the slope consists of the material which the eroding agent carries—*i.e.*, in alluvial slopes if the carrying agent is water. However, the picture thus envisaged may be the same for the erosion of almost any bedrock slope. The mechanism of such erosion can be explained as follows: In general, a relationship exists between the speed of flow and the carrying capacity of the water flowing over the slope. If the water carries less material than is its capacity, it takes on more material from the slope underneath and thus makes it recede. Conversely, if the flow is slowed down, the carrying capacity decreases, and therefore material will be deposited. The latter process is called slope development by accumulation.

A model for the slope build-up by accumulation has been discussed by the writer (Scheidegger, 1959). It was shown that accumulative slopes should be essentially concave (this was taken as an *a priori* assumption in a theory of the development with time of accumulative slopes proposed by Strahler [1952]) and flatten out with time, as they are represented by an error-function complement. This theory can be extended to the case where material transport effects removal of mass. The equation of the slope is always an error-function complement: the transport of mass effects erosion on one side of the center of the slope and accretion on the other side (Fig. 5).

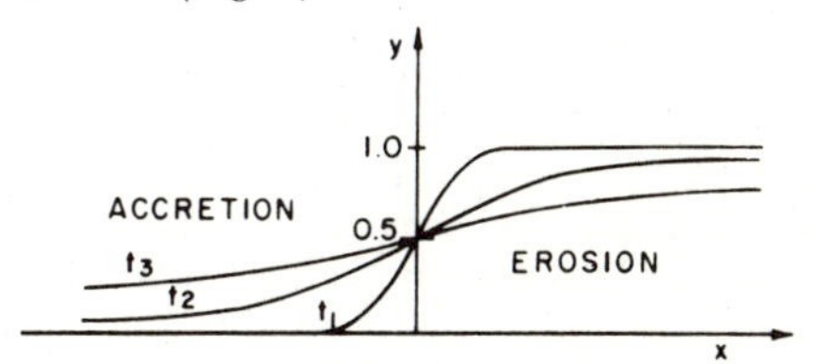

FIGURE 5.—SLOPE DEVELOPMENT BY ACCUMULATION AND EROSION

3. VARIATIONS OF EXPOSURE: LINEAR THEORY

The theories of weathering reviewed above are concerned with the shape of a rocky core beneath a pile of debris. However, Bakker and his followers also applied this theory to the shapes of visible slopes. To uncover the slopes the debris upon them must somehow be removed by the action of some eroding medium. It then appears doubtful that the original shape of the slopes existing beneath the debris could remain unaffected.

In the theory of central rectilinear slope recession, an attempt was made to take into account variations of the rapidity of degradation. However, this was done by a postulate regarding the geometry of the slope without justifying it by a proper model of mass transport. One might therefore want to analyze systematically the effect of various assumptions regarding the rapidity of degradation on the shape of the slope and hope that the various shapes encountered in nature could be obtained directly from appropriate postulates regarding the process of degradation.

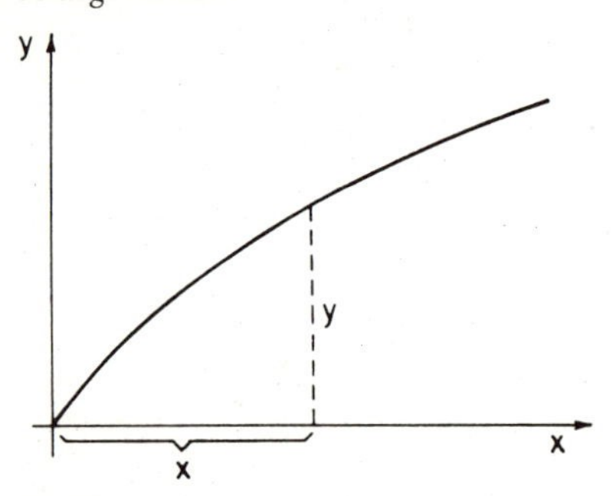

FIGURE 6.—GENERAL GEOMETRY OF A SLOPE

If the slope material is homogeneous, variations of the rapidity of degradation will be due to variations of exposure of the rock to the elements of the weather. The following are possibilities:

CASE 1: The degradation is independent of the slope and proceeds at an equal rate at any exposed portions of the slope.

CASE 2: The degradation is proportional to the height of the point under consideration above a certain base level. This could be justified by the observation that in certain areas precipitation increases with height.

CASE 3: The degradation is proportional to the steepness of the slope. This could be justified by noting that the degradation is due to the exposure of the slope. The steeper the slope, the faster the debris will be removed. Thus, the steeper slopes will generally be more exposed than less steep ones.

Let us investigate how degradation affects the shape of slopes in the above three cases. In order to do this, we assume that the lowering of the slope per unit time at any given point is proportional to a constant (case 1), the height of the slope (case 2), or the slope (case 3). Thus, denoting the height by y, the location by x, we have (*cf.* Figure 6):

$$\frac{\partial y}{\partial t} = - \mathit{const.}\ \phi \tag{3.1}$$

with

$$\text{CASE 1:} \quad \phi = 1 \tag{3.2}$$

$$\text{CASE 2:} \quad \phi = y \tag{3.3}$$

$$\text{CASE 3:} \quad \phi = \partial y/\partial x. \tag{3.4}$$

It will turn out that cases 1 and 2 are quite unrealistic. However, they correspond in essence to the models envisaged by Bakker and followers and therefore must be followed up. One could also envisage other models, such as setting $\phi = 1/(\partial y/\partial x)$, but this probably would be no improvement over case 3, as it would mean that a horizontal slope would degrade infinitely fast, but a vertical one not at all.

It is obviously always possible to change the time scale so that the constant in (3.1) can be set equal to 1. Thus, one has a partial differential equation to solve: the original shape of the slope represents the arbitrary function that enters into the solution of every partial differential equation. In the three cases under consideration, the solution is very easily obtained

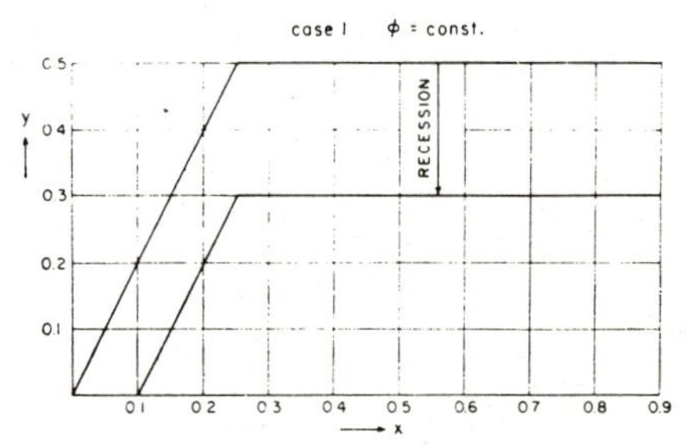

FIGURE 7.—SLOPE RECESSION IN CASE 1 OF THE LINEAR THEORY

CASE 1: The differential equation is

$$\frac{\partial y}{\partial t} = - 1 \tag{3.5}$$

with the initial condition $y = f_o(x)$. The solution is

$$y = f_o(x) - t. \tag{3.6}$$

This represents the case of equal slope recession. The slope retains its shape and simply moves downward. (See Figure 7.) If the slope is rectilinear initially, then one has parallel slope recession downward, as expected.

CASE 2: The differential equation is

$$\frac{\partial y}{\partial t} = -y \qquad (3.7)$$

with the initial condition $y_o = f_o(x)$. The solution is

$$y = f_o(x)e^{-t}. \qquad (3.8)$$

At any time, all slope heights, therefore, are reduced proportionately. (*See* Figure 8.) If the slope is rectilinear to begin with, then this

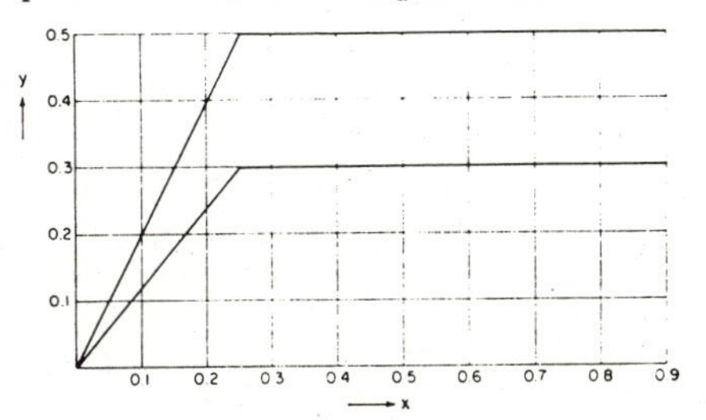

FIGURE 8.—SLOPE RECESSION IN CASE 2 OF THE LINEAR THEORY

represents central slope recession. The latter has, thus, been given a clear physical justification.

CASE 3: The differential equation is

$$\frac{\partial y}{\partial t} = -\frac{\partial y}{\partial x}. \qquad (3.9)$$

With the initial condition $y_o = f_o(x)$, the solution is

$$y = f_o(x - t). \qquad (3.10)$$

This solution signifies that any given slope profile will wander to the right with time. (*See* Figure 9.) If the slope is rectilinear initially, this means parallel slope recession, as in case 1. Thus, parallel rectilinear slope recession can occur in case 1 as well as in case 3.

The various cases discussed above are those that have been treated in the earlier work of Bakker and followers. During its recession, a rectilinear slope remains rectilinear; the development is either parallel or central, depending on the model that is chosen. The physical conditions leading to central slope recession are not very satisfactory, as it is artificial to assume that the degradation is proportional to the height of the slope above a certain base level. It is much more natural to assume that the rapidity of degradation is proportional to the slope itself which, according to the above discussion, leads to parallel slope recession.

The basic shape of the slope remains unaltered in all three cases treated above. The hope that a variation of exposure would change the slope shapes is therefore not fulfilled in the above mathematical models.

4. VARIATIONS OF EXPOSURE: NONLINEAR THEORY

The mathematical models of slope development discussed so far are basically simple, but important conditions obtaining in nature have been neglected, and the calculated slope profiles therefore appear far too simple.

A serious oversimplification was made in the models when the vertical lowering of the slopes was set proportional to some expression which was either a constant, equal to y or equal to $\partial y/\partial x$. One should allow for the fact that weathering acts normal to the slope, so that

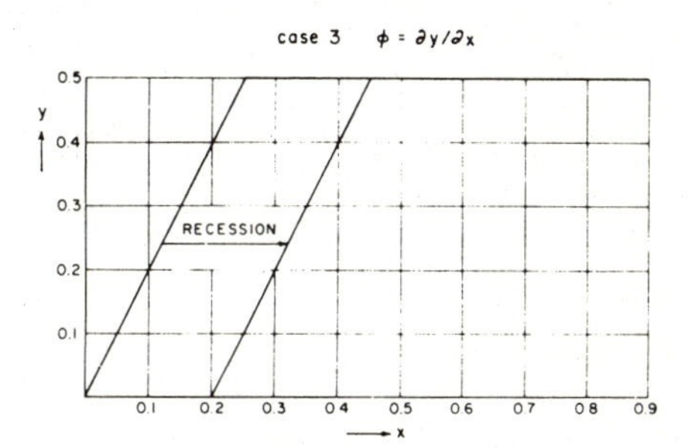

FIGURE 9.—SLOPE RECESSION IN CASE 3 OF THE LINEAR THEORY

the vertical lowering is then represented by the vertical effect of the denudation action (the latter is taken as proportional to a constant, y or $\partial y/\partial x$, according to the case under consideration) which is directed normally against the slope.

In order to investigate the effect of this assumption, consider Figure 10:

$$\Delta\, ABC = \Delta\, DAE;$$

hence

$$\textit{distance } AB = \textit{distance } AD$$
$$dx = \textit{distance } BC = \textit{distance } AE.$$

However, by Pythagoras' theorem

$$\textit{distance } AB = \sqrt{dx^2 + dy^2}.$$

Furthermore

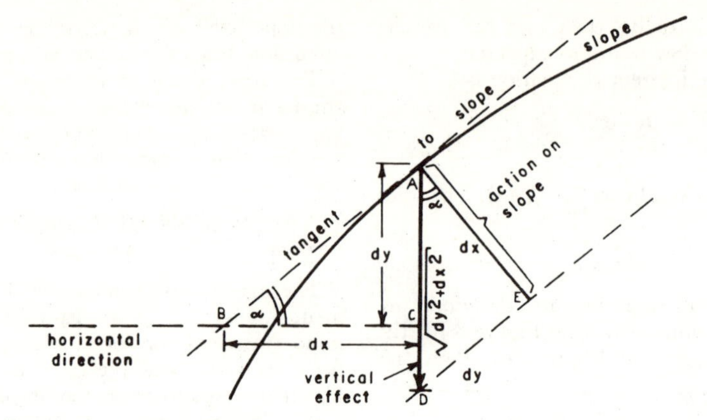

FIGURE 10.—VERTICAL EFFECT OF WEATHERING ACTION NORMAL TO THE SLOPE

$$\frac{\textit{action on slope}}{\textit{effect on slope}} = \frac{\textit{distance } AE}{\textit{distance } AD}.$$

Hence

$$\frac{\textit{action on slope}}{\textit{effect on slope}} = \frac{dx}{(dx^2 + dy^2)^{1/2}},$$

or

$$\textit{effect on slope} = \frac{\sqrt{dx^2 + dy^2}}{dx} \textit{ action on slope};$$

thus

$$\frac{\partial y}{\partial t} = -\sqrt{1 + \left(\frac{\partial y}{\partial x}\right)^2} \cdot \phi, \quad (4.1)$$

if ϕ is the action on the slope, as before. Thus, ϕ must be one of the expressions (3.2), (3.3), or (3.4) corresponding to the three possible cases under consideration.

The improved "new" differential equation (4.1) of slope development differs from the old one in a very fundamental regard: it is non-linear. Easy solutions of the new equation can therefore no longer be obtained. The problem for the three cases was therefore tackled with the help of an electronic computer (the Burroughs 205 Datatron of the California Institute of Technology) by approximating the differential equation by a difference equation. As always with nonlinear hyperbolic partial differential equations, the choice of the steps in the approximation procedure is critical. The steps for Δx and Δt must be consistent with the domain of influence defined by the net of characteristics (*see e.g.*, Collatz, 1951), but this is merely a necessary, not a sufficient, condition for achieving stability for the solution.

In all cases considered, the development of a long straight slope bank was studied (in profile). The original height of the bank was assumed to be equal to 0.5 (arbitrary) scale unit of y and the original slope at one end of the slope equal to 2. The co-ordinate x varies from 0 to 1 in 100 steps. The original slope, thus, has the shape shown in Figure 11. The procedure adopted in the individual cases was as follows:

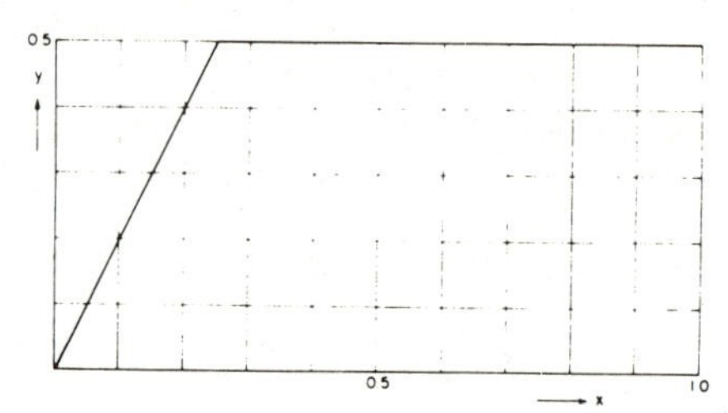

FIGURE 11.—ORIGINAL SLOPE BANK

CASE 1: The differential equation is

$$\frac{\partial y}{\partial t} = -\sqrt{1 + \left(\frac{\partial y}{\partial x}\right)^2} \quad (4.2)$$

which was approximated by the following difference equation:

$$\left(y_{t_{m+1}} - y_{t_m}\right)\Big|_n = -\sqrt{1 + \left(\frac{y_n - y_{n-1}}{x_n - x_{n-1}}\right)^2_m} (t_{m+1} - t_m). \quad (4.3)$$

The characteristics are (using, *e.g.*, the formula given by Collatz, 1951, p. 241):

$$dt/ds = 1 \quad (4.4)$$

TABLE 1.—SLOPE RECESSION IN CASE 1 OF THE NONLINEAR THEORY

x \ Time	5	10	15	20	25	30	35	40	45
0.00	0.00000	0.00000	0.00000	0.00000	0.00000	0.00000	0.00000	0.00000	0.00000
0.02	0.00000	0.00000	0.00000	0.00000	0.00000	0.00000	0.00000	0.00000	0.00000
0.04	0.00000	0.00000	0.00000	0.00000	0.00000	0.00000	0.00000	0.00000	0.00000
0.06	0.00000	0.00000	0.00000	0.00000	0.00000	0.00000	0.00000	0.00000	0.00000
0.08	0.04820	0.00000	0.00000	0.00000	0.00000	0.00000	0.00000	0.00000	0.00000
0.10	0.08820	0.00000	0.00000	0.00000	0.00000	0.00000	0.00000	0.00000	0.00000
0.12	0.12820	0.01639	0.00000	0.00000	0.00000	0.00000	0.00000	0.00000	0.00000
0.14	0.16820	0.05639	0.00000	0.00000	0.00000	0.00000	0.00000	0.00000	0.00000
0.16	0.20820	0.09639	0.00000	0.00000	0.00000	0.00000	0.00000	0.00000	0.00000
0.18	0.24820	0.13639	0.02458	0.00000	0.00000	0.00000	0.00000	0.00000	0.00000
0.20	0.28820	0.17639	0.06458	0.00000	0.00000	0.00000	0.00000	0.00000	0.00000
0.22	0.32820	0.21639	0.10458	0.00000	0.00000	0.00000	0.00000	0.00000	0.00000
0.24	0.36820	0.25639	0.14458	0.03278	0.00000	0.00000	0.00000	0.00000	0.00000
0.26	0.40820	0.29639	0.18458	0.07278	0.00000	0.00000	0.00000	0.00000	0.00000
0.28	0.42811	0.33639	0.22458	0.11278	0.00098	0.00000	0.00000	0.00000	0.00000
0.30	0.45000	0.37630	0.26458	0.15278	0.04098	0.00000	0.00000	0.00000	0.00000
0.32	0.45000	0.39999	0.30458	0.19278	0.08098	0.00000	0.00000	0.00000	0.00000
0.34	0.45000	0.40000	0.34302	0.23278	0.12098	0.00917	0.00000	0.00000	0.00000
0.36	0.45000	0.40000	0.35000	0.27273	0.16098	0.04917	0.00000	0.00000	0.00000
0.38	0.45000	0.40000	0.35000	0.29994	0.20097	0.08917	0.00000	0.00000	0.00000
0.40	0.45000	0.40000	0.35000	0.30000	0.24006	0.12917	0.01736	0.00000	0.00000
0.42	0.45000	0.40000	0.35000	0.30000	0.25000	0.16914	0.05736	0.00000	0.00000
0.44	0.45000	0.40000	0.35000	0.30000	0.25000	0.19969	0.09736	0.00000	0.00000
0.46	0.45000	0.40000	0.35000	0.30000	0.25000	0.20000	0.13684	0.02556	0.00000
0.48	0.45000	0.40000	0.35000	0.30000	0.25000	0.20000	0.15000	0.06554	0.00000
0.50	0.45000	0.40000	0.35000	0.30000	0.25000	0.20000	0.15000	0.09900	0.00000
0.52	0.45000	0.40000	0.35000	0.30000	0.25000	0.20000	0.15000	0.10000	0.00355
0.54	0.45000	0.40000	0.35000	0.30000	0.25000	0.20000	0.15000	0.10000	0.05000

$$\frac{dx}{ds} = \frac{\partial y/\partial x}{\sqrt{1 + (\partial y/\partial x)^2}} \tag{4.5}$$

with s equal to the arc length on the characteristics. This yields as a necessary condition for stability:

$$\Delta t < \Delta x \sqrt{1 + 1/(\partial y/\partial x)^2} \tag{4.6}$$

with

$$\Delta t = t_{n+1} - t_n;\ \Delta x = x_n - x_{n-1}.$$

Since the right-hand side of this inequality is always larger than 1, one can safely set

$$\Delta t = \Delta x. \tag{4.7}$$

The result of carrying out the approximation procedure is shown in Figure 12; some of the numerical values are tabulated in Table 1.

It is evident that there is a difference if the present case be compared with the analogous one of the linear theory. The recession is now no longer straight downward but partly sideways. At the same time, the sharp edge becomes rounded.

CASE 2: The differential equation is:

$$\frac{\partial y}{\partial t} = -y\sqrt{1 + \left(\frac{\partial y}{\partial x}\right)^2}. \tag{4.8}$$

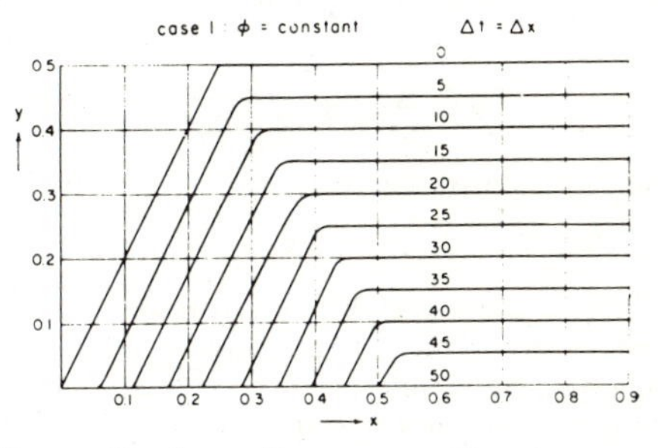

FIGURE 12.—SLOPE RECESSION IN CASE 1 OF THE NONLINEAR THEORY

TABLE 2.—SLOPE RECESSION IN CASE 2 OF THE NONLINEAR THEORY

x \ Time	20	40	60	80	100	120	140	160
0.00	0.00000	0.00000	0.00000	0.00000	0.00000	0.00000	0.00000	0.00000
0.02	0.02705	0.01983	0.01518	0.01188	0.00944	0.00758	0.00612	0.00496
0.04	0.05410	0.03968	0.03036	0.02366	0.01889	0.01516	0.01224	0.00992
0.06	C.08115	0.05952	0.04554	0.03565	0.02833	0.02274	0.01836	0.01488
0.08	0.10820	0.07936	0.06073	0.04753	0.03778	0.03032	0.02448	0.01985
0.10	0.13525	0.09920	0.07591	0.05941	0.04722	0.03790	0.03060	0.02481
0.12	0.16230	0.11904	0.09109	0.07130	0.05667	0.04547	0.03672	0.02977
0.14	0.18935	0.13888	0.10627	0.08318	0.06611	0.05305	0.04284	0.03473
0.16	0.21640	0.15718	0.12145	0.09506	0.07555	0.06063	0.04896	0.03969
0.18	0.24345	0.17856	0.13664	0.10695	0.08500	0.06821	0.05508	0.04465
0.20	0.27050	0.19840	0.15182	0.11883	0.09444	0.07579	0.06120	0.04961
0.22	0.29755	0.21824	0.16700	0.13071	0.10389	0.08337	0.06732	0.05458
0.24	0.32460	0.23808	0.18218	0.14259	0.11333	0.09095	0.07344	0.05954
0.26	0.35165	0.25792	0.19736	0.15447	0.12278	0.09853	0.07956	0.06450
0.28	0.37854	0.27778	0.21254	0.16636	0.13222	0.10611	0.08568	0.06946
0.30	0.40242	0.29757	0.22773	0.17824	0.14166	0.11368	0.09180	0.07442
0.32	0.40895	0.31706	0.24289	0.19012	0.15111	0.12126	0.09792	0.07938
0.34	0.40895	0.32434	0.25790	0.20199	0.16055	0.12884	0.10404	0.08434
0.36	0.40895	0.33449	0.27075	0.21367	0.16996	0.13641	0.11015	0.08930
0.38	0.40895	0.33449	0.27358	0.22265	0.17884	0.14382	0.11620	0.09423
0.40	0.40895	0.33449	0.27358	0.22353	0.18262	0.14893	0.12117	0.09857
0.42	0.40895	0.33449	0.27358	0.22353	0.18264	0.14923	0.12193	0.09963
0.44	0.40895	0.33449	0.27358	0.22353	0.18264	0.14923	0.12193	0.09963

The difference equation approximating this is:

$$y_{t_{m+1}} - y_{t_{m_n}}$$

$$= -\sqrt{1+\left(\frac{y_n - y_{n-1}}{x_n - x_{n-1}}\right)^2}\Bigg|_m (t_{m+1} - t_m) y_n. \quad (4.9)$$

The equations for the characteristics are:

$$dt/ds = 1 \quad (4.10)$$

$$\frac{dx}{ds} = y \frac{\partial y/\partial x}{\sqrt{1+(\partial y/\partial x)^2}} \quad (4.11)$$

which lead to the condition

$$\Delta t \leqslant \Delta x \frac{1}{y}\sqrt{\frac{1}{(\partial y/\partial x)^2}+1}\,. \quad (4.12)$$

Since y is always smaller than 0.5, it is safe to set

$$\Delta t = 2\Delta x. \quad (4.13)$$

The results obtained by this approximation procedure are shown graphically in Figure 13. Some of the values are tabulated in Table 2. In presenting these results, the time steps have been measured in units of Δx.

The slope recession is now no longer "central," as was the case in the linear theory. One also has a rounding of the top edge.

CASE 3: The differential equation is

$$\frac{\partial y}{\partial t} = -\frac{\partial y}{\partial x}\sqrt{1+\left(\frac{\partial y}{\partial x}\right)^2}. \quad (4.14)$$

This is approximated by the difference equation:

$$y_{t_{m+1}} - y_{t_{m_n}} \quad (4.15)$$

$$= -\frac{y_n - y_{n-1}}{x_n - x_{n-1}}\sqrt{1+\left(\frac{y_n - y_{n-1}}{x_n - x_{n-1}}\right)^2}_m (t_{m+1} - t_m).$$

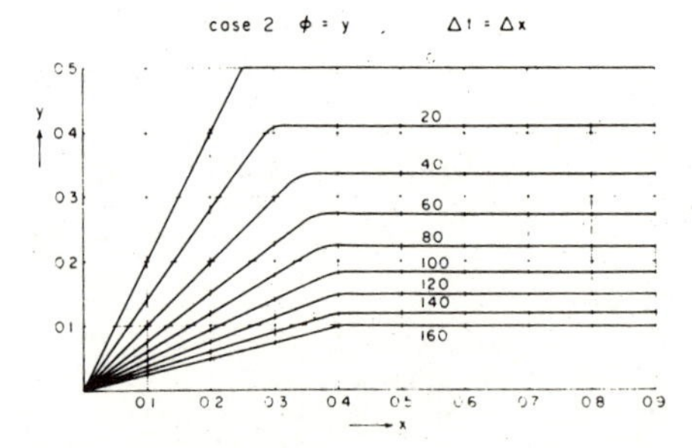

FIGURE 13.—SLOPE RECESSION IN CASE 2 OF THE NONLINEAR THEORY

TABLE 3.—SLOPE RECESSION IN CASE 3 OF THE NONLINEAR THEORY

x \ Time	40	80	120	160	200	240
0.00	0.00000	0.00000	0.00000	0.00000	0.00000	0.00000
0.05	0.00678	0.00029	0.00001	0.00000	0.00000	0.00000
0.10	0.04147	0.00809	0.00084	0.00004	0.00000	0.00000
0.15	0.10149	0.03310	0.00893	0.00135	0.00010	0.00001
0.20	0.18209	0.07370	0.02978	0.00938	0.00184	0.00020
0.25	0.27681	0.12775	0.06227	0.02774	0.00976	0.00228
0.30	0.37640	0.19405	0.10486	0.05557	0.02645	0.01010
0.35	0.47564	0.27160	0.15663	0.09163	0.05121	0.02556
0.40	0.49999	0.35905	0.21701	0.13515	0.08302	0.04810
0.45	0.50000	0.45392	0.28559	0.18563	0.12119	0.07687
0.50	0.50000	0.49994	0.36196	0.24277	0.16527	0.11125
0.55	0.50000	0.50000	0.44558	0.30623	0.21496	0.15817
0.60	0.50000	0.50000	0.49973	0.37592	0.27005	0.19528
0.65	0.50000	0.50000	0.49999	0.45158	0.33036	0.24445
0.70	0.50000	0.50000	0.50000	0.49992	0.39576	0.29815
0.75	0.50000	0.50000	0.50000	0.50000	0.46612	0.35628
0.80	0.50000	0.50000	0.50000	0.50000	0.49997	0.41873
0.85	0.50000	0.50000	0.50000	0.50000	0.50000	0.48487
0.90	0.50000	0.50000	0.50000	0.50000	0.50000	0.49999
0.95	0.50000	0.50000	0.50000	0.50000	0.50000	0.50000

The equations for the characteristics are:

$$dt/ds = 1 \qquad (4.16)$$

$$\frac{dx}{ds} = \sqrt{1 + \left(\frac{\partial y}{\partial x}\right)^2} + \frac{(\partial y/\partial x)^2}{\sqrt{1 + (\partial y/\partial x)^2}} \qquad (4.17)$$

which yield the condition

$$\Delta t \leqslant \frac{\Delta x}{\sqrt{1 + \left(\frac{\partial y}{\partial x}\right)^2} + \frac{(\partial y/\partial x)^2}{\sqrt{1 + (\partial y/\partial x)^2}}}. \qquad (4.18)$$

The machine was programmed to have the last condition always satisfied. Starting with $\Delta t = \Delta x$, the computer would halve the time steps until the inequality was satisfied and then proceed with the calculation. At the beginning it was necessary to use

$$\Delta t = \tfrac{1}{8}\,\Delta x. \qquad (4.19)$$

Later it was possible to increase the time steps to $\Delta x/4$. However, for the presentation of the results in Table 3 and Figure 14, $\Delta t = \Delta x/8$ was chosen as time unit.

The results obtained in the last case are eminently reasonable. The originally straight slope eats its way into the bank. The toe becomes very broad, the head remains relatively sharp, and thus the slope assumes a concave over-all appearance. Simultaneously, the average inclination becomes less and less, and as time progresses, the steep bank will eventually yield to a very gentle slope.

Of the three models discussed, the last is thus the most reasonable. However, comparison of an actual slope in nature will, in every case, nail down the physical conditions that produced it.

5. EFFECT OF ENDOGENETIC PROCESSES ON SLOPE DEVELOPMENT

As mentioned in the Introduction, endogenetic and exogenetic geodynamic processes may occur at the same time. Thus, mountains may be built up while degradation is taking place, or oceanic islands may sink into the sea concurrently with but independently of exogenetic planation. An analysis of the effect of superposition of the two types of processes (endogenetic and exogenetic) on slope development may therefore be of some interest.

The models of slope development discussed here can be modified to describe external ef-

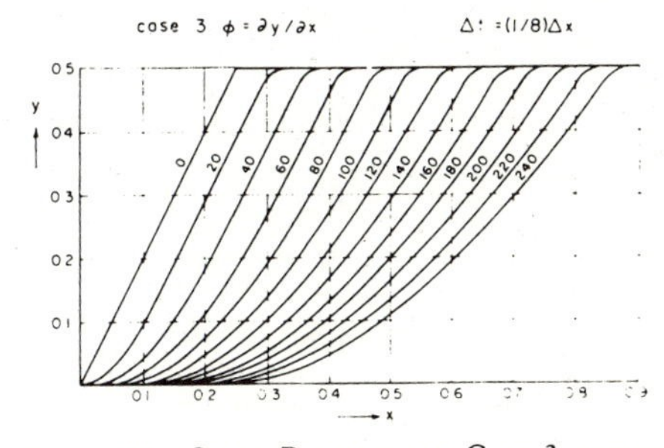

FIGURE 14.—SLOPE RECESSION IN CASE 3 OF THE NONLINEAR THEORY

TABLE 4.—DEVELOPMENT OF AN ENDOGENETICALLY DECREASING SLOPE

x \ Time	10	20	30	50	70	90	110
0.00	0.00000	0.00000	0.00000	0.00000	0.00000	0.00000	0.00000
0.02	0.00843	0.00244	0.00069	0.00005	0.00000	0.00000	0.00000
0.04	0.02844	0.01193	0.00436	0.00063	0.00007	0.00000	0.00000
0.06	0.05636	0.02644	0.01298	0.00282	0.00047	0.00005	0.00000
0.08	0.08817	0.04637	0.02525	0.00730	0.00175	0.00028	0.00003
0.10	0.12082	0.06992	0.04087	0.01404	0.00432	0.00099	0.00017
0.12	0.15350	0.09560	0.05895	0.02276	0.00827	0.00246	0.00057
0.14	0.18618	0.12221	0.07903	0.03318	0.01349	0.00497	0.00141
0.16	0.21887	0.14880	0.10023	0.04503	0.01984	0.00795	0.00279
0.18	0.25155	0.17551	0.12189	0.05800	0.02714	0.01186	0.00472
0.20	0.28423	0.20221	0.14369	0.07177	0.03525	0.01641	0.00715
0.22	0.31691	0.22892	0.16551	0.08601	0.04400	0.02152	0.01001
0.24	0.34960	0.25562	0.18732	0.10005	0.05320	0.02707	0.01325
0.26	0.38218	0.28233	0.20915	0.11502	0.06268	0.03298	0.01681
0.28	0.40600	0.30885	0.23096	0.12958	0.07232	0.03913	0.02062
0.30	0.40851	0.32959	0.25248	0.14414	0.08201	0.04543	0.02461
0.32	0.40854	0.33363	0.26870	0.15864	0.09173	0.05179	0.02872
0.34	0.40854	0.33380	0.27246	0.17208	0.10141	0.05819	0.03290
0.36	0.40854	0.33380	0.27273	0.17993	0.11066	0.06458	0.03710
0.38	0.40854	0.33380	0.27273	0.18182	0.11760	0.07080	0.04131
0.40	0.40854	0.33380	0.27273	0.18206	0.12064	0.07601	0.04543
0.42	0.40854	0.33380	0.27273	0.18207	0.12141	0.07890	0.04906
0.44	0.40854	0.33380	0.27273	0.18207	0.12154	0.07987	0.05144
0.46	0.40854	0.33380	0.27273	0.18207	0.12156	0.08008	0.05245
0.48	0.40854	0.33380	0.27273	0.18207	0.12156	0.08011	0.05274
0.50	0.40854	0.33380	0.27273	0.18207	0.12156	0.08012	0.05280
0.52	0.40854	0.33380	0.27273	0.18207	0.12156	0.08012	0.05280

fects, simply by introducing an additional function F into the basic differential equation (4.1). The latter then becomes:

$$\frac{\partial y}{\partial t} = -\sqrt{1 + \left(\frac{\partial y}{\partial x}\right)^2}\,\phi + F. \qquad (5.1)$$

Many possibilities exist for the choice of F. If F were simply taken as a constant, no significant effect on the curves calculated earlier would be obtained. To get any reasonable results, it is therefore necessary to assume that F is a function of x and y and thus to set

$$F = F(x, y). \qquad (5.2)$$

In the present connotation, two cases were investigated. In the first case we assumed an endogenetic decrease of the slope and therefore set:

$$F = -\,const.\,y. \qquad (5.3)$$

Only the possibility

$$\phi = \frac{\partial y}{\partial x} \qquad (5.4)$$

has been analyzed because the corresponding model of degradation appears to be the most reasonable one. Using (5.3) would presumably give a good picture of a body of mass (such as a volcanic island rapidly thrown up) sinking because of the action of isostasy. The speed of sinking is then proportional to the height of

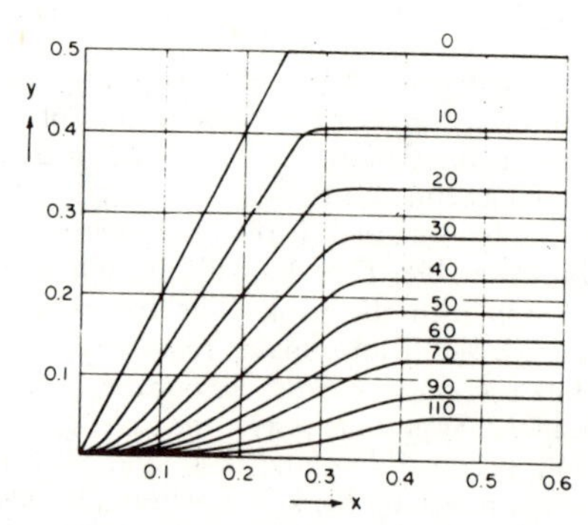

FIGURE 15.—DEVELOPMENT OF AN ENDOGENETICALLY DECREASING SLOPE

Table 5.—Development of an Endogenetically Increasing Slope

x \ Time	10	20	30	40	50	60
0.00	0.00000	0.00000	0.00000	0.00000	0.00000	0.00000
0.02	0.01141	0.00464	0.00193	0.00079	0.00032	0.00013
0.04	0.03794	0.01989	0.01123	0.00634	0.00348	0.00181
0.06	0.07610	0.04487	0.02894	0.01923	0.01274	0.00822
0.08	0.12190	0.07848	0.05422	0.03912	0.02861	0.02079
0.10	0.17042	0.11995	0.08647	0.06543	0.05061	0.03931
0.12	0.21917	0.16841	0.12534	0.09776	0.07831	0.06336
0.14	0.26793	0.22256	0.17051	0.13587	0.11144	0.09264
0.16	0.31669	0.28073	0.22174	0.17958	0.14979	0.12691
0.18	0.36545	0.33958	0.27868	0.22874	0.19325	0.16605
0.20	0.41421	0.39901	0.34088	0.28323	9.24170	0.20992
0.22	0.46297	0.45844	0.40763	0.34294	0.29506	0.25845
0.24	0.51173	0.51788	0.47774	0.40772	0.35325	0.31157
0.26	0.56047	0.57732	0.54963	0.47741	0.41623	0.36923
0.28	0.60288	0.63676	0.62203	9.55179	0.48392	0.43138
0.30	0.60944	0.69607	0.69448	0.63051	0.55626	0.49798
0.32	0.60950	0.74031	0.76693	0.71303	0.63318	0.56899
0.34	0.60950	0.74295	0.83938	0.79850	0.71461	0.64438
0.36	0.60950	0.74297	0.90148	0.88582	0.80045	0.72411
0.38	0.60950	0.74297	0.90567	0.97394	0.89054	0.80816
0.40	0.60950	0.74297	0.90568	1.06111	0.98469	0.89649
0.42	0.60950	0.74297	0.90568	1.10402	1.08262	0.98906
0.44	0.60950	0.74297	0.90568	1.10456	1.18387	1.08582
0.46	0.60950	0.74297	0.90568	1.10456	1.28716	1.18674
0.48	0.60950	0.74297	0.90568	1.10456	1.34700	1.29173
0.50	0.60950	0.74297	0.90568	1.10456	1.34777	1.40072
0.52	0.60950	0.74297	0.90568	1.10456	1.34777	1.51136
0.54	0.60950	0.74297	0.90568	1.10456	1.34777	1.62360
0.56	0.60950	0.74297	0.90568	1.10456	1.34777	1.64452
0.58	0.60950	0.74297	0.90568	1.10456	1.34777	1.64454
0.60	0.60950	0.74297	0.90568	1.10456	1.34777	1.64454

the mass above a certain base level; this is expressed by Equation (5.3).

The present case has been solved on the electronic computer, starting with the usual original slope bank shown in Figure 11. The time steps, at the beginning at least, had to be chosen as follows:

$$\Delta t = \Delta x/8 \tag{5.5}$$

which corresponds to Equation (4.19), since the characteristics are the same as in (4.14). The constant in (5.3) was chosen equal to 16; thus:

$$F = -16y. \tag{5.6}$$

The results obtained in this manner are shown graphically in Figure 15 and numerically in Table 4.

The second case investigated in the present study corresponds to that analyzed above, but with a reversed sign. Thus we set

$$F = 16y. \tag{5.7}$$

This yields a slope whose height is increasing; the endogenetic rate of increase is proportional to the height already reached. This may correspond to conditions obtaining in recent orogenetic belts that are still active. At the beginning, one could take the same time steps as in (5.5), but as the calculation went along, these had to be shortened to fulfill the conditions for stability imposed by the characteristics. Of course, only the possibility for ϕ represented by (5.4) was considered. The results obtained are shown graphically in Figure 16 and numerically in Table 5. In these presentations, time is measured in units Δt, as given in (5.5), although the steps, as mentioned above, were shortened in the latter stages of the computation.

The results show that the superposition of an endogenetic displacement apparently does not materially affect the character of the slope profiles that will develop. An originally straight slope bank will essentially become concave at the toe and convex at the head; the toe will be much broader than the head. This is somewhat at variance with the ideas of Penck (1924), who postulated that "waxing" and "waning" developments would lead to characteristically different slope forms. However, whereas the third model of slope development without the interference of endogenetic effects (embodied in the differential equation 4.14) is probably substantially correct, the particular choice of the function F in (5.6) and (5.7) is probably too oversimplified. Further experimentation with other possibilities might yet produce Penck's slope types and thereby ascertain the physical conditions that produce them.

6. CONCLUSION

Each theoretical model of slope formation proposed in this paper is the result of particular, specific physical conditions. If the various results be compared with actual slopes observed in nature, then such a comparison will provide a means to ascertain the true physical conditions that have caused the observed slope. The task of observation of slopes falls in the domain of field geomorphology, a branch of the earth sciences with which the writer is not very familiar. He hopes, therefore, that people qualified to do this will search for the various types of slopes and make the proper comparisons.

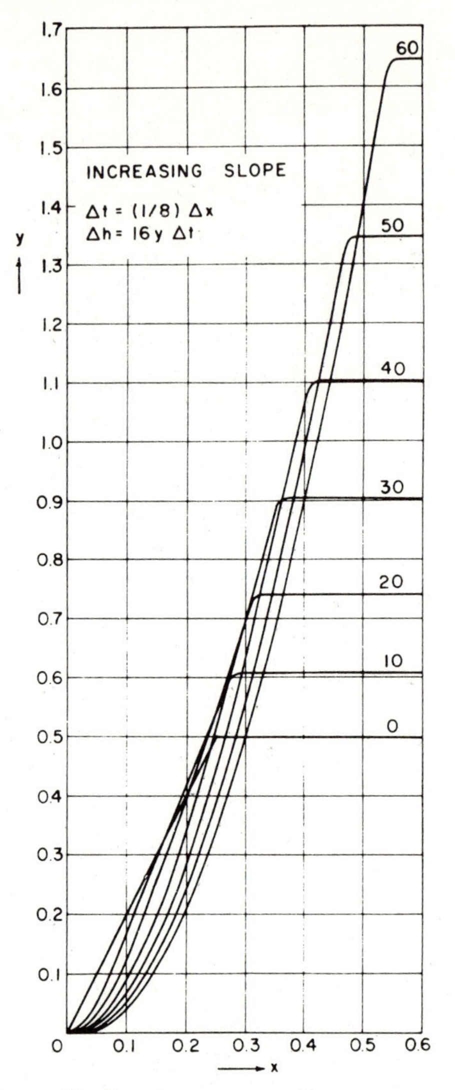

FIGURE 16.—DEVELOPMENT OF AN ENDOGENETICALLY INCREASING SLOPE

REFERENCES CITED

Bakker, J. P., and Le Heux, J. W. N., 1946, Projective-geometric treatment of O. Lehmann's theory of the transformation of steep mountain slopes: K. Nederl. Akad. Wetens. Proc., v. 49, p. 533-547

—— 1947, Theory of central rectilinear recession of slopes, I and II: K. Nederl. Akad. Wetens. Proc., v. 50, p. 959-966, 1154-1162

—— 1950, Theory of central rectilinear recession of slopes III and IV: K. Nederl. Akad. Wetens. Proc., v. 53, p. 1073-1084, 1364-1374

—— 1952, A remarkable new geomorphological law: K. Nederl. Akad. Wetens. Proc., Ser. B, v. 55, p. 399-410, 554-571

Bakker, J. P., and Strahler, A. N., 1956, Report on quantitative treatment of slope recession problems: Cong. Int. Géog., Rio de Janeiro, 1er Rapp. Comm. étude des versants, 12 p.

Collatz, L., 1951, Numerische Behandlung von Differentialgleichungen: Berlin, Springer Verlag, 458 p.

Lehmann, O., 1933, Morphologische Theorie der Verwitterung von Steinschlagwänden: Vierteljahrsschr. Schweiz. Natf. Gesellsch. Zürich, v. 78, p. 83-126

Looman, H., 1956, Observations about some differential equations concerning recession of mountain slopes: K. Nederl. Akad. Wetens. Proc., Ser B, v. 59, p. 259-284

Penck, W., 1924, Geomorphologische Analyse: ein Kapitel der physikalischen Geologie: Stuttgart, Verlag von Engelhorns Nachf., 283 p.

Scheidegger, A. E. 1958, Principles of geodynamics: Berlin, Springer Verlag, 280 p.

—— 1959, Hydraulic effects in geodynamics: Geologie und Bauwesen, v. 25, p. 3-49

Smith, K. G., 1958, Erosional processes and landforms in Badlands National Monument, South Dakota: Geol. Soc. America Bull., v. 69, p. 975-1008

Strahler, A. N., 1952, Hypsometric (area-altitude) analysis of erosional topography: Geol. Soc. America Bull., v. 63, p. 1117-1142

—— 1956, Quantitative slope analysis: Geol. Soc. America Bull., v. 67, p. 571-596

Van Dijk, W., and Le Heux, J. W. N., 1952, Theory of parallel rectilinear slope recession: K. Nederl. Akad. Wetens. Proc., Ser. B, v. 55, p. 115-129

Manuscript Received by the Secretary of the Society, June 23, 1959
California Institute of Technology, Division of Geological Sciences, Contribution No. 935

Reprinted from *Zeitsch. Geomorph.*, **9**, 88–101 (1970)

10

A comparison of theoretical slope models with slopes in the field

by

FRANK AHNERT, College Park, Maryland

with 14 figures

Zusammenfassung. Mit dieser Arbeit soll an Hand von Feldbeobachtungen an rezenten Hängen der Wert theoretischer Modelle der Hangentwicklung überprüft werden, die anläßlich des IGU-Kongresses 1964 vorgeführt wurden.

Auf den Modellhängen variiert der hangab gerichtete Massentransport mit dem Sinus des Hangwinkels, und der Anfall an Verwitterungsmaterial wird durch die Dicke der Verwitterungsdecke reguliert. Hangabwärts-Bewegung und lokale Verwitterung werden somit als Determinanten der Veränderung der Mächtigkeit der Verwitterungsdecke auf dem Hangprofil angesehen.

Feldbeobachtungen, zusammen mit Profilmessungen, Probenentnahme und Messungen der Verwitterungsdecke mit einem tragbaren Refraktionsseismographen, wurden an einigen Hängen im Gneisgebiet von North Carolina durchgeführt. Aus der multiplen Regressions-Analyse ergab sich eine Zunahme der Mächtigkeit der Verwitterungsdecke mit der Entfernung vom höchsten Punkt des Hanges, und eine Abnahme mit dem Wachsen des Sinus des Hangwinkels. Die Korrelation ist sehr gut (multipler Korrelationskoeffizient R = O. 873). Um die Feldbefunde mit denen des theoretischen Modells zu vergleichen, wurde auch für die letzteren die gleiche Regressionsanalyse durchgeführt.

Die berechneten Gleichungen waren einander sehr ähnlich:

Feldbefunde: $C = 67{,}55 + 1{,}52\,d - 156{,}04 \sin \alpha$
Modellhang: $C = 68{,}77 + 1{,}16\,d - 110{,}07 \sin \alpha$
(C = Mächtigkeit in % der mittleren Mächtigkeit,
d = Entfernung vom höchsten Punkt des Hanges in % seiner Gesamtlänge).

Im weiteren wird die Ähnlichkeit zwischen den beiden Hängen diskutiert und analysiert. Es ergibt sich, daß der theoretische Hang ein vernünftiges Modell rezenter Hänge in der Natur darstellt.

Summary. The purpose of this study is to test the validity of theoretical models of slope development that were presented at the 1964 IGU Congress, by means of field observations on actual slopes. On the model slopes, the downslope waste transport varies with the sine of the slope, and the rate of waste supply by bedrock weathering is regulated by the thickness of the waste cover. Downslope movement and local weathering thus are seen as determinants of the variation of waste cover thickness along the slope profile.

Field observations, including profile surveys, collection of samples, and measurements of waste thickness with a portable refraction seismograph, were carried out on several slopes in an area of gneiss in North Carolina. By multiple regression analysis, it was found that the waste cover thickness varies positively with the distance from the top of the slope, and negatively with the sine of the slope. This correlation is very high (multiple correlation coefficient $R = 0.873$). For comparison of the field data with the waste cover data of the theoretical model, the same regression analysis was carried out for the latter.

The estimating equations obtained are very similar:

Field data: $C = 67.55 + 1.52\,d - 156.04 \sin \alpha$
Model slope: $C = 68.77 + 1.16\,d - 110.07 \sin \alpha$
(C = thickness in % of mean thickness, d = distance from top in % of total slope length).

The similarity between the model slope and the field slopes is further discussed and analyzed. It supports the contention that the theoretical model is reasonably representative of actual slopes in the field.

Résumé. Le but de cette étude est de vérifier la validité de modèles théoriques de l'évolution des versants, qui ont été présentés au Congrès de l'U.G.I. en 1964, à l'aide d'observations effectuées sur des versants réels.

Sur les versants des modèles, l'importance du transport des produits meubles varie avec le sinus de la pente, et la vitesse de la formation de ces produits par l'altération est contrôlée par l'épaisseur du manteau d'altération. Ablation et altération locales sont donc considérées comme déterminant l'épaisseur du manteau d'altération en chaque point du versant.

Des observations sur le terrain, comprenant l'élaboration de profils, la collecte d'échantillons, la mesure de la profondeur de l'altération avec un séismographe ont été effectuées sur une série de versants dans une région de schistes cristallins de la Caroline du Nord. L'analyse par regression multiple des résultats a montré que l'épaisseur du manteau d'altération varie avec la distance entre le sommet et le point du versant considéré et en raison inverse du sinus de la pente. La corrélation est très élevée (coefficient de corrélation multiple, $R = 0{,}873$). Afin de comparer les données du terrain avec celles du modèle, ces dernières ont été traitées suivant le même processus d'analyses que les premières. Les équations obtenues pour l'un ou pour l'autre sont assez semblables:

Terrain: $C = 67{,}55 + 1{,}52\,d - 156{,}04 \sin \alpha$
Modèle: $C = 68{,}77 + 1{,}16\,d - 110{,}07 \sin \alpha$
(où C est l'épaisseur du manteau d'altération en % de l'épaisseur moyenne
d la distance du sommet en % de la longueur totale du versant.)

Cette similitude de formule a été discutée et analysée avec plus de détails. Elle suppose que le modèle théorique choisi est représentatif pour les pentes réelles.

One of the problems in the study of slope development is the difficulty to verify a hypothesis by field observation. A theoretical assessment of slope development requires the inclusion of slope-transforming processes, in particular of waste production by weathering, and of the downslope transport of the waste. The rates at which these processes occur need to be quantitatively defined, at least in relative terms, as functions of other slope properties; the rate of mass-wasting, for example, can be defined as a function of the slope angle.

Despite recent improvements of measuring techniques whose application has made an increasing amount of data available (YOUNG 1960, RAPP 1961, RUDBERG 1964, SCHUMM 1964, LEOPOLD, WOLMAN & MILLER 1964, EMMETT 1965), direct field measurements of slope processes still contain large error margins which limit their usefulness as a basis for theoretical models. Only in part are these error margins due to the instrumentation, in the sense that the rods, test pillars, pits, spikes and washers etc. that are used for the observation of processes interfere to varying extent with the process they are meant to measure. More importantly, the common quasi-continuous processes of weathering and downslope transport are usually too slow to be accurately measured in a limited period of observation. Most of the rapid processes, on the other hand, occur too rarely and too discontinuously in any one location to permit reasonably accurate determination of their rates over the long periods of time required for slope development.

Because of these difficulties, theories of slope development tend to be deductive rather than inductive–based on reasonable assumptions that appear to be in general accord with observational evidence, rather than directly on specific field observations. This rule applies to all theoretical models of slope development that have been designed so far (W. PENCK 1924, O. LEHMANN 1933, BAKKER & LE HEUX 1947, 1950, SCHEIDEGGER 1961, YOUNG 1963, AHNERT 1964, 1966). Corroboration of the theoretical models has customarily been sought by comparison between the surface profiles of the model slopes with the surface profiles of slopes in the field. Critics of this approach have pointed out that a given profile shape can result from a variety of process combinations (LEOPOLD, WOLMAN & MILLER, 1964, pp. 500–503), so that a coincidence of profile shapes by itself is insufficient evidence for the validity of the model.

Some theoretical slope models show not only the transformation of surface profiles, but also the spatial and temporal variations of waste cover thickness on the model slope. These variations are directly caused by variations in the rates of waste production and of downslope waste transport–i. e. of processes that constitute essential components of the theories on which the models are based. A comparison of the distribution of waste cover thickness on slopes in the field with that on model slopes can therefore provide a valuable additional criterion for assessing the model's reliability.

An opportunity to carry out morphological field work in the crystalline Appalachians of North Carolina was used to make a series of measurements of waste cover thickness on slopes, for a comparison with a slope model that was designed earlier. The results of this comparison are presented in this paper.

Characteristics of the Model

The model in question has been described in detail elsewhere (AHNERT 1966), so that a summary of its essential characteristics will suffice here. It simulates the development of slope profiles in response to weathering and mass-wasting.

The rate of waste production by bedrock weathering at any point of the model slope is assumed to decrease with increasing thickness of the waste cover at that point. Two different versions of this functional relationship were used, namely, a linear decrease and an exponential decrease of waste production. The model slopes developed with the exponential function bear a closer resemblance to the studied field slopes and will therefore be used for the comparison here. Their rate of waste production is defined by the formula

(1) $$W_C = W_0 e^{-C}$$

where W_0 is the rate of waste production on bare bedrock, and W_C the rate of waste production under a waste cover of thickness C; e is the basis of natural logarithms. The waste moves downslope at a velocity that is proportional to the sine of the slope angle.

During any given time unit, the waste cover thickness at each point on the model slope is determined by

(a) the waste thickness C present at the beginning of that time unit,
(b) the amount A of waste arriving from upslope,
(c) the amount of waste W supplied by local weathering, and
(d) the amount of waste R being removed in a downslope direction, so that

(2) $$C' = C + A + W - R$$

where C' is the waste cover thickness at the end of the time unit in question. It can be shown that under these conditions the rates of waste supply (from upslope and by local weathering) and of waste removal at each slope point tend towards a state of dynamic equilibrium (GILBERT 1877, JAHN 1954, AHNERT 1954, 1966, 1967).

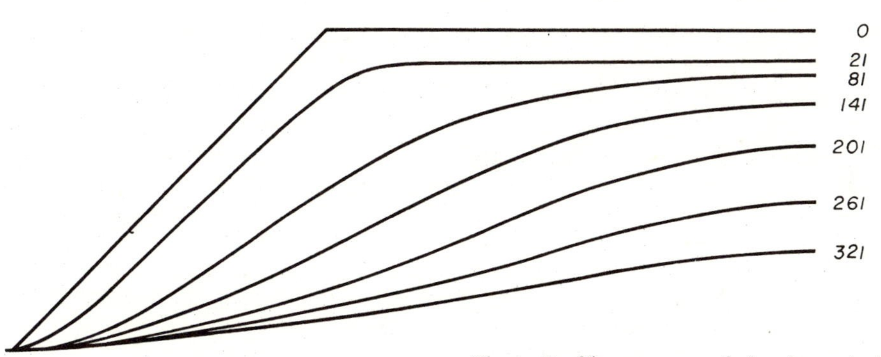

Fig. 1. Profile sequence of the theoretical model. The numbers on the right indicate elapsed time units (computer iterations).

The ingredients of the theoretical model, fed into a computer, produce a developmental sequence of slope profiles (fig. 1) from an arbitrary initial surface to a subdued gentle slope. The profiles of figure 1 are bedrock profiles; however, the computer printout also lists the waste cover thickness present at each slope point, so that there is no problem in drawing any one profile with its waste cover (fig. 2).

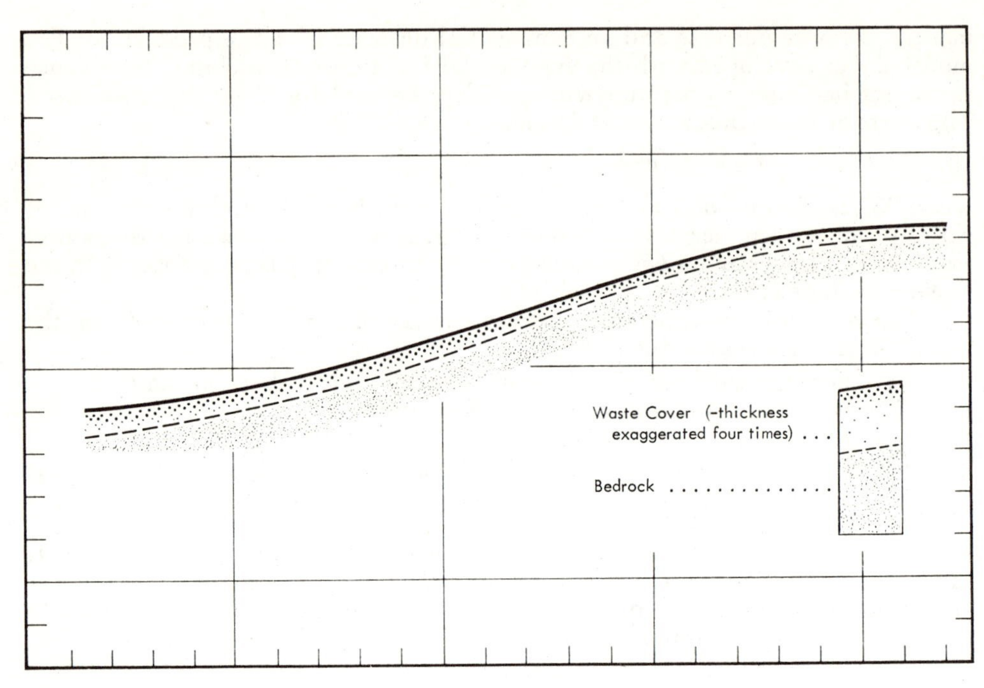

Fig. 2. Model profile after 241 time units (computer iterations).

Characteristics of the Field Slopes

The model had been designed with the tacit assumptions that the bedrock was homogeneous and that there was neither downcutting by a stream at the slope foot nor a significant climatic change during the time of slope development. The slopes in the field that were to be used for an attempt to verify the model by comparison had to match these conditions as closely as possible. Massive crystalline rocks tend to be more homogeneous over greater distances than sedimentary rocks. Areas of relative tectonic stability, at some distance from the sea, are least likely to be affected by recent stream incision. Finally, the condition of climatic stability in humid-temperate regions is best fulfilled in areas that lie equatorward of the zone of known Pleistocene periglacial effects. The crystalline Appalachians of North Carolina meet these conditions rather well. The slopes that were selected lie on gneiss, at elevations below 4,000 ft (1,200 m) and at a latitude of about 35°50″N. None of them borders directly on streams that are actively cutting down. Profiles that ranged from 250 ft. (75 m) to 1,300 ft. (390 m) in length were surveyed with tape and hand level. On five of these profiles, the waste cover thickness was determined by means of a portable refraction seismograph (type Terrascout, rented from Soiltest, Inc., Evanston, Illinois) at regular intervals along the slope profile. Failure of one of the instrument's transistorized

Fig. 3. Profiles of field slopes.

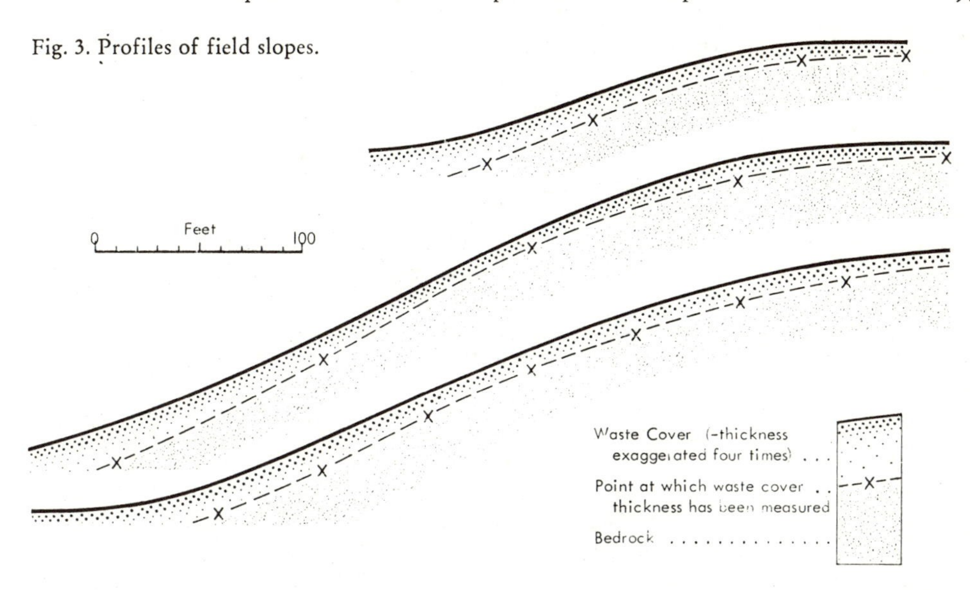

modules during the field work unfortunately prevented extension of these measurements over a greater number of slopes. The two longest profiles showed irregular variations of slope angles and of the waste cover thickness that are probably due to local variations of rock resistance. They had to be eliminated from the evaluation. The short profiles that were retained are shown in fig. 3.

The seismic measurements on these slopes reveal a pronounced subsurface discontinuity at which the wave velocity increases from less than 1,000 ft./sec. (300 m/sec.) to more than 1,600 ft./sec. (480 m/sec.), and then remains remarkably constant at this higher value to the depth limit of sounding (13 ft. or 4 m). The highest velocity that was found below the discontinuity amounted to about 2,400 ft./sec. (720 m/sec.). These values indicate that the bedrock underneath the waste cover is weathered, since fresh bedrock transmits shock waves at considerably higher velocities. Road cuts in the area indeed show deep *in situ* weathering of the bedrock.

A preliminary inspection of the thickness data suggested two tendencies of thickness variation along the profile:

(a) a decrease of waste cover thickness with increasing slope angle, and
(b) an increase of waste cover thickness with increasing distance from the top of the slope.

For quantitative evaluation of these two tendencies, multiple regression analysis offers the most suitable method. The multiple regression equation has the general from

$$C = a + b \cdot d + c \cdot \sin \alpha \quad (3)$$

(C = thickness, d = distance from the top, α = local slope angle)

7*

The combined sets of data for the three profiles of fig. 3 yield the estimating equation

$$C\ (\text{ft.}) = 1.9\ (\text{ft.}) + .00733\ d\ (\text{ft.}) - 1.81\ (\text{ft.}) \cdot \sin\alpha \quad (4)$$

or in metric terms

$$C\ (\text{m}) = .58\ (\text{m}) + .00733\ d\ (\text{m}) - .55\ (\text{m}) \sin\alpha\ ,$$

with a multiple correlation coefficient $R = .860$ (fig. 4). According to equation (4), the waste cover thickness increases with increasing distance d from the top of the slope, presumably because weathering adds material to the waste as it moves downslope; and the thickness decreases with increasing sine of the local slope angle, presumably because the waste moves faster on the steeper parts of the slope.

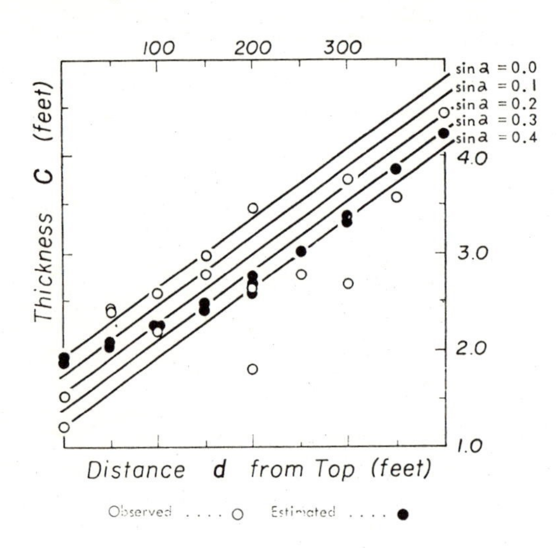

Fig. 4. Relationship between waste cover thickness C, distance from the top d, and sin α, for the field slopes:
$C\ (\text{ft.}) = 1.9\ (\text{ft.}) + .00733\ d\ (\text{ft.}) - 1.81\ (\text{ft.}) \sin\alpha$
$R = .860$

Comparison with the Model

For comparison of field data with the model, a conversion of the length units becomes necessary, as the model has arbitrary length units instead of feet or meters. This is accomplished by converting the values of C from feet into percentages of the mean thickness for each slope, and by converting the values of d into percentages of the total length of each slope. The estimating equation for the field data then becomes

$$C\ (\%) = 67.55\ (\%) + 1.51\ d\ (\%) - 156.04\ (\%) \sin\alpha \quad (5)$$

with $R = .873$. Figure 5 isolates the relationship between observed and estimated thicknesses.

From the series of model profiles, a profile similar to the field slopes was singled out for comparison. After the model data have been converted into per-

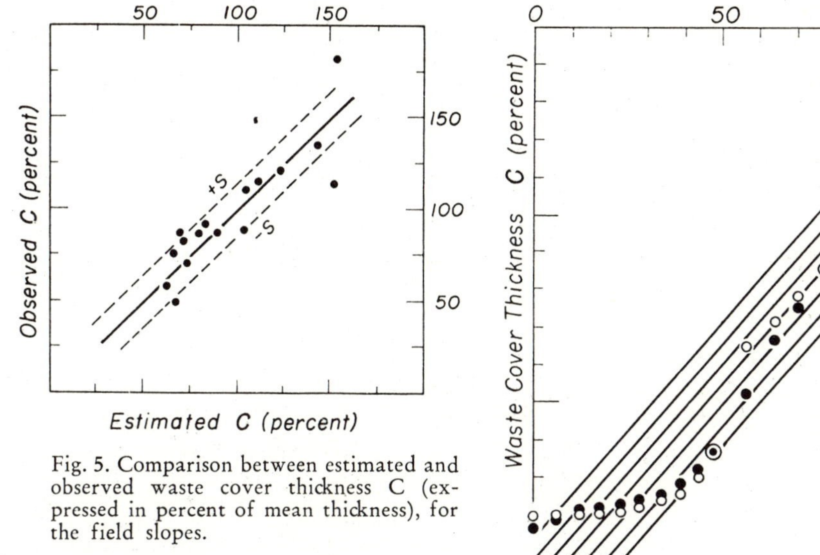

Fig. 5. Comparison between estimated and observed waste cover thickness C (expressed in percent of mean thickness), for the field slopes.

Fig. 6. Relationship between waste cover thickness C, distance from the top d, and sin α, for the model slope:
$C\ (\%) = 68.77\ (\%) + 1.16\ d\ (\%) - 110.07\ (\%) \sin \alpha$
$R = .993$

centages in the same manner as the field data, multiple regression produces the estimating equation

$$C\ (\%) = 68.77\ (\%) + 1.16\ d\ (\%) - 110.07\ (\%) \sin \alpha \qquad (6)$$

for the model, with a near-perfect multiple correlation coefficient $R = .993$. The correspondence between observed and estimated values is evident in fig. 6. The model equation (6) bears a very close resemblance to the equation (5) that was derived from field data. In order to obtain a more precise quantitative assessment of the similarity between these two equations, one can predict the waste cover thicknesses on the field slopes by means of the model equation (6), simply by entering the values of d and sin α from the field data into the model equation (fig. 7). Such use of one set of data in the estimating equation derived from another set of data is, of course, statistically unorthodox. The mean of observed values is not identical with the mean of estimated values, and the sum of residuals differs therefore from zero. Nevertheless, by using standard correlation methods, one can compute a multiple correlation coefficient that describes the degree of correspondence between model predictions and field observations (fig. 8). This coefficient is $R = .861$, very similar to that derived solely from the field data (equations (4) and (5)).

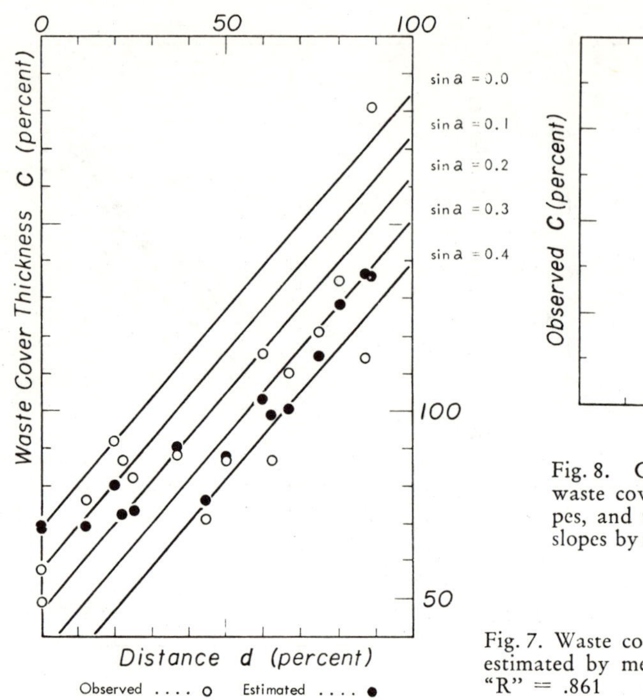

Fig. 8. Comparison between observed waste cover thickness C on the field slopes, and thickness estimated for the field slopes by means of the model equation (6).

Fig. 7. Waste cover thickness of the field slopes estimated by means of the model equation (6). "R" = .861

Thus the similarity between the model slope and the field slopes becomes fully apparent. The model had been designed on the basis of assumptions regarding processes on slopes–not as a statistical scheme to match waste cover thicknesses observed on slopes in the field. The close agreement between the properties of the model slope and those of the field slopes indicates that the theoretical model is very probably a valid representation of conditions and processes on real slopes.

Other Observations

Besides measurements of waste cover thickness, the field work included collection of waste samples at regular intervals along the profiles. Mechanical analysis of these samples has shown no clearly discernible trends, except that the proportion of fines (particles smaller than .062 mm) has a vague tendency to increase slightly downslope, and another tendency, equally vague, to decrease with increasing slope angle in the upper part of the slope. Multiple correlation analysis of these tendencies yielded a coefficient R = .255, too low to be of any great significance. These findings are in agreement with the observations of FURLEY (1968) in England, who found no clear trend in the proportion of fines on slopes covered by non-calcareous waste.

A somewhat better correlation with slope parameters was obtained for the quartz content of the coarse fraction (particles larger than .062 mm) of the waste samples. Determination of quartz grains in percent of the total number of grains was made for each of 52 waste samples from seven slope profiles. On six of these profiles, the quartz content is relatively high at the top of the slope, then drops markedly in the upper slope zone where the greatest increase of slope angle occurs. Farther down along the profile the quartz content tends to rise again, but beyond 300 ft. from the top of the slope it varies without much noticeable relationship to either the distance along the profile or the slope angle. Multiple correlation of the data for the upper 300 ft. of the profiles, with quartz content as the dependent variable, and distance from the top and sin α as the determining variables, produced R = .505 for six slopes, and R = .760 for five slopes, after elimination of the profile that contained the largest residuals (fig. 9). Of the two determining variables, the sine of the slope is far more

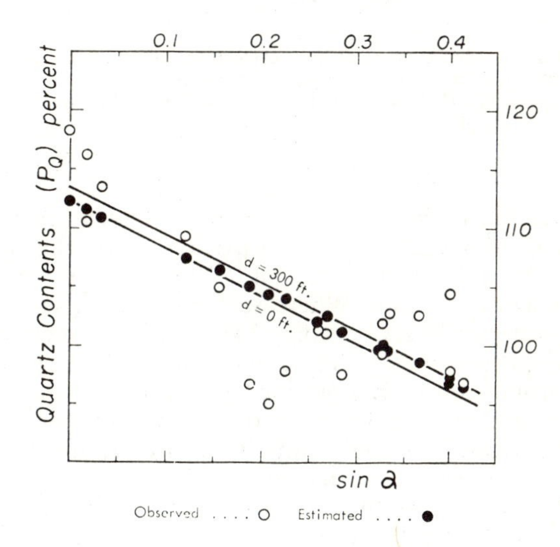

Fig. 9. Relationship between quartz contents P_Q in the coarse fraction of waste samples, distance d from the top, and sin α, for five slopes:
P_Q (%) = 112.27 (%) + .0032 d (ft.) — 40.3 sin α
R = .760

determinative than the distance from the top. This becomes also apparent when separate Spearman rank correlations are carried out for these two variables. On the upper 300 ft. of six profiles, the Spearman rank correlation coefficient between quartz content and sin α is R = –.554; it rises to R = –.825 when the rank correlation is limited to the upper 100 ft. of these six profiles (fig. 10). On the other hand, the rank correlation between quartz content and distance from the top computed for samples between 100 ft. and 300 ft. from the top of these same six profiles, has the coefficient R = .364, and that only after two samples (out of 17) with the largest residuals have been eliminated from the computation (fig. 11).

There is an apparent contradiction between the results of the analysis of waste cover thickness distribution and those of the waste sample analysis. The

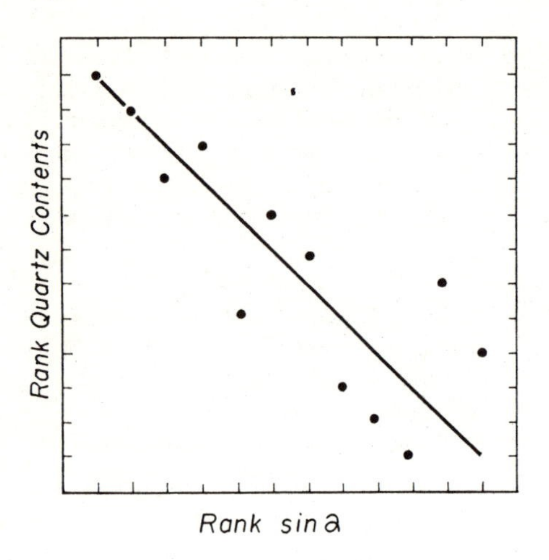

Fig. 10. Rank correlation between quartz contents of coarse fraction and sin α, on upper 100 ft. of six profiles. R = —.825

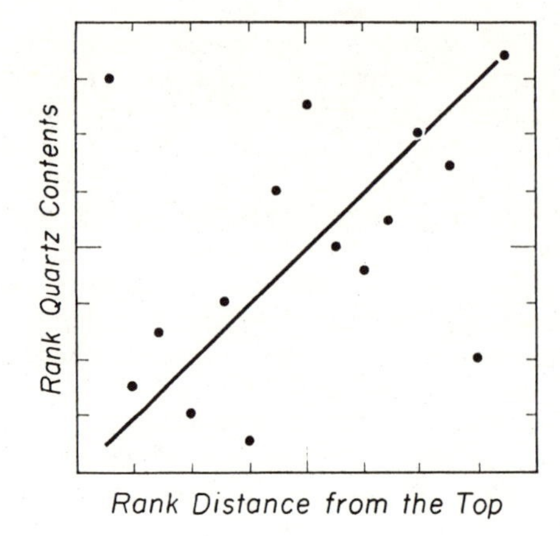

Fig. 11. Rank correlation between quartz contents of coarse fraction and distance from the top, for samples between 100 ft. and 300 ft. from the top of six profiles. R = .364

general increase of waste cover thickness from the top to the foot of the slopes indicates that the waste distribution on the slopes is caused by slow downslope mass movement, presumably creep, in conjunction with rock weathering underneath the waste. Where such movement is present, one would expect that the waste itself also becomes progressively more weathered as it travels along the slope profile and that, consequently, there would be an increase in both the proportion of fine-grained material and the proportion of quartz grains in the coarse fraction, from the top to the foot of the slopes. Instead, there is very little variation in the silt and clay contents, and the quartz contents vary much more pronouncedly in dependence of the slope angle than in dependence of the distance from the top. In other words, the composition of the waste samples points more towards *in situ* weathering of the waste than to downslope mass movement. A possible explanation might be that the mass movement was active in the past, but is either very weak or entirely absent today and that the waste now weathers in place except for minor surface wash and illuviation. Since there are no signs of cryoturbation, that past mass movement probably was creep, and may have occurred at a time when freeze-thaw cycles were more numerous, and frost was more severe, than at present. Thus the climate of the area may have been less stable than was originally expected.

It is possible to make a very crude estimate of the mean velocity of downslope waste movement that would be required to account for the downslope increase of waste cover thickness as observed on the field slopes. According to equation (4), the waste cover thickness increases, on a constant-angle slope, by $\triangle C = .00733d$ over any downslope distance d. Hence, during the time in which the waste moves along a profile segment of the length d, the total amount

Fig. 12. Addition of waste by weathering during the time in which the waste moves over the distance d (see equations [7] and [8]).

$$\frac{d\Delta C}{2} = \frac{.00733\ d^2}{2} = .00367\ d^2 \quad (7)$$

is added to the waste cover cross-section. This addition is shown by the shaded triangle in figure 12. The mean amount of waste addition, or mean depth of waste added by weathering, during this time is then

$$\frac{\Delta C}{2} = \frac{.00367\ d^2}{d} = .00367\ d \quad (8)$$

In order to obtain the time needed for the waste to move over the distance d–and thus, to obtain the velocity of waste movement–information about the rate of weathering is required. To measure this rate directly is impossible. However, a rough approximation may be attempted by equating the rate of weathering with the rates of denudation derived from large river basins. Figure 13 demonstrates the quantitative relationship between mean denudation rates and sine of mean slope for 19 major mid-latitude drainage basins (the denudation rates are taken from CORBEL, 1959, LEOPOLD, WOLMAN & MILLER, 1964, and others).

If it be assumed that the relationship between denudation rate and sine of mean slope also holds true for the relationship between the rate of weathering

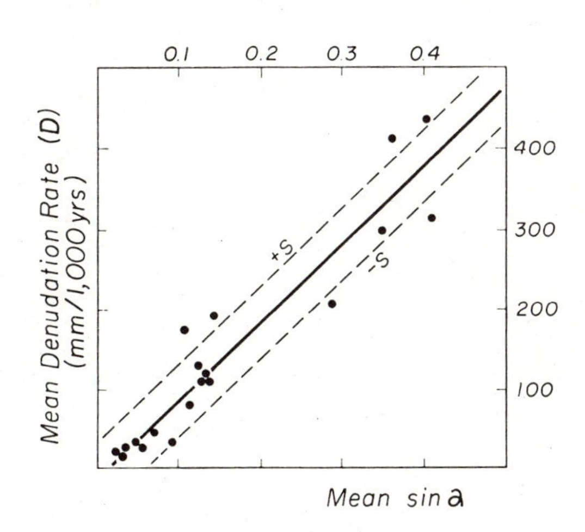

Fig. 13. Relationship between mean denudation rate D and sine of mean slope for 19 mid-latitude river basins:
D (mm/1,000 years) = 959.6 sin α — 4.4
R = .951

and the sine of local slope, one can combine the regression equation of figure 13 with the equation for weathering (equation 8) so that the velocity V is

(9) $$V = .00367\ d\ (\text{mm}/1{,}000\text{yrs}) = 959.6 \sin \alpha - 4.4\ (\text{mm}/1{,}000\ \text{yrs})$$

from which follows

(10) $$V = 261 \sin \alpha - 1.2\ (\text{mm/year})$$

or

(11) $$V = 10.3 \sin \alpha - .05\ (\text{inches/year})$$

On a six-degree slope ($\sin \alpha = .1$), the velocity would be approximately one inch (25 mm) per year, on a twelve-degree slope ($\sin \alpha = .2$), two inches per year, and on an eighteen-degree slope, three inches per year (fig. 14). However, not all of the weathered bedrock is incorporated in the waste cover. In humid warm-temperate climates such as the one prevailing here, dissolved load often makes up more than half of the total stream load. If one assumes that fifty percent of the weathered material is removed from the slope in solution, the velocity figures drop to half their previous values, for example, on an eighteen-degree slope, to approximately 1.5 inches (38 mm) per year. If waste movements at these velocities were taking place at present, deformations of the surface should be common, but such deformations were not observed. It appears safe to conclude that the present distribution of waste cover thickness on these slope profiles is the result of mass movements that do not occur today.

Past attempts by YOUNG (1960) and by LEOPOLD, WOLMAN & MILLER (1964, p. 351–352) to measure the rates of creep in humid warm-temperate climates have had generally negative results. GERLACH (1967) also found no significant creep on vegetation-covered slopes in the Carpathians. The values they obtained lie within the error margins of the instrumentation; furthermore, these values are too small to account for the development of valley side slopes during the

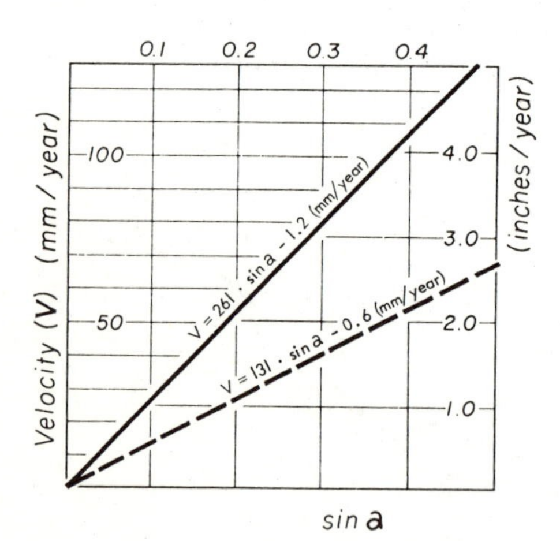

Fig. 14: Estimated mean velocity of mass-wasting required to explain the observed distribution of waste cover thickness.

Pleistocene in many areas where such slope development is known to have occurred. These measurements, as well as the composition of waste samples described here, suggest that creep as a quasi-continuous movement is virtually absent in humid warm-temperate climates, and that the slopes in such areas owe their shape and their distribution of waste cover thickness to more intensive denudational processes of the past.

Acknowledgements

The research presented in this paper was financially supported by the U. S. Army Research Office, Durham, North Carolina. Mr. LELAND ADAMS assisted in the field work, and Mr. RICHARD KESEL carried out the mechanical analysis of waste samples; both are graduate students of geography at the University of Maryland.

References

AHNERT, F. (1954): Zur Frage der rückschreitenden Denudation und des dynamischen Gleichgewichts bei morphologischen Vorgängen. – Erdkunde, **8**: 61–64.
– (1964): Quantitative Models of Slope Development as a Function of Waste Cover Thickness. – Abstracts of Papers, 20th Internat. Geogr. Congr., London, p. 118.
– (1966): Zur Rolle der elektronischen Rechenmaschine und des mathematischen Modells in der Geomorphologie. – Geogr. Z., **54**: 118–133.
– (1967): The Role of the Equilibrium Concept in the Interpretation of Landforms of Fluvial Erosion and Deposition. – L'Evolution des Versants, Université de Liège, pp. 23–41.
BAKKER, P., & J. W. N. LE HEUX (1947, 1953): Theory of Central Rectilinear Recession of Slopes. – Kon. Ned. Akad. v. Wetensch., Proceedings, **50**, 8 and 9; **54**, 7 and 8.
CORBEL, J. (1959): Vitesse de l'Erosion. – Z. f. Geomorph., **3**: 1–28.
EMMETT, W. W. (1965): The Vigil Network: Methods of Measurement and a Sampling of Data Collected. – Publ. 66, Int. Ass. of Sc. Hydrol.
FURLEY, P. A. (1968): Soil Formation and Slope Development. 2. The Relationship between Soil Formation and Gradient Angle in the Oxford Area. – Z. f. Geomorph., **12**: 25–42.
GERLACH, T. (1967): Evolution actuelle des versants dans les Carpathes. – L'Evol. des Versants, Université de Liège, pp. 129–138.
GILBERT, G. K. (1877, 1880): Report on the Geology of the Henry Mountains, U. S. G. S., 2nd ed. 1880.
JAHN, A. (1954): Denudacyjny bilans stoku. – Csas. Geogr., **25**: 38–64.
LEHMANN, O. (1933): Morphologische Theorie der Verwitterung von Steinschlagwänden. – Vierteljschr. d. Naturf. Ges. Zürich, **78**: 83–126.
LEOPOLD, L. B., M. G. WOLMAN & J. P. MILLER (1964): Fluvial Processes in Geomorphology.
PENCK, W. (1924): Die morphologische Analyse.
RAPP, A. (1961): Recent Development of Mountain Slopes in Kärkevagge and Surroundings, Northern Scandinavia. – Geog. Ann., **42**: 67–200.
RUDBERG, S. (1964): Slow Mass Movement Processes and Slope Development in the Norra Storfjäll Area, Southern Swedish Lappland. – Z. f. Geomorph., Suppl. Bd. **5**: 192–203.
SCHEIDEGGER, A. (1961): Theoretical Geomorphology.
SCHUMM, S. A. (1964): Seasonal Variations of Erosion Rates and Processes on Hillslopes in Western Colorado. – Z. f. Geomorph., Suppl. Bd. **5**: 215–238.
YOUNG, A. (1960): Soil Movement by Denudational Processes on Slopes. – Nature, **188**: 120–122.
– (1963): Deductive Models of Slope Evolution. – Nachr. d. Akad. d. Wiss. Göttingen, II. Math.-Naturw. Kl., **5**: 45–66.

Address of the author: Dr. F. AHNERT
Department of Geography, University of Maryland, 20742 – College Park, Md./U.S.A.

Form

III

Editors' Comments on Papers 11 Through 15

Description of hillslope form is in itself a complicated task, but description, whether qualitative or quantitative, is just the beginning, and the final objective is to understand cause and effect relationships between process and form.

In this section a number of papers are reproduced which deal primarily with hillslope form, and more specifically, with inclination and shape. It is inevitable that a competent researcher will discuss both form and process, and hence some difficulty in categorizing the sections was experienced.

The first contribution is an excerpt from Kirk Bryan's well-known U.S. Geological Survey report on the geology of southern Arizona. Bryan's work is very similar to that of Gilbert; he reports his observations, makes measurements, and attempts to develop an explanation for the phenomena he sees. While he was at work in the desert, others were becoming bogged down with academic discussions centering on Davis's ideas of landscape evolution and peneplanation.

The two major topics discussed by Bryan subsequently received a great deal of attention from other geologists. For example, Roth (1956) and others have discussed the roles of temperature change and water on desert weathering of rocks, perhaps laying more stress on the presence of water than did Bryan. His classification of slopes into cliffy, boulder-controlled, and rainwashed is a precursor of the recognition of limiting, repose, and threshold slopes (Young and Carson, Papers 13 and 14). Melton (1965) has discussed and criticized some aspects of Bryan's report, using data gathered in and near Bryan's field area.

Although there had been one or two instances in which actual measurement of slopes had been made (for example, Tyler, 1875), studies of hillslopes up to the 1940s depended largely on qualitative observation in the field and deductive reasoning. Even the classic "Geology of the Henry Mountains" and "Les Formes du Terrain" were of this nature. However, in the late 1940s and early 1950s a number of researchers, apparently independently, realized that the use of data gathered by actual measurement in the field would allow long-held ideas to be decisively tested.

Three men deserve special mention: Fair, working in South Africa; Savigear, working in Britain; and Strahler, working in America. The contributions by Fair and Savigear are included in a later section of this volume; that of Strahler follows the excerpt from Bryan's report.

Strahler's long paper is worthy of careful consideration. The use of statistics and carefully prepared experimental design make it a noteworthy methodological contribution, and moreover, the close relationship between slopes and the drainage system is pointed out. Hillslopes are thus shown to be but one part of a geomorphic unity—the watershed. Strahler applies the concept of dynamic equilibrium, introduced by Gilbert and discussed by Hack (1960), to his study area. He employs accurate measurement of slope morphology to attack problems which are basically insoluble when one's only tool is deductive reasoning. Thus, the Davis–Penck controversy is discussed, and valley asymmetry examined. The use of the techniques of measurement and statistical analysis was a great advance which has revitalized geomorphology.

Young's paper expands on the idea of characteristic and limiting slope angles introduced by Bryan, and discusses the occurrence of multimodal distributions of slope angles in a given area. He suggests that only angles of between 25° and 29° are intrinsic features of slope evolution; other peaks are explicable in terms of local erosional history. Carson, on the other hand, feels that there are several frequently found slope angles, which are functions of the weathering characteristics of debris. His conclusions tend to support a model of hillslope evolution which includes an element of slope replacement, although both parallel retreat, associated with replacement from below, and decline are thought to occur. Carson's work is especially noteworthy for its extensive application of techniques used in soil mechanics and engineering (Carson and Kirkby, 1972).

Melton's paper is another fine example of careful experimental design and hypothesis testing. The many discussions of the effect of aspect on hillslope form and inclination are reviewed, and the effect of microclimate demonstrated. Melton feels that erosional environment is the final link between microclimatic factors and landforms; clearly this must be taken into account in any discussion of hillslope development and evolution.

DEPARTMENT OF THE INTERIOR
ALBERT B. FALL, Secretary

UNITED STATES GEOLOGICAL SURVEY
GEORGE OTIS SMITH, Director

Bulletin 730—B

11

EROSION AND SEDIMENTATION IN THE PAPAGO COUNTRY, ARIZONA

WITH A SKETCH OF THE GEOLOGY

BY

KIRK BRYAN

Contributions to the geography of the United States, 1922
(Pages 19-90)
Published June 8, 1922

WASHINGTON
GOVERNMENT PRINTING OFFICE
1922

MOUNTAIN SCULPTURE.

MOUNTAIN SLOPES.

The mountains of the Papago country rise from the surrounding plain with startling abruptness. To the traveler approaching them the mountain slopes stand like a wall, without transition from the plain on which he travels.[38] Only in a distant view is it possible to realize that the surrounding plain rises gradually on all sides toward the mountains, which stand up like jagged ornaments on the ridgepole of a low-gabled roof.

This appearance is due to contrast in angle of slope between the mountains and the plain. The angles of the mountain slopes range from 15° to almost 90° with the horizontal; those of the plain from 1° to 6°. Between these slopes there is usually no region of transition, either of intermediate slopes or of low foothills. In many ranges slopes that average 25° to 30° rise directly from the plain to the crest of the mountains. The factors which produced the mountain slopes must then differ radically from those which produced the plain.

CONDITIONS DETERMINING MOUNTAIN SLOPES.

Mountains composed of rock of any one type have a relatively constant angle of slope, either at the mountain border or in the side walls of canyons. Even small isolated hills have similar slopes. These conditions have been stated by Lawson[39] as follows: "(1) That the hard-rock slopes of desert ranges which shed large spalls are steep, while those which shed small fragments have a low angle;

[38] Hornaday, W. T., op. cit., pp. 38–39 (a humorous statement of these facts).

[39] Lawson, A. C., The epigene profiles of the desert: California Univ. Dept. Geology Bull., vol. 9, p. 29, 1915.

(2) that ranges composed of hard rock, which are thus naturally steep, maintain their steepness as long as the rock slopes endure."

The resistance of rock to weathering is the dominating condition which determines the angle of slope, but the presence of large spalls is only one of the results of resistance to weathering, though in many types of rocks there is a significant relation between the size of spalls and the angle of slope, as noted hereafter. Resistance of rock to weathering seems to be divided into two phases—resistance to the detachment of large blocks and resistance to direct weathering into small particles. Both kinds of resistance involve mechanical and chemical processes of weathering, but in the Papago country, because of the arid climate, mechanical processes are dominant.

MECHANICAL PROCESSES OF EROSION

The mountain slopes are ordinarily barren of soil, the bedrock presenting an uneven surface, the lowest parts of which are covered by a layer of broken rock fragments but recently parted from the parent rock. The projections of the bedrock are large or small, according to the spacing of joints. Weathering proceeds along the joint planes, and the projections tend to be detached and then lie on the surface as blocks or boulders or, as Lawson speaks of them, "rock spalls." In many places the surface is completely mantled by blocks and boulders. The production of blocks of this character is accomplished in part by the weakening of cement and widening of joints by solution but largely by expansion and contraction of the rock under the influence of changes in temperature.

Daily as well as seasonal changes are very large, the mean daily air temperature in January, the coldest month, being 54.7° and that in August, the warmest month, being 90.1°. At mid-day the temperature of the soil and rock is frequently higher than that of the air, whereas at night the rock cools nearly to air temperature.

The depth to which changes in temperature are effective in disrupting rock can only be estimated. The water of wells dug wholly in rock has the temperature of the surrounding rock, from whose crevices and seams it seeps. Water standing only 59.5 feet from the surface in Al John's well at Gibson, on the east flank of the Little Ajo Mountains, had a temperature of 67.5° on October 10, 1917, and doubtless represents the mean annual air temperature for that locality, as it is only 1.6° below the mean annual air temperature at Ajo for the years 1915 to 1918. In the vicinity of this well seasonal changes of temperature do not penetrate perceptibly below 60 feet. However, the well is in an interstream area of gently rolling country, and it is not unlikely that on the crests of the ridges insolation is effective to greater depths. In the valleys, where the water table is nearer the surface, the reverse is probably true and insolation is effective to depths of about 25 feet, the depth at which water ordinarily stands in wells.

The daily changes in temperature produce effects much more striking and probably more effective in the breaking up of rock than seasonal changes. The effects may be summed up under three heads—rupture and spalling, exfoliation, and granular disintegration.

Rupture appears to be due to the expansion of rock under the influence of a rise in temperature beyond the elastic limit of the material, doubtless followed by sudden cooling as suggested by Walther.[40] It is most often observed in fine-grained brittle rock, such as siliceous lava. The action affects blocks of rock lying loose on the surface which are from 1 to 2 feet through in the direction of minimum thickness. The rupturing takes place on sharply defined but irregular planes and apparently in a single action, for the broken surfaces, though all manifestly younger than the exterior of the block, have a uniform color, corresponding to the time they have been exposed to the weather, but show no marks of weathering along initial cracks. Similar rocks suffer also from the spalling off of slivers, which in many places are scattered about like the quarry refuse left by Indians in making stone arrowheads. Though this process is most easily observed in loose blocks of rock, it undoubtedly affects the bedrock also, and thus in its ultimate effect it grades into exfoliation and serves to break up the plates formed by that process.

Exfoliation is a process by which successive surface sheets of rock are parted from the underlying mass. The surface of a rock mass, being in contact with the air; changes in temperature from hot to cold or from cold to hot more rapidly than the interior. As a consequence of the resultant changes in volume, the surface shell is either too large or too small for the interior. Shearing strains are set up between the shell and the interior and a crack forms. The thickness of the shell varies according to the area and form of the surface exposed to the air and the texture of the rock. When first exposed to this process rock masses are commonly angular and bounded by joint planes. The first places attacked are the corners and edges, where heat is most readily absorbed and radiated. The first parting plane is commonly 3 to 4 inches below a corner and thence approaches the surface in all directions, so that the first fragment to break off is a four-sided piece whose lower surface is concave. With the splitting off of successive sheets, each of which is slightly thicker at the protuberances of the rock, the remaining mass approaches a spherical shape. In the Papago country exfoliation sheets from one-fourth of an inch to 2 inches thick have

[40] Walther, Johannes, Das Gesetz der Wüstenbildung, p. 29, Berlin, 1900.

been observed on granite, from an inch to 2 feet thick on arkosic conglomerate, and of almost paper thinness on certain felsitic lavas. To exfoliation may be attributed the rounded bosslike forms that occur on most outcrops of the Tertiary arkosic conglomerates.

* * * * * * *

Exfoliation sheets from 6 inches to 2 feet or more in thickness are recognized only on rocks that have few joints and very massive structure. It seems likely that these sheets are produced by seasonal changes in temperature. In rocks with more closely spaced joints yielding takes place along existing fractures or by the reopening of older lines of weakness. Thus in such rocks the curved forms of exfoliation are not produced.

Granular disintegration is due to the unequal rates of expansion of different minerals whereby a rock composed of more than one mineral is subject to internal strains with every change in temperature. These strains tend to separate the different mineral grains one from another and thus to reduce the rock to a loose rubble. The process is best observed in coarse-grained granite and is probably the most effective type of weathering in this rock.

CHEMICAL PROCESSES OF EROSION.

The freshness of rock débris, the lack of soil, and the absence of chemical deposits all testify to the preponderance of mechanical over chemical action in the Papago country. Though chemical action is not large in amount, however, evidence of its presence is widespread, and its total effect is likely to be underestimated.

Chemical action is dependent on the presence of water and is most active in regions of circulating water, abundant organic matter, and high temperature. Few of the rocks of this region are ever saturated with water or are even wet frequently, yet water circulates in the joints and adheres to the walls of all cavities and pore spaces for a considerable time after each wetting. It is also possible that moist air in the cracks of the rocks may be effective in chemical work on the side walls. Prospect holes and other excavations show few joints which are not widened by solution or along which rock decay is not evident. Even loose boulders are marked by concentric bands of color showing solution and redeposition, particularly of the iron minerals. In the mines at Ajo copper sulphides are oxidized to depths of 20 to 150 feet.[41]

[41] Joralemon, I. B., The Ajo copper-mining district, Ariz.: Am. Inst. Min. Eng. Bull. 92, pp. 2011–2028, 1914.

Rock débris seldom shows grains of hornblende or biotite mica, for these minerals are largely destroyed by chemical action before the rock breaks up. In certain localities in the Papago country granite surfaces are pitted by the removal of these minerals and subsequent mechanical disintegration is facilitated. The feldspars, on the other hand, are very fresh and clear and rarely have the dull-reddish opaque appearance which they assume in the residual soils of the Sierra Nevada and central New Mexico. Solution of the glassy matrix of lavas is also a fairly rapid process and probably plays a large part in the removal of lava blocks on talus slopes.

GRADES OF MOUNTAIN SLOPES.

Mountain slopes range in steepness from vertical cliffs to grades up which it is easy to ride a horse. Slopes can be divided into three groups—cliffy slopes, boulder-controlled slopes, and rain-washed slopes. The limiting angles and characteristic rocks of these groups are shown in figure 7. Each group has certain common characteristics, although the dividing lines between them are not very sharp.

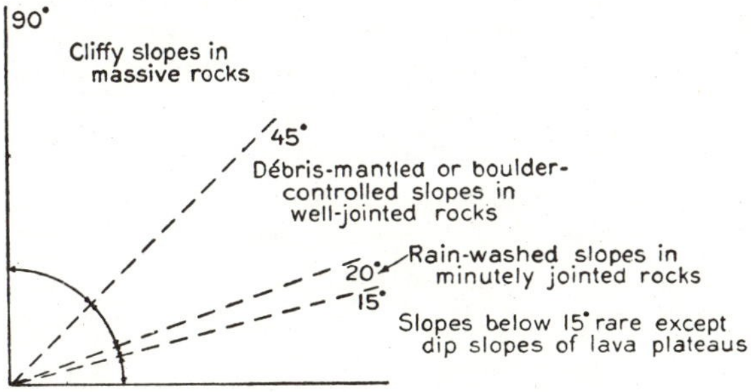

FIGURE 7.—Diagram to show the range of mountain slopes.

CLIFFY SLOPES.

Mountain slopes inclined between 45° and 90° from the horizontal are developed on granite, granite gneiss, massive lava beds, intrusive fine-grained porphyritic rocks, and arkosic conglomerate. These cliffy slopes are stable under the existing conditions of erosion, and the mountains composed of suitable rocks grow smaller in size but maintain the same angle of slope until they are totally reduced. The rocks are massive, with widely spaced joints, and the steepness of slopes seems to be related to the sparse jointing, for near by may be found the same rocks with more closely spaced joints undergoing erosion on gentler slopes. In sparsely jointed rocks joint blocks are not easily dislodged, and the mountain slope recedes either by undermining at the base or simply by surface disintegration and weathering of the rock wall. When at long intervals blocks are dislodged, they have been so weakened by weathering that in falling to the base they break into fine rubble. Thus no talus forms at the foot of these slopes. Slopes of this type on arkose are shown in Plate X, *A*, on massive lava flows in Plate X, *B*, and on granite in Plate XII, *A*.

BOULDER-CONTROLLED SLOPES.

Mountain slopes inclined between 20° and 45° from the horizontal are characteristic of most granites, granite gneisses, and horizontally bedded lava flows. The angle of the common mountain slope is between 30° and 35°. The steeper slopes are usually interrupted by cliffs, and the gentler ones are interrupted by slopes of the next class.

Mountain slopes of this type composed of bedded lava flows consist of a cliff with talus below, successive cliffs with intervening talus, successive cliffs with smooth intervening slopes developed on tuff, continuous talus, or talus gullied and dissected. Very thick and massive lava flows resist the dislodgement of joint blocks and form cliffs. Whether these cliffs lie at the top or midway of the slope, they simply retard the recession of the mountain slope and increase its average steepness. The processes on the intervening talus slopes are similar to those described below.

Smooth slopes on tuff recede more rapidly than the cliffs above them, and the undermined blocks roll down over them. Only where such slopes are short stretches intervening between two steeper slopes are they free of rock waste. Where they are bare, the processes of erosion on the tuff are those described for rain-washed slopes (p. 46).

Bedded lavas under the attack of weather usually break into joint fragments from 2 to 6 feet in diameter, which, although easily dislodged, are comparatively resistant to disintegration and hence form a talus of rock waste that gradually mantles the whole mountain slope. The grade of the slope then becomes the angle of repose of the average-sized joint fragment (Pl. XI, *A*). Further erosion takes place by removal of the rock waste and formation of a new talus.

On certain mountain slopes part of the talus is removed by the formation of gullies, as shown in Plate XI, *B*. The mountain here illustrated consists of a cap of lava on a base of granite, and because of the difference in color the gullied talus is easy to photograph. Other mountains similar in size and composed wholly of lava have like gullies with like triangular areas of unremoved rock waste. Figure 8 shows the distribution of gullied and ungullied mountains of flat-lying lavas of approximately the same size and height observed in the Papago country. Many others of both kinds exist, and only those observed are included.

No local lowering of base-level is evident at the foot of these mountains, and the formation of the gullies can be ascribed only to the work of unusually heavy local rains. Such rains are well known under the term "cloud-bursts," and if disintegration of the boulders that form the talus and the bedrock on which they lie had proceeded until an unstable condition had been reached a single cloud-burst would be competent to produce the gullies. Weathering and the normal rains combine to dislodge new fragments, which will in time fill the gullies and restore the original unbroken mantle of rock waste.

Confirmation of this explanation is obtained from the talus slopes of the ungullied mountains. The greater part of the rock waste has been in position for a long time. The boulders are cracked, exfoliated, and pitted by the solution and removal of the ferromagnesian minerals. The under surfaces of the boulders are coated with calcium carbonate derived from solution of the minerals. Bushes grow

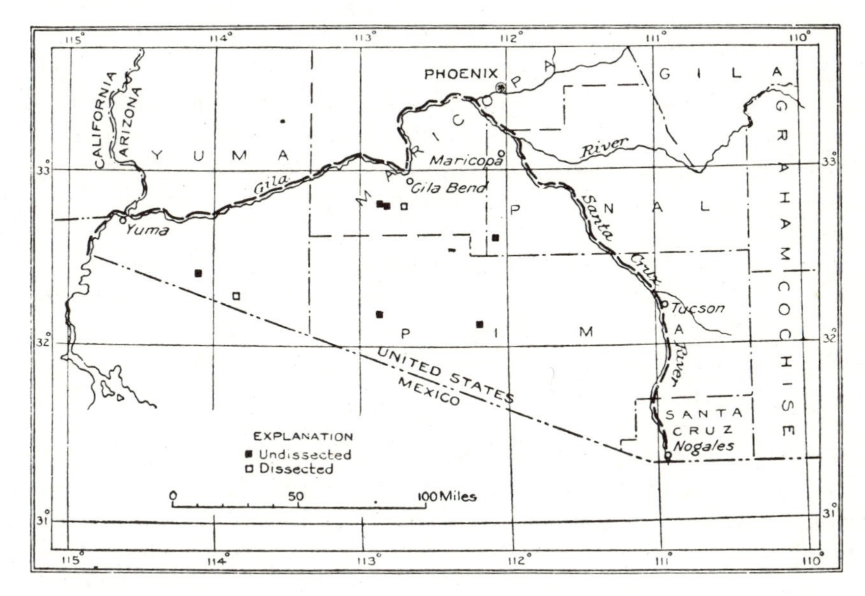

FIGURE 8.—Sketch map of the Papago country, Ariz., showing distribution, so far as observed, of gullied and ungullied mountains of approximately the same size.

between the boulders. Of these the most significant in indicating the stability of the slopes is the palo verde, *Parkinsonia microphylla*, whose age on the slopes of Tumamoc Hill was found to range from 10 to 400 years.[42] In contrast to this ancient rock waste strings and lenses of fresh rock waste occur on all slopes. The blocks are sharp-angled and more closely packed than those of the older rock waste and occupy positions corresponding to the gullies previously described.

The retreat of a talus slope on horizontal lavas takes place by the slow production of the fragments that compose the talus blocks, by their decay, and by the creep of the rock waste from top to bottom of the slope. This slow process is occasionally interrupted by the

[42] Shreve, F., Establishment and behavior of the palo verde: Plant World, vol. 14, p. 293, 1911.

catastrophic removal of large amounts of rock waste by great storms, exposing a fresh rock surface from which rock waste is again gradually produced and the talus is replaced.

Granite slopes at angles between 20° and 45° also have a mantle of rock waste, although it is by no means so nearly complete as the talus on lava slopes, particularly on the steeper slopes. The talus on granite slopes commonly consists of a layer of boulders only, and in many places scattered boulders and patches of boulders between protuberant knobs of the bedrock seem to determine the angle of the slope. The granite boulders range from 10 feet down to 1 foot in diameter, and all sizes may be found on a single slope, but in a general way the size of the boulders is proportional to the grade of the slope. Large boulders mantle steep slopes, and small boulders gentle slopes. As the size of the boulder is determined primarily by the spacing of joints, fine-grained granite and most granite gneisses, which usually have closely spaced joints, yield smaller boulders and consequently produce gentler mountain slopes than coarse-grained granite and gneiss.

Many of the granite boulders have a brown or blackish color from the so-called "desert varnish," which is associated. in granite at least, with the deposition of limonite in the outer 1 or 2 inches of the rock. This outer crust is in many boulders a shell which covers a completely disintegrated interior so soft and crumbling that the minerals may be picked apart with the fingers. Many boulders also are cracked, and some are completely split, many of them in two or three directions. Still others show the work of exfoliation. On many mountain slopes the boulders are so weather beaten and ancient in appearance that it seems that they could not possibly have been produced by any process now in action, but that, laid on the slope in some ancient time, they have ever since been slowly rotting and disintegrating.

The bedrock on which the boulders lie is covered with a loose film of small fragments made up of the mineral grains that once formed the granite. The fragments slip under the foot, and new pieces crumble from the bedrock continually. Disintegration of the bedrock proceeds most rapidly along joints but is effective everywhere. Every rainstorm sets trains of this fine débris moving down the slope. As the fine débris is removed the boulders roll down to find a new lodgment either lower on the slope or at its base. In this movement many of the boulders, already disintegrated within the outside crust, are shattered into fragments. Under normal conditions few of them reach the bottom and no accumulation of boulders takes place there.

As the bedrock disintegrates and rain washes the débris away more rapidly along joint planes protuberances of the bedrock are formed,

80477°—22——3

which consist of the most compact rock between the most widely spaced joints. The protuberances are cut loose by the same process that formed them, and a new crop of boulders comes into existence.

By these slow but continuous processes the mountain front recedes, but the same great storms that affect the slopes of lava mountains fall on granite slopes also. The process of recession is doubtless hastened by these storms, but no direct evidence of their effect has been obtained.

RAIN-WASHED SLOPES.

Mountain slopes at angles less than 20° from the horizontal are rare in the Papago country. They are developed on the least-resistant rocks, which, probably occurring in relatively small amount, have been largely removed by erosion. The mountains are composed almost wholly of the resistant rocks, which thus dominate the mountain slopes and keep them at a relatively high angle, the gentle slopes occurring on hills and as parts of mountains.

Closely jointed gneiss, schist, phyllite, and felsite form gentle slopes. The closely spaced cracks absorb and retain more rain than those developed in rock of other types. These rocks usually contain also a larger percentage of soluble material, either of ferromagnesian minerals in the gneiss and schist or of glass in the felsite, and consequently chemical action takes place more readily when they are exposed to the weather. The processes of mechanical action, because of the smaller size of the joint fragments and smaller grain of the component minerals, tend to produce finer rock débris. This finer débris is readily moved by rain wash on a flatter slope than that on which corresponding material derived from granitic rocks can be moved. The joint fragments, being smaller, are also more easily moved and instead of falling are undermined and carried away by rain wash.

Tuff and shale also develop gentle slopes. The products of the weathering of these rocks are fine and rather easily produced, so that the angle of the slope is determined largely by the grade on which rain wash can transport débris. Certain tuffs, however, are so compact that they weather slowly and hence form steep slopes, the angle being determined wholly by the rapidity with which particles may be detached from the matrix.

CANYON CUTTING.

In the previous sections the recession of slopes has been considered. If rocks were absolutely homogeneous so that no irregularities were formed on slopes in which rain water could be concentrated, then mountains would be eroded wholly by slope recession. But when water is concentrated and flows in a stream it erodes the rock by corrasion at a rate faster than that of slope recession, and to this fact is due much of the diversity of mountain topography and all the larger features of mountain sculpture in the Papago country.

Corrasion is wear by water in a stream and the rock débris which it transports upon the stream bed, and it is assisted by corrosion or solution of the rock so far as this may be accomplished by the passing water. The stream therefore cuts downward in a narrow groove, and in homogeneous material the walls are essentially vertical. Slumping and the processes of slope formation previously described operate to widen the cut so that in time each stream flows in a rather broad valley. As there is a limiting grade below which a given quantity of water can not transport a given quantity of material, corrasion

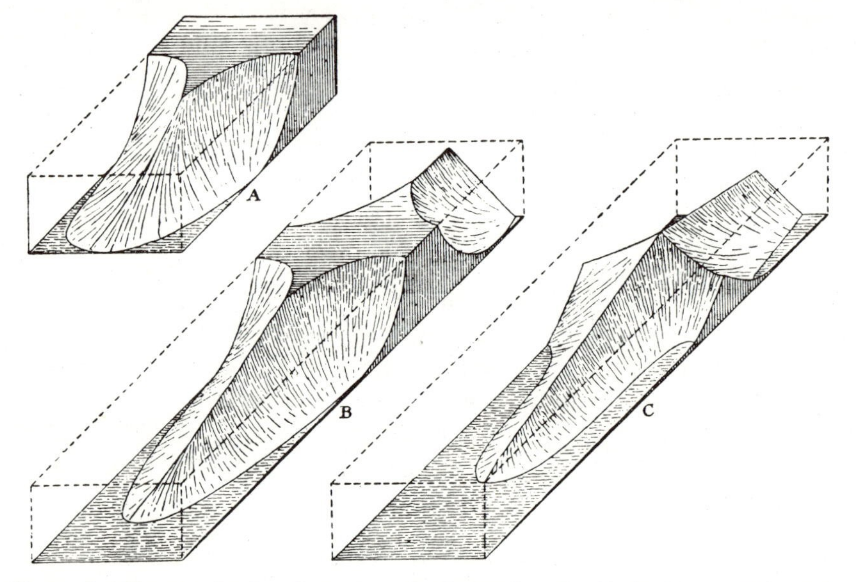

FIGURE 9.—Diagram showing three stages in the erosion of a block of the earth's crust to form the sierra type of mountain.

of the bed can proceed only until this grade is reached, and further lowering is prevented by the deposition of sediment. A stream which has reached this grade in any part is said to be graded.

The way in which canyon cutting produces the sierra type of mountain under the climatic conditions of the Papago country is shown in figure 9. In block A an original rectangular block of rock has been affected by erosion so that the face of the block has receded to the angle of 45°, assumed for the purpose of the diagram to be the slope normal to this type of rock. During the same time canyons whose side walls have the same slope of 45° have been cut into the block. In block B slope recession has carried back the point of the spur or ridge between the canyons, and the side walls of the canyon have also been eroded by the same process. Corrasion has lengthened the canyon. At the mouth of the canyon and at the foot of the ridge a small pediment has been formed. By the recession of the canyon walls the crest of the ridge has been carried below the level of the original block. The ridge, then, has a slope which is a function of the original length of the block, the length of the canyons, and the angle of the mountain slope. In block C the canyons from opposite sides of the block have met, and the head of the canyon now assumes the grade of the mountain slope. From this time on stream erosion is at a minimum and the mountains decrease in size by slope recession and the lateral cutting of the streams on the canyon floors. It is obvious that if blocks similar to block C are placed on either side of it and others beside them, the typical sierra with its crenulated ridge and projecting spurs will be reproduced.

HEADWATER BASINS.

Although the erosive power of mountain streams is great when they are in flood, floods are so infrequent in the Papago country that widening of the valleys takes place slowly. The steep grades necessary to transport débris prevent the formation of meanders, and lateral cutting is at a minimum. The lower section of a mountain canyon, once the stream is brought to grade, is in consequence relatively stable in cross section, and the side walls recede by the slow processes of slope recession.

At the entrance of a tributary, however, there are two spurs, each bounded by one slope of the main canyon and one of the tributary. Here the recession of mountain slopes is doubly active. Near the divide several tributaries usually unite to form the main stream, and the corrugated surface between them affords the optimum conditions for slope recession. Hence, at such junctions headwater basins are formed.

If two or more streams develop such basins on either side of the divide, the divide may be reduced long before the rock mass lying between the lower portions of the streams is reduced. The production of headwater basins is accelerated, especially at the beginning of the process, by the more abundant rainfall which is characteristic of the higher parts of a range.

Development of these headwater basins is a common phenomenon in the Papago country, and part of one is shown in Plate XIII, *A*. Many mountains consist of groups of isolated hills, more or less irregular in size and height, scattered on plains, that rise to low divides which lie in the position of the original mountain crests and constitute merged headwater basins, or mountain pediments.

A. AJO PEAKS, ARIZ., FROM THE EAST.

Cliffy slopes of Penasco Peak developed on massive Tertiary conglomerates and dissected pediment in foreground. Exfoliation and niches on bosslike hill at the right.

B. CLIFFY SLOPES OF MASSIVE LAVA FLOWS ON THE WEST SIDE OF THE AJO MOUNTAINS, ARIZ.

Montezuma head in the left center.

A. GROWLER MOUNTAINS SOUTH OF BATES WELL, ARIZ., FROM THE WEST.

Unbroken, talus-controlled slopes developed on slightly tilted lavas.

B. LAVA-CAPPED MOUNTAIN IN THE CABEZA PRIETA MOUNTAINS, ARIZ., FROM THE WEST.

Gullied lava talus and, to the right, mountain slope on granite with talus one boulder deep.

A. PASS WEST OF TULE TANK, ARIZ.

Niches in a granite mountain slope of the cliffy type due to massive jointing.

A. HEADWATER BASIN AND DISSECTED PEDIMENT IN THE SAND TANK MOUNTAINS, ARIZ.

Tilted lavas of Jack-in-the-Pulpit, on the extreme right, rest on the crystalline rocks in the foreground.

[AMERICAN JOURNAL OF SCIENCE, VOL. 248, OCTOBER 1950, PP. 673-696]

American Journal of Science

OCTOBER 1950

12

EQUILIBRIUM THEORY OF EROSIONAL SLOPES APPROACHED BY FREQUENCY DISTRIBUTION ANALYSIS

ARTHUR N. STRAHLER

PART I

ABSTRACT. In the quantitative analysis of erosional landforms one of several form-elements requiring measurement is slope angle of valley walls. Field sampling of slope angles in several mature regions differing widely in lithology, relief, vegetation, climate and soils has yielded data suited to frequency distribution analysis. Determination of means, estimated standard deviations, skewness and normal distribution fitness have provided a quantitative basis for slope description and significance tests.

Distributions show a wide range of means, but are symmetrical and have low dispersions. This is taken as evidence of a prevailing condition of form-equilibrium accompanying a steady state in an open system of erosion and transportation. Comparison of valley-wall slopes with adjacent channel gradients reveals a strong, positive correlation, indicating a high degree of adjustment among component parts of a drainage system.

Tests for significance of differences in slope means of three localities in the Verdugo and San Rafael Hills showed that (1) factors other than lithology exert major control over observed differences in slope angles, (2) directional exposure has no significant effect upon slope angles, and (3) slopes left to weathering, sheet wash and creep without stream corrasion at the base have reclined in angle.

INTRODUCTION

A LARGE proportion of the earth's land surface is composed of erosional landforms produced by channel erosion in drainage networks operating in conjunction with raindrop and sheet-runoff erosion and mass gravity wasting on the contributing slopes. In a mature stage of development the landscape consists of an intricate combination of channels, slopes and divides. Although the same unit form is not exactly repeated throughout this type of topography, an observer cannot fail to recognize that the forms are approximately the same throughout an area having a uniform lithology, geologic history, climate, soil and vegetation. Examples may be cited from the intricately dissected California Coast Ranges; from the monotonously repetitious landforms of the dissected Appalachian Plateaus; or from the marl hills of badlands in northern Arizona.

Although our understanding of erosional landforms is still in a stage of generalized verbal description, some attempts have been made to separate the various elements of form from the topographic complex. Robert E. Horton (1945) has analyzed drainage network composition in terms of drainage density, stream frequency, stream number, stream order, and bifurcation and length ratio.

Despite the apparent complexity of a quantitative analysis, it is possible to define and measure the various form-elements which comprise a fluvially dissected landscape. This is the first step in an over-all investigation aimed at relating quantitatively the influence of causative factors to form characteristics. Slope is one of the form-elements and is the principal topic of this paper.

In considering quantitatively the element of slope in an erosional landmass it is possible to measure and compare (1) forms of slope profiles as mathematical curves, (2) lengths of slopes from divide to base (scale-aspect of the topography), and (3) angles of slope with respect to the horizontal (gradient-aspect of the topography). Because the principal concern of this paper is with the third measure, the first two are discussed only briefly.

In a maturely dissected region of homogeneous rock the common slope profile, when taken along an orthogonal line with respect to the contours, tends to be straight in its middle and lower parts (fig. 1). At the upper end the profile becomes convex-up, passing into the rounded divide. Where streams are effectively corrading a channel at the slope base the profile continues to the base as a nearly straight line (plate 1). Concavity at the base, so commonly shown in elementary textbooks, seems to occur only where accumulations of slopewash or slide-rock lie at the slope base, or where structural control is present. Straightness of profile has been noted by Lawson (1933) in the California Coast Ranges and by the writer in that same province as well as in many other places in the United States. It may be seen in the bold mountains of the Great Smoky mass as well as in small-scale badland forms in weak clays. The common illusion of generally concave-up slopes comes from viewing profiles of lateral spurs or other lines which are not true orthogonals.

Length of slope, a function of the general scale of the drainage network, is related to the infiltration capacity and resistivity of the surface to corrasion and other factors and is therefore left to be treated elsewhere in quantitative considerations of drainage density and its variations.

Slope steepness is a particular form-element which lends itself to analysis by the standard methods of frequency distribution statistics. Even the most casual geological observer has noted that slope steepness varies widely from one region to another, but that it seems fairly constant within the limits of a small area of uniform relief and rock composition. Thus slopes may be said to reflect a steady state in the rate of removal of debris and the rate of supply of debris. Slopes seem to tend toward a constant angle which is related to the particular conditions of

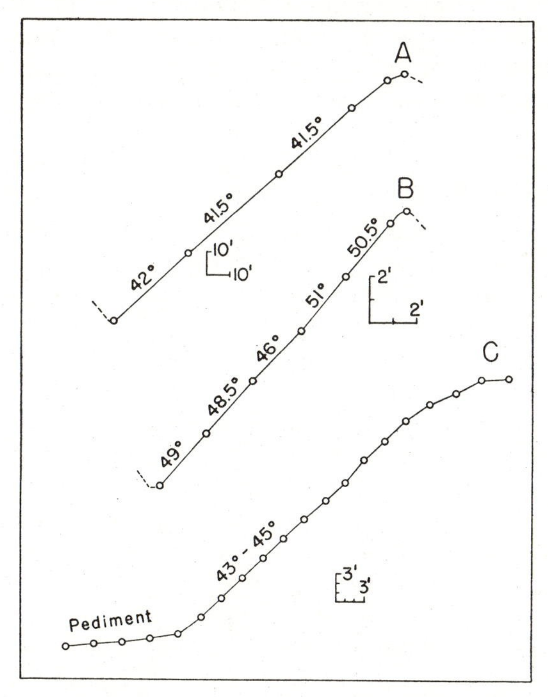

Fig. 1. Profiles of three valley-wall slopes surveyed in the field. (A) Verdugo Hills, Los Angeles Co., California. (B) Dissected clay dump, Perth Amboy, N. J. (C) Petrified Forest National Monument, Ariz.

soil, bedrock, vegetation, climate and relief, and which is integrally adjusted with conditions of channel gradient and stream regimen.

The concept of equilibrium has long been applied to graded streams and their associated slopes but the nature of this equilibrium and its basic similarities with other systems of equilibrium in nature seem not to have been fully examined. A graded drainage system is perhaps best described as an open system in a steady state (von Bertalanffy, 1950) which differs from a closed system in equilibrium in that the open system has import and export of components. An example of an open system in physics might be the flow of heat through a conducting solid. When a steady state is attained the temperature at a given point remains constant while the flow of heat continues. In biology the metabolism of a stable organism is best described as operating in a steady state requiring an exchange of materials with the environment and a continuous building up and breaking down of materials within the organism. Open systems consume energy to maintain a steady state, whereas closed systems require no energy for the maintenance of equilibrium. Still another property of the open system is that a disturbance in the flow of materials or energy will cause a readjustment to take place until a time-independent steady state is reestablished.

In a graded drainage system the steady state manifests itself in the development of certain topographic form characteristics which achieve a time-independent condition. (The forms may be described as "equilibrium forms".) Erosional and transportational processes meanwhile produce a steady flow (averaged over periods of years or tens of years) of water and waste from and through the landform system. Potential energy of position is transformed into kinetic energy of water and debris movement or heat. Over the long span of the erosion cycle continual readjustment of the components in the steady state is required as relief lowers and available energy diminishes. The forms will likewise show a slow evolution.

Applied to erosion processes and forms, the concept of the steady state in an open system focuses attention upon the relationships between dynamics and morphology. With this concept in mind, the first steps in measurement and description of slope forms typical of the steady state can be undertaken.

ACKNOWLEDGMENTS

The writer wishes to thank Dr. Howard Levene of the Department of Mathematical Statistics, Columbia University, for critically reading the paper and suggesting improvements. Field work in 1947 was supported by the Kemp Memorial Fund of Columbia University and in 1948 and 1949 by Grant 525-48 from the Penrose Bequest of the Geological Society of America.

GENERAL PRINCIPLES

Method of Obtaining Field Data.—Slope sampling requires decision as to (1) type of data wanted, (2) rules of sampling to be followed in the field, and (3) size of sample required.

In the present investigation it was required that only the steepest part of a profile line of slope be measured; furthermore, that the profile line must follow the true down-slope line from divide to base of slope (the orthogonal line with respect to contours). Following this rule the observer is prevented from having a choice as to what part of a profile to select; the use of apparent, rather than true maximum slopes is avoided. Thus any two or more observers should obtain the same angle of slope from any given point in the field, subject only to errors of measurement.

There is a more important and fundamental reason for using the steepest part of a slope profile. It is postulated that in mature topography a slope is the morphological manifestation of a steady state in which the forces which remove material are so adjusted to the resistive properties of the surface as to provide a steady supply of debris to the streams. The steepest part of a slope will then reflect the maximum angle which can be maintained, and is an indicator of the relative effectiveness of the opposed forces in the equilibrium relationship. Because the slope at its upper part declines in angle to become horizontal on the divide, angles less than the maximum have little significance and their use would bring in a broad element of subjective sampling. As the ultimate aim in geomorphology must be a dynamic understanding, it is essential that those form-elements be measured which can be most easily related to causative forces.

In field sampling, only valley-wall slopes leading down into the sides of a channel were used. Slopes leading into the extreme head of a ravine were not used because of the element of strong convergence present. Slopes down the axes of spurs, where lateral divides descend to a main stream, were not used because these are not true orthogonals. Drainage basins of the first and second, sometimes third orders were used. Avoidance of higher order master streams eliminates the reading of slopes abnormally steepened by strong lateral corrasion of a large stream.

Within each area an attempt was made to take slope readings at approximately equally spaced positions along the contour. In the case of fine-textured badlands these were spaced only 5 to 10 feet apart, or about one-fourth to one-half of the prevailing slope length. In medium-textured topography with slope lengths of 100 to 300 feet the spacing was 100 to 150 feet. This raises the question of how close together any two profile lines can lie before they can be considered as one statistical item in a frequency distribution, rather than as two separate items. The problem exists because the sampling is from a continuous surface which varies in orientation from place to place, rather than from a population of discrete pieces or objects, as for example, a clastic sediment. Because all movement of detritus on a slope takes place along the true down-slope profile lines, one would not have to move far laterally along a slope to encounter relatively independent threads of debris movement. With the spacing used by the writer successive profiles were separated by what seemed to be a relatively broad belt of slope, equivalent to one-half to one-third of the slope length itself; hence no mutual interference seemed likely.

Slope angles were read to the nearest degree or half degree with Abney hand level or Brunton compass. A class interval of 2° or more is desirable in the frequency distribution analysis, because the range of error for any item is almost a half degree. On mountain or hill slopes the steepest down-slope line was determined by rolling boulders or pebbles down-slope, or by selecting the maximum of a range of readings aimed down-slope through a lateral arc. When a two-man team is used, each man takes a position on the slope line. The instrument man then reads the slope to the eye-level of his partner. In this way minor ground irregularities and brush can be overcome. When one is working alone, he can put up a perpendicular rod with the eye-level marked, or make a mark at eye-level on a convenient tree trunk or bush. Slopes were read over lengths equivalent to one-fourth or one-fifth of the slope length so as to avoid minor irregularities made by boulders or clumps of vegetation. As most slopes are relatively straight or steepen toward the base the measurement is taken near the base. Some slopes have a basal concavity due to local accumulations of slopewash or talus; the slope is then read above the zone of accumulation. In the case of small-scale badlands a 3-foot board was laid on the slope and the measurement taken as with dip of a bedding plane.

It is desirable to read the azimuth of each slope measurement so as to permit study of the effects of direction of exposure on slope angle. Readings should be located on the field map by a station number corresponding to the notebook record. This permits statistical analysis between individual basins or groups of basins.

Sample sizes taken by the writer ranged from 20 to 200 readings. Subsequent statistical analysis has shown that a sample of 50 slopes is adequate to reveal the characteristics of the distribution in a homogeneous region of 6 to 12 first-order drainage basins; while a sample of 25 will closely fix the mean.

The use of topographic maps as sources of slope data has been investigated and tests run between field and map readings of the same slopes. The statistical analysis of this procedure is treated in a later section of this paper. Regarding collection of data from maps the following conclusions have been reached: Maps on a scale of 1:24,000 or 1:25,000 published by the U. S. Geological Survey and the Army Map Service in recent years will give reasonably accurate results where topography is coarse-textured and the relief strong. These maps are prepared by photogrammetric methods and comply with National Standard map accuracy requirements. It was found that maps of a scale of 1:31,680 in Los Angeles County are unsuited to slope sampling because the topographic texture is too fine for the map scale and contour interval; *i.e.*, slopes are too short to be represented accurately under the cartographic conditions. Maps of the scale 1:62,500 are generally beyond the limits of reasonable accuracy and should be entirely avoided for slope sampling.

General Characteristics of Slope Frequency Distributions.— The methods of slope sampling described above were applied to several areas differing widely in lithology, soil, relief, vegetation, and climate in order to obtain some idea of the range of characteristics present and the degree of homogeneity within each area. Figure 2 shows several histograms drawn on a common abscissa to show comparative positions on the slope scale 0° to 90°. For each distribution the following measures are given:

N Number of items in the sample.

$\overline{X}$ Arithmetic mean.

s Estimated standard deviation.

Cumulative frequency percentage distributions for the same samples are shown in figure 3. Both step-graphs, showing individual classes, and smoothed curves are shown.

With regard to general distribution properties, two points are especially noteworthy: (1) Dispersions are low, considering the range of angle possible, giving evidence of a high degree of homogeneity in the populations sampled. (2) Distributions are generally symmetrical, the principal exception being sample D. Inasmuch as sampling was carried out according to previously set rules and no readings were discarded

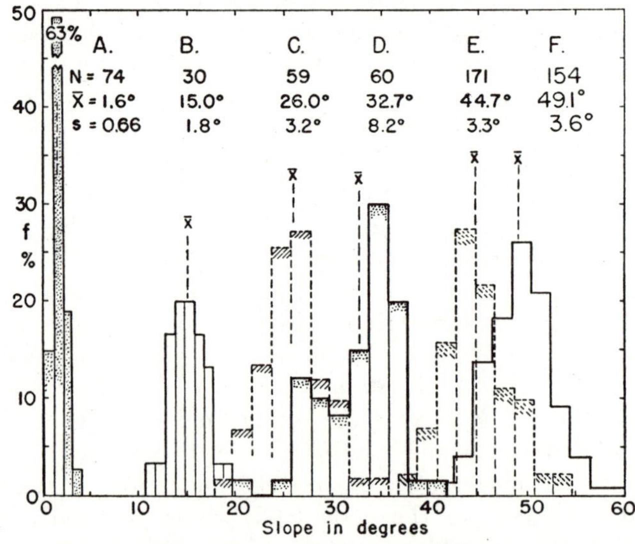

Fig. 2. Histograms of six slope frequency distributions. (A) Steenvoorde, France, (B) Rose Well gravels, Arizona, (C) Bernalillo, N. M., Santa Fe formation, (D) Hunter-Shandaken area, Catskill Mts., N. Y., (E) Kline Canyon area, Verdugo Hills, Calif., (F) Dissected clay fill, Perth Amboy, N. J. All data except (A) from field readings.

arbitrarily, either during the field sampling or during the processing of the data, the low dispersions and symmetry are interpreted as an indication that within a given small region where conditions of lithology, climate, soil, vegetation and relief are uniform, slopes tend to approach a certain equilibrium angle, appropriate to those controlling factors.

Slope Distributions and the Normal Curve of Error.—The problem of whether the distribution of maximum slopes in a selected region follows the normal curve of error was considered worthy of analysis in view of the possibility that the distribution is inherently skewed and is better represented by a logarithmic-normal curve, or some other variety of normal curve, such as one in which the abscissa is on a scale of tangents or log-tangents.

Three methods were used in analyzing the form of the distributions: (1) A plot on arithmetic probability paper. (2) A fitted normal curve and χ^2 test of goodness of fit. (3) Determination of significant skewness and kurtosis based on the third and fourth moments of the distribution.

A rough test of normal distribution may be had by plotting cumulative percentage frequencies on arithmetic probability paper. This paper is constructed to show the normal curve of

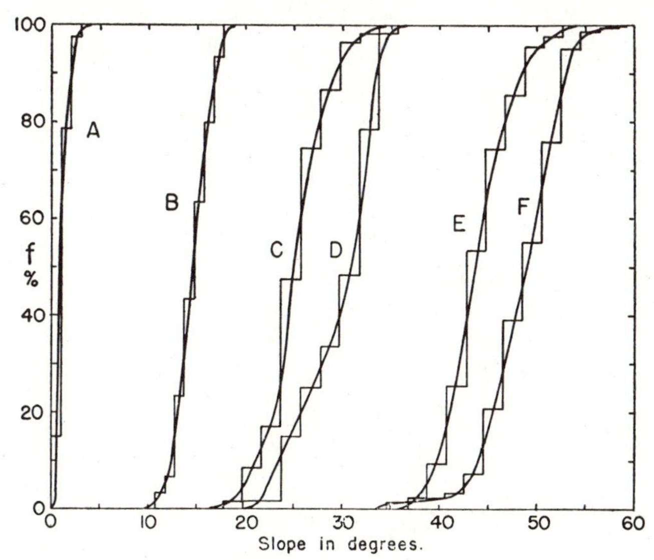

Fig. 3. Cumulative percentage frequency distributions. Same data as in figure 2.

error (cumulative) as a straight line. If the distribution appears reasonably straight one can proceed to fit a normal curve to the data. An additional feature of the cumulative curve on probability paper is that the slope is proportional to dispersion, hence the differences in dispersions between two normal distributions can be noted at a glance.

Figure 4 shows the data of figure 3 plotted on arithmetic probability paper. End classes containing only one item have been combined to compensate for the influence of scarcity. With the exception of sample D and the lower tail of sample F a reasonable straightness prevails, suggesting that the distributions are well described by normal curves, which may now be fitted to the data.

Fitting of the normal curve to the slope data of sample C, an area of dissected Santa Fe formation near Bernalillo, New Mexico, is illustrated in figure 5.

Goodness of fit has been determined by use of a χ^2 test in which theoretical frequencies and observed frequencies are compared as follows: (Croxton and Cowden, 1946, p. 286)

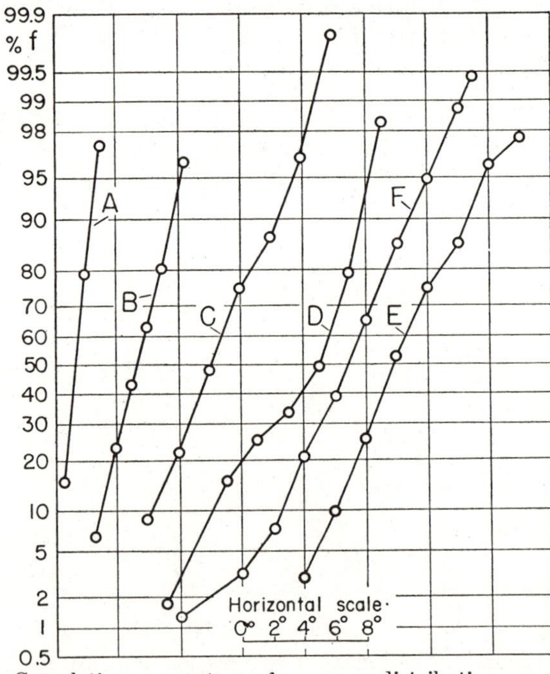

Fig. 4. Cumulative percentage frequency distributions on arithmetic probability paper. Same data as figures 2 and 3.

$$\chi^2 = \Sigma \frac{(f - f_c)^2}{f_c} = 0.631$$ where

f = observed frequency

f_c = expected frequency

The value *n*, representing the number of degrees of freedom, is obtained by subtracting the three degrees of freedom lost in the curve-fitting process from the original number of classes, seven, and is therefore equal to four. By consulting a prepared table we find for these values of χ^2 and *n* a probability, P, of approximately 0.97. Thus a normal curve of error is an excellent description of the distribution and, if the distribution is in fact normal, we should expect to find this good a fit or better in only 6 times out of one hundred such samples because of chance variations due to sampling only.

Tests of significant skewness and kurtosis were carried out on the raw data of three frequency distributions of slopes in the Verdugo and San Rafael Hills, Los Angeles County, California[1].

The purpose of the investigation was to determine if the various amounts of skewness and kurtosis visible in the three distributions are of such an order as to be commonly ob-

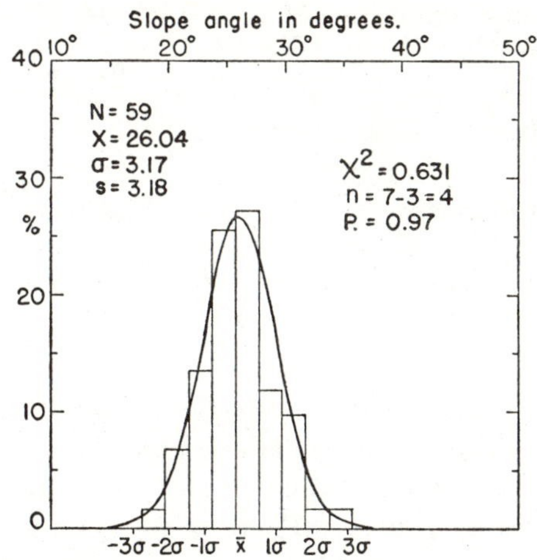

Fig. 5. Normal curve of error fitted to a frequency distribution of slope angles from the dissected Santa Fe formation near Bernalillo, New Mexico. Slopes measured in field.

[1] The analysis was carried out under the direction of Mr. Richard Ostheimer, Instructor in Statistics, Columbia University. Moments were computed from the raw data in order to remove possible influences of grouping.

tained by chance variations in sampling of a normal population, or whether they are significant of actual departures from normal in the population (fig. 6).

Skewness was determined from the value of β_1, a measure of skewness based upon the third moment of the frequency distribution:

$$\beta_1 = \frac{N\ (\Sigma x^3)^2}{(\Sigma x^2)^3}$$

Table 1 shows the results and conclusions for samples A, B, and C of figure 6. Samples B and C showed no significant skewness; sample A is doubtful of interpretation. Kurtosis was determined by the value of β_2, a measure of kurtosis based upon the fourth moment of the frequency distribution:

$$\beta_2 = \frac{N\ \Sigma x^4}{(\Sigma x^2)^2}$$

TABLE 1

Summary of analysis of skewness in three sample slope frequency distributions.

DETERMINATIONS MADE	Area A	Area B	Area C
β_1 (Distribution symmetrical when β_1 is 0; Right skewness when β_1 is +; left skewness when β_1 is −.)	+0.153	—0.005	—0.103
Probability of sample β_1 at least this large, if β_1 of universe is 0.	0.02 to 0.10	>0.10	>0.10
g_1 (Normally distributed measure of skewness. Symmetry when $g_1 = 0$.)	0.394	0.070	0.340
$\frac{g_1}{\sigma_{g_1}}\left(\frac{x}{\sigma}\text{ of a normal curve}\right)$...........	2.12	0.27	0.75
Probability of sample g_1 at least this large if g_1 of universe is 0.	0.02	0.39	0.23
Is the skewness significant?	?	No	No

Table 2 shows the results and conclusions for samples A, B and C. None of the samples shows significant kurtosis although the probability shows a considerable range.

The general conclusion which may be stated from the analysis of β_1 and β_2 is as follows: There is no reason to doubt that frequencies of each of the three areas are normally distributed. A possible exception is area A, in which the sample has right skewness, but the skewness is of questionable significance.

Law of Constancy of Slopes.—The tendency of slopes to group closely about a mean value may be taken as an expression of a morphologic law relating slope to other form factors. This law may be stated as follows: Within an area of essentially uniform lithology, soils, vegetation, climate and stage of development, maximum slope angles tend to be normally distributed with low dispersion about a mean value determined by the combined factors of drainage density, relief and slope-profile curvature.

TABLE 2

Summary of analysis of kurtosis in three sample slope frequency distributions.

DETERMINATIONS MADE	Area A	Area B	Area C
β_2 (Distribution normal, mesokurtic, when $\beta_2 = 3.0$. Leptokurtic, or peaked, when $\beta_2 > 3$. Platykurtic, or flattened, when $\beta_2 < 3$.)	2.992	3.707	2.381
Probability of sample β_2 differing by at least this much from a population β_2 of 3.0	0.05	0.05	0.05
g_2 (Normally distributed measure of kurtosis. Mesokurtic when $g_2 = 0$.)	0.027	0.819	0.485
$\frac{g_2}{\sigma_{g_2}}\left(\frac{x}{\sigma}\text{ of a normal curve}\right)$...........	0.07	1.62	0.55
Probability of sample g_2 at least this large, if g_2 of universe is 0.	0.47	0.05	0.29
Is there significant departure from normal, mesokurtic form?	No	No	No

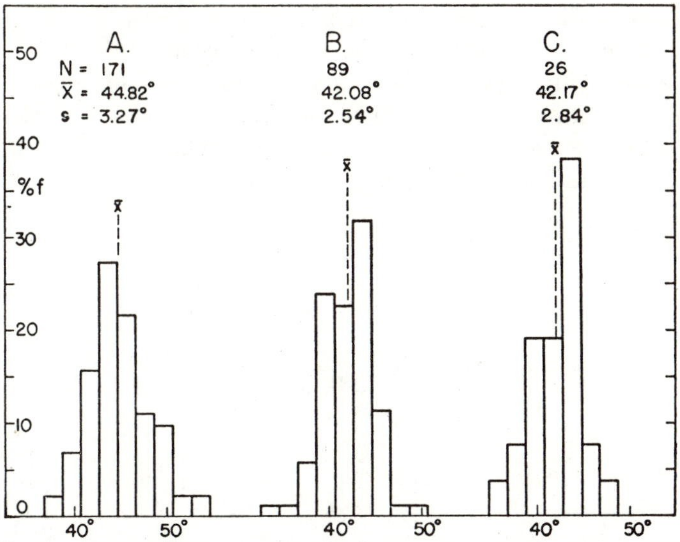

Fig. 6. Histograms of three slope frequency distributions from the Verdugo and San Rafael Hills, California. Location of areas A, B, and C is shown in figure 10. See tables 1, 2, 3, 4, 5, and 6 for summaries of data analysis.

If drainage density (Horton, 1945, p. 248),

$$D_d = \frac{\Sigma L}{A}$$ where ΣL = sum of stream lengths (projected onto a horizontal plane) A = area.

then the mean horizontal distance from divide to channel is roughly one-half of the reciprocal of drainage density:

$$H = \frac{1}{2D_d}$$ where H = horizontal distance from stream to divide.

Now, if the spacing of streams is uniform throughout and the length of contributing slopes is similarly uniform, slope steepness can vary only as relief or profile curvature are varied (fig. 7).

Given two areas of the same drainage density, the area having the lower relief must have lower slopes, assuming profile curvature characteristics to be similar (fig. 7A). Given two areas of similar relief, but of different drainage density, slope must be less steep where drainage density is less (fig. 7B). If divides are broadly convex and only the basal fraction of the slope is straight, a steeper maximum angle is required than in a region of narrow divides and long, straight slopes, even though relief and drainage density are the same in both areas (fig. 7C). Because divide curvature is a factor readily influenced by stage of development or by structural conditions, it can be minimized only by the use of mature areas which behave as if lithology and structure were homogeneous.

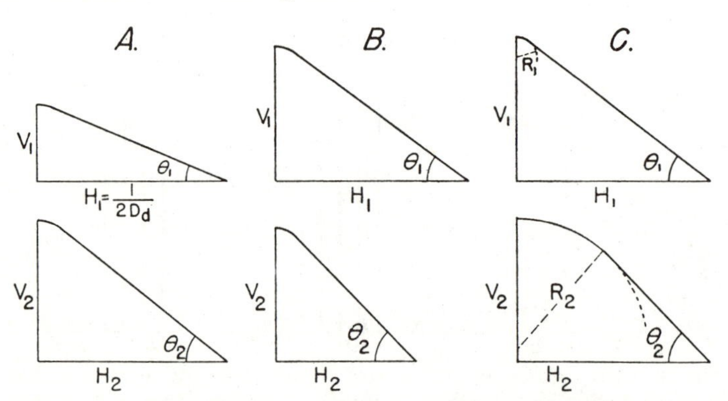

Fig. 7. Relationship of relief (V), slope (θ) and horizontal distance between divide and stream (H). Radius of divide curvature, R, is a variable diagram *C*.

Slope is thus seen to be proportional to the ratio between relief and one-half of the reciprocal of drainage density, assuming constant profile curvature characteristics. From figure 1 it is apparent that profiles are similar in form for both the fine-textured badlands on dissected clay fill near Perth Amboy (A) and for the medium-textured Verdugo Hills in Los Angeles County (B). Both have long, straight slopes and only a small curature radius at the divides (plates 1 and 2). On the basis of detailed profile plotting of slopes in the field and plane table mapping it appears that relief in the first area is about 20 times the second, but drainage density is only 1/20 as great, so that the slope angles are of the same general order of magnitude (fig. 2, E, F). If land slopes were perfectly straight from divide to base, the relationship of relief to drainage density would be approximately

$$\tan \theta = \frac{V}{\frac{1}{2D_d}} = 2VD_d$$

where θ = slope angle

V = relief (average vertical distance between divide and channel)

D_d = drainage density

For example, it was found in the Perth Amboy locality that the mean of a large number of measurements of vertical distance between channel and divide was approximately 6.0 feet, or 0.00114 miles. Drainage density sampled from several basins was 500 (that is, 500 miles of channel per square mile of area, projected upon a horizontal plane). Then

$$\tan \theta = \frac{V}{\frac{1}{2D_d}} = \frac{0.00114}{0.00100} = 1.14$$

$$\text{and } \theta = 48.7^\circ$$

Note that the mean of slope readings in this area as shown in figure 2, sample F, was 49.1^0. One would not normally expect such good agreement between calculated and observed slope angles, and because the range of error has not been investigated, the coincidence may be misleading.

Relations of Slopes to Channel Gradients.—Davis (1909, p. 266-268) clearly stated the concept that the slopes of a mature landmass are graded, as are the streams. We may restate this principle by saying that slope profiles are in equilibrium with the channel profiles to which the slopes contribute their debris.

For a given area free of systematic structural control, but subject to uniformly controlling factors of climate, vegetation, soil and stage of development, all morphological characteristics tend to approach a time-independent form. Ground and channel gradients, as well as drainage density achieve a form best adapted to maintaining a steady state in the removal of debris. This is simply an extension of Playfair's Law to include all form-aspects of the topography.

If the type of adjustment described above actually exists, one would expect the angle of slope of valley walls to vary systematically with gradient of channel at the slope base. Steep ground slopes would be expected to correspond with steep channel gradients; low ground slopes with low stream channel gradients. This relationship has been stated by R. E. Horton (1945, p. 285) who uses the definition

$$\text{slope ratio} = \frac{S_c}{S_g} \quad \begin{array}{l} \text{where } S_c = \text{channel slope} \\ \text{and } \quad S_g = \text{ground slope} \end{array}$$

Ground slope can be determined by the field methods previously described in this paper. For each slope reading the channel gradient at the base of the slope would need to be measured. This was not done in the field by the writer, but the general trend of the slope ratio was obtained from large-scale topographic maps, including two which were specially mapped for the present study. Figure 8 shows the means of ground slopes and channel slopes plotted on log-log paper. Each point represents means of a large number of measurements of ground slopes and channel slopes. Because samples were taken from maps and the correlation is believed to have wide inherent spread of values, a curve was fitted by inspection. We might say that, if these data are representative of general conditions, ground slope varies as somewhat less than the first power of channel slope, and that the slope ratio tends to range from 1/5 or 1/4 for low values of slopes, up to about 1/2 in the higher values.

There are several uncertain elements in the foregoing analysis. Data from a greater number of regions would be expected to modify the estimating equation considerably. Substitution of careful field measurements for map measurements would increase the confidence in the data. It is further obvious that channel slope steepens rapidly toward the head of a valley, whereas ground slope tends to remain nearly constant along the

Plate 1. Topography of Verdugo Hills. Cabrini Canyon, south side of Verdugo Hills, Los Angeles Co., Calif. Slope in foreground faces northwest, had chamise-chaparral cover prior to burning. Canyon is about 200 feet deep.

Plate 2. Topography of Perth Amboy badlands. Badlands in clay-sand backfill, Perth Amboy, N. J. Relief in this photograph is 20 to 30 feet. This topography is similar in form-characteristics to that shown in plate 1.

sides of the valley, as the frequency distribution studies show. To minimize this variable, which can easily vitiate the results, readings were made along second-order stream basins, near the point where they are formed by the junction of two first-order streams (Horton, 1945, p. 281). This assures that all data come from the same relative position in the drainage system. Further standardization is desirable.

There is a suggestion from the data of figure 8 that climate has a profound influence on the slope ratio. Areas 7 and 9, which lie to the right of the line of the estimating equation, and area 8 which is on the line, have typical bare ground surfaces of badlands, or are lightly covered by chaparral, as in the Mt. Gleason area (7). Here one might expect a given slope of valley side to supply proportionately more detritus to the channel than a comparable slope in vegetation-clothed regions

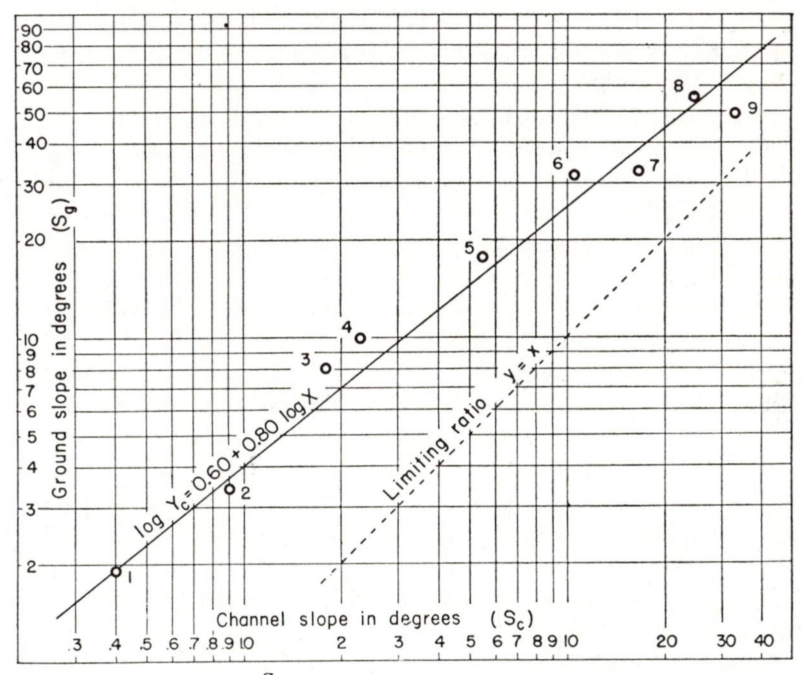

Fig. 8. Slope ratio, $\frac{S_c}{S_g}$, for nine maturely dissected regions. 1. and 2. Grant, La., 3. Rappahannock Academy, Va., 4. Belmont, Va., 5. Allen's Creek, Ind. 6. Hunter-Shandaken, N. Y. 7. Mt. Gleason, Calif. 8. Petrified Forest, Ariz. 9. Perth Amboy (clay fill), N. J. All data from U. S. G. S., A. M. S. or special field maps.

in the humid climates. The thinly vegetated areas would require relatively steeper channel gradients for the transportation of the greater quantity of detritus. In the two cases lying to the right of the line the channel slope is high in proportion to ground slope. While this could easily be merely the result of chance sampling, it is in agreement with the deduced behavior. On the other hand, well-vegetated slopes in the humid climate would tend to supply less detritus because of the increased surface resistivity and infiltration capacity, and would be associated with relatively lower channel slopes. As if to verify this deduction, points 3 through 6, representing humid regions, lie to the left of the line of the estimating equation. As this could readily be the result of chance, the observation merely makes more intriguing the prospect for a more adequate investigation of this phase of equilibrium relationships.

Comparison of Slope Data of Maps with those of Field Observations.—The problem of reliability of slope readings taken from maps in comparison with readings made at the same locations in the field is an important one, inasmuch as the labor and cost of the first method is only a fraction of the second. The writer was inclined to suppose that the most recent maps of the Army Map Service on a scale of 1:25,000 and of the U. S. Geological Survey on a scale of 1:24,000 made by photogrammetric methods would give results equal to ground observations.

For test purposes the Shandaken and Hunter, N. Y. quadrangles of the Army Map Service were selected. In the field the writer walked along the side slopes of five drainage basins, following trails marked on the maps. The maps have been prepared from air photographs. Planimetric detail seemed remarkably good when checked in field use. The drainage basins are large, the slopes long and of relatively simple, smooth character. A total of 60 field readings was taken, or about 12 in each basin. These readings were spaced from 50 to 150 yards apart. A frequency distribution diagram of these data is shown in figure 9, lower half.

In the laboratory slopes were measured from the contours along the same stretches of the trails. Approximately the same number of readings, 58, was taken, and while these are sampled from zones corresponding to those in the field, they are equidistantly spaced and do not coincide exactly with the field readings. A frequency distribution diagram of these readings is shown in figure 9, upper half. A comparison of statistical measures of the two distributions is given in table 3.

TABLE 3

Summary of field and map data of five drainage basins, Hunter and Shandaken Quadrangles, N. Y.

DRAINAGE BASIN	FIELD DATA $\bar{X}$		MAP DATA $\bar{X}$	
Hunter Brook	27.7°		26.9°	
Becker Hollow	30.0°	$N = 60$	30.7°	$N = 58$
		$\bar{X} = 32.69°$		$\bar{X} = 30.72°$
Taylor Hollow	32.8°		28.3°	
		$s = 8.24°$		$s = 6.20°$
Notch Lake Hollow	33.4°	$s_{\bar{x}} = 1.07°$	31.9°	$s_{\bar{x}} = 0.81°$
Shanty Hollow	35.5°		34.4°	

Examination of the histograms and summary table shows that the field distribution has a higher degree of irregularity

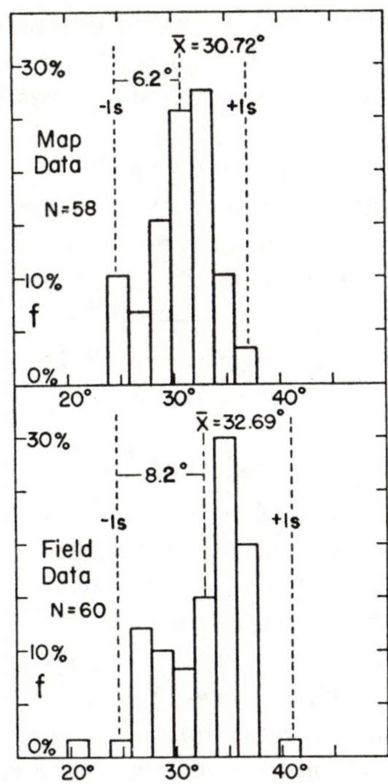

Fig. 9. Histograms comparing field and map samples of slope angles from the same drainage basins. Shandaken and Hunter, N. Y. quadrangles, A. M. S. 1:25,000.

than the map distribution. Dispersion is greater in the field data, with an estimated standard deviation a full 2° greater and with two end classes which are not present in the map data. It appears that use of the map tends to reduce the dispersion and remove irregularities. This might be expected if the drawing of contours in preparation for reproduction results in increasing the degree of parallelism present in the original contours.

The field data are strongly bi-modal because, of the five drainage basins used, two have low but similar means; the other three are relatively high. Note that while the map data also show a bi-modal tendency, it is greatly reduced; the sequence of means is not in quite the same order as in the field data. The greater homogeneity of the map data is again apparent.

Of greatest interest is the difference in means, a whole 2°. We might at first suppose that the processes of mapping and map drafting have caused a reduction in slopes, through the process of drawing the contours evenly spaced on the map, even where steeper and less-steeper portions of the slope are actually present. This would tend to lower the angle read from contours alone.

Are the field and map means significantly different? To test the difference, the ratio of the observed difference to the standard error of the difference (Croxton and Cowden, 1946, p. 319) was computed as follows:

$$\frac{x}{\sigma} = \frac{\bar{X}_1 - \bar{X}_2}{\sigma_{\bar{x}_1 - \bar{x}_2}} = 1.462$$

On the normal curve this value of $\frac{x}{\sigma}$ lies at such a position that the area under the tail of the curve is 0.07. The difference is therefore not significant but the level of significance is not convincing. We can make the following statement: Even if the map is an exact replica of the field topography, we should expect a difference in means of any 2 samples of this size to be two degrees (1.97° exact) or more in 15 out of 100 times because of chance variations in sampling alone. Hence we cannot say that the map is inaccurate. Even if a very significant difference were present, it would not necessarily mean that the map were at fault. Differences resulting from the map and field methods of reading slopes might cause the observed differences in means.

In conclusion, we may say that while a fairly close agreement exists between field slope data and those taken from the best available topographic maps, the inconsistencies are of too great an order to permit significance studies between small groups of field and map data. It is advisable to use only the one type of data, preferably the field observations where these can be had.

A Tentative Classification of Graded Erosional Slopes.—As a result of frequency distribution analysis and field study of slopes in a wide range of localities, the classification of erosional slopes into three general groups seems feasible:

(1) *High-cohesion slopes.* Slopes underlain by cohesive, fine-textured materials such as clays or clay-rich soil or by strong, massive bedrock, such as granite, schist or gneiss tend to have means in the range 40 to 50° and sometimes higher in regions of strong relief. These slopes are higher than the angle of repose of the loose, dry fragments of the same materials, so that taluses of lower surface angle may accumulate at the slope base where material rolls or slides down the slope and is not swept away by streams. These steep slopes are subject to occasional rapid flowage and sliding movements of the soil or surficial layer following torrential rains. Where the slope has a thick soil of weathered, clay-rich material resting on a resistant, massive crystalline rock, a debris-avalanche may strip off the entire layer, exposing a slope of bare rock (Bryan, 1940, p. 262-263; Freise, 1938).

Examples of high-cohesion slopes are found in badlands, such as the Petrified Forest, New Mexico, or in eroded clay dumps along the Raritan River, New Jersey. In the Verdugo and San Rafael Hills of the California Coast Ranges are examples of slopes of similar steepness where a soil layer with moderate vegetation rests on deeply decomposed diorite or metasediments. Bailey (1948, p. 304) has described stable vegetated 60-degree slopes underlain by fine-textured soils with well-developed profile. He attributed the high angle to binding and protective action of vegetation, principally grass. Examples where bare, hard, massive crystalline rock forms the slope were seen in the Blue Ridge in North Carolina.

Thus steepness of slope, alone, without regard to scale of

slope length or drainage density, can exist under a wide variety of lithologic conditions. There is the common property of maintaining a slope angle above that of the angle of repose of coarse fragmental materials because the slopes possess a high degree of cohesion or strength. All seem to require also that a steep-gradient stream system in process of vigorous corrasion be present.

Bryan has pointed out (1940, p. 263) that for steep slopes of the tropical regions there is nothing to indicate that the slopes should flatten with time. We might reason that, so long as a steady state prevails in which the streams are corrading their channels and constantly removing the detritus from the slope base, conditions of vegetation, climate and lithology remain constant, the slopes would maintain a fairly constant angle in equilibrium with the other factors. However, just as graded streams gradually regrade their profiles to lower gradients with generally decreasing land relief, the contributing slopes might be expected to be gradually reduced in gradient, the system maintaining a steady state through any given short period of time.

(2) *Repose slopes.* Slopes which shed loose, coarse rock particles may be controlled by the angle of repose of those materials provided that vigorous stream corrasion into the slope base promotes the maintenance of the maximum or near maximum repose angle. Meyerhoff (1940, p. 251) seems to have included this type of slope in the term *gravity slope.* Bryan (1922, p. 43) terms certain desert mountain slopes *boulder-controlled slopes.* He states that they are commonly in the range 30-35° and that the grade of the slope is the angle of repose of the average-sized joint fragment.

Repose slopes are illustrated by certain drainage basins in the Catskill Mountains (see figure 9) where a coarse layer of angular sandstone and conglomerate mantles slopes whose mean was found to be 32° in five first-order drainage basins.

The term *talus-slope,* referring to slopes of slide rock accumulation made by accretion of particles rolled or dropped from a higher source, should not be confused with repose slopes here described. The latter are essentially erosional, having only a thin veneer of loose particles over the parent bedrock.

Erosional repose slopes are observable in gravel pits where glacial-fluvial sands and gravels are being excavated. Left undisturbed, a fresh vertical cut into a gravel bank will recline to a straight slope equivalent to the angle of repose of the loosened particles, but composed of the almost bare, undisturbed, bedded sands and gravels. On this slope, which merges into a true talus at the base, particles will just slide or roll when dislodged from the parent mass. Laboratory analysis of angles of repose in fragmental materials (Van Burkalow, 1945) shows that repose slopes have distribution characteristics similar to the valley slopes described in the present paper.

Slide-rock slopes might be expected to maintain a constant angle as a region is progressively denuded, provided that the streams maintain a vigorous corrasion which saps the valley walls and maintains the maximum angle of the slide-rock fragments.

(3) *Slopes reduced by wash and creep.* Where channel corrasion is greatly reduced and channel gradients are low, due to onset of a late-mature stage, slopes are not maintained at the maximum angle by basal corrasion. Here sheet-erosion, rain beat and creep reduce the slope to angles well below the angle of repose. Bryan (1922, p. 46) terms these slopes *rain-washed slopes* and states that they are developed on fine-grained materials, readily subject to removal by rain-wash. Meyerhoff (1940, p. 251) terms these *wash slopes* and states that they seldom exceed 15°. He regards them as secondary forms which replace the steeper varieties by encroachment at the base. In the W. Penck system these would be classed as *Haldenhänge*.

Examples of slopes reduced by wash and creep were found to have means ranging from 20° to as low as 1½° (see fig. 2, A, B). It is postulated here that the slopes were reduced gradually in angle as the relief of the region diminished and stream gradients declined. Some evidence favoring a gradual decline of slope angle rather than replacement of a steep slope by a distinctly lower slope is discussed in a later section.

Summary Statement.—The application of standard methods of frequency distribution statistics to study of slope angles is of value in two principal respects: (1) It provides a standardized quantitative method of describing the characteristics of maximum slope angles within a small area, although it does not rectify fundamental errors of judgment made in the sampling process. (2) It provides a tool, through the method of significance of difference of sample means, of evaluating the results of field measurements and thereby prevents the drawing of unwarranted conclusions from numerical data. At the same time it may give confidence in certain conclusions based upon truly significant differences in sample means.

The measurement and analysis of real landforms is an inductive method, contrasting strongly with the purely deductive projective-geometric treatment of slopes used by Lawson (1915) and Putnam (1917) and recently revived by Bakker and Le Heux (1946). The latter method lays great stress on precision attained by setting up initial assumptions and deducing the slope profile to be expected. Whether the initial assumptions or deduced profile development bear any resemblance to real landforms is dubious at best, because only one process, out of several acting together, is selected for analysis.

(*To be continued*)

[American Journal of Science, Vol. 248, November 1950, Pp. 800-814]

EQUILIBRIUM THEORY OF EROSIONAL SLOPES APPROACHED BY FREQUENCY DISTRIBUTION ANALYSIS

ARTHUR N. STRAHLER

PART II

SIGNIFICANCE TESTS APPLIED TO SLOPE PROBLEMS IN THE VERDUGO AND SAN RAFAEL HILLS, CALIFORNIA

Introduction—Studies of the significance of differences in sample means have applications in the attack upon a number of fundamental geomorphic problems involving factors which affect equilibrium of graded slopes. Data collected in the Verdugo and San Rafael Hills, Los Angeles County, California, have been examined with a view to determining (1) if differences in underlying rock type are associated with differences in slope angles, (2) if differences in directional exposure to sunlight and other meteorological factors produce differences in slope angles, and (3) if slopes decline in angle when left to weathering and erosion processes not accompanied by basal erosion and removal.

Description of the Area.—For initial research three small areas in the Verdugo and San Rafael Hills were studied (figure 10). These lie in the southern side of a single NW-SE trending Figure 10.
Coast Range block of moderate relief (Miller, 1934). Conditions of climate, vegetation, relief and tectonic history are essentially the same throughout, but lithologic factors are not. The easternmost area is underlain by a relatively homogeneous mass of Wilson diorite; the western two areas by the metasediments and small intrusive bodies comprising the San Gabriel formation. Despite marked variations in bedrock composition and structure in the latter formation, the entire southern side of this range appears homogeneous in topographic aspect.

The field work was concentrated upon measuring slope angles in several categories, in measuring and plotting detailed topographic profiles from ravine floors to divides, and in determining the number and location of the smallest discrete channels of the drainage network. Information was plotted on photostatically enlarged copies of the U. S. Geological Survey topographic maps on a scale of 1:24,000 corrected on the basis of air photograph study and field inspection (fig. 11).

Topography.—Mountain topography of the Verdugo and San Rafael Hills is in a mature stage of development. The drainage pattern is roughly dendritic, although the predominant trend of streams is southwesterly, from the major divide in which the streams head, to the sloping fan surfaces which encroach from the south upon the ragged foothill belt of the mountain base. Although there are many irregularities of bedrock composition and structure, these seem to have rather minor influences upon the placement of minor streams and divides.

Slopes are steep (plate 1), with a tendency to straightness of profile (figs. 1 and 12). Divides tend to be smoothly rounded throughout, a form emphasized by the fact that firebreaks are made and maintained by bulldozer along the major divides. Serrate crests and craggy peaks, often seen in desert moun-

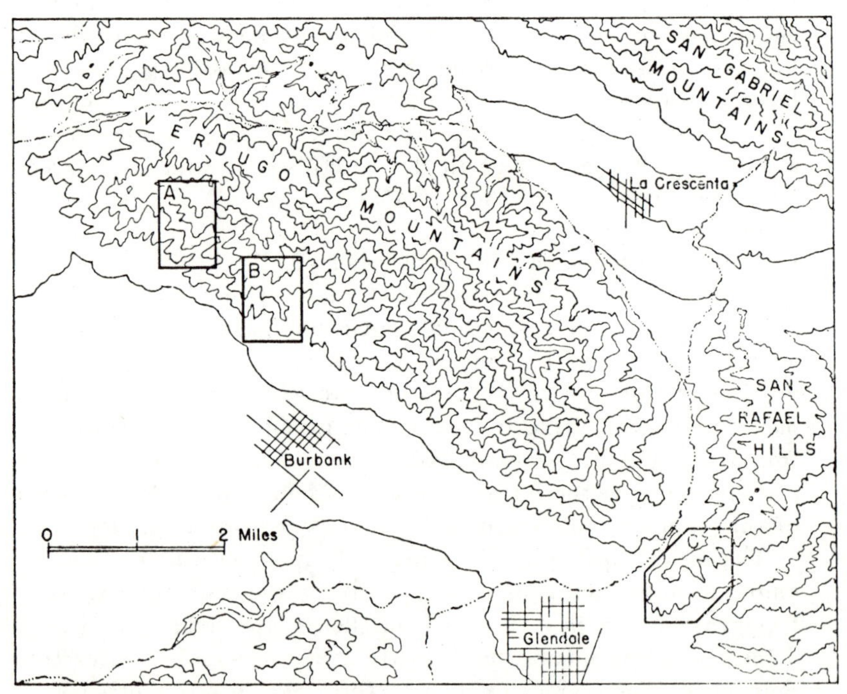

Fig. 10. Outline map of Verdugo and San Rafael Hills, Los Angeles County, California.

tain ranges and glaciated mountains are absent here. Relief is not great, despite the steep slopes. The areas studied have less than 1000 feet total relief while most slope profiles measured in the field have less than 100 feet of difference in elevation from divide to channel.

Ravines may be subdivided into two classes: (1) The smaller ravines and the upper ends of the larger ones are V-shaped in transverse profile. The enclosing walls have slopes in the range 40° to 45° for the most part. The stream channel, dry most of the time, is narrow and is cut in bedrock. Locally, however,

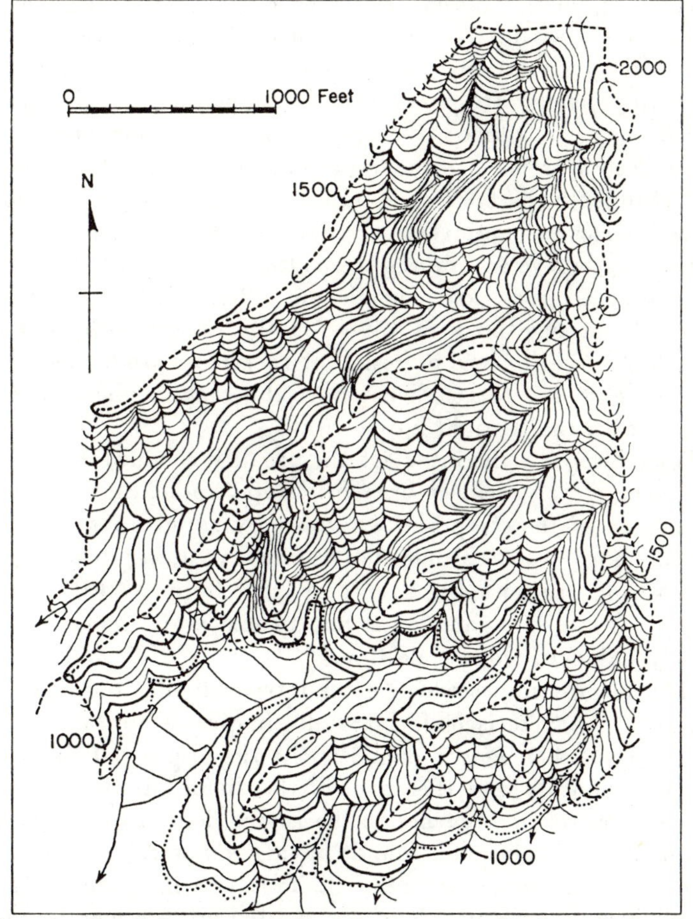

Fig. 11. Detailed map of Kline Canyon area, Verdugo Hills, Calif. (area A of figure 10.) Enlargement of U. S. G. S. map, 1:31,680, corrected in field.

the channel is choked with debris which has slid, rolled or flowed from the steep walls. (2) The lower courses of most of the larger canyons are graded and have a flat alluvial floor ranging from a few feet to one or two hundred feet wide. To what extent the flat is due to lateral corrasion in bedrock, rather than to alluviation by flood- and mud-flow of a formerly V-shaped valley is not easily determined. In the areas studied, a period of recent channel trenching showed alluvial and mud-flow fill. Streams strongly impinge on the base of the steep canyon walls, undercutting the bedrock. Where the stream has temporarily moved away from the slope base steep cones of talus have been built, causing a smoothed, concave-up basal slope of accumulation to be present.

Ravine heads are steep-walled, funnel-like amphitheaters which may be termed "hoppers" because their converging slopes feed detritus into the uppermost end of the stream channel. Hopper walls tend to be nearly straight in profile and to make the form of an inverted part-cone. Most hopper walls are very steep, above 45° in angle. Evidences of rolling and sliding of weathered rock are conspicuous. Bedrock exposure comprises from one-third to three-fourths of the hopper walls. Locally a layer or lens of resistant rock may produce a nearly vertical cliff, but for the most part the rock surfaces conform closely with the straight hopper walls.

Some hopper walls intersect the broadly rounded divide with a sharp, clearly defined break in slope, indicating the rapid expansion of the steep hopper slopes by rapid mass movement into a more stable profile determined by slower processes of creep and rain-wash. Elsewhere the hopper walls merge smoothly into the rounded divide profile and suggest a more nearly balanced slope condition.

The lateral slopes which extend from divide to ravine bottom are characteristically straight in profile for a considerable part of their length (fig. 12). At the upper end the profile merges smoothly into the curve of the divide; at the lower end the straight profile reaches to the ravine floor. In some ravines a composite profile was found, having a lower, or inner, segment steeper than that above it. This was interpreted as the result of a recent epicycle of accelerated erosion locally affecting some of the canyons.

Between-Area Differences in Slope.—Significance of differences in means of three areas was examined for the purpose of determining a possible influence of bedrock on slope. Areas A (Kline Canyon area) and B (Stough Park area) are both underlain by the San Gabriel formation whereas area C (San Rafael Hills) is underlain by the Wilson diorite. Should the means of the first two areas prove to be similar, but to differ significantly from the third, the difference may possibly be due to rock type. Should the results fail to show any significant differences, one might conclude that these particular varieties in bedrock have no noticeable influence. Should the first two areas, both in the same type of rock, differ significantly, but one of them to be the same as the third, we might conclude that factors other than rock type influence slope more strongly.

Histograms of three frequency distributions, one from each area are shown in figure 6. In an earlier section on general distribution characteristics of slopes an analysis of significant skewness and kurtosis was discussed (tables 1, 2). Differences

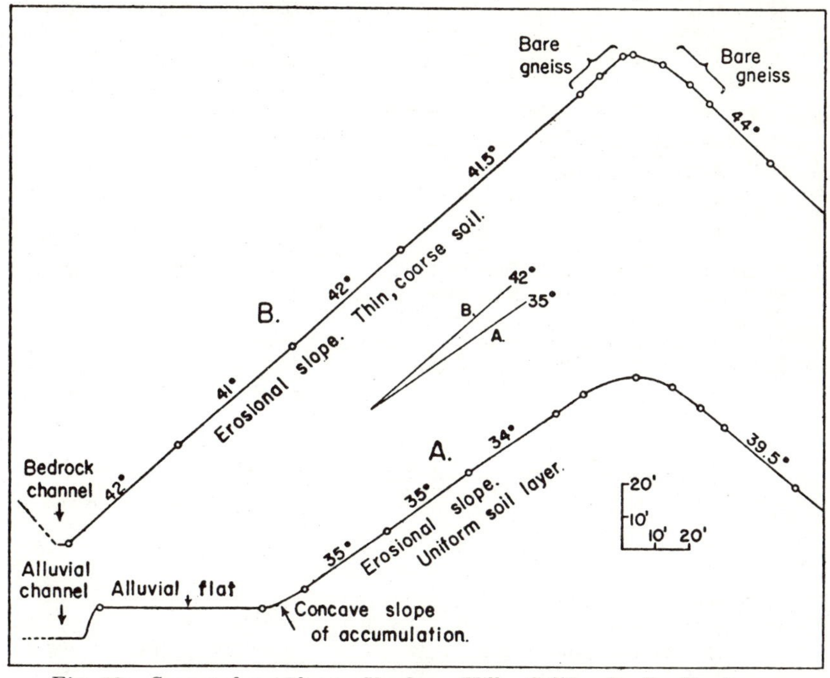

Fig. 12. Surveyed profiles in Verdugo Hills, California. Profile A, representing the type in sample A of figure 14, has escaped basal cutting for a long time. Profile B, representing the type in sample B of figure 14, is being actively corraded at the base.

in sample size are largely the consequence of the prevailing conditions of vegetation at the time of the field work. Area A had been recently burned over and presented an ideal condition for slope work. Area B had a light cover of shrubs following recent burns. Area C was densely overgrown with a chaparral of many years' development.

Tables 4, 5, and 6 give the essential data for the between-area analysis. Tests were made for significance of differences in estimated standard deviations and means.[1] Tests of sig-

TABLE 4

Data used in determinations of significance of differences in standard deviations and arithmetic means (tables 5 and 6.).

DETERMINATIONS	Area A	Area B	Area C
N Number of observations in sample	171	89	26
$\overline{X}$ Arithmetic mean of sample	44.825°	42.084°	42.173°
σ Standard deviation of sample	3.257°	2.522°	2.780°
s^2 Unbiased estimate of population variance	10.669°	6.433°	8.039°

$$s^2 = \frac{N}{N-1}\sigma^2$$

TABLE 5

Tests for significance of differences in standard deviations.

DETERMINATIONS	A and B	A and C	B and C
F (Ratio of larger s^2 to smaller s^2.)	1.658	1.327	1.250
$F = \frac{s_1^2}{s_2^2}$			
n_1 (Degrees of freedom in Sample 1.)	170	170	25
n_2 (Degrees of freedom in Sample 2.)	88	25	88
P. (Ratio of area in two tails of F-distribution to area under entire curve.)	0.02	$>.40$	$>.40$
Is there a significant difference in standard deviations?	Yes (?)	No	No

TABLE 6

Tests for significance of differences in arithmetic means.

DETERMINATIONS	A and B	A and C	B and C
t Where $t = \frac{\overline{X}_1 - \overline{X}_2}{s_{\overline{x}_1 - \overline{x}_2}}$	6.90	3.92	0.15
n Degrees of freedom	258	195	113
P Ratio of area in two tails of t distribution to area under entire curve	$<.001$	$<.001$	.80 to .90
Are the differences in means significant?	Yes	Yes	No

[1] Analysis was performed by Mr. Richard Ostheimer, Instructor in Statistics, Columbia University.

nificance of differences in estimated standard deviations were required by the differences between area A and the other two areas. As indicated in table 5 there is probably a significant difference between deviations of areas A and B, but not between A and C, or between B and C. Inasmuch as the level of significance in the first case is not minute, we cannot rule out the possibility that they are really not significantly different. Certainly any influence of difference in bedrock fails to show strongly in the estimated standard deviations.

Differences in means are treated in table 6. The results are clearcut in terms of probabilities and satisfactory significance levels. Whereas areas A and C, and areas A and B, are significantly different, the differences in B and C are definitely not significant, with a probability of 80 to 90 per cent that this difference or greater may be due to chance sampling variations alone. Because the two areas of similar rock differ significantly in slope, whereas two of the unlike areas do not differ significantly in slope, we may conclude that some factor (or factors) other than bedrock has a controlling influence on slope differences. These factors might be undetected differences in climate, vegetation, soils, geologic (tectonic) history, or sampling differences due to unintentional changes in sampling method. The conclusions are thus negative, even though clear, and serve to emphasize the complexity of the type of problem being treated.

A possible explanation for a lack of any strong bedrock influence may be the deep weathering of bedrock in this region. Decomposition due to alteration of feldspars and ferromagnesian minerals has proceeded to depths of tens of feet, so that highway grading in bedrock may be done with power shovels. Landforms are thus little influenced by differences in physical characteristics of the unaltered rock.

Influence of Differences in Exposure of Slopes.—Throughout the areas studied there is a marked difference in the appearance of slopes facing south through southeast to east and those facing north through northwest to west. This difference is obvious both in vegetation and soil. In the field one is apt to conclude that systematic differences in slope angle are also present because of the fact that one set of slopes affords better footing for climbing than the other. Whether or not a slope angle difference actually exists is here investigated by the method of differences in sample means.

All of the areas studied are clothed in a chaparral vegetation representing the natural flora of the region as it develops in various stages following fire. Two distinct types of vegetation are recognizable; that on the drier south-facing slopes, and that on the relatively protected north-facing slopes.

On the south-facing slopes is a light growth, 3 to 5 feet high, termed *chamise-sage* by J. S. Horton (1941). On the south flank of the Verdugo and San Rafael Hills this consists largely of *Adenostema fasciculatum* (Chamise). There are in addition *Eriogonum fasciculatum* (Buckwheat), *Salvia mellifera* (Black sage), *Artemesia californica*, and *Yucca whipplei*. Chamise seems to have excluded almost all other plants except the yucca on foothill slopes which have escaped burning for many years. Slopes burned over recently (3 to 5 years ?) may consist of almost pure stands of *Eriogonum fasciculatum* or *Salvia mellifera* or a mixture of the two, interspersed with some yucca. *Rhus laurina* occurs throughout the chamise-sage areas, making conspicuous dense clumps 8 to 10 feet high and most commonly attaching to rocky parts of the steeper slopes. According to measurements by J. S. Horton (1941) in the San Dimas Experimental Forest, the average percentage of ground occupied by plants is only about 24 per cent, while 76 per cent is open ground. Although no measurements were taken within the areas studied by the writer, it seems likely that there is a similar proportion of open ground. Viewed from opposite hillsides a chamise-sage slope shows a considerable amount of exposed soil, weathered rock mantle, or bedrock.

In contrast with the chamise-sage association is the *chamise-chaparral* association (J. S. Horton, 1941) of north-facing slopes. This is a denser, higher growth, some 8 to 15 or 20 feet high, composed largely of stunted trees. Horton's measurements show that only about 52 per cent of open ground area is present. Viewed from a distance little or no bare ground may be visible. A considerable amount of litter, made up of dead leaves and twigs, is present. In the area studied by the writer, the chamise-chaparral contained such species as *Prunus ilicifolia, Rhamnus crocea, Quercus dumosa* and *Rhus integrifolia*, intermixed with chamise.

Protection of ground by both of these chaparral associations is considerable, but would appear to be much greater where the chamise-chaparral association is present. On the chamise-sage slopes individual plants catch considerable quantities of twigs and leaves on the up-slope side of the plant base. This in turn holds soil and rock particles, making a miniature terrace. When the slope is swept by fire, much of the debris is released to slide and roll downslope into the ravine bottoms where it accumulates in small talus cones. The bare ground between individual plants or clumps of plants is readily attacked by raindrop beat (splash erosion) and sheet-runoff, which in places has cut small rill channels several inches into the weathered rock mantle. In dry weather these grooves serve as chutes down which loosened particles roll or slide. The activities of rabbits, coyotes, deer and other animals which live in large numbers on chaparral slopes serve to release much fragmental rock and soil material throughout the long, dry Mediterranean summer.

Nature of the soil cover is likewise contrasted on the two types of slopes. The N-NW-W chamise-chaparral slopes commonly have a thick, dark-brown soil through which no bedrock crops out. A maze of fine rootlets binds the surficial part of the soil together so that free sliding and rolling of particles is minimized. The thick soil is given readily to flowage movements; the scars typical of such movements are conspicuous on burned-over slopes. On the S-SE-E slopes with a light chamise-sage cover, soil is thin. Outcrops are numerous and make up a large portion of the surface area of slopes over 40° or 42°. In the absence of soil held by rootlets there is much free rolling and sliding of weathered rock particles. Both because of the greater proportion of outcrops and because of the greater amount of sliding of detrital fragments, these slopes give the impression of being the steeper. They are more difficult to ascend or descend on foot than the soil-covered slopes (where both are freshly burned over) because footing is poorer in the loose unstable detritus on S-SE-E slopes.

In order to determine if a true and systematic asymmetry does exist between the two sets of slopes, a test was set up using slope angles previously gathered without regard for the particular problem of asymmetry. All ravines having simple, opposed valley walls were used; eight in all. These ranged in trend from N-S to ENE-WSW and showed contrasting slope conditions as previously described. In all cases the ravine was sharply V-shaped with a narrow stream channel occupying the bottom. Thus the deepening of the ravine would cause approximately equal basal corrasion of both valley walls. Due to the complexity of bedrock, which included metasediments and small intrusive bodies, it is not certain that control by rock structure and composition differences is absent. No such relationship suggested itself, however.

Fifty-two readings were analyzed from slopes facing in a westerly or northwesterly direction; 53 from slopes facing southerly or southeasterly. The equality in number of readings was wholly fortuitous, because all available readings were used and none arbitrarily discarded. In general the distribution of readings was even throughout and opposite sides of the same ravine received almost equal sampling.

Of the first group, facing W, NW or N, the arithmetic mean was 44.54°; for the second group, facing S, SE or E, the arithmetic mean was 44.56° (fig. 13). Note that standard deviations differ markedly even though means are almost identical.

A test of significance of difference in means shows

$$\sigma_{\bar{x}_1 - \bar{x}_2} = 0.631$$

$$\frac{x}{\sigma} = 0.03168$$

The area under the two tails of the normal curve is 97.2 per cent. Hence the probability of obtaining this great a difference in means or greater due to sampling variations alone would be about 97 out of 100, if samples were repeatedly drawn from a slope population uniform throughout. It is nevertheless possible that a true difference exists due to exposure but that the two samples happened to have almost identical means. If we take the observed difference, plus or minus 2.6 times the standard error of the difference of the means ($.02 \pm 2.6 \times 0.631$) and we state that the true difference lies somewhere in this range of values, then, if we regularly proceed according to such a rule, in the long run 99 per cent of the statements will be true.

This example serves as an illustration of the value of systematic quantitative measurement and testing in investigating what seems upon casual field observation to be a difference in slope angles due to exposure to meterological elements. The eye alone cannot detect significant small differences and may, on the other hand, be deceived into registering a slope difference where none exists.

Parallel versus Declining Slope Retreat.—A fundamental problem in geomorphology is whether slopes of a fluvially dissected landmass retreat in parallel planes accompanied by replacement with a set of lower basal slopes, as Penck (1924) has held, or whether they decline to progressively lower angles as the landmass relief is lowered, as Davis (1909, p. 268-269) has held.

Under the equilibrium theory, outlined in earlier pages, slopes maintain an equilibrium angle proportional to the channel gradients of the drainage system and are so adjusted as to permit a steady state to be maintained by the process of erosion and transportation under prevailing conditions of climate, vegetation, soils, bedrock and initial relief or stage. Thus both slopes and streams are graded. In any one small area where homogeneity prevails slopes tend to cluster about a mean value with relatively low dispersion. As the landmass is reduced

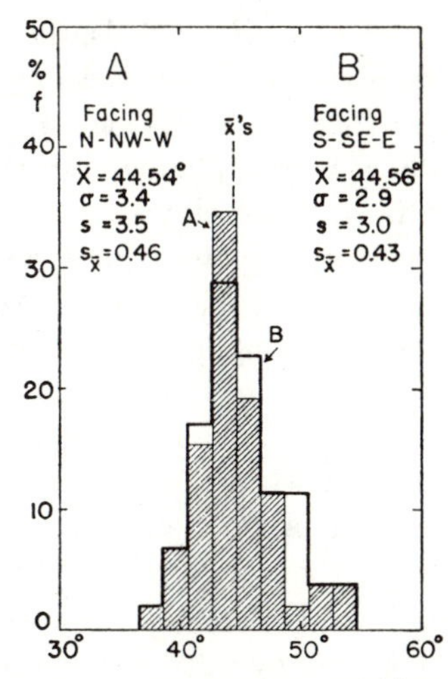

Fig. 13. Histograms of two samples superimposed for comparison. Sample A (cross-hatched) represents the more densely vegetated, sheltered northerly-facing slopes. Sample B (heavy line) represents thinly covered chamise slopes with southerly exposure. Means are almost identical, hence are indicated by one line.

both slopes and stream gradients are reduced, being slowly and continuously regraded to maintain approximate equilibrium. As the correlation of stream gradients with slopes (fig. 8) suggests, the decline of stream gradient is accompanied by slope reduction. This concept of maintenance of a steady state by slow readjustment is essentially Davis' concept of landmass development in the normal cycle.

Under the Penck (1924) theory of landmass development an initially steep slope or wall (*Böschung*) such as the valley wall of a youthful canyon, might be expected to retreat in successive parallel planes. So long as the stream is down-cutting at a constant rate, no other elements of slope would be present and uniform development would prevail. With diminishing rate of down-cutting by the stream there will begin to develop at the base of the steep slope a gentle slope (*Haldenhang*) which expands in area as the steep slope retreats from the valley axis. The *Haldenhang* is, in turn, replaced by still lower slopes (*Subhaldenhängè*). This concept fails to correlate stream gradients with slopes, as the theory of equilibrium forms in a steady state demands. Once formed, a steep slope (*Böschung*) would retreat at a constant angle regardless of changes of gradient of the associated stream.

In the Verdugo and San Rafael Hills the prevailing conditions are those of steep-walled, V-shaped canyons, or canyons with narrow alluvial valley flats. If the topography is developing by parallel slope retreat, as under the Penck theory, there might not yet be any development of *Haldenhänge* because of the rapid valley deepening and widening. It would thus be difficult to draw conclusions as to the correctness of these theories on the basis of general topographic aspect. There is, however, a possible means of determining the validity of Penck's assumption that a steep slope retreats in parallel planes once it has been formed and is no longer subject to direct basal stream action. At various places along the narrow alluvial flats in some of the canyons the stream is vigorously undercutting the bedrock at the slope base on one side, but has not recently been active on the opposite side, where there is an accumulation of soil and rock fragments in the form of talus or basal sheet-wash apron. The slope above this zone of accumulation has been free to waste back by processes of mass movement and sheet erosion, but without any removal or undercutting at the slope base. If retreat has been parallel, slopes of this type should maintain a mean angle the same as for the slopes vigorously corraded at the base; if retreat has been of the declining type, these slopes should be significantly lower than the latter group.

During the field work measurements were made of erosional slopes which had basal accumulations of slopewash or talus and had, therefore, for a greater or lesser period of time, no vigorous corrasion at the base. In all 33 measurements of this class of slopes in the Kline Canyon area were made. These slopes were located among the slopes being actively corraded, of which the sample of 171 readings has already been described (sample E in figures 2, 3 and 4). The larger sample includes none of the smaller, but the two are intermixed as far as geographic distribution is concerned. Figure 14 shows the histograms of the two frequency distributions plotted on the same coordinates. The mean of the protected slopes is 38.2°; that of the basally corraded slopes 44.8°, a difference of 6.6°. Estimated standard deviation of the first group (2.70°) is somewhat lower than the second (3.27°) but both are of the same general order of magnitude. It is obvious, in view of the significance levels obtained in previously described tests, that a difference in means of 6.6° is significant beyond doubt. A test shows that

$$\frac{x}{\sigma} = 11.83$$

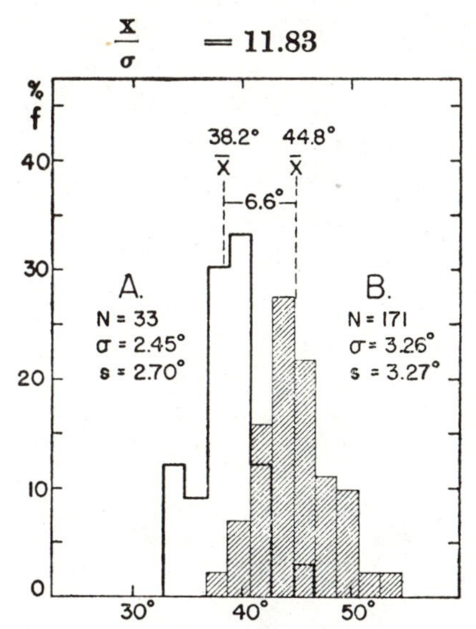

Fig. 14. Histograms of two samples superimposed for comparison. Sample A, protected slopes. Sample B, basally corraded slopes.

Tables of areas under parts of the normal curve fail to include this high a value. If the value were 5.0 the probabilities of obtaining this great difference or greater by chance variations in sampling from the same population would be on the order of one in one million.

We thus conclude that the slopes which have been protected from recent basal cutting have significantly lower angles. The statistical analysis does not explain this difference, but in view of the fact that conditions of climate, vegetation, soils, bedrock, and tectonic history seem to be essentially the same for both samples it is concluded that the one group of slopes declined in angle during the period when only sheet-runoff, creep and other mass-wasting processes operated on the slope. Some suspicion, at least, is thus cast upon the Penck assumption of parallel retreat, while some support is given Davis' scheme of declining slopes.

Summary Statement.—The method of evaluating the significance of differences in sample means appears to have applications in the solution of fundamental problems of geomorphology. The method serves as a check upon the reliability of conclusions drawn from samples. In the three examples described above, the second and third illustrate opposite extremes of significance level which give considerable confidence to the investigator in drawing conclusions. The first case illustrates an inconsistent relationship among three sample means relative to a possible causative factor and points to control by other factors not under consideration.

Although the specific results of this investigation may prove of little consequence in the light of future studies, some value may come through increasing the range of techniques available in the attack upon basic geomorphic problems.

References

Bailey, R. W., and Craddock, Geo. W., 1948. Watershed management and sediment control: Federal Inter-Agency Sedimentation Conference Proc., Bureau of Reclamation, U. S. Dept. Interior, Washington, pp. 302-310.

Bakker, J. P., and Le Heux, J. W. N., 1946. Projective-geometric treatment of O. Lehmann's theory of the transformation of steep mountain slopes: K. Akad. Wetensch. Amsterdam Proc., vol. XLIX, no. 5, pp. 533-547.

Bryan, Kirk, 1922. Erosion and sedimentation in the Papago Country, Arizona: U. S. Geol. Survey Bull. 730, pp. 19-90.

———, 1940. The retreat of slopes: Assoc. Am. Geographers Annals, vol. 30, pp. 254-268.

Croxton, F. E., and Cowden, D. J., 1946. Applied general statistics, Prentice-Hall, Inc., New York.

Davis, W. M., 1909. Geographical essays, Ginn & Co., Boston.

Freise, F. W., 1938. Inselberge und Inselberglandschaften im granit- und gneisgebiete Brasiliens: Zeitschr. Geomorphologie, Band 10, pp. 137-168.

Horton, J. S., 1941. The sample plot as a method of quantitative analysis of chaparral vegetation in southern California: Ecology, vol. 22, pp. 457-468.

Horton, R. E., 1945. Erosional development of streams and their drainage basins; hydrophysical approach to quantitative morphology: Geol. Soc. America Bull., vol. 56, pp. 275-370.

Lawson, A. C., 1915. The epigene profiles of the desert: California Univ., Dept. Geol. Sci., Bull., vol. 9, pp. 23-48.

———, 1933. Rain-wash erosion in humid regions: Geol. Soc. America Bull., vol. 43, pp. 703-724.

Meyerhoff, H. A., 1940. Migration of erosional surfaces: Assoc. Am. Geographers Annals, vol. 30, pp. 247-254.

Miller, W. J., 1934. Geology of the western San Gabriel mountains of California. Pub. U. C. L. A. in Math. and Physical Sci., vol. 1, University of California Press, Berkeley.

Penck, Walther, 1924. Die morphologische analyse, J. Engelhorns Nachf., Stuttgart.

Putnam, T. M., 1917. Mathematical forms of certain eroded mountain sides: Am. Math. Monthly, vol. 24, pp. 451-453.

Van Burkalow, A., 1945. Angle of repose and angle of sliding friction: an experimental study: Geol. Soc. America Bull., vol. 56, pp. 669-707.

Von Bertalanffy, Ludwig, 1950. The theory of open systems in physics and biology: Science, vol. 111. pp. 23-28.

DEPARTMENT OF GEOLOGY
COLUMBIA UNIVERSITY
NEW YORK, N. Y.

Reprinted from *Zeitsch. Geomorph.*, **5**, 126–131 (1961)

13

Characteristic and limiting slope angles

By

ANTHONY YOUNG, Zomba

With 1 figure

In studies of slope development it has frequently been suggested that certain angles are of particular significance; various terms have been used to refer to these, but two main concepts are found. These will be defined, and field evidence relating to them presented.

DE LA NOE & DE MARGERIE (1888, pp. 44—46) discussed in general terms the angles of slope characteristic of clays, marls, limestones, sandstones, and granite. SIMPSON (1953, p. 17), describing the Lyn drainage system of Exmoor, south-west England, stated that a gradient of 1 in $1^1/_2$ ($33^1/_2^\circ$) "appears to be the characteristic hanging slope developed in this area when a hillside is undercut". SAVIGEAR (1956) has drawn attention to the concept, and tabulated characteristic angles for seaward slopes in south-west England, obtained from analysis of surveyed profiles; these are the angles which most commonly occur on these slopes, intermediate angles being absent or less developed. The term will be defined in this sense: *characteristic angles of slope are those which most frequently occur, either on all slopes, under particular conditions of rock type or climate, or in a local region.*

Distinct from this is the concept of limiting angles. GÖTZINGER (1907) distinguished between erosion slopes and denudation slopes *(Erosionsböschungen, Abtragungsböschungen)*. The former, of which an angle of 45° is given as an example, are produced by river undercutting; when undercutting becomes less rapid these develop into scree slopes, which are colonized by vegetation at angles of 30°—35°; this transforms them into denudation slopes, the subsequent form of which is due to denudational processes. This classification implies a change in both the nature of the ground surface and the processes which act upon it at a certain angle. A similar concept to this had previously been given by MAW (1866). BRYAN (1925), writing of the arid regions of Arizona, distinguished "cliffy slopes" at 90°—45°, "debris-mantled or boulder-controlled slopes" at 45°—20°, chiefly 30°—35°, and "rain-washed slopes" at 20°—15°. The minimum angles upon which certain forms of mass-movement take place were discussed by PENCK (1924, pp. 98, 103, 105, 110). CHALLINOR (1931) used the term "limiting angle" to refer to the maximum angle which a rock cliff will maintain when undercut at the base; this angle is dependent upon the geology. In civil engineering the term "stable slope" is used to describe a slope not subject to rapid mass-movements; in clay strata this depends both upon the angle and the height of the slope (SKEMPTON [1953]). The angle of repose for loose material has frequently been discussed in relation to scree slopes (e. g. VAN BURKALOW [1945]). SAVIGEAR (1952), describing slopes in South Wales, gave the maximum angle on which a continuous debris cover is retained, in sandstones and marls of the Old Red Sandstone, as 32°. In south-west England SAVIGEAR (1956) found that slopes of 40° and above are formed by bare bedrock, whilst those below 32° (37° on gritstone) possessed a regolith cover. TRICART (1957) has suggested the existence of "*seuils de fonctionnement*", threshold angles below which certain denudational processes do not operate.

This concept has been used in describing both the nature of the slope surface (e. g. bare, scree-covered, regolith-covered), and the denudational processes to which it is subjected (e. g. rapid mass-movements). It is clear in practice which is being referred to, therefore a definition which includes both of these, and which is in accordance with the normal meaning of the words, will be adopted: *limiting angles of slope are those that define the range within which particular types of ground surface occur, or particular denudational processes operate.* They may be applied to all slopes, or restricted to a rock type, climate, or local region. It is normally clear without the need for explicit statement whether the upper or lower limits are being referred to.

Defined in this way, characteristic and limiting angles are separate concepts, although it is probable that causal relations will be found to exist between them.

Field evidence

In considerations of the course of slope evolution, a fundamental question is whether particular characteristic angles are of intrinsic importance; that is, whether the nature of slope retreat is such that certain angles are formed more frequently, or remain unaltered for longer periods of time, than others. Data relevant to this has been drawn from 30 profiles of valley sides, from 3 areas of Britain. These are a small river basin, the Heddon, in North Devon; an area in Central Wales lying between the rivers Rheidol and Ystwyth; and the basin of the Upper Derwent in the Southern Pennines (fig. 1a). Previous geomorphological work in these areas has been summarized by BALCHIN (1952), CHORLEY (1958), BROWN (1950), and SISSONS (1954). The profiles consist of measurements of angle and distance along a line from interfluve crest to valley floor in the direction of maximum slope; they were surveyed with Abney level and tape, reading angles to the nearest half-degree (SAVIGEAR [1952]). In order to investigate characteristic angles, the measured lengths along the ground surface at each half-degree were summed. Random fluctuations were smoothed by computing the running means of three half-degrees; for example the value taken for $3^1/_2$° is the mean of all measured lengths at 3°, $3^1/_2$°, and 4° on the profiles. These values are shown in figure 1b—f.

The Heddon basin is underlain by two rock types, shales of the Ilfracombe Beds and sandstones of the Hangman Grits, both of Devonian age and dipping mainly at 40°—45°; the area is unglaciated. Figure 1b shows data for slopes on the shales. Angles of up to 10° are the most common; there is then a rapid fall

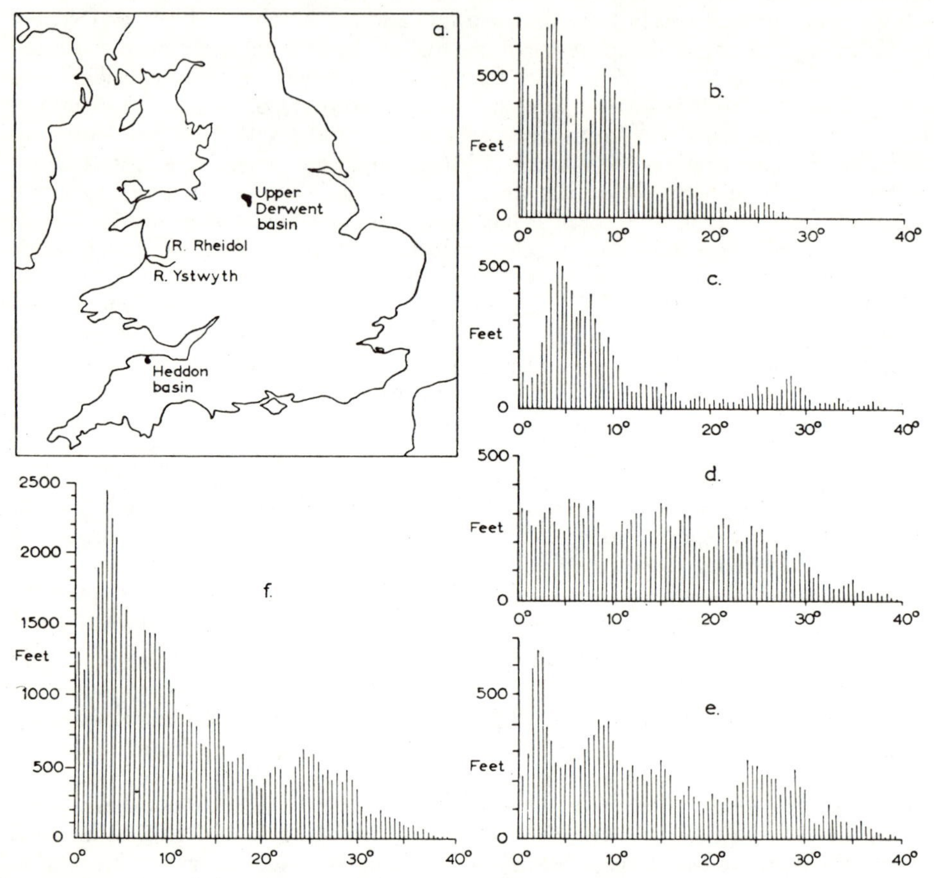

Fig. 1. a. Location of the areas in which slope profiles were surveyed. b—f. Total distances at each angle on the profiles; the distances are the running means of three half-degree totals. b. The Heddon basin, profiles on shales. c. The Heddon basin, profiles on sandstones. d. Central Wales, between the rivers Rheidol and Ystwyth. e. The Upper Derwent basin. f. All areas; note the difference in scale

in frequency up to 14°, followed by a slower fall to 28°. Peaks in frequency occur at 3°—4° and 9°—10°, with smaller peaks at 16° and 23°—25°. On the sandstones (fig. 1c) there is a rise in frequency up to 4°, a fall to 11°, followed by short distances at all angles up to 34° and very short distances to 38°. Peaks occur at 4°—5° and 28°—29°, with a slight peak at 13°—15°. Other geomorphological observations indicate that the valleys of the area have undergone three epicycles of erosion; the first of these reached an advanced stage, but the two later rejuvenations occurred only relatively recently. The overall angle distributions on both rock formations reflect this morphological history, the slopes at less than 10° corresponding to the extensive interfluve areas formed in the first long epicycle. The peaks in frequency on the two formations may be correlated as follows:

Shales (Ilfracombe Beds):	3°—4°	9°—10°	23°—25°
Sandstones (Hangman Grits):	4°—5°	13°—15°	28°—29°

This correlation includes all of the four main peaks, and two of the three minor ones. In each pair the angles are higher on the sandstones, which are more resistant to weathering. This suggests that these three pairs of angles correspond to the three epicycles of erosion in the area, each being represented by a slightly higher angle on the more resistant rock.

Figure 1d shows the distribution for the area in Central Wales. This is underlain by intensely folded shales of Silurian age, and has been glaciated. The overall distribution shows that all angles up to 25° are common, followed by a decline in frequency to 32°, and short distances at up to 39°. These angles above 32° do not occur on the shales of the Ilfracombe Beds, which are lithologically very similar; they are probably due to excessive slope steepening caused by glaciation. Peaks in frequency are not well marked, occurring at 6°—8°, 15°, 21°—22°, and 25°. The third area, in the Southern Pennines, is formed on almost horizontal sandstones and shales of the Millstone Grit Series, of Carboniferous age. Figure 1e shows that all angles up to 30° are common, above which there are short distances at up to 40°. Peaks are well defined, at 2°, 8°—10°, and 24°—25°. The first of these represents erosion surface remnants, practically unmodified by later youthful dissection. The second and third again appear to correspond with successive rejuvenations that have affected the area.

The above peaks in frequency, for each of the three areas, include every angle from 2° to 29°, with the exceptions of 11°—12°, 17°—20°, and 26°—27°; this wide distribution suggests that each peak is related to the local morphological history. The exception to this is that in each area the most recent epicycle of erosion is represented by a maximum at 25°, together with at 28°—29° on the sandstones of the Heddon basin; this may be a feature intrinsic to slope development, under the conditions of rock and climate found in the three areas. These conclusions are confirmed by the distribution for the three areas combined, shown in figure 1f. From a maximum at 2°—4° this declines, at a decreasing rate, to 40°; the main interruption to an otherwise regular decline is that produced by a peak at 24°—26°, with adjacent moderately high values at 26°—29°. This may also be related to the characteristic angle of 26°—29° distinguished by SAVIGEAR (1956) for seaward slopes in south-west England.

Observations of limiting angles for different types of ground surface were carried out on steep slopes subject to rapid basal erosion in the Heddon and Upper Derwent basins. The results may be summarized as follows:

Discontinuous soil cover, with projecting rock outcrops:	41°—49°
Continuous soil, with bare gashes in the vegetation cover:	36°—40°
Continuous soil and vegetation cover, with terracettes:	33°—36°
Smooth soil and vegetation cover, no surface irregularities:	below 33°

This may be compared with the results of Savigear (1956), who found bare rock on more than 40°, and irregular surface detail above 32°; and MacGregor (1957), who stated that at 40° terracettes and crescentic gashes are well marked, whilst at 25° they are absent. There is substantial agreement between these results, indicating that the limiting angle for a continuous soil cover is approximately 40°, and for a smooth soil and vegetation cover, without scars or terracettes, slightly below 32°.

Conclusions

If it is assumed that the presence of ground surface irregularities indicates relatively rapid denudation and slope retreat, then the sharp fall in frequency at 30° shown by the angle distribution for all profiles (fig. 1f) may be explained. Slopes of 30°—40° are produced as a result of rapid basal erosion, by a river or other agency, undercutting a slope; but they are relatively short-lived forms, and soon after the basal erosion slackens or ceases they are transformed into slopes of less than 30°. This can be related to the group of characteristic angles at 25°—29°. Such slopes are not necessarily the first to be produced following rejuvenation in an area; but they are the first to be developed which are not relatively rapidly reduced by denudation to a gentler angle. The extent to which this change takes place by slope decline or parallel retreat is not indicated by the data presented here.

The conclusions from this evidence relate to valley slopes formed on Palaeozoic sedimentary rocks under a humid temperate climate. They are firstly, that the majority of the characteristic slope angles of an area are related to local morphological history, and are not intrinsic features of slope development. Secondly, that an exception to this is a slope which forms at between 25° and 29°; this is the first to be developed, following a period of slope steepening by basal erosion, which retains its angle for a relatively long period of time.

Acknowledgements

A modified form of this paper was read at the 126th meeting of the American Association for the Advancement of Science, Chicago, December 1959. The field work on which it is based was assisted by a grant from the University of Sheffield Research Fund.

References

Balchin, W. G. V., 1952: The erosion surfaces of Exmoor and adjacent areas. — Geogr. J., **118**, p. 453—477.

Brown, E. H., 1950: Erosion surfaces in North Cardiganshire. — Trans. Inst. Brit. Geogr., **16**, p. 49—66.

Bryan, K., 1925: The Papago Country, Arizona. — U.S. geol. Surv. Wat. Supply Pap. **499**, 213 pp.

Challinor, J., 1931: Some coastal features of North Cardiganshire. — Geol. Mag., **68**, p. 111—121.

Chorley, R. J., 1958: Aspects of the morphometry of a "poly-cyclic" drainage basin. — Geogr. J., **124**, p. 370—374.

De la Noë, G., & de Margerie, E., 1888: Les formes du terrain. — Paris, Surv. géogr. de l'Armée, 205 pp.

GÖTZINGER, G., 1907: Beiträge zur Entstehung der Bergrückenformen. — Geogr. Abh., **9**, H. 1, 174 p.
MACGREGOR, D. R., 1957: Some observations on the geographical significance of slopes. — Geography, **42**, p. 167—173.
MAW, G., 1866: Notes on the comparative structure of surfaces produced by subaërial and marine denudation. — Geol. Mag., **3**, p. 439—451.
PENCK, W., 1924: Die morphologische Analyse; translated by H. Czech and K. C. Boswell, 1953, Morphological analysis of land forms. — London, Macmillan, 429 pp. (The page references given in this paper are to the translation.)
SAVIGEAR, R. A. G., 1952: Some observations on slope development in South Wales. — Trans. Inst. Brit. Geogr. **18**, p. 31—51.
—, 1956: Technique and terminology in the investigation of slope forms. — Union géogr. Int., lr. rap. Com. ét. versants, Amsterdam, p. 66—75.
SIMPSON, S., 1953: The development of the Lyn drainage and its relation to the origin of the coast between Combe Martin and Porlock. — Proc. Geol. Ass., Lond., **64**, p. 14—23.
SISSONS, J. B., 1954: The erosion surfaces and drainage system of South-West Yorkshire. — Proc. Yorks geol. Soc., **29**, p. 305—342.
SKEMPTON, A. W., 1953: Soil mechanics in relation to geology. — Proc. Yorks geol. Soc., **29**, p. 33—62.
TRICART, J., 1957: Mise au point: l'évolution des versants. — Inform. géogr., **21**, p. 108—116.
VAN BURKALOW, A., 1945: Angle of repose and angle of sliding friction: an explanatory study. — Bull. geol. Soc. Amer., **56**, p. 669—707.

Reprinted from *Inst. British Geog. Special Pub.*, **3**, 31–47 (1971)

An application of the concept of threshold slopes to the Laramie Mountains, Wyoming

14

M. A. CARSON
(*Assistant Professor of Geography, McGill University, Montreal*)

MS received 5 December 1969

ABSTRACT. This paper examines certain aspects of the development of hillside slopes in the Laramie Mountains, Wyoming. Survey work showed that many slopes in the area contain straight sections; the frequency distribution of the angles of these straight slopes shows a pronounced modal group at 25°–28°, and upper and lower limits at 33° and 18° respectively. Conventional slope stability analysis suggests that these three angles correspond to angles of limiting stability (threshold slopes) at different stages of weathering of the debris mantle. These results reinforce a model of hillslope evolution developed earlier in a humid temperate area, and offer a general explanation for the similarity of characteristic angles of straight slopes on debris derived from the weathering of strong, well-jointed rocks in different climates.

SLOPE studies by the writer over the last 5 years in several areas of England[1] led to the conclusion that most straight hillsides in a humid temperate area stand at a critical angle of stability determined by the nature of the debris on the slope. In particular, it has been indicated[2] that angles of slope at about 26° are extremely common in areas where the underlying rock mass weathers into debris comprising a mixture of rock rubble and coarse soil particles. This has tentatively been explained in terms of the angle of stability of this type of debris when subjected to a particular pattern of pore-pressure distribution within the mantle.

It is expected that the model of hillslope evolution developed from this research may not be applicable to more arid areas for many reasons. Soil erosion, for instance, may proceed so rapidly that soil particles are washed away as soon as they are produced in such areas, and there may be no chance for a mixed mantle of rock rubble and soil-size material to develop. In addition, the concept of a fully saturated debris mantle subjected to positive pore pressures during prolonged rainstorms seems rather improbable in a drier environment. Nevertheless, reports by B. P. Ruxton, M. A. Melton and G. Robinson,[3] among many, have indicated that, even in semi-arid areas, straight hillsides cut into rocks which weather into a taluvial[4] mantle, commonly occur at angles of about 26°. This apparent similarity in the steepness of slopes in areas of strong well-jointed rock prompted the writer to examine the relevance of the model in a drier area. The Laramie Mountains were chosen since their debris-mantled slopes in a semi-arid environment were similar in many ways to those previously examined by the writer in more humid areas, and since the character of the bedrock and associated debris enabled comparisons to be made with the findings of Melton,[5] who undertook the first quantitative studies of mantle stability on slopes in arid areas.

THE FIELD AREA

The area comprises the Pre-Cambrian core of the Laramie Mountains, Wyoming, extending about 200 km north of the interstate highway linking Laramie and Cheyenne, as shown in Figure 1. The geology of the area is fairly well known, although the lithology of the Pre-Cambrian core is extremely complex and mostly undivided. On the western side of the mountain chain lies the Laramie Basin where Quaternary alluvial deposits mask Tertiary sediments, and to the east are the Tertiary sediments of the Denver Basin. Slope profiles were surveyed within valleys and on hills inside the area of the old mountain core and, in addition, in small valleys cut into the Tertiary sediments flanking the eastern side of the mountain chain. On Figure 1, a distinction is made between these two sets of slopes.

The slopes within the Pre-Cambrian core are rock slopes mantled by a thin veneer of debris derived from the weathering of the underlying rock mass. Most of the bedrock is plutonic in origin and the most extensive individual rock type is granite. There is, however, considerable lithological variety within the crystalline core and, in many places, metamorphism has produced granitic-gneiss from the parent granite. Much of the area east of the mountains and immediately flanking them comprises weakly consolidated and cemented material closely resembling terrace deposits. This material is intricately dissected by streams draining to the North Platte and Laramie rivers. The second set of slopes corresponds to profiles in the valleys of these streams. This material has been interpreted by M. N. Denson[6] as the coarse mountainward facies of the Upper Miocene and Pliocene Ogallala Formation. These deposits occur extensively in the area west of Wheatland and the small area west of Esterbrook; these areas are the only parts of the Ogallala that were sampled.

Although the landscapes of the Pre-Cambrian and Ogallala areas are very different in scale, appearance and slope geometry (Fig. 2), straight slope sections in both areas are similar in terms of steepness and mantle character. This is perhaps not unexpected since the material of the Ogallala Formation is essentially the product of weathering in the crystalline core, transported by streams to the edge of the mountain chain. In the context of this particular study, therefore, the differences in landscape of the mountain chain and foothill areas are probably irrelevant.

It is inappropriate here to consider the regional geomorphology of the Laramie area. The geomorphological history of the area has been studied by many workers and an excellent summary is given by D. V. Harris.[7] It is generally believed that the Laramie Range escaped major modification by glaciers during the Pleistocene. The present climate may be described as semi-arid. Wheatland, at an elevation of 1469 m, records an average precipitation total of 317 mm, over three-quarters of which occurs in the summer months between April and September. The frost-free period at Wheatland averages about 130 days and the annual average range in temperature is from $-27°C$ to $38°C$.

SLOPE PROFILE SURVEY

Because of the scant map coverage of the area, it was impossible to adopt an entirely random sampling design for the selection of slopes to survey. Moreover, previous work has indicated that a sampling design based on maps is usually impractical on many grounds and an element of subjectivity always remains at the small scale. The whole area accessible to a four-wheel drive vehicle was inspected during the summer months of 1968 and slopes

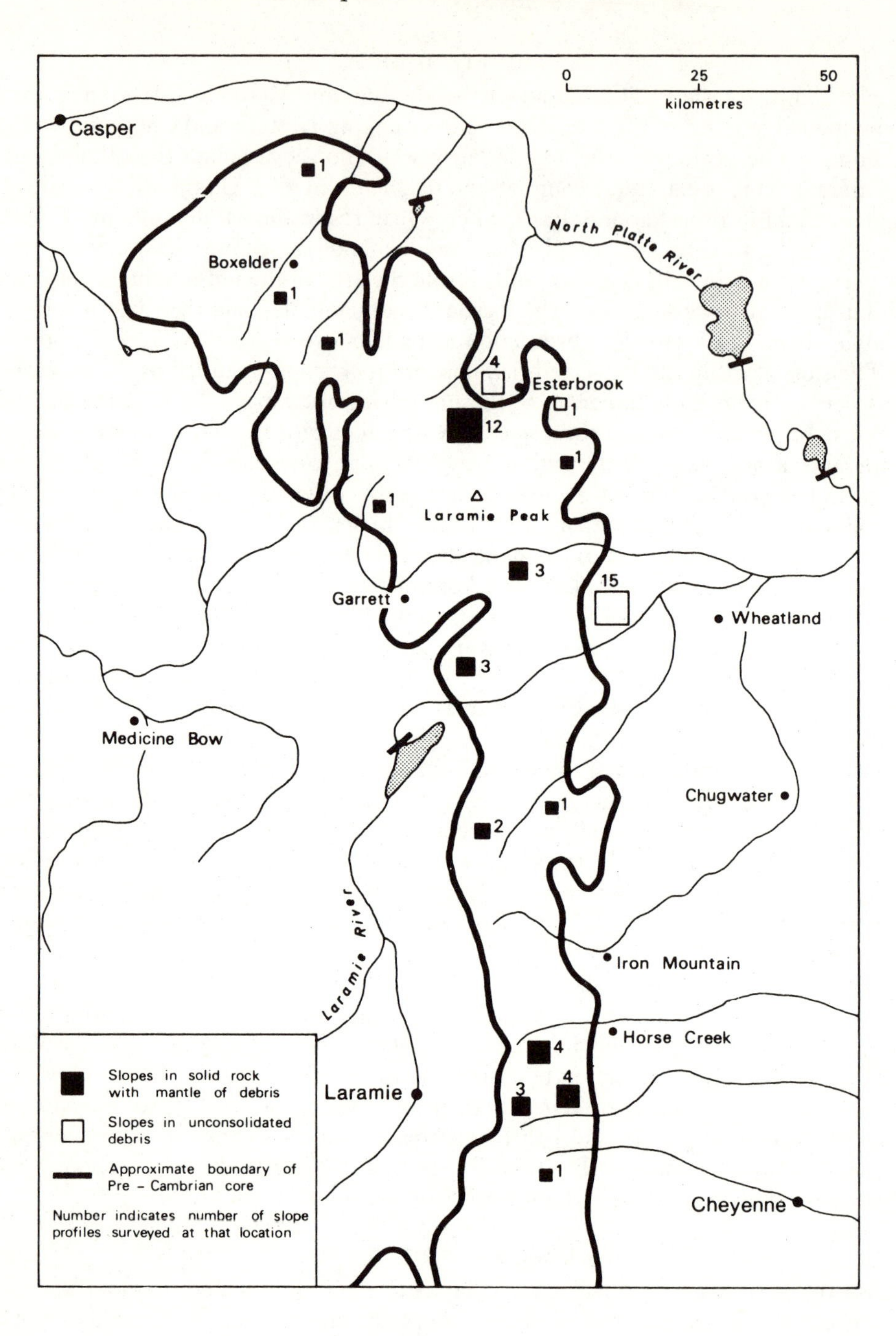

FIGURE 1. The field area

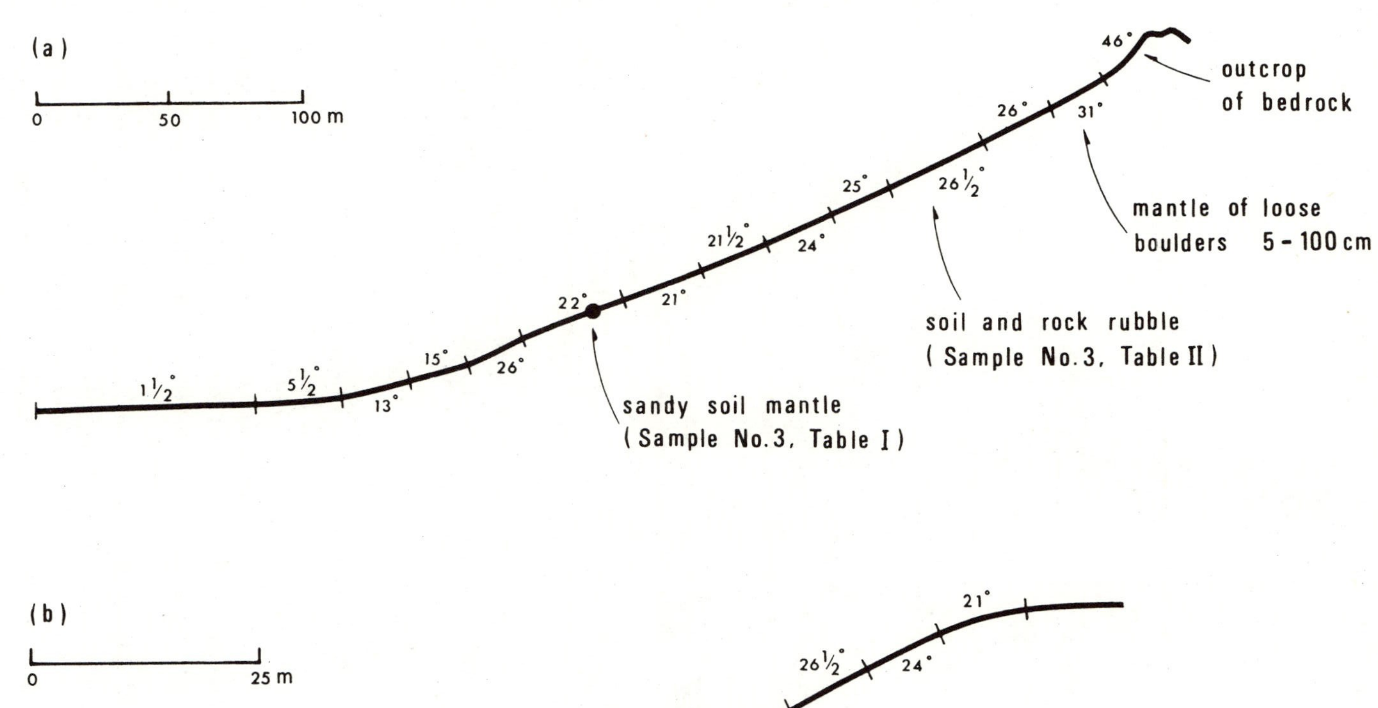

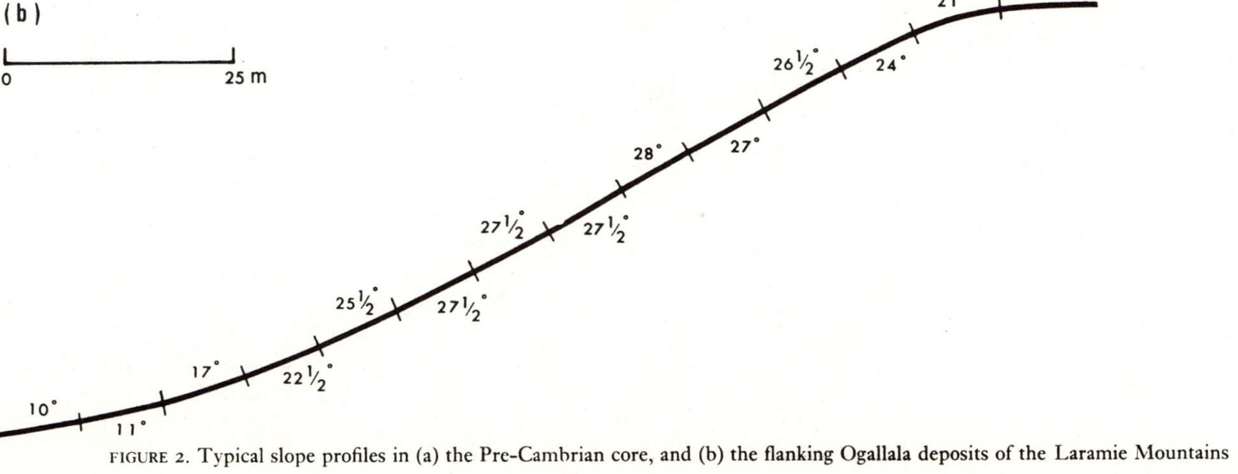

FIGURE 2. Typical slope profiles in (a) the Pre-Cambrian core, and (b) the flanking Ogallala deposits of the Laramie Mountains

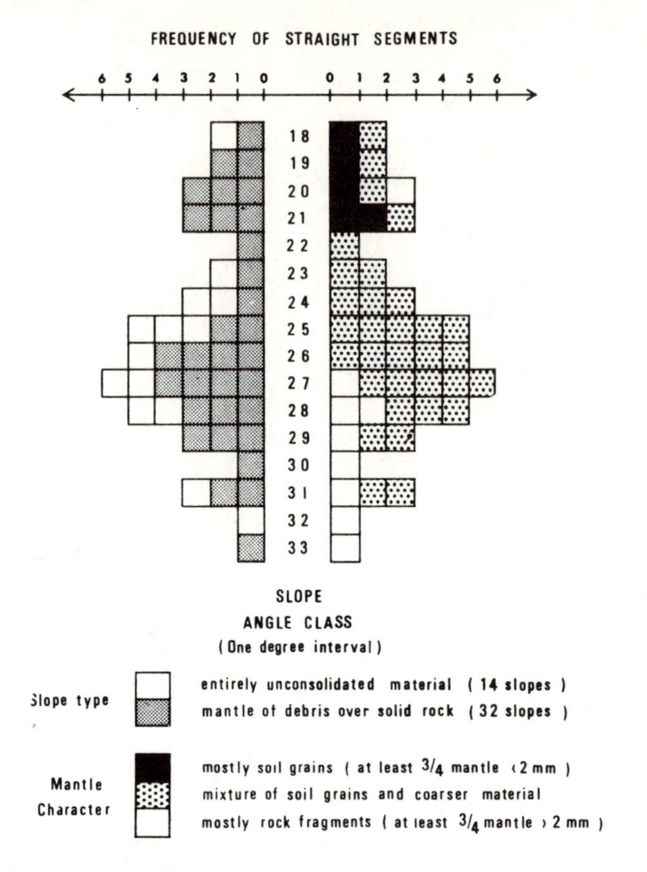

FIGURE 3. The frequency distribution of angles of straight slopes in the Laramie Mountains

were selected wherever straight sections appeared to be well developed. Some slopes selected contained, in fact, more than one straight section, and in other instances actual survey showed that the slope was entirely convexo-concave. Measurements were made with an Abney Level traversing the line of maximum slope from the base to the top of the slope, and readings usually were taken over lengths of 10 m.[8]

In all, fifty-eight slope profiles were surveyed, and inspection of these profiles indicated the existence of forty-six straight[9] segments. The location of the surveyed slopes is shown in Figure 1. The frequency distribution of straight segments, by slope angle, is depicted in Figure 3, where the slope type and mantle character are indicated for slopes in each class of slope angle. The total length of all surveyed sections in each slope angle class is given in Figure 4, together with the frequency distribution of maximum slope angles. The term 'maximum slope angle' denotes the angle of the steepest surveyed section on each profile. The majority of the slope profiles were surveyed within the main mountain chain rather than in the area of the flanking Ogallala deposits.

The main features of interest in the left-hand histogram of Figure 3 are the major grouping at 25°–28°, and the upper and lower limits on the distribution of angles at 33° and 18° respectively. The interesting aspect of the right-hand histogram of Figure 3 is the

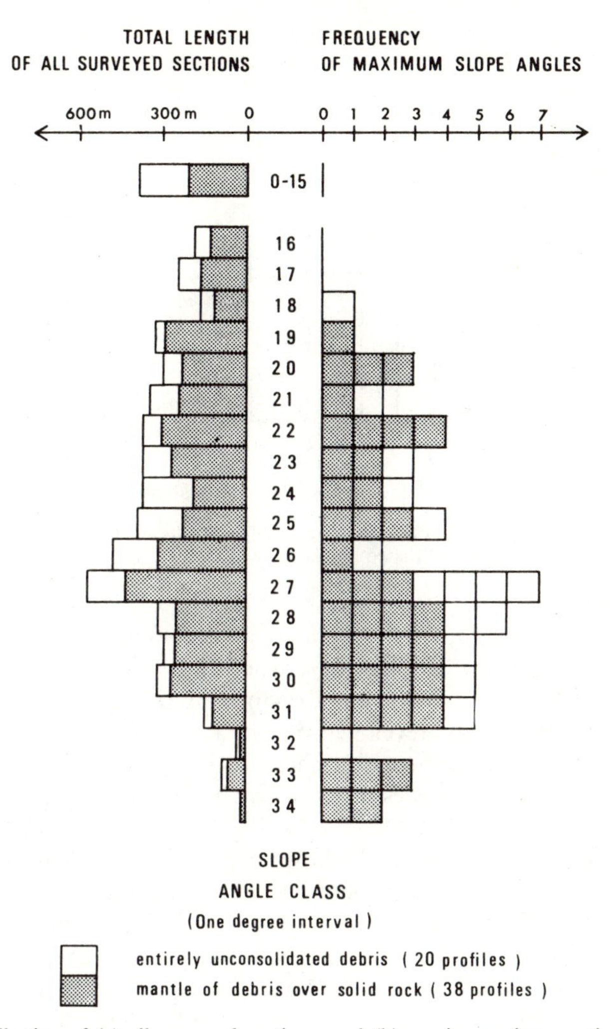

FIGURE 4. The distribution of (a) all surveyed sections, and (b) maximum slope sections, by slope angle

definite change in mantle character with slope angle: most of the high-angle slopes are rock rubble slopes and most of the low-angle slopes possess a more weathered mantle of genuine soil. The major slope group at 25°–28° contains straight slopes which are mostly mantled by a mixture of rock rubble and soil debris in an intermediate stage of breakdown.

It is difficult to assess the statistical reliability of this particular sample of the hillslopes in the Laramie Mountains; indeed, the impression of trimodality in the frequency distribution is statistically insignificant on the basis of 46 slopes. This paper is, however, not so much concerned with the total character of the frequency distribution as with the modal grouping at 25°–28° and the limits at 33° and 18°. The field data[10] indicate that, in these respects, the sample distribution may be regarded as representative of the sampled area.

The distribution of straight-slope angles in Figure 3 closely resembles the pattern

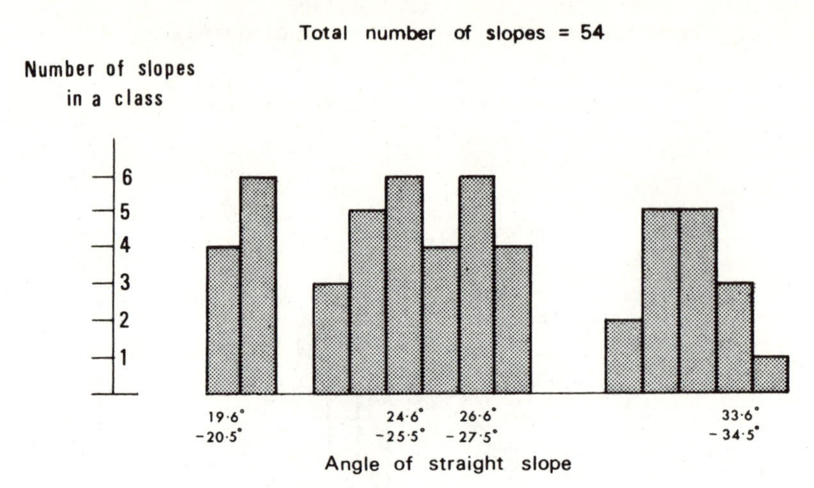

FIGURE 5. The frequency distribution of angles of straight slopes in parts of Exmoor and the southern Pennines, England (based on Carson and Petley, 1970)

observed by M. A. Carson and D. J. Petley[11] in the southern Pennines and northern Exmoor. These data are shown in Figure 5 where a predominant grouping at 24°–26° occurs; the upper and lower limits on the distribution are 34° and 19° respectively. The data of Figure 5 relate to slopes developed on gritstones and shales, rather than granite, but much of the weathering sequence of these rocks is similar. The early products of weathering are joint-controlled pieces of rock which eventually break down into soil-size debris.

These observations also agree with those of Melton[12] for granitic rock in Arizona. In a sample of twenty-seven measurements, more than 80 per cent of the slope angles occurred between 19° and 35°, with a mean value of 26·6°. The minimum angle measured by Melton was 12° and the largest was 37°, so that the limits of his distribution are not identical to those of the Laramie data. The observations in Figure 3, however, relate only to the angle of *straight* slopes, whereas Melton's data refer to the maximum slope angle on any type of profile.

It is unfortunate that there are only a few studies on the scale of Melton's work in granitic country available for comparison. Many early observations[13] suggested that debris-mantled granite slopes stand at angles near to 30°, but there is little quantitative measurement to support this. Ruxton,[14] in the Sudan, described slope profiles (Fig. 6a) on granite which closely resemble the profile in Figure 2a. Similar profiles, on dolerite-sandstone rather than granite, have been described by Robinson[15] (Fig. 6b) in the Great Fish river basin, South Africa. The existence of a middle slope at 24°–27° in these two cases is particularly noticeable.

The results of this work therefore seem to agree with the observations of previous workers in semi-arid areas. The persistent recurrence of slope angles near to 26° on rocks which weather into a taluvial mantle is especially interesting. No satisfactory explanation of this fact has yet been offered, although A. Young[16] hinted that it might be connected with the stability of the debris mantle. The next section of this paper is therefore con-

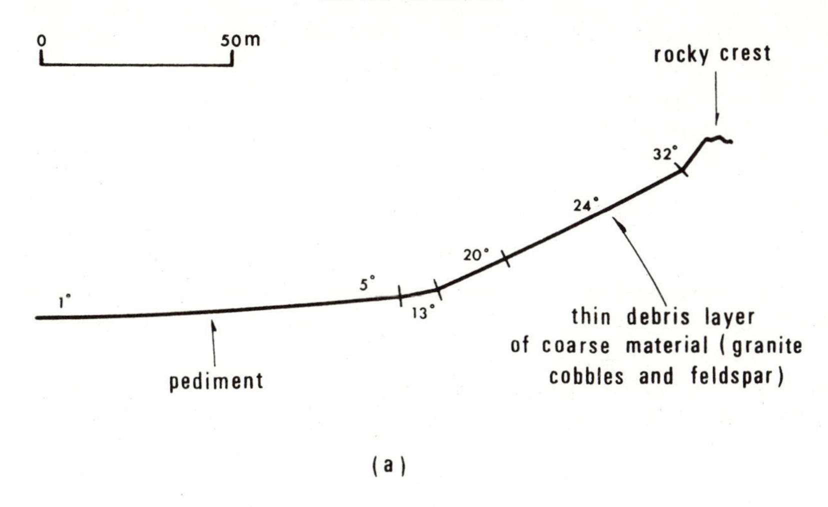

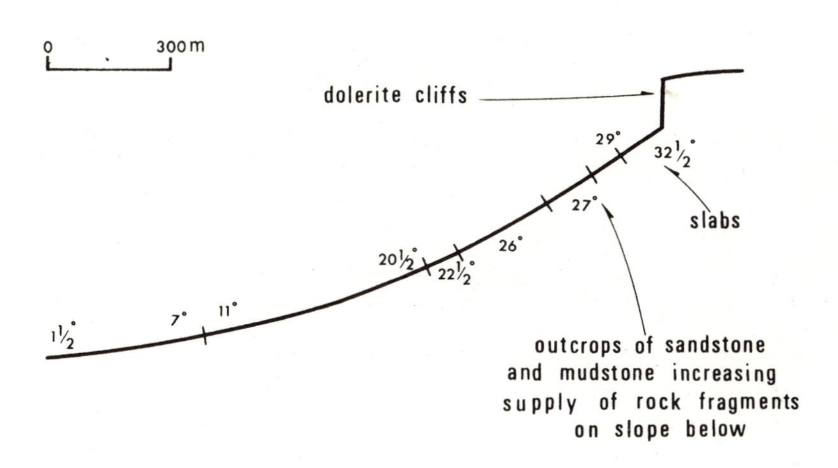

FIGURE 6. Typical hillslope profiles in (a) granitic rock at Jebel Balos, Sudan (based on Ruxton, 1958), and (b) the Great Fish river basin, South Africa (based on Robinson, 1966)

cerned with an examination of the nature of the debris mantle on these slopes from the viewpoint of slope stability.

ANALYSES OF DEBRIS MANTLES

The steepest slopes in the area, at angles close to 33°, are mantled by rocky talus, with fragments ranging in size from 2 cm to 200 cm. Soil may exist in pockets among the boulders, but never comprises more than 25 per cent of the material. These talus slopes are most common in the main mountain chain and, usually, correspond to scree slopes beneath cliffs in the granite at the top of the slope. Soils engineers and geologists have long known that the angle of repose of this type of material is about 35°, so that it may be inferred that the upper limit to the distribution of straight-slope angles is determined basically by the angle of repose of the first products of weathering of the granite. The modal group at 25°–28°, and the lower limit at 18° are not so readily explicable, however, and samples of debris were taken from these slopes and their shear strengths examined.

Samples of debris were taken from nine different straight slopes: four from slopes with angles in the range 19° to 23° and five from slopes at angles of 26° and 27°. All these samples were examined to determine the percentage of humus, of the mantle coarser than 2 mm and of silt-clay in the material finer than 2 mm. This information is presented in Tables I and II. In addition, the shear strengths of three samples from the 19°–23° group and three samples from the 26°–27° group were investigated.

TABLE I

Properties of samples taken from the mantle on slopes ranging from 19° to 23°

Sample number	*Slope angle*	*h*	*sd*	*w*	*γ*	*c′*	*ϕ′*
1	19	7·1	68·0	25·0	1900	0	33·1
2A	20	5·4	77·8				
2B	20	4·3	88·5				
2C	20	2·8	87·7				
3	22	3·7	33·0	22·1	1860	0	32·0
4	23	4·8	55·5	23·4	1880	0	33·6

Key
h: humus content (per cent by weight)
sd: sand fraction (> 0·06 mm) as percentage of material finer than 2 mm
w: per cent moisture content of sample during shear test
γ: bulk specific weight of sample during test (kg/m³)
c′: cohesion
ϕ′: angle of internal friction

Shear-strength tests on the material from the gentler slopes were undertaken in the standard way[17] using 6 cm direct-shear apparatus. The sandy nature of this material made it impossible to extract undisturbed samples in the field, but this is not thought to be a serious problem; it has been demonstrated[18] that, in general, the *residual* shear strength of natural soils, as distinct from the peak strength, is probably independent of stress history and should, therefore, be unaffected by disturbance. The samples were recompacted to their natural bulk unit weight (1900 kg/m³) prior to testing.

The condition of the soil samples is only one item which affects the meaning of the test results. Equally important is the degree to which these samples represent the mantles from which they were taken. In this study, this was not a serious issue. Mantles are usually thin in the area (20–65 cm) so that there was little variance with depth. On one of the three slopes, samples were taken about 5 cm beneath the humic layer; in the other two cases the location of the samples within the mantle profile is indicated in Figure 7. Variability along the length of the straight slope was small.

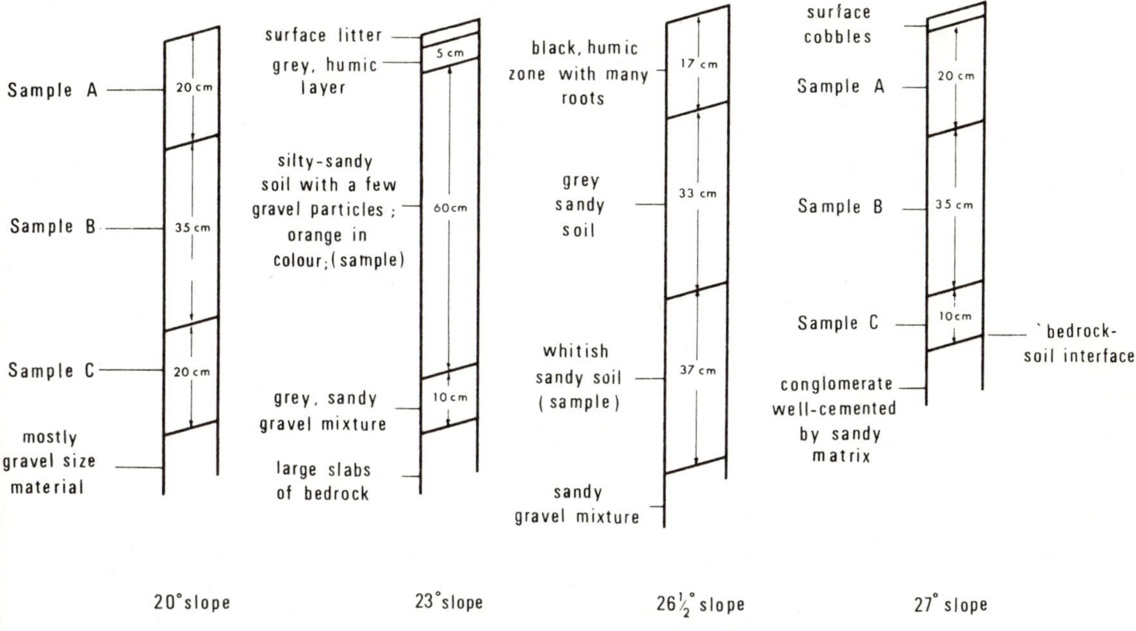

FIGURE 7. Debris mantle profiles indicating origin of soil samples

Tests were carried out at a displacement rate of 0·0007 cm/s and may be considered fully drained; the parameters obtained therefore relate to effective stresses. The effective normal stress ranged from 0·2 kg/cm^2 to 0·6 kg/cm^2, and although this might seem rather small and close to the limits of sensitivity of the apparatus, abundant data exist emphasizing the importance of testing material within the limits of field values of the normal stress. This, therefore, was the chief factor determining the limits of the normal stress values. The moisture content of the samples during the tests varied between 22 and 25 per cent.

A specimen shear strength–displacement curve is shown in Figure 8. All tests revealed the same tendency to move slowly towards an ultimate residual strength value rather than to produce an initial peak and then decrease in value towards the residual state. It is probable that this is because of the low initial bulk density of the samples. All three samples showed zero cohesion, as might be expected, and the ϕ' values averaged 33·0°. These values (Fig. 9) are very consistent and not entirely unexpected since it is well known that the angle of repose of sand is close to 35°. The results are similar to the recent data presented by T. C. Kenney[19] for the shearing resistance of small grains of 'massive' minerals.

It was not possible to test the shear strength of the coarser material using the conventional techniques of soil mechanics, since no apparatus large enough to accommodate the bigger rock pieces was available. The negligible clay content of these samples may, however, be taken as indicating minimal cohesion, and the problem is essentially that of deriving an estimate of the angle of internal friction. This was determined as follows. Debris was placed in a box (20 × 25 × 35 cm) hinged at one end to a horizontal platform. The debris was recompacted as far as possible to field bulk density. As the box is lifted

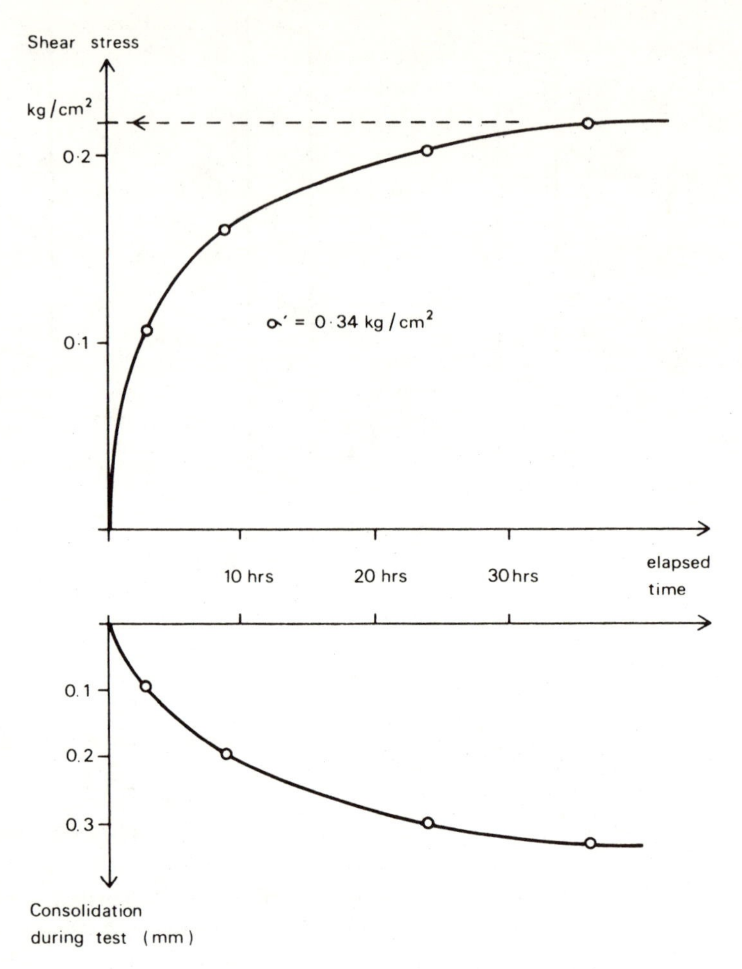

FIGURE 8. Typical shear stress–displacement curve during 6 cm direct shear test of sample No. 3 from low-angle slope

above the platform, the sample is tilted and, eventually, at a critical angle of stability, it fails. Since the sample is packed in a completely dry state, loose sand grains on the surface begin to roll before the sample as a whole fails, and identification of the actual instant of failure is a little subjective. Nevertheless the results of eight tests on each sample were consistent; they are summarized in Table II. The gravel fraction (coarser than 2 mm) of the samples varied between 50 and 61 per cent; the estimated angle of internal friction ranged between 43·1° and 44·1°. The tests were undertaken with samples at bulk unit weight of about 1800 kg/m³ (see Note on p. 48).

Notwithstanding the apparent crudeness of the testing procedure, the results compare well with other data obtained through tests with more sophisticated apparatus, and in many ways it can be argued that the mode of testing is appropriate. The Laramie data agree well with more precise work from engineers[20] working on coarse debris in connection with the building of earth dams (Table III). The test data on the coarse samples

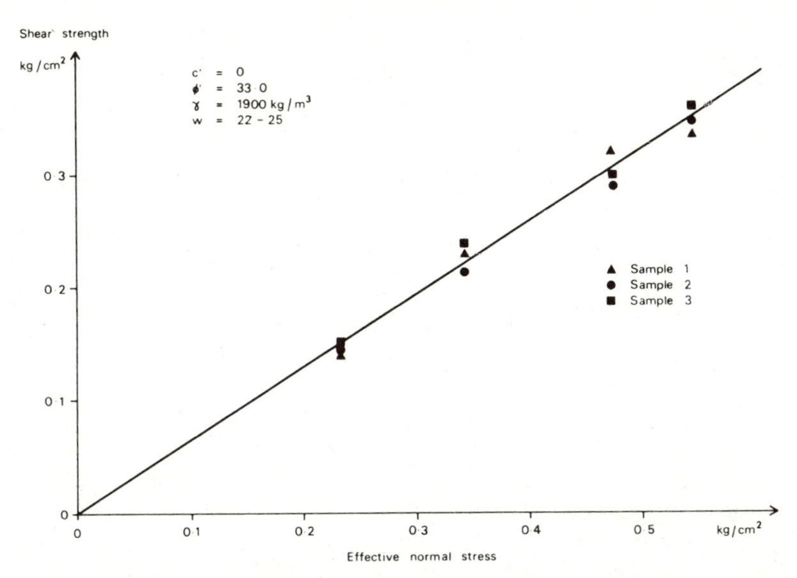

FIGURE 9. Summary of 6 cm direct shear tests on samples from slopes in the 19°–23° range

are therefore believed to be a reasonable indication of the angle of internal friction of this material.

STABILITY ANALYSES OF THE TWO SETS OF SLOPES

The factor of safety of a hillslope, that is the ratio of shear strength to shear stress along the critical potential failure surface, is not constant over time. The most important reason for this is that the effective normal stress on the potential failure surface fluctuates in magnitude according to changes in the moisture content of the soil. The effective normal stress is usually given by[21]

$$\sigma' = \sigma - u \tag{1}$$

TABLE II

The properties of samples taken from the mantle on slopes at angles of 26°–27°

Sample number	*Slope angle*	*g*	*sd*	*γ*	*φ′*
1	26	65	89·2		
2	26	44	75·3		
3	26·5	70	57·8		
4A	27	61	84·9	1720	43·2
4B	27	51	80·5		
4C	27	50	73·4	1800	43·1
5	27	53	74·2	1850	44·1

Key
g: per cent of sample, by weight, coarser than 2 mm
sd and *γ*: as defined in Table I
φ′: estimate of angle of internal friction

TABLE III

Angles of internal friction on sandy-gravel samples obtained in recent engineering studies

Source	*Material*	*g*	*sd*	γ	ϕ'
Lowe, 1964	terrace cobble-gravel	70	45–85	1900–2170	41–44
Queiroz, 1964	Furnas Dam gravel			1840–2320	43
Schultze, 1957	slaty-greywacke	85	85	1600–1900	43
Hall and Gordon, 1963	silty-sandy gravel	70	80	2350–2400	45·5

Key: as in Table II

where σ is the total normal stress and u is the pore pressure on the failure surface. Above a water table, pressures are negative relative to atmospheric pressure, and below it they are positive. It is easy to appreciate, therefore, that fluctuations in the level of the water table, or the creation of temporary perched water systems, will markedly affect the factor of safety of a hillslope. Since geomorphologists are concerned with slopes on a long timescale, it is necessary to assume the maximum pore pressures that are likely to occur in the mantle in a slope stability study. A. W. Skempton and F. A. DeLory,[22] working on the long-term stability of slopes in the London Clay, suggested that this condition is attained when the water table occurs at the ground surface and sub-surface flow is parallel to the ground slope. This model gives a pore pressure–depth relationship described by

$$u = \gamma_w . z . \cos^2\theta \qquad (2)$$

where γ_w is the unit weight of water, z is the depth beneath the surface, and θ is the angle of the surface slope. This relationship is identical to the pattern which would develop with a temporary water system perched above the soil–rock interface and with saturated flow parallel to the ground slope. Previous work[23] has suggested that this is applicable to valley slopes on Exmoor and in the Pennines. In the discussion of slope stability in the Laramie Mountains it was, therefore, assumed that the largest pore pressures during the history of the hillslope are given by equation 2.

The maximum stable angle of a hillside depends not only on the effective shear strength parameters, but also on the geometry of the incipient failure surface. On straight hillsides which are essentially rock slopes mantled by weathered debris, the failure surface is confined to the slope mantle and may be considered almost planar and parallel to the ground surface. The condition of limiting stability on an infinite planar failure surface is given by[24]

$$\gamma . z . \sin\theta . \cos\theta = c' + (\gamma . z . \cos^2\theta - u) . \tan\phi' \qquad (3)$$

and, since the cohesion intercept c' is zero here, this is easily solved to give the maximum angle of slope θ, where

$$\gamma . z . \tan\theta = (\gamma . z - u/\cos^2\theta) . \tan\phi' \qquad (4)$$

Substituting for u from equation 2, we obtain

$$\tan\theta = \frac{(\gamma-\gamma_w)}{\gamma}\tan\phi' \qquad (5)$$

Insertion of values of γ and ϕ' into equation (5) for the Laramie slopes will now provide a quantitative estimate of how the actual slopes compare with the threshold slope angles for these two mantles. Tests in the field indicated that γ is approximately equal to 1900 kg/m³. Assuming γ_w is about 998 kg/m³, equation (5) then reduces to

$$\tan\theta = 0{\cdot}48\tan\phi' \qquad (6)$$

On the slopes at about 26°, ϕ' was shown to be about 44°; substituting this value into equation (6) gives a value of θ equal to 25½°, which provides remarkable agreement with the actual angle of slope. It should be stressed that equation (5) is very sensitive to the value of γ, and actual determination of the bulk unit weight of taluvial material is not easy, so that the γ value indicated above probably contains a small amount of error, although it agrees with other figures for this type of material. In the case of slopes at 19°–23°, insertion of $\phi' = 33°$ into equation (6) produces a value of θ equal to 17½° which, again, is close to the actual angles of slope from which the samples were taken.

In summary, the major aspects of the frequency distribution of straight-slope angles in the Laramie Mountains may be explained in terms of the soil mechanics properties of the debris mantles masking the slopes. The *upper limit* is probably the angle of repose of the first, coarse products of weathering of the rock mass; the *lower limit* appears to coincide approximately with the threshold angle of slope of the mantle in its most weathered condition, when subjected to particular moisture conditions; and the *modal group* of slopes seems to be linked to the stability of the mantle in an intermediate stage of weathering.

Admittedly, many questions remain unanswered. Why, for instance, is the frequency distribution of slope angles peaked, when the weathering of the mantle proceeds in a gradual and continuous manner? And what are the geometrical changes associated with the development of slopes at 18° from slopes at 33°? These questions are touched upon in the concluding section.

THE EVOLUTION OF STRAIGHT SLOPES

The study of slope development through frequency distribution analysis is not new. It was employed by A. N. Strahler[25] in his classic paper on slopes in various parts of the U.S.A., and also by Young[26] working in the United Kingdom. Neither, however, could offer explanations for the basic properties of the distribution. The first real attempt to explain the frequency distribution of slope angles in an area was by Melton[27] in his study of slope angles in the southern Arizona desert. Melton noted that 'Hillslopes of less than 28·5° on both granite and volcanic rocks are highly stable, whereas those between 28·5° and 36° are increasingly unstable . . . debris-covered hillslopes above 36° are rare. The two stability limits, 28·5 and approximately 36°, are related to the frictional properties of the rock itself.' Melton attempted to relate the two limiting angles of slope to the static and sliding coefficients of friction of cobbles, resting on a planar slab of mica schist. In these experiments he discovered that the average coefficients of static and sliding friction were, respectively, 0·675 and 0·52, corresponding to arctangent values of 34° and 27·6°. The agreement between these values and the actual limiting angles in the field is extremely good and, in this way, Melton offers an explanation for the two limiting values.

Melton also points out that 'the 28·5° stability limit is a characteristic limit for a variety of classes of hillslopes that are covered by coarse debris'. There can be little doubt now that the prevalence of slopes mantled by coarse debris at angles between 25° and 29° is related to the stability of this type of mantle, but whether Melton's particular explanation is the correct one is debatable. It is clearly difficult to justify a stability analysis based purely on frictional strength, when the shear strength of the material under study is a result of *interlocking* as well as plane friction. Melton's laboratory tests measure only the plane frictional element of the debris. The correct shear-strength parameter which should be used in the stability analyses of these slopes is not the plane friction coefficient, but the angle of *internal* friction which integrates the interlocking and frictional strength components.

The evidence from numerous soils engineers, as well as that reported directly from this research, indicates that the angle of internal friction of gravel–soil mixtures is about 42°–45°. This value is substantially higher than the limiting angle of slope with which this type of debris is associated. This disparity can only be reconciled through the existence of positive pore pressures in the mantle acting to dissipate some of this frictional resistance. The data above suggest that the model employed in other areas appears to be valid, even in the semi-arid environment of the Laramie Mountains. Since this is, *prima facie*, rather surprising, a closer examination of the problem follows.

In semi-arid areas such as southern Wyoming, the true water table is usually far beneath the ground surface, but it is possible that, during long storms, shallow debris mantles might become saturated and perched water systems form for a short period. The shallowness of debris in these areas, of course, means that the amount of water needed to saturate the mantle above the more impermeable bedrock may be much less than in humid temperate areas. A hypothetical case should make this clear. A mantle 60 cm deep with a void ratio of 0·5, a fairly typical value for this material, will accommodate 20 cm of water when fully saturated. Assuming that, in periods between storms, about 20 per cent of the void space contains capillary water, only 16 cm of water are needed to saturate the mantle completely.

Storms capable of producing 16 cm of percolation into these debris mantles clearly do not occur every year, but they might be expected to occur with sufficient frequency, on a geological timescale, that they must be considered in the stability analysis of slopes in these areas. Unfortunately, there is little information available on the frequency of prolonged rainstorms of moderate intensity in the Laramie area. D. L. Yarnell's data[28] on short, high-intensity storms in the U.S.A. indicate that in the Denver–Cheyenne area, the maximum 100-year 24-hour rainfall averages about 110 mm, and storms with a return period of 1000 years might be expected to produce 24-hour rainfall amounts approaching 160 mm. Such figures hardly provide unequivocal support for the model developed above, although they do indicate that more prolonged storms could saturate these waste mantles fairly frequently on a geological timescale. On the other hand, assuming that more moist conditions prevailed during parts of the Pleistocene, it may be that these slopes are unrelated to present-day soil moisture conditions, and are relicts of former times. Much more research is necessary, however, in order to settle this point.

The importance of sub-surface flow on slopes has only recently been recognized by earth scientists[29] and its role in semi-arid areas has rarely been considered. Over 10 years ago, however, Ruxton[30] pointed to the temporary existence of sub-surface flow, during storms, on granite slopes in the Sudan, and R. F. Hadley and B. N. Rolfe[31] noted that micro-instability on slopes in parts of the Great Plains of the U.S.A. appeared to result from the temporary build-up of water pressure in the soil mantle during rainstorms.

Even accepting the role of pore pressures in determining the threshold slope of a taluvial mantle at about 26° in semi-arid areas, there remains the problem of explaining the peaked character of the slope angle distribution at this particular value, rather than a rectangular distribution with a continuum of values between its limits. An explanation for this has already been offered elsewhere;[32] it may be briefly summarized as follows. Notwithstanding the gradual nature of the weathering of rock rubble into soil, the change in the angle of internal friction during this process is much more abrupt. The angle of internal friction is about 35° for rock rubble, approximately 43° for a wide variety of taluvial mantles, and about 35° for a fully reduced sandy soil. The change between these three conditions of the mantle is apparently quite sudden. Over most of the weathering sequence, mantles appear to behave as taluvial material and, on the basis of the model above, this would account for the peaked character of slope angle distributions at about 26°.

No comment has so far been made on the mode of slope profile change, whether retreat or decline, from the upper to the lower limit of the slope angle distribution. Both on Exmoor and in the southern Pennines it has been noted[33] that both decline and retreat occur depending on the thickness of the waste mantle. In the Laramie Mountains, the multisegment character of many slopes suggests that the 33° slope retreats and is replaced by a 26° slope and that, in turn, the 18° slope emerges as a result of the retreat of the 26° slope. One might attempt to support this through reference to the 'gaps' in the frequency distribution at 30° and 22°, but a much larger sample would be needed to establish the statistical significance of these apparent breaks in the histogram.

It is beyond the scope of this paper to consider whether these hillslopes ever become completely stable at about 18° or whether subsequent retreat produces an ever-extending pediment surface. The purpose of the paper is not to speculate on the complete development of slopes in the Laramie Mountains, but to examine the relevance of a previously developed model to certain aspects of slope evolution in the area. The model appears to agree with field data and indicates a possible, albeit rather surprising, answer for observations, not only in the plutonic area of the Laramie Mountains, but in areas of rock which generally weather to produce coarse debris mantles on hillslopes.

ACKNOWLEDGEMENTS

This research was supported by Grant A 5111 of the National Research Council of Canada as part of a general project on the long-term stability of natural slopes. I am grateful to Mr A. N. Dolman and Mr P. A. W. Smith of McGill University for assistance in the field and in the laboratory respectively. A grant towards the cost of illustrations was awarded by the National Research Council of Canada.

NOTES

1. M. A. CARSON, 'The evolution of straight debris-mantled hillslopes', Unpubl. Ph.D. thesis, Univ. of Cambridge (1967)

2. M. A. CARSON and D. J. PETLEY, 'The existence of threshold hillslopes in the denudation of the landscape', *Trans. Inst. Br. Geogr.* 49 (1970), 71–95

3. B. P. RUXTON, 'Weathering and subsurface erosion in granite at the Piedmont angle, Balos, Sudan', *Geol. Mag.* 95 (1958), 353–77; M. A. MELTON, 'Debris-covered hillslopes of the Southern Arizona Desert—con-

sideration of their stability and sediment contribution', *J. Geol.* 73 (1965), 715–29; G. ROBINSON, 'Some residual hillslopes in the Great Fish River Basin, S. Africa', *Geogrl J.* 132 (1966), 386–90

4. The term 'taluvial' is used to describe waste mantles which comprise a mixture of coarse rocky rubble and more weathered material of soil-size particles. It was originally used for this purpose by C. K. WENTWORTH, 'Soil avalanches on Oahu, Hawaii', *Bull. geol. Soc. Am.* 54 (1943), 53–64, as a hybrid from talus and colluvium.

5. M. A. MELTON, op. cit. (1965)

6. M. N. DENSON (United States Geological Survey, Denver, Colorado), written communication (1968)

7. D. V. HARRIS, 'Geomorphology of Larimer County, Colorado' in *Geology of the Northern Denver basin and adjacent uplifts* (ed. P. J. KATICH and D. W. BOLYARD, 1963), Rocky Mountain Association of Geologists

8. The only exception to this procedure was when distinct breaks of slope occurred at distances of less than 10 m.

9. The designation of whether or not a section of a profile is straight follows the policy outlined at length by M. A. CARSON, op. cit. (1967), 101–4.

10. Many spot checks, without detailed survey, were made on the angles of straight slopes besides those included in Figures 3 and 4. These reinforce the sample data in terms of modal and limit values. Slopes above about 33° do not possess a continuous debris cover. Some slopes occur at angles less than 18° (see M. A. MELTON, 'Intra-valley variation in slope angles related to microclimate and erosional environment', *Bull. geol. Soc. Am.* 71 (1960), 133–44) but these do not include straight segments.

11. M. A. CARSON and D. J. PETLEY, op. cit. (1970)

12. M. A. MELTON, op. cit. (1965)

13. K. BRYAN, 'Erosion and sedimentation in the Papago Country, Arizona', *Bull. U.S. geol. Surv.* 730 (1922), 19–90

14. B. P. RUXTON, op. cit. (1958)

15. G. ROBINSON, op. cit. (1966)

16. A. YOUNG, 'Characteristic and limiting slope angles', *Z. Geomorph.* 5, 2 (1961), 125–31

17. A. W. SKEMPTON and A. W. BISHOP, 'The measurement of the shear strength of soils', *Géotechnique* 2 (1950), 90–108

18. A. W. SKEMPTON, 'The long-term stability of clay slopes', The Rankine Lecture, *Géotechnique* 14 (1964), 77–101

19. T. C. KENNEY, 'The influence of mineral composition on the residual strength of natural soils', *Proc. geotech. Conf. Oslo* 2 (1967), 123–9

20. J. LOWE, 'Shear strength of coarse embankment dam materials', *Proc. 8th int. Congr. Large Dams* 3 (1964), 745–61; L. QUEIROZ, 'Geotechnical properties of weathered rock and behaviour of Furnas rockfill dam', *Proc. 8th int. Congr. Large Dams* 1 (1964), 877–90; E. SCHULTZE, 'Large-scale shear tests', *Proc. 4th int. Conf. Soil Mech.* 1 (1957), 193–9; E. B. HALL and B. B. GORDON, 'Triaxial testing with large-scale high pressure equipment', *Symposium on laboratory shear testing of soils, Ottawa, ASTM STP 361* (1963), 315–28

21. I. K. LEE and I. B. DONALD, 'Pore pressures in soils and rocks' in *Soil mechanics—selected topics* (ed. I. K. LEE, 1968), 58–81

22. A. W. SKEMPTON and F. A. DELORY, 'Stability of natural slopes in London Clay', *Proc. 4th int. Conf. Soil Mech.* 2 (1957), 378–81

23. M. A. CARSON and D. J. PETLEY, op. cit. (1970)

24. A. W. SKEMPTON and F. A. DELORY, op. cit. (1957)

25. A. N. STRAHLER, 'Equilibrium theory of erosional slopes approached by frequency distribution analysis', *Am. J. Sci.* 248 (1950), 673–96 and 800–14

26. A. YOUNG, op. cit. (1961)

27. M. A. MELTON, op. cit. (1965)

28. D. L. YARNELL, 'Rainfall intensity-frequency data', *U.S. Dept. Agric. Misc. Publ.* 204 (1935)

29. Saturated sub-surface flow on slopes has recently been reported by R. Z. WHIPKEY, 'Subsurface stormflow from forested slopes', *Bull. int. Ass. scient. Hydrol.* 10 (1965), 74–85; R. P. BETSON, J. B. MARIUS and R. T. JOYCE, 'Detection of saturated interflow in soils with piezometers', *Proc. Soil Sci. Soc. Am.* 32 (1968); and C. A. TROENDLE, *U.S. Dept. Agric. Forest. Serv.* (Virginia) in several unpublished reports.

30. B. P. RUXTON, op. cit. (1958)

31. R. F. HADLEY and B. N. ROLFE, 'Development and significance of seepage steps in slope erosion', *Trans. Am. geophys. Un.* 36 (1955), 792–804

32. M. A. CARSON and D. J. PETLEY, op. cit. (1970); in addition to the evidence presented in that paper, experimental data given by W. G. HOLTZ and H. T. GIBBS, 'Shear characteristics of pervious gravelly soils as determined by triaxial shear tests', *Conv. Am. Soc. civil Engnrs, San Diego, California* (9–11 Feb. 1955) and D. B. SIMONS, 'Study of the angle of repose of non-cohesive materials' (Mimeo., Dept. of Civil Engineering, Colorado State University, Fort Collins, Colorado) strongly support these ideas.

33. This point has been discussed briefly by the writer in 'Models of hillslope development under mass failure', *Geogrl Analysis* 1 (1969), 76–100, and in more detail in the Ph.D. thesis cited in note 1.

BULLETIN OF THE GEOLOGICAL SOCIETY OF AMERICA
VOL. 71, PP. 133-144, 2 FIGS. FEBRUARY 1960

INTRAVALLEY VARIATION IN SLOPE ANGLES RELATED TO MICROCLIMATE AND EROSIONAL ENVIRONMENT

BY MARK A. MELTON

15

ABSTRACT

East- or west-trending stream valleys that have erosional slopes have been reported to be (1) asymmetrical with the north-facing slopes steeper, (2) asymmetrical with the south-facing slopes steeper, and (3) symmetrical, with both sides of equal mean slope angle. A series of measurements of angles of erosional slopes, taken in three east-trending, low-gradient valleys in the Laramie Range, Wyoming, shows that north-facing slopes there tend to be 4.42° steeper than opposed south-facing slopes; slope angle is further affected by nearness to the channel. In valleys that have channel gradients greater than 6° and greatly differing vegetation density across the valley, measurements of slope angles definitely show valley symmetry. The interpretation given is that unless the channel has been maintained against the base of the north-facing slope by greater slope wash from the south-facing slope, vegetation differences and resulting rates of slope erosion alone do not produce asymmetric valleys. Valley asymmetry that results from a variety of basic causes can be attributed to a single mechanism, asymmetric lateral corrosion by the stream.

Difference in frost action on north- and south-facing valley sides is rejected as a cause of valley asymmetry in the areas studied because (1) the asymmetry is opposite that reported in tundra regions elsewhere, and (2) the degree of asymmetry in two widely separated areas does not reflect differences in the degree of frost activity in the two areas.

CONTENTS

TEXT

ILLUSTRATIONS

TABLES

INTRODUCTION AND ACKNOWLEDGMENTS

The question of whether the orientation of a valley side sufficiently affects the processes of erosion and mass transport, through the agency of differences of microclimate, to cause perceptible differences in the declivity of the steepest, straight portion of approximately mature, erosional slopes in homogeneous material has received the attention of geologists for at least 50 years. The question is inherently a statistical one; because we cannot measure the steepness of every valley-side slope, it is necessary to make estimates of population parameters on the basis of limited sample data, and we wish to control as much as possible the accuracy of these estimates. This is best done by the methods of statistical analysis. Procedures of sampling, measurement, and analysis needed for a scientific attack on the problem of valley asymmetry are relatively simple, but only two systematic, quantitative studies have been conducted (Emery, 1947; Strahler, 1950), and the results of these two studies do not agree. No conclusive quantitative evidence has yet been adduced in support of any hypothesis.

Where it is due to differences in microclimate, the difference in angle between north- and south-facing slopes, if any, is usually small in comparison to the mean slope angle; the importance of a study of this phenomenon, then, lies in the possibility of verifying predictions of the manner of operation and effectiveness of certain gradational processes.

The conclusions that are reached below are based on a large number of slope-angle measurements taken by the writer in Arizona and Wyoming, and by Prof. A. N. Strahler in California. Their validity is limited in the strictest view to those and similar areas. However, a wide variety of rock types and climatic zones is represented in the combined sample, and the writer believes that the conclusions will prove to be generally applicable in the subhumid to arid regions of the northern hemisphere, middle latitudes. Similar information obtained in more humid regions and in the Arctic might provide a very interesting counterpart to this study.

The Department of Geology of the University of Chicago supplied the necessary support for field expenses and assistants. Dr. Wm. H. Kruskal of the Department of Statistics, University of Chicago, offered some very helpful advice in planning the sampling program. Roger Melton of Stillwater, Oklahoma, and C. J. Acker of the University of Arizona assisted in the field during the summers of 1957 and 1958, respectively. The writer is indebted to Prof. A. N. Strahler, Columbia University, and to Prof. John P. Miller, Harvard University, for critically reading the manuscript and for offering a number of helpful comments. Any errors are of course the responsibility of the writer alone.

Theoretical Consideration of Causes of Valley Asymmetry

On the question of valley asymmetry there are only three geometrical possibilities; each has been supported fairly recently in print:

(1) North-facing slopes tend to be steeper than south-facing slopes in the drier portions of the North Temperate Zone, if other factors are equal. This appears to be the consensus of geomorphologists at present and is supported in the paper by Emery (1947) and in many earlier works. (It is generally believed that the difference increases northward.) At least three plausible explanations for this can be given:

(a) Gilbert's "law of divides," states that the declivity at a point on a slope is less in proportion as the quantity of runoff over it is greater (Gilbert, 1880, p. 110). The runoff and the erosion rates on north-facing slopes are less than that on corresponding south-facing slopes because of greater vegetation density and soil development and therefore higher infiltration rate and surface resistance to the eroding stress of running water. South-facing slopes accordingly tend to be eroded more rapidly, with resulting lowered slope angle (Emery, 1947, p. 67–68).

(b) The susceptibility to erosion is greater on south-facing slopes for the reasons given above, so that a greater quantity of debris is washed down into the subjacent channel from the north, thereby shifting the channel and stream of water against the toe of the north-facing slope. The south-facing slope continues to be lowered at a greater rate than the north-facing slope, which is kept steep by being undercut frequently. The toe of the south-facing slope is buried and built up to some extent, which further reduces the over-all angle. The difference in slope angles increases until another agency, presumably rate of surface creep, enters the picture and the difference in slope angles becomes steady.

(c) In regions in which the January mean temperature is 32°F. or less, snow and ice tend to remain on north-facing slopes for longer periods than on corresponding south-facing slopes; this protects the north-facing slope and allows more rapid erosion on south-facing slopes. The result is that in regions that have the requisite temperature regime, north-facing slopes are steeper than south-facing slopes (Russell, 1931).

Of course, mechanisms (1a, b, c) are not necessarily mutually exclusive. A major contention of this paper is that mechanism (b) has by far the greatest effect in most arid to subhumid regions where slope differences exist because of differences in microclimate across valleys.

(2) That north-facing slopes tend to be less steep than opposed south-facing slopes has been suggested by von Engeln (1942, p. 143) without restriction and more recently by Büdel (1953, p. 255) as well as by a number of earlier writers (Smith, 1949, p. 1503) for tundra and "fossil tundra" areas. As this is primarily a study of fluvially controlled regions, no attempt is made to discuss completely periglacial processes that might produce valley asymmetry. Two of the proposed mechanisms are:

(a) In periglacial regions a south-facing slope, receiving much more heat by insolation, will be frozen less frequently than the opposed north-facing slope. Consequently, corrasion of the base of the south-facing slope by the stream will be more frequent and severe and will tend to produce steeper south-facing slopes. If modern processes have not obliterated the asymmetry, it will remain as an indication of a former frigid episode in now temperate regions (Smith, 1949, p. 1503).

(b) In periglacial regions, early spring snow melt-water causes greater erosion and debris movement on north-facing slopes because frozen subsoil under the north-facing slope reduces the infiltration capacity. A greater amount of debris will be brought down into the south side of the subjacent channel, causing the stream to shift against the base of the south-facing slope and steepen it (Büdel, 1953, p. 255).

Geologists who have observed Arctic and periglacial terrain from air photographs or in the field do not agree that south-facing slopes are steeper or that these mechanisms are of greater importance than others that tend to produce steeper north-facing slopes (F. A. Melton, Don Currey, J. Corbel, personal communications). Qualitative reasoning is of little further use, and quantitative work is needed to resolve the issue.

(3) The hypothesis that north-facing slopes on the average have the same declivity as south-facing slopes is supported by data collected by Strahler in the Verdugo Hills and San Rafael Hills, California (1950, p. 806–810). On the basis of his sample of 105 slope measurements, the 99 per cent confidence interval for the true difference in mean slope angles, Δ, is

$$-1.62° \leq \Delta \leq +1.66°,$$

where Δ is the true mean angle of north-facing slopes minus the true mean angle of south-facing slopes, in the Verdugo and San Rafael hills. This interval includes zero and therefore supports the hypothesis of no consistent difference in mean slope angles. All slopes measured in this sample were in sharp, V-shaped ravines with narrow, steep channels, affording approximately equal opportunity for corrasion of the toe of either side of the ravine (Strahler, 1950, p. 809). Strahler found that slopes that had been protected from recent basal corrasion by an accumulation of slide-rock and other debris averaged 6.6° less steep than slopes that had recently been corraded (p. 812–813). Statistically, this difference in sample means is extremely improbable if the difference in population means does not exist.

All three situations described above could not be true in general. It is important, then, to see whether any one is true in general, or whether the particular situation in any given case depends on the existence of certain conditions of stream regimen and climate. Otherwise, the conflicting viewpoints will only serve to emphasize that much of the published thought on slope development is not based on sound facts and is of less interest to the scientist than to the historian of science.

Neither Strahler's nor Emery's work is conclusive: Strahler's sample is too small to provide a good chance of detecting actual small differences in slope angles on the order of 1.0°–1.5° (see below). The sample taken by Emery was from a region of tilted strata with a regional southerly dip; the scarp slopes facing north would introduce valley asymmetry even if differences in microclimate had no effect (Emery, 1947, p. 67).

Analysis of Slope Angles in the Laramie Mountains, Wyoming

An area in the southern part of the Raggedtop Mountain quadangle, Wyoming, was selected for study on the basis of the presence of a number of similar, east-trending valleys that have low gradients, fairly homogeneous, deeply weathered granite bedrock in which no influence of structural planes on valley-side slope angles could be discerned, distinct difference in density of vegetation on north-facing and south-facing slopes, and absence or rarity of extensive modern gullying or channel trenching. An alluvial fill was present in each valley of third or higher order. Three separate valleys were selected for detailed slope sampling.[1] At most points on the channels, the gradients measured 0.5°–1.5°; the maximum recorded was 3.5°. In each valley, at the mouth of every extensive tributary, a small alluvial fan has been built outward into the main valley by the deposition of debris eroded during periods of runoff and left as the water is lost into the fill. These fans appear to be still growing. The region lies between 7500 and 8000 feet above sea level. When the writer visited the three valleys none contained water; the streams are presumably intermittent or ephemeral.

The mean percentage of bare-soil area on north-facing slopes was 26 per cent; on south-facing slopes it was 41 per cent; this is a difference of means of 15 per cent. Measurements of slope angles were taken at points along segments of each valley such that channel gradient, width of the main valley, and the mean slope angle appeared approximately uniform from end to end.

After some preliminary work, the experimental design for a single valley was chosen (Fig. 1) to allow consideration of the effects of the alluvial fans at the mouths of tributaries on nearby valley-side slope angles. In Figure 1 and Table 1, F_o stands for absence of alluvial fans in the neighborhood of the slope, F_f stands for facing an alluvial fan, *i.e.*, opposite a tributary entering from the other side; F_a stands for above an alluvial fan, *i.e.*, on the same side of the main channel as an entering tributary and close enough to the junction so that the toe of the slope intersects part of the fan. The six situations are mutually exclusive, and each determines what is called here the erosional environment of the slopes. From four to six slope measurements in each of the six situations were made per valley, and the average is recorded in Table 1 in the appropriate cell. Thus, a three-way factorial design was obtained; each valley is considered a "block."

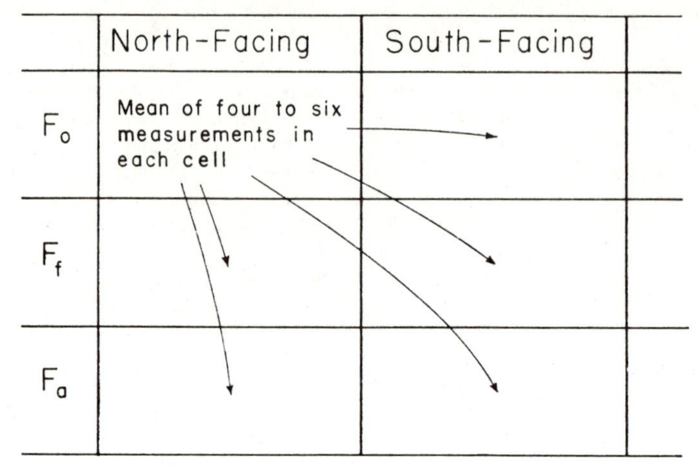

FIGURE 1.—TWO-WAY FACTORIAL DESIGN FOR A SINGLE VALLEY

The analysis of variance of the data in Table 1 is presented in full in Appendix I. Conclusions supported by the analysis are:

(1) No first-order interaction is significantly different from zero at the $\alpha = .05$ level of significance; we may assume that the quantitative effects of orientation and erosional environment within each valley are essentially additive. That this is true could be determined only by quantitative methods (Appendix IA).

(2) The column effects differ significantly from zero; north-facing slopes average $2 \times 2.2\bar{1}° = 4.4\bar{2}°$ steeper than south-facing slopes (Appendix IB).

(3) The slope angles at a distance from any tributary alluvial fan, *i.e.*, in the F_o erosional

[1] The location of the mouth of each of the three valleys is:
(1) NW1/4 sec. 21, T. 15 N., R. 71 W.
(2) SW1/4 sec. 16, T. 15 N., R. 71 W.
(3) W1/2 sec. 6, T. 15 N., R. 71 W.

environment, tend to be the same as the average of angles in the F_f and F_a erosional environments. The important conclusion is that the slope angles above an alluvial fan are reduced by the same amount that slope angles facing an alluvial fan are steepened (Appendix IB).

(4) The means of all slopes within each of these particular valleys differ significantly, although the greatest difference, between valley 2 and valley 3, is less than 2.5°. This tells something about the power of this analysis to detect differences in slope-angle means.

These results reaffirm the observations of a number of workers in geomorphology and clearly imply that valley-side slopes are sensitive indicators not only of the intensity of gradational processes operating on their surfaces but in addition display the effects of the proximity and activity of nearby channels, that is, their erosional environment. Therefore, generalizations about slope processes and evolution must include considerations of erosional environment as well as of climate, vegetation, and rock and soil type. The interpretive geomorphologist must also consider possible past erosional environments as well as processes no longer active on the surfaces of slopes. A number of recent discussions of slope development have ignored this seemingly obvious point, as did the majority of earlier works.

In the present case, the high degree of association of slope angles with specific erosional environments (Appendix IB) is evidence against the hypothesis that frost action, ancient or modern, is primarily responsible for

TABLE 1.—MEAN SLOPE MEASUREMENTS FOR EACH VALLEY, ARRANGED ACCORDING TO EROSIONAL ENVIRONMENT

	North-facing	South-facing	Totals ($T_{i..}$)	Means ($\bar{X}_{i..}$)	Effects (A_i)
F_o	1. 19.12 2. 18.38 3. 21.00	1. 14.75 2. 12.38 3. 15.33	100.96	16.83	+0.01 (=16.83 − 16.82)
F_f	1. 19.38 2. 20.25 3. 23.17	1. 16.44 2. 15.50 3. 16.33	111.07	18.51	+1.69 (= 18.51 − 16.82)
F_a	1. 16.88 2. 14.50 3. 18.62	1. 13.50 2. 13.25 3. 14.00	90.74	15.12	−1.70 (= 15.12 − 16.82)
$T_{.j.}$	171.29	131.48	302.77		
$\bar{X}_{.j.}$	19.03	14.61		16.82 (Grand mean)	
B_j	+2.2$\bar{1}$ (= 19.03 − 16.82)	−2.2$\bar{1}$ (= 14.61 − 16.82)			

Blocks

Totals	Means	Effects
$T_{..1}$ = 100.06	$\bar{X}_{..1}$ = 16.68	C_1 = −0.14 (= 16.68 − 16.82)
$T_{..2}$ = 94.25	$\bar{X}_{..2}$ = 15.71	C_2 = −1.11 (= 15.71 − 16.82)
$T_{..3}$ = 108.46	$\bar{X}_{..3}$ = 18.08	C_3 = +1.26 (= 18.08 − 16.82)

the asymmetry of the valleys in the region studied.

Interaction of Channel Gradient with Slope Orientation in Producing Valley Asymmetry

In view of the statistical evidence for the existence of valley asymmetry presented above, why was no asymmetry detected in the statistical study conducted by Strahler? One possibility is that Strahler's sample contained too few measurements to provide an adequate chance of detecting a real, though small, difference in slope-angle means. The power of a t test to detect several alternative values of the true $\bar{\theta}_N - \bar{\theta}_S = \Delta$, based on Strahler's data (1950, p. 810), is presented in Table 2, assuming that $\alpha = .05$ and d.f. $= 103$ (Pearson and Hartley, 1951, p. 115). The probability of detecting a true difference of 3.0° is very good, but a difference of 0.5° or 1.5° might be undetected.

Possibly no difference in slope-angle means for north- and south-facing slopes exists in the Verdugo and San Rafael hills, in spite of the considerable difference in vegetal cover across the valleys (Strahler, 1950, p. 807). The conditions described seem ideal for the operation of agencies (1a) or (1b). However, the channel gradients are generally rather steep in the portion of the valleys having narrow channels cut in bedrock (Strahler, 1950, p. 802, map). The slopes are about four times steeper than in the Laramie Mountains, and the region is much less subject to frost action. Further work in regions that combine some of these properties was necessary.

The writer developed the following working hypothesis to explain why in some cases no difference in opposed slope angles exists, where a considerable difference in microclimate and vegetation density does exist: a steep channel gradient in bedrock allows the same runoff that brings detritus into the channel to sweep the channel clean. Such material is carried immediately out of the steep segment of the valley, in some cases as a mudflow, and is deposited either beyond the mouth of the basin on an alluvial fan or in the channel where the gradient is sufficiently lower and water is absorbed or evaporated. This prevents the accumulation of an asymmetric valley fill in the steep segment that would act to shift the channel against the base of the north-facing slope, thereby corrading it but not the base of the south-facing slope. As nothing in this would prevent the modification of the valley-side slopes by differing runoff and slope erosion rates alone (1a or c), if no difference in angles is found, then this agency acts too feebly to produce an effect. Thus, the factor of the channel gradient itself, which should be symmetric in its effect on the adjacent slopes, hypothetically would interact with and nullify the factor of slope orientation when the gradient is steep.

Table 2.—Power of Strahler's (1950) Test for Difference in Slope-Angle Means Against Four Alternatives

$\lvert\Delta°\rvert$	$\phi = \frac{1}{2}\Delta\sqrt{n/S_p}$	Power
0.5	0.393	.12
1.0	0.786	.34
2.0	1.572	.62
3.0	2.36	.92

Analysis of Slope Angles in Southern Arizona

As a check on the argument presented above, and to eliminate a possible effect of slope angle on asymmetry, the writer conducted a program of sampling in valleys that have mean slope angles that range from approximately the same as those in the Laramie Mountains area to somewhat steeper but have much steeper channel gradients, roughly comparable with those in the Verdugo and San Rafael hills. Eight areas, in the Lower Sonoran, Upper Sonoran, and Canadian climatic zones were selected as follows: Care was taken to avoid sources of valley asymmetry not related to differences in microclimate on the slopes. Areas that have pronounced structural planes, modern, accelerated slope erosion, winding channels, or differences in height of divides above the channel were avoided. The valleys sampled appeared to be representative of the region. In every steep-gradient valley of the southern Arizona sample, the channel contained at most only a thin bed load; no alluvial fans were present at the mouths of tributaries.

After deciding upon a valley segment to be included in the sample, the writer and assistant established points along the channel at random. Near these points the channel gradient, γ, valley-side slope angles, θ, and the percentage of bare-soil area, b, were measured. The error in measurement of the slope angles was probably no greater than ½°, and the error in measurement of the percentage of bare-soil area was probably no greater than 5 per cent.

The topographic variable treated is $\Theta = \theta_N - \theta_S$; the θ's were measured in pairs, opposite each other across the valley. The reason for taking paired differences instead of the quotient, as did Emery (1947, p. 62), becomes apparent when we examine the various terms that make up each slope measurement:

$$\theta_N = \bar{\theta}_{reg} + e_v + e_p + e_n + e_b + \epsilon,$$
$$\theta_S = \bar{\theta}_{reg} + e_v + e_p + e_s + e_b + \epsilon,$$

where θ_N is the measured angle on north-facing slopes, θ_S is the measured angle on south-facing slopes, $\bar{\theta}_{reg}$ is the regional mean slope angle, e_v is the quantitative effect of the particular valley studied, e_p is the effect of the particular position within that valley where the slope pair was measured, e_n and e_s are the effects of facing north and facing south, respectively, e_b is any consistent observer's bias, ϵ is the random error. Some of these terms could be zero, of course. When we take the difference, $\Theta = \theta_N - \theta_S$, the common terms drop out (*i.e.*, the regional mean, any peculiarity of the valley studied, any correlation between north- and south-facing slopes because of position within the valley, and the observer's bias), leaving $\Theta = e_n - e_s + \epsilon$, which is the quantity we wish to examine. In most cases the "random" errors are not strictly random but are positively correlated for neighboring measurements taken within a short span of time; thus the ϵ term will probably be reduced somewhat by taking differences. The quotient θ_N/θ_S would still contain all the extraneous terms in a complex and unmanageable form. The frequency distribution of Θ is very nearly symmetrical, and the cumulative distribution gives essentially a straight-line plot on arithmetic probability paper; this suggests strongly that Θ is normally distributed.

In the southern Arizona sample, the difference in percentage of bare-soil area across valleys, $\bar{b}_s - \bar{b}_n$, ranging between 15 and 45 per cent, was about 25 per cent for most valleys. Channel gradients ranged from 6° to 23° with minor modes at 8° and 15°. Slope-angle measurements ranged from 15° to 37°, with the majority lying between 19° and 30°. Forty-seven pairs of slope measurements taken were used; a number were discarded because the gradient of the intervening channel was less than 6°. The slope data are summarized as follows:

$$\bar{\Theta} = -0.025\bar{5}°$$
$$s_\theta = 1.97° \quad \text{(standard deviation of the } \Theta\text{'s)}$$
$$N = 47.$$

A paired t test shows:

$$t = \frac{0 - (-0.025\bar{5})}{1.97/\sqrt{47 - 1}} = 0.089, \text{ d. f.} = 46.$$

The value of t is so small that it can be expected far more than 50 per cent of the time by chance alone on repeated sampling and testing in the same way, if the population mean of Θ is zero. The 95 per cent confidence interval for the true mean value of Θ is

$$-0.61° \leq \bar{\Theta} \leq +0.55°.$$

The power of this paired t test against specified alternatives to the null hypothesis, assuming that $\alpha = .05$, is given in Table 3. We may thus be fairly confident that the true difference in the valley-side slope angles in steep-gradient valleys, if not zero, is not much greater than 1.0°, whereas it might reasonably be expected to be 4° or greater on the basis of comparison with the sample from the Laramie Mountains. As most of the actual slope measurements of this sample were in the same range of values as the measurements in the Laramie Mountains, the possibility that the lack of valley asymmetry is due to steeper slopes is eliminated.

The southern Arizona sample can be broken down on the basis of gross climatology to see what effect that has on valley asymmetry. Table 4 summarizes the data according to life zones. These means appear to support in part the hypothesis of Russell (1931, p. 484) that areas where the January mean temperature is 32°F. or less have steeper north-facing slopes, because of snow accumulation (1c). We might further conclude that the asymmetry is reversed in the hot desert areas. The analysis of these data (Appendix II) shows, however, that this amount of difference among means could be expected rather frequently by chance alone where no difference among population means exists. Despite the rather small sample size and the uneven distribution of measurements among the climate classes, the power of this analysis against departures of Θ from zero of 1° is .65; thus we may be fairly confident in rejecting the hypothesis that gross climatology in the range Lower Sonoran–Canadian affects valley asymmetry to any appreciable degree. Many field observations confirm that in valleys that have low-gradient channels in the Lower Sonoran zone, north-facing slopes are generally at least 4° or 5° steeper than opposed south-facing slopes where an asymmetric fill is present, and the difference can be greater.

Table 3.—Power of Paired t Test

$\lvert\Delta°\rvert$	ϕ	Power
0.5	0.871	≈.3
1.0	1.742	.67
2.0	3.483	>.99

Table 4.—Sample Parameters of Θ According to Life Zone

	Lower Sonoran	Upper Sonoran	Canadian
$\bar{\Theta}$	−0.37°	0.09°	0.33°
N	15	26	6
s_θ	1.98°	2.14°	1.03°

As further illustration of the effect of channel gradient on slope-angle differences, a series of paired measurements was taken in a single east-trending valley, located in the Campo Bonito quadrangle, Arizona (N½ sec. 2, T. 11 S., R. 16 E.), near the climatic transition between the Lower and Upper Sonoran life zones. The valley is incised into an alluvial fan. Values of Θ plotted against channel gradient show a gradual increase in valley asymmetry as the gradient decreases downstream (Fig. 2); Θ probably approaches zero closely for $\gamma > 6°$.

The evidence strongly favors the conclusion that unless an asymmetric valley fill is maintained, differences in vegetation across a valley, acting through differences in rates of slope erosion, will not produce differences in valley-side slope angles and thus that the erosional environment of a slope, *i.e.*, the proximity and erosional activity of the channel, if any, at its base, is of greater importance to the slope angle than the gradational processes acting on the slope surface *per se*. This is further emphasized by the observation of Strahler's (1950, p. 812–813) that debris-protected slopes average more than 6° less in angle than slopes not so protected, and by the commonly observed fact that slopes on the outside of channel bends are considerably steeper than slopes above the inside of channel bends or above straight channel segments. Asymmetry can exist even in steep-gradient valleys, but only if the base of one slope facet is corraded more strongly than that of the opposed slope. This effectively rules out the possibility that the lack of asymmetry in spite of a considerable difference of microclimate and vegetal density can be due simply to the youthful condition of the valley or to the greater rate of channel downcutting in relation to slope erosion, as was suggested by Emery (1947, p. 68), who has also noted the lack of asymmetry in steep valleys. In steep-gradient valleys the conditions are not favorable for the maintenance of a continuous asymmetric fill, which if present would cause asymmetric basal corrasion. Thus, valley asymmetry that is the result of a variety of basic causes can be attributed to a single mechanism, asymmetric basal corrasion.

Perhaps the degree of valley asymmetry observed in areas studied here is the result of factors no longer active in these areas. One possibility is that the differences in slope angles developed solely because of differences in vegetation density and of erosional intensity on the slope surfaces (1a), and that a valley fill accumulated only after a change toward a generally drier climate, which may also have caused the initiation of the tributary channels. This hypothesis does not adequately explain why asymmetry does not always exist in valleys that have steep gradients where microclimatic differences exist and is confuted by the data from the Laramie Mountains. It would require a very clumsy *ad hoc* hypothesis to explain the observed variation in slope angles if the slopes developed entirely before the tributary alluvial fans were built. Further, in both study areas, the writer noted a rough correlation between the degree of asymmetry in a valley and the relative size of the valley fill it contained. The slopes in these areas show a remarkable adjustment to present erosional environments and must have been at least modified with the inception of present conditions.

The hypothesis that the observed valley asymmetry (or lack of it) is a result of differences in frost action or length of snow cover alone during Wisconsin times is untenable for areas similar to those examined. Were frost action the main cause, a much greater asymmetry should occur in the Laramie Mountains same. (Compare Table 1, F_o row, with Fig. 2.) The direction of asymmetry is opposite that reported in tundra regions, where frost action is

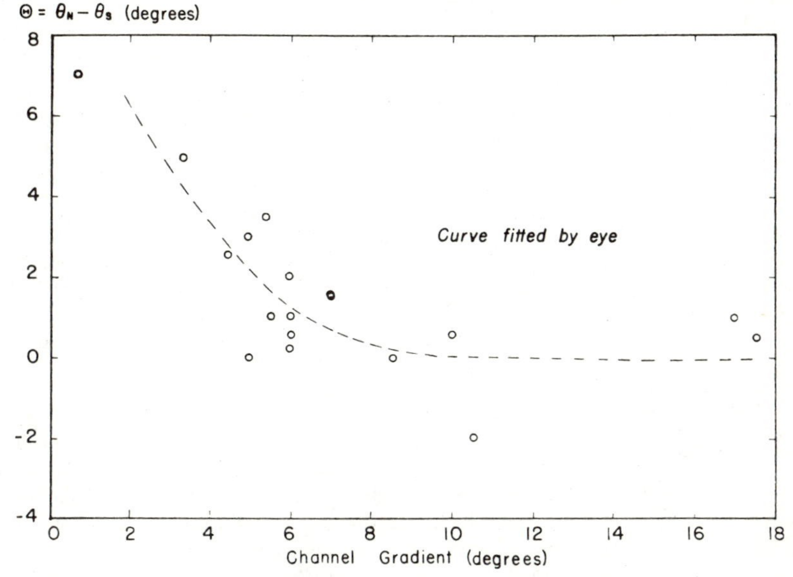

FIGURE 2.—RELATION BETWEEN CHANNEL GRADIENT AND VALLEY ASYMMETRY WITHIN A SINGLE VALLEY

TABLE 5.—COMPARISON OF POTENTIAL FROST ACTIVITY IN SOUTHERN ARIZONA AND SOUTHEASTERN WYOMING

All figures are from Visher (1954). Because of the small scale of the maps and the generality of the sources of data, these figures should be considered only as rough comparisons.

Measure of frost activity	Southern Arizona	Southeastern Wyoming
Normal annual number of days with mean temperature of 20° to 30°	0	>100
Normal annual number of days continuously below freezing	<1	≈45
Normal annual number of nights with frost	<30	≈180
Average annual number of alternations of freezing and thawing	≈25	≈130
Average depth in inches of frost penetration	<3	>20

than in the Lower Sonoran zone of southern Arizona. Table 5 compares modern frost intensity and the frequency of alternation of freeze and thaw in the two areas. The difference may have been greater during the Wisconsin. In neither area is the soil of a type to be subject to much frost heaving. In spite of the considerable difference in potential frost activity, in valleys that have low-gradient channels on the order of 1° – 2°, and away from extraneous factors such as channel bends and alluvial fans, the degree of asymmetry is nearly the morphologically important, although this fact cannot be given much weight.

EFFECT OF EROSIONAL ENVIRONMENT ON DOWNSLOPE PROFILES

The literature contains much discussion of the interpretation of straight, convex, concave, and sigmoidal downslope profiles. The following three points, which are based on purely qualitative observations of many hundreds of individual hills and valleys located in Arizona.

Utah, New Mexico, Colorado, Wyoming, South Dakota, Iowa, and Illinois, are presented to point out that the concept of the erosional environment of a slope facet has some relevance to this subject:

(1) Erosional slopes of valleys in homogeneous rocks in which there is no flood plain, and slopes of isolated hills that have channels maintained at their base, normally have straight profiles (Strahler, 1950, p. 681).
(2) A hill in homogeneous rock that is isolated from other hills and channels normally has concave slopes on all sides, presumably because of the accumulation of debris that has been washed down from a higher, erosional segment, progressively burying the toe more deeply, since no stream is present to transport the debris away (Lawson, 1932, p. 720).
(3) Most convex slopes, on careful examination, appear to be composites of two or three facets that have straight profiles, representing successive entrenchments followed by partial regrading and some smoothing of the corners. In most, the lower portion is straight.

Evidence that erosional environment is the chief determining factor of the figure of the downslope profile, and not regional climate or type of lithology (if homogeneous), can be found in many places where a hill on one side faces an open plain and on the other faces a valley and has an active stream impinging on its base. On most, the profile facing the plain is concave, and the profile facing the valley is straight. These relations are well displayed along the Black Canyon Highway (Arizona Highway 69) between Phoenix and Bumble Bee. Exceptions to these generalizations exist but are unusual in the subhumid to arid regions examined, and they can be explained by special conditions of rock structure or changes in stream regimen.

Sauer (1930, p. 368) has reported the existence of isolated hills in the Chiricahua area of Arizona that have concave, denudational slopes from summit to base. The concavity of such hillslopes cannot be the result of accumulation of debris near the base, as bedrock is present below a very thin cover of slide-rock. The writer examined the "hills of circumdenudation" mentioned by Sauer and many others that are similar and invariably found that an inhomogeneity in lithology existed; *i.e.*, at the summit of each hill there was a layer of rock, more resistant to weathering and erosion than the rock lower down, that retarded erosion of the summit relative to the lower slopes, thus producing the concavity. This relation appears to be so invariant that the writer believes it can be used with confidence to predict inhomogeneities that are not discernible from a distance, provided the hill slopes are entirely erosional.

Summary

The inferences drawn from the foregoing discussions must be that in arid to subhumid temperate climates, excluding the extremes in dryness and humidity[2], (1) low channel gradients favor the development of valley asymmetry in east- or west-trending valleys, the north-facing slopes become steeper as the channels and streams are moved against their toes by filling of debris from the south-facing slopes; (2) steep channel gradients in V-shaped valleys favor more symmetric development of valley sides, in spite of possibly marked differences in microclimate and vegetal density across the valley and the implications of Gilbert's Law of Divides; (3) most cases of valley asymmetry can probably be attributed directly to asymmetric basal corrasion, which can result from a variety of conditions within the valley; (4) both the maximum slope angle and the figure of the downslope profile are functions of the erosional environment of the slope, although the latter is strongly influenced also by rock inhomogeneities; (5) the effects on slope angle of several components of the erosional environment of the slope are additive.

References Cited

Büdel, J., 1953, Die "periglazial"-morphologischen Wirkungen des Eiszeitklimas auf der ganzen Erde: Erdkunde, Bd. 7, p. 249–266
Emery, K. O., 1947, Asymmetric valleys of San Diego County, California: Southern Calif. Acad. Sci. Bull., v. 46, pt. 2, p. 61–71
Gilbert, G. K., 1880, Report on the geology of the Henry Mountains: 2d. ed., Washington, D. C., Govt. Printing Office, 170 p.
Lawson, A. C., 1932, Rain-wash erosion in humid regions: Geol. Soc. America Bull., v. 47, p. 703–724
Pearson, E. S., and Hartley, H. O., 1951, Charts of the power function for analysis of variance tests, derived from the non-central F-distribution: Biometrika, v. 38, p. 112–130
Russell, R. J., 1931, Geomorphological evidence of a climatic boundary: Science, v. 74, p. 484–485

[2] Excluding the dry extreme because no vegetation will grow on either slope; excluding the humid extreme both because no data from such areas were included, and because presumably sufficient moisture is available on all slopes for a complete vegetal cover. Further work is needed in these areas.

Sauer, C. O., 1930, Basin and range forms in the Chiricahua area: Univ. Calif. Pub. in Geog., v. 3, p. 339–414
Smith, H. T. U., 1949, Physical effects of Pleistocene climatic changes in nonglaciated areas: Geol. Soc. America Bull., v. 60, p. 1485–1516
Strahler, A. N., 1950, Equilibrium theory of erosional slopes approached by frequency distribution analysis: Am. Jour. Sci., v. 248, p. 673–696, 800–814
Visher, S. S., 1954, Climatic atlas of the United States: Cambridge, Mass., Harvard Univ. Press, 403 p.
von Engeln, O. D., 1942, Geomorphology: N. Y., The Macmillan Co., 655 p.

Dept. of Geology, University of Arizona, Tucson, Ariz.

Manuscript Received by the Secretary of the Society, October 27, 1958

APPENDIX I.—ANALYSIS OF VARIANCE OF SLOPE MEASUREMENTS COLLECTED IN THE LARAMIE MOUNTAINS, WYOMING

A. Analysis of Variance—Test for Interactions

Source of variation	Sums of squares	Degrees of freedom	Mean squares	F
Row effects				
1. $\frac{1}{2}(F_f + F_a) - F_o$	.00034	1	.00034	
2. $F_f - F_a$	34.44	1	34.44	
Column effects	88.05	1	88.05	
Blocks effects	17.01	2	8.50	
Row × column interaction	4.18	2	2.09	2.21 (2, 4)*
Row × block interaction	1.23	4	0.31	0.33 (4, 4)†
Column × block interaction	3.82	2	1.91	2.02 (2, 4)*
Residual	3.78	4	0.945	
Total	152.51	17	....	....

B. Analysis of Variance—Test for Main Effects

Source of variation	Sums of squares	Degrees of freedom	Mean squares	F
Row effects				
1. $\frac{1}{2}(F_f + F_a) - F_o$	.00034	1	.00034	.00031 (1, 12) (n s)
2. $F_f - F_a$	34.44	1	34.44	31.89 (1, 12) (.001)
Column effects	88.05	1	88.05	81.53 (1, 12) (.001)
Block effects	17.01	2	8.50	7.87 (2, 12) (.01)
Pooled	13.01	12	1.08	
Total	152.51	17	....	....

* $F_{.05}(2, 4) = 6.94$
† $F_{.05}(4, 4) = 6.39$

APPENDIX II.—ONE-WAY ANALYSIS OF VARIANCE OF SLOPE MEASUREMENTS TAKEN IN SOUTHERN ARIZONA, ACCORDING TO LIFE ZONE

Analysis of Variance

Source of variation	Sums of squares	Degrees of freedom	Mean squares	F
Among means	2.86	2	1.43	0.36 (2, 44)*
Within groups	174.95	44	3.98	
Totals	177.81	46	....	....

* $F_{.05}(2, 44) = 3.21$

Evolution

IV

Editors' Comments on Papers 16 Through 21

Geomorphologists have always been concerned to a major extent with changes in hillslope morphology over time, but they are hampered by the brief span of geologic time available for observation of actual slope evolution. In certain areas, such as in badlands or earth-fill areas, it is possible to measure temporal changes in hillslope form, but, more frequently, it is assumed that sampling in space is equivalent to sampling through time (the ergodic hypothesis). Thus, a series of hillslope profiles arranged in terms of their apparent age may provide information on evolution. The first three papers in this part, by Fair, Savigear, and Koons, rely primarily on the latter approach; Schumm's study of badland slopes uses both.

Fair's conclusions have some relevance to other articles in this volume (Bryan, Paper 11, and King, Paper 6). For example, he presents evidence which essentially confirms the concept of boulder-controlled slopes, giving specific data regarding slope angles and associated debris sizes. Fair also observes that a caprock is necessary for parallel retreat to occur. In a comparison of Karroo and Natal slopes, Fair (1948, p. 79) states that "these observations reflect the fine adjustment of landscape to even small changes of process, and bring out clearly the considerably wider contrasts between regions of climatic divergence."

At about the same time Savigear had also discovered, in Britain, the value of accurate survey in geomorphic research. Measurement of slope angles along profiles in South Wales enabled Savigear to conclude that slopes retreat parallel to themselves with unimpeded removal of debris from their foot, and decline with impeded removal. Savigear's field situation was superior to that of Fair because progressive isolation of a cliff face from attack by the sea had occurred, and profiles illustrating progressive stages in the erosional development of cut cliffs could be selected. However, unlike Fair, Savigear largely disregards the erosion processes which modify slopes.

Koons' paper combines field observation and measurement with simple laboratory experiments to explain differences in cliff profiles. He concludes that cliff retreat is not a continuous process; rather, a threshold of stability is exceeded at intervals, and collapse occurs. A long period of stability then ensues. Other papers discussing scarp retreat in the Colorado Plateau are those by Bradley (1963), and Schumm and Chorley (1964, 1966).

Schumm's study of small-scale slopes uses the technique of serial profiles to show that both parallel and declining retreat may occur in badlands. The very high rate of erosion permits measurement of hillslope erosion, which supports the conclusions drawn from the serial profiles. Thus a combination of form and process studies provides conclusive evidence that parallel and declining slope evolution occurs under these limited conditions.

The next paper in this part, by Everard, was selected in order to emphasize that the progress of hillslope evolution may be strongly affected by nongeomorphic factors, especially climatic change and human activity. In Cyprus, post-glacial climatic change, deforestation, and grazing prevent an accurate assessment of the "normal" processes of erosion and landscape evolution. It is clear that the Davis–Penck–King controversies are meaningless when such factors are ignored.

Finally, Jahn's study is included because it considers hillslopes as three-dimensional surfaces, and shows that much valuable data may be obtained from other research areas, in this case from soil science, agronomy, and agricultural engineering (see also Smith and Wischmeier, Paper 28). Jahn amplifies the point made by Everard, that man's activities may have a decisive influence on landscape development. Cultivation is especially important, since its role may be to greatly accelerate and intensify naturally occurring erosional processes.

Reprinted from *Trans. Geol. Soc. South Africa,* **50**, 105–119 (1948)

Slope Form and Development in the Interior of Natal, South Africa.

By T. J. D. Fair.

16

[Plates XIII-XIV].

ABSTRACT.

Slope elements typical of a sub-humid region are described, and the following factors tending to produce or modify these are discussed :—

(*a*) Horizontal structure with alternate hard and soft strata.

(*b*) A sub-humid climate with a summer rainfall of the convectional type.

(*c*) A base-level that is stable or slowly falling.

CONTENTS.

INTRODUCTION.

From the point of view of slope form and development, the chief characteristics of the interior of Natal are that it is an externally-drained region, it is almost sub-humid, and it possesses over the greater part of its extent a base-level that is stable or slowly falling. The region lies behind a line of elevated country which extends from Louwsburg in the north-east, through Ceza, Babanango and Nkandhla to the Karkloof spur*. It is occupied by the basins of the Tugela, Buffalo, Blood, Umfolosi and Pongola Rivers and reaches inland to the foot of the Great Escarpment. Geomorphologic accounts of the area, which is included in Wellington's zone 9C (1946, p. 79), are contained in papers by Dixey (1942, p. 167), King (1944, p. 270) and Fair (1944).

The virtually horizontal Ecca and Beaufort series of the Karroo System are the basement rocks of the region. They consist of thick bands of sandstone alternating with shales, and together with sills of Karroo dolerite, which cap nearly every hill in the region, impose a strong structural control upon landscape. Base-level has remained comparatively stable to permit the production of an imperfectly planed erosion surface, now at an elevation of some 4,000 feet. A younger cycle (Pleistocene) (King, 1947) is encroaching up the major rivers.

* See Topographical Map of the Union of South Africa, 1 : 500,000, Sheet 2, Natal. Published by the Union Irrigation Dept., 1936.

The climatic regime, of which the outstanding feature is the concentration of 80–90 per cent. of the rainfall in the summer season, is illustrated by the accompanying figures :—

Mean Annual Rainfall.	Summer Per Cent.	Average Rainy Days per Year.	Average Intensity per Rainy Day.	Mean Jan. Max.	Mean July Min.	Mean Annual Range.
25–35 in.	80–90	67	·45 in.	80–90°F.	30–45°F.	40–55°F.

Convectional downpours of high intensity (up to 4 inches per hour) are accompanied by rapid run-off. Mist is absent and high summer temperatures (Wellington, 1934) favour rapid evaporation. Grass is the chief vegetative cover but thorn bush commonly occurs in drier areas of less than 30 inches of rain per annum. At higher altitudes above 4,500 feet on divide areas and along the Great Escarpment conditions are humid due to mists and to a higher and better distributed rainfall.

SLOPE ELEMENTS.

Four elements may constitute a slope, though all may not be present in a single slope (Wood (1942), and King and Fair (1944)). They are the upper convex element or waxing slope, the outcrop of bare rock or free face, the talus or detrital slope and the lower concave slope flattening towards the valley, known by various names of which waning slope, wash-slope and pediment are common. Detailed measurement of hillslope profiles in the field has been the chief method of study adopted by the writer in order to determine the form of slopes as the cycle progresses and the factors tending to produce or modify these slopes.

THE FREE FACE.

The free face is the youngest element in any slope cycle, but it may persist into the mature landscape provided detritus is removed from the foot of the slope at a rate more rapid than it is supplied. Such a condition may be fulfilled if the hill is composed of rock of considerable resistance to weathering. In Natal a free face on hillsides overlooking the 4,000-feet surface is disappointingly rare, and imposing sandstone cliffs of a height of 100 feet or more are confined to those areas where youthful Pleistocene dissection has occurred, as, for example, along the eastern slopes of Dumbe, Hlobane and Enyati Mountains. In the pediplaned (King, 1947) region cliffs of more than 25 feet in height occur most commonly where small streams head up into hillsides. For the most part, the capping sills of dolerite, in places 200 to 300 feet thick, often shed, on weathering, quantities of spheroidal boulders which litter hillsides and often obscure the underlying sandstone bands.

Further, as most hills have retreated beyond the effective reach of lateral planation by large streams, this efficient mechanism for the removal of waste at the foot of hills is absent which, in turn, guarantees the persistence of the free face provided rock type is suitable. Nevertheless, slopes are remarkably steep and, while the processes of transportation at the foot of hills may not be able to remove the debris at a rate greater than that of supply, they can, at least, keep pace with it.

Provided conditions are favourable, the middle Ecca sandstones give the finest cliffs, followed by the middle Beaufort sandstones. The dolerite, unless columnar jointed, seldom gives free faces but rather an untidy bouldery slope.

THE TALUS OR DETRITAL SLOPE.

The angles of talus slopes on practically all dolerite hills and scarps in the region are governed by the angle of repose of the dolerite boulders, and the fact that steep slopes are maintained as scarps retreat suggests that the factors governing their initial appearance are not changed. As the Great Escarpment retreats, dolerite sills appear in the landscape at all elevations above local base-level. The higher the elevation the more easily are boulders dislodged and the larger is their size. As size of detritus determines mainly the degree of declivity of the slope, it is the hills, which are highest with respect to local base-level that have the steepest taluses.

On all talus slopes boulders vary in size from 4 inches to 4 feet across but any particular slope is determined by the predominance of a particular size of boulder. The following are the approximate ranges of angles of repose for the detritus of talus slopes of dolerite or dolerite mixed with sandstone :—

Size of Detritus	Angle of Repose
3-4 feet across	30-35 degrees
2-3 feet across	23-29 degrees
1-2 feet across	19-23 degrees

All taluses decrease in declivity downslope because :—

(i) Detritus decreases in size, and

(ii) the increasing accumulation of soil washed down from above causes free talus to pass into bound mantle, so that boulders lie at angles less than the maximum angle of repose for their size.

On a talus ranging from an angle of some 35 degrees to about 19 degrees, the rate of change of declivity varies from 5 to 8½ degrees per 100 feet, depending not only on the height of the hill top above local base-level but also on the type of waste predominating on the slope. For both dolerite and sandstone waste, the upper angle of the talus is much the same but the dolerite, though easily dislodged from its parent sill, is mechanically resistant and does not shatter easily on falling as do blocks of sandstone. The result is that taluses composed solely of Beaufort or Ecca sandstone decrease in angle more rapidly downslope than predominantly dolerite boulder slopes and consequently pass at a gentler angle from talus to pediment.

The profile of the underlying bedrock, whether it is shale or sandstone, provided it is adequately covered by talus, follows closely that of the talus slope. In certain instances, however, where a dominant sandstone band or dolerite sill is absent the hillside is made of a series of small free faces with intervening rocky slopes, and the slope as a whole is governed by the occurrence of these small outcrops (see Fig. 1). Slopes under these conditions are not as steep as those dominated by a thick cap-rock, and, therefore, much of the steepness of hillsides in the interior of Natal must be ascribed to the structural control exerted by such a cap-rock.

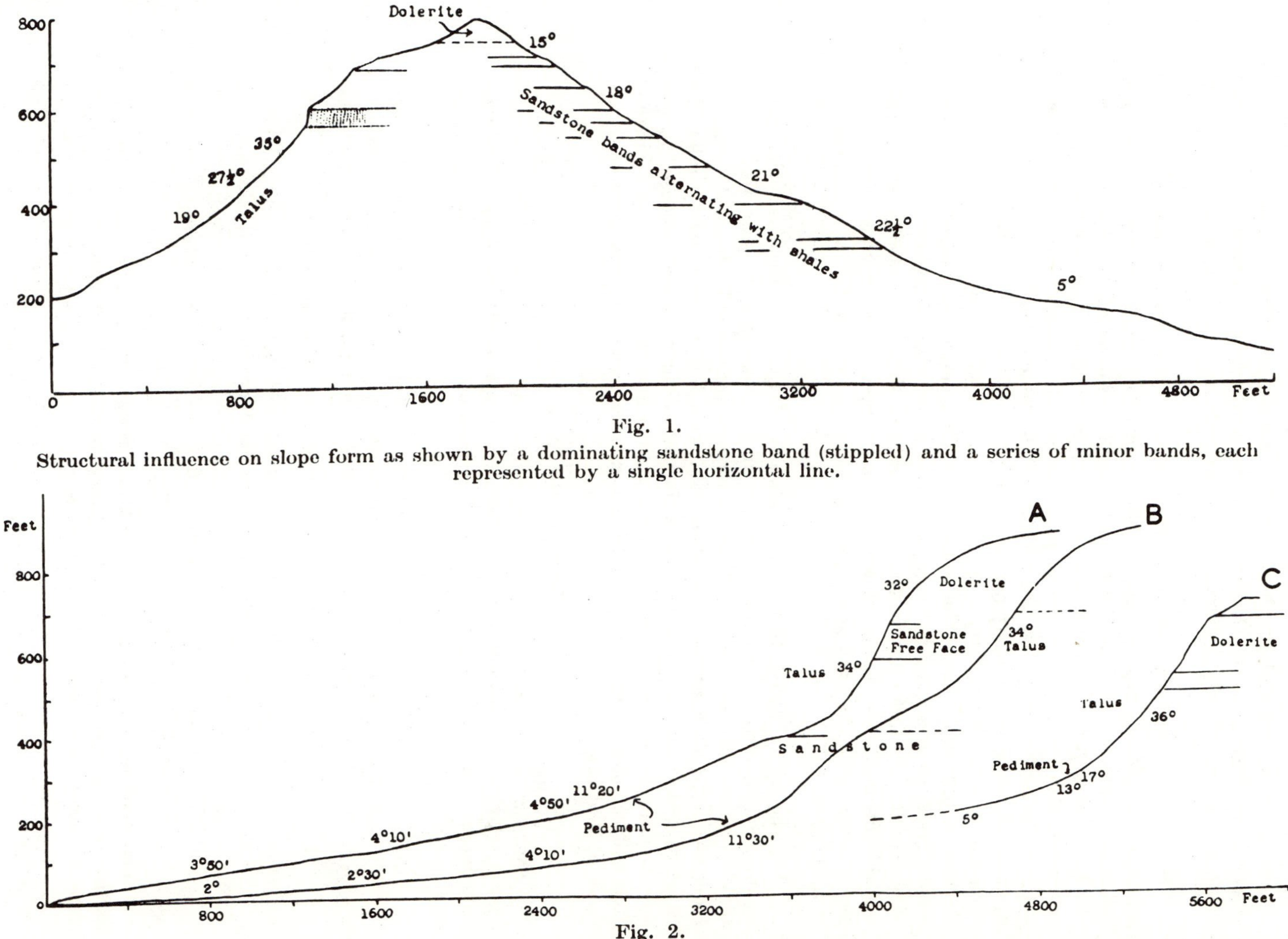

Fig. 1.

Structural influence on slope form as shown by a dominating sandstone band (stippled) and a series of minor bands, each represented by a single horizontal line.

Fig. 2.

Three profiles of a middle Beaufort sandstone scarp, capped with dolerite. A and B are broken by intermediate ledges. C is a profile of the scarp seen in Plate XIV, Fig 1.

It has been stated above that the minimum angle for true boulder-controlled slopes is about 19 degrees. Below this angle occurs a zone of transition between talus and pediment, which joins the two elements in a graceful curve (see Fig 2C and Plate XIV, Fig 1). In this zone boulders of the talus disappear beneath the accumulation of soil washed down from above. Boulders do not lie at their maximum angles of repose and the movement downslope of waste under the influence of gravity gradually gives way to the removal of fine waste by sheetwash. It is a zone in which neither gravity nor sheetwash are the dominant processes, but one in which waste is prepared for transportation on the lower angle of the pediment. Thus the rate at which waste is comminuted determines in no small measure the sharpness of angle between talus and pediment. Apart from the relationship between the rate of supply of debris and its removal at the foot of hills and scarps, the weathering characteristics of dolerite are highly important in determining the comparatively sharp change from talus declivity to pediment. As free talus on the upper parts of slopes, the mechanically resistant dolerite boulders weather slowly but, downslope, as they become increasingly buried under soil, they are kept moist and weather chemically at a rapid rate to smaller sizes.

On the other hand, there are examples of talus slopes passing without an intervening zone to the pediment. Boulders, one foot in diameter, resting at 14 degrees end abruptly and are replaced immediately by pediments at 4 degrees. All such profiles are associated with low rocky dolerite bults, the slopes of which consist solely of free, unburied boulders, which, exposed in this way, remain mechanically resistant from top to bottom of the slope. There is thus a direct change from boulder to soil and such sharp knicks are not unlike those of arid regions.

THE PEDIMENT.

Sharply rising scarps and inselberge* with well-developed pediments are characteristic of the scenery of the 4,000-feet surface (see Plate XIII). The wide occurrence of pediments is the result of certain outstanding features common to the region :—

(i) Base-level is stable or very slowly falling and the retreat headwards of knickpoints associated with the younger Pleistocene cycle is effectively retarded by formidable mechanically-resistant dolerites.

(ii) Much of the area is planed across weak Karroo shales.

(iii) Climate causes the formation of a surface hardpan which considerably aids run-off, and rainfall of the thunderstorm type is dissipated first as sheetwash and only subsequently does it collect into stream channels. This wash is wholly capable of transporting comminuted detritus from the foot of the talus slopes.

The pediment is essentially a rock-cut surface covered by 15 to 20 feet, at the most, of fine waste. In longitudinal profile, though gradients vary from slope to slope, the pediment rises very gently from local base-level to about 5

* Not to be confused with the inselberge of savanna landscapes.

degrees after which there is a rapid increase to a maximum angle varying from 7 to 13 degrees below the foot of the talus slope. The following examples serve to illustrate the gradients of some typical pediments :—

(*a*) 3° to 5°50′ over 4,700 feet : 5°50′ to 8°50′ over 200 feet.
(*b*) 2°50′ to 5°10′ over 2,800 feet : 5°10′ to 9°20′ over 300 feet.
(*c*) 1° to 4°50′ over 2,600 feet : 4°50′ to 12°40′ over 500 feet.
(*d*) 2° to 5° over 2,250 feet : 5° to 11°30′ over 500 feet.
(*e*) 4° to 4°50′ over 2,150 feet : 4°50′ to 13° over 650 feet.
(*f*) 1°10′ to 5° over 1,350 feet : 5° to 12°30′ over 300 feet.

Whereas the upper part of the pediment flattens at the rate of 1 to 2½ degrees per 100 feet, the lower part does so at 2 to 20 minutes per 100 feet so that here, too, the contrast between hillside and plain is emphasized. Though it rises to a maximum angle of 13 degrees the pediment is essentially a slope of less than 5 degrees over by far the greater part of its length.

To determine the upper limit of the pediment with precision is as difficult as setting a limit to the lower part of the talus slope. The upper parts of pediments thinly covered with free loose boulders are as common as the lower parts of talus slopes completely covered with soil. However, it is clear that the maximum angles of individual pediments fall roughly into two groups, viz., 11 to 13 degrees and 4 to 9 degrees. In the first group are pediments associated with major dolerite and sandstone scarps at the foot of which the smooth curve of the transitional zone joins the talus to pediment. In the second, are pediments cut on shale at the foot of low dolerite bults, described above, but these are less conspicuous elements in the landscape.

In transverse profile parallel to the foot of the scarp, the pediment is broken by small streams, both ephemeral and perennial, spaced at intervals varying from ¼ to 1 mile. Between the streams the transverse profile is straight. The spacing of these shallow stream channels decreases as the drainage area becomes smaller, so that in most instances inselberge are surrounded by pediments unbroken by such streams (see Plate XIII).

Any account of pediment evolution involves both the method of scarp retreat as well as the processes operative on the pediment itself.

Scarp Retreat. The fact that inselberge and scarps overlooking the 4,000-feet surface have retreated considerable distances from their initial positions and have retained their steep sides, is sufficient evidence to postulate a parallel retreat of slopes. Admittedly, these unconsumed masses are structural features but, unless material is removed from the foot of the hillsides at a rate not out of proportion to that supplied, then such parallel retreat cannot occur.

There are two possible means whereby scarps may be kept steep ; by lateral planation at the foot of the scarp (Johnson, 1940, p. 232) or by the removal of weathered waste by water-work.

In considering lateral planation, the work of three stream types must be examined. They are major through-flowing rivers, perennial tributaries, and wet-weather gullies and ravines on the sides of hills. Apart from a narrow zone at the foot of the Great Escarpment the major rivers play almost no part in keeping scarps steep as the latter have retreated beyond the range of planation of such rivers. Base-level is slowly falling at a rate sufficient to retard lateral planation by rivers already well supplied with water, so that floodplains of more than a mile in width are uncommon. The function of perennial tributaries is primarily to break up major divides into smaller units, e.g., the Biggarsberg (King, 1944, p. 271), and they, too, achieve little lateral planation either in the embayments or where they debouch from the scarp fronts. Instead, responding to the slowly falling base-level, they cross the pediment in shallow channels. In local instances, where they do swing laterally against scarps, they are not sufficiently closely-spaced to be responsible for wearing back scarps as a whole.

In arid regions, Howard (1942, p. 100-2) notes the ability of small closely-spaced streams in gullies and ravines to achieve lateral planation at their bases. With reference to the Book Cliffs, Utah, he says, " The steep gradients of the ravines, their close spacing, and the weak sediments in which they are cut, all suggest that streams are primarily responsible for the recession of the scarp and for the sloping surface between the scarp and the pediment beyond." In a somewhat similar fashion, though applied on a larger scale, King (1944, p. 278) describes the retreat of the well-watered Natal Drakensberg as follows, " Irregularities in the outline of the crest are related to the heads of Natal streams, and spurs project into Natal for, commonly, three to five miles. For a scarp which is being driven back principally by headstream erosion, these spurs are remarkably short. Their distal extremities appear to be removed by widening of the river valleys rather than by wasting under weathering. The contrast between the steep headwater streams and the rivers of the lower valleys is remarkably abrupt . . . "

Provided then, that such gullies are sufficiently closely-spaced whether they be occupied by the ephemeral floods of arid regions or the perennial streamlets of humid regions, it is possible that lateral corrasion may be achieved at their base. In the sub-humid interior of Natal such gullies occur but well-mantled talus slopes do not favour the collection of water into channels and, hence, they tend to be rather widely spaced. Many of these gullies at the foot of scarps are continued across the pediment as shallow channels, others spread out fanwise as concentrated flow is transformed to sheetwash. These latter do achieve a certain amount of lateral planation but the evidence of undercut banks is so limited and the distance between the gullies sufficiently wide, that lateral corrasion cannot be considered the major process keeping scarps steep. Finally, it has already been stated that as the drainage area of inselberge becomes smaller such gullies disappear entirely.

Thus it is considered that, while large unconsumed divides may be considerably broken up into smaller units by large perennial streams, the retreat of hillsides is accomplished primarily by weathering together with the efficient removal of the debris by sheetwash at the foot. The movement downhill of the easily dislodged dolerite boulders under gravity is greatly facilitated by sheetwash and rillwash on the talus slope itself. That overgrazed talus slopes may suffer donga erosion (see Plate XIV, Fig. 1) is sufficient testimony to the power of sheetwash and the fact that small gullies on hillsides do no more than notch the crests, shows that weathering keeps pace fairly well with gully-head erosion.

At the foot of the hillslope the comminution of detritus has been shown to be comparatively rapid, and that sheetwash is wholly capable of transporting it is demonstrated by the rapid flattening of the pediment. Further, the fact that the transitional zone between talus and pediment is characteristic of almost all the slopes in the region is strong evidence that lateral corrasion at the foot of hills is wholly subordinate to the retreat of slopes by weathering. Thus weathering down the whole slope and continuous removal of fine debris by sheetwash at the base, aided by favourable conditions of structure, explains the retreat of slopes with constant declivity in this sub-humid region.

As scarps retreat the grass-covered pediment evolves as a true wash-slope, generally, but not everywhere, channelled at intervals by shallow stream courses. On weak shales and particularly on the weak old granite of the north-eastern part of the region round Paulpietersburg, these streams swing laterally giving discontinuous segments slightly below the general level of the pediment and with slightly smaller gradients.

Dongas. At this stage some aspects of the relation of dongas to the pediment and the talus slope (King and Fair, 1944) may be usefully described, as illustrating not only modification of processes which result from disturbance of natural conditions, but also the power of sheetwash on the pediment. Slope dongas, often more than 20 feet deep, are of two types: those that cross the whole length of the pediment, or will do so as headward erosion proceeds, and those, less in number, that occur at the head of the pediment only. Both are due, as is well known, to the change from sheetwash to concentrated water flow, but certain interesting responses to changes of load are evident.

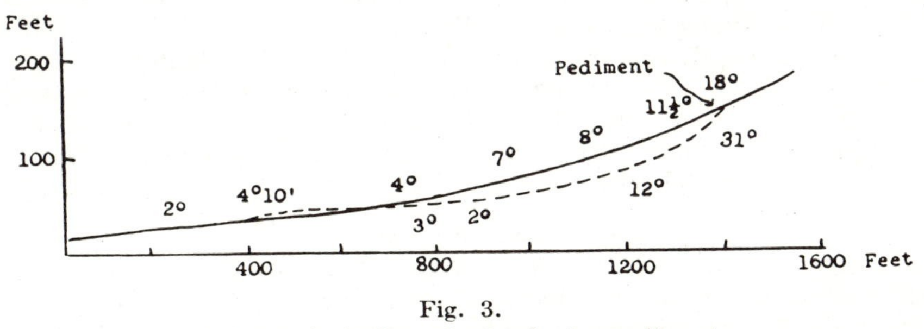

Fig. 3.
A donga incised at the head of a pediment.

The first type, once their heads penetrate to the foot of the transitional zone between talus and pediment, often aggrade considerably throughout much of their length. The second type, generated initially at the head of the pediment, are almost immediately overcome by an excess of fine waste from the transitional zone so that no sooner is this picked up by the ephemeral stream than it is spread out at the mouth of the donga as a very low fan with a slightly increased gradient. Thus concentrated flow changes to sheetwash, a reversal of processes usually associated with the upset of natural conditions. The gradients of a typical donga, the pediment into which it is incised and the associated fan are illustrated in Fig. 3.

Further, the soil transported by wash from the mouth of such dongas can easily be traced ½ mile and more down the pediment on gradients of less than 3 degrees in spite of a cover of tall grass. These facts serve to show the great quantity of fine detritus that is available for delivery to the pediment by decay of talus waste and the ability of sheetwash to transport it on comparatively low gradients.

Regrading of Pediments. One of the fundamentals of slope development in externally drained regions with graded rivers is that local base-level may be stable or, and this is probably commoner, it may be slowly falling. Certainly, the latter is the case so far as the 4,000-feet surface is concerned, for it is being developed westward, its inner edge growing as the Great Escarpment retreats and its outer margins being encroached upon by the younger Pleistocene cycle. Continual regrading of pediments by streams, but more widely by sheetwash, has been in progress, a process easily accomplished in shale areas.

The result of such regrading is that the height of scarps between their crests and the head of pediments is not reduced by progressive pedimentation at the rate they would be if base-level was stable. Hence, the downward extent to which backwearing of the hillside can proceed, determined by the gradient given to the pediment by its sheetfloods (Davis, 1933) is increased, and skylines in which the pediment reaches smoothly to the top of major divides, more than, say, 500 feet above local base-level, are absent.

A structural factor is also involved here, of course, for the pediment rises highest up hillsides capped by only thin dolerite sills or sandstone bands and underlain by weak shales (see Fig. 4B). The upper parts of such features, nevertheless, still retreat parallel to themselves provided the cap persists but generally they do not stand high above local base-level. Certainly, apart from the effects of continuous regrading of pediments, stream spacing is not sufficiently wide to allow for the disappearance of scarps (parallel to through-flowing rivers) by progressive pedimentation, but continuous regrading does keep retreating scarps higher than they would be were local base-level stable. It must add considerably to the prominence of retreating major cyclic scarps (King, 1947) running roughly parallel with the coast.

LOWERING OF DIVIDES.

Under conditions of parallel retreat as here described, divides are lowered as opposing hillsides meet one another along sharply rounded crests (see Fig. 4A), and the closer that through-flowing streams are spaced, the sooner will lowering be achieved. As the supply of debris from above becomes less and less, the talus slope assumes angles of decreasing declivity until it finally disappears. The pediment then merges into the convex or waxing slope of the summit and subdued rounded forms result. This is the only method whereby the talus slope is caused to disappear in this region for, due to the comparative close spacing of streams governing local base-level and the slow fall of base-level, progressive pedimentation of the scarps of structural plains does not occur.

The highest angle that can be found on the sides of such subdued forms depends primarily on the upper angle of the pediment on the original slope before disappearance of the talus slope, for in this region the ever-decreasing talus is replaced by the upward growth of the pediment rather than by the downward

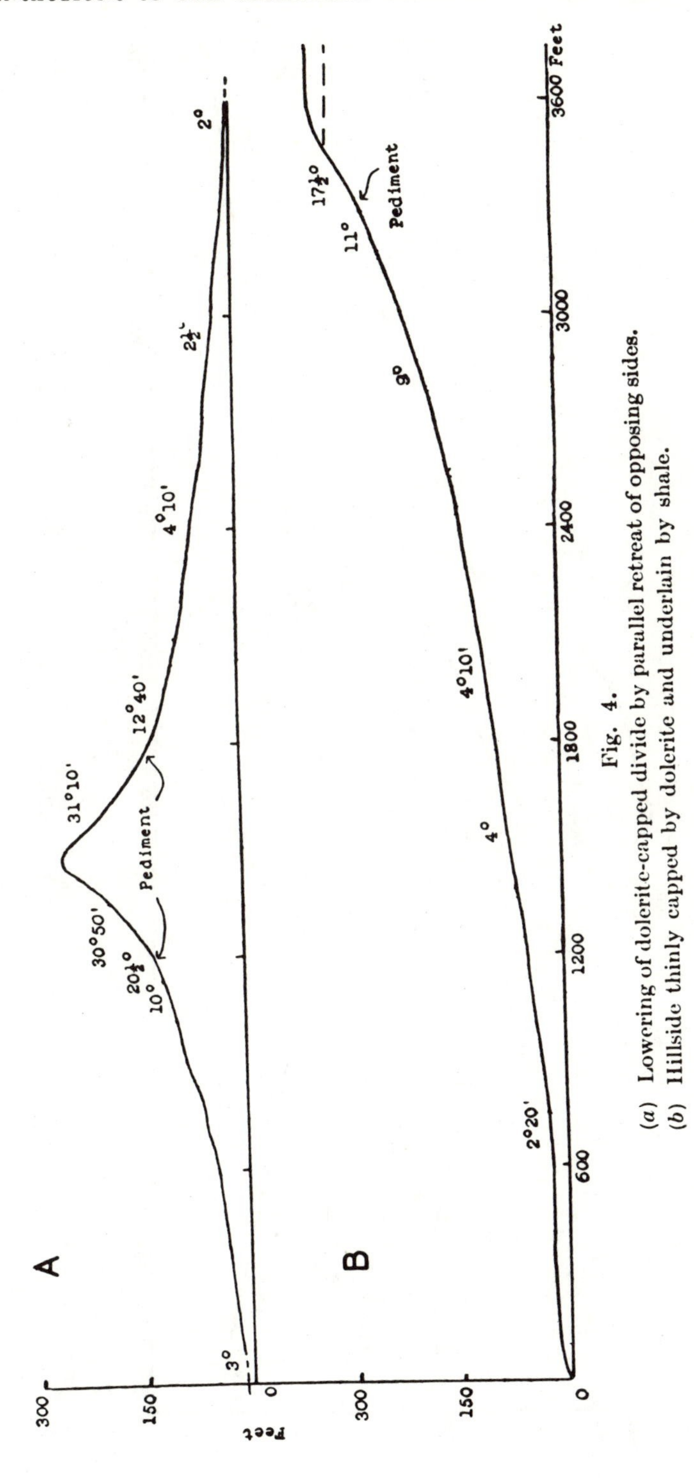

Fig. 4.

(*a*) Lowering of dolerite-capped divide by parallel retreat of opposing sides.
(*b*) Hillside thinly capped by dolerite and underlain by shale.

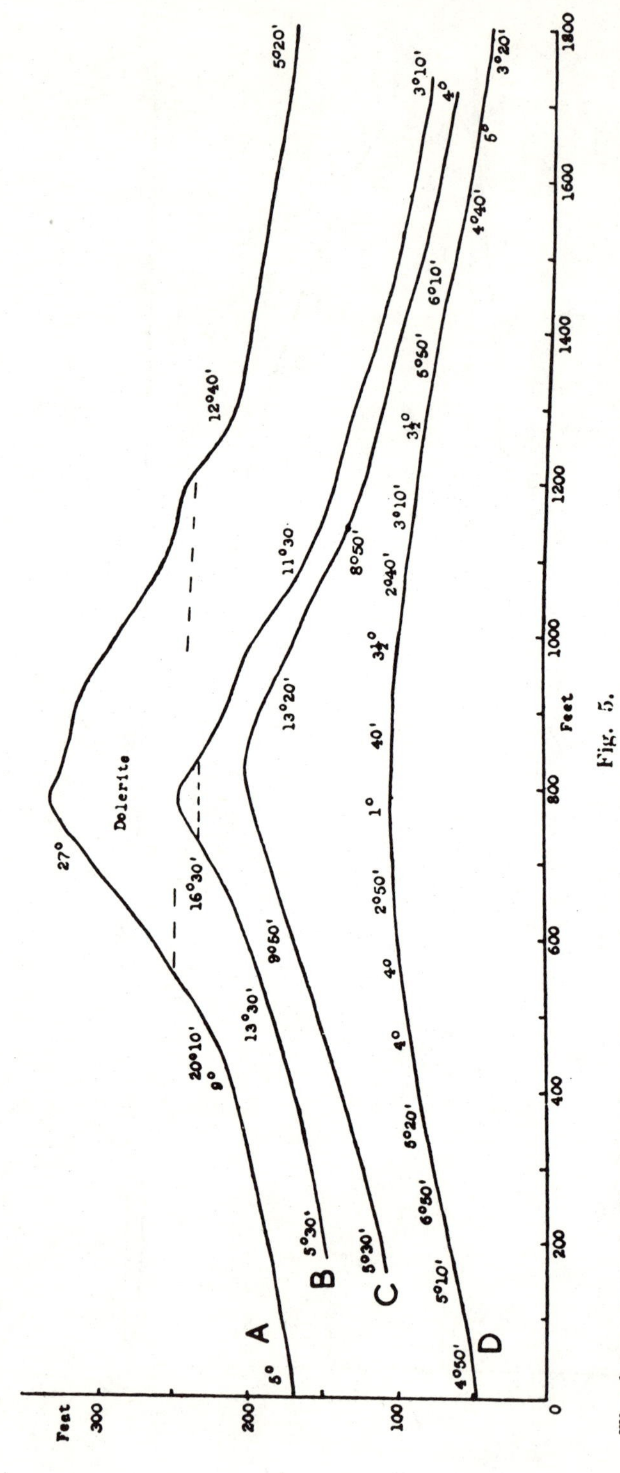

Fig. 5.

The later stages in the lowering of a dolerite-capped sandstone divide. The heights of the profiles are not relative to one another. On the right-hand side profiles extend another 2,000 feet to local base-level.

growth of the waxing slope. Thus transport of fine debris is accomplished by waterwork rather than by creep and as soon as the pediment meets the waxing slope there results a mound-like form on which concave lower slopes are dominant over convex crests. At this stage, these subdued forms are clearly intermediate between the predominantly rounded forms of mature humid landscapes and the sharp crests and concave lower slopes of arid regions.

As continued reduction proceeds, two further changes in slope form are clearly discernible (see Fig. 5). First, both the area and the radius of curvature of the convex crest increase but at no time does it become greater proportionately than the concave slope. Secondly, a straight slope between summit convexity and pediment concavity becomes more noticeable. This is no new slope element but merely a zone of change between slope profiles

That these swells should consist predominantly of concave slopes is understandable in a region in which sheetwash is highly active, but the increasing area of the convex summit as the top of the swell is lowered and declivities are reduced from a maximum of 13 degrees to less than 3 degrees, requires explanation. Similar features in humid regions have been dealt with by Lawson (1932). The following are characteristic of these subdued forms :—

(i) Soil depth (A Horizon) is remarkably uniform, being about 1 to 1½ feet.
(ii) Grass cover is uniform.
(iii) Effects of rainwash between grass tufts may be observed over the whole swell.
(iv) Occasionally small rock bands may interfere with the smoothness of the slopes but in most instances structure is wholly subordinate to process at this stage.

When the waxing slope and the pediment are first joined along a maximum slope of some 13 degrees, it is possible that soil creep is operative and plays some part at least in rounding the low swell in the same way that weathering and soil creep along the tops of scarps rounds off their edges. But as the swell is lowered and declivities decrease, it is difficult to assign to soil creep, if it operates at all on such low declivities, a part superior to sheetwash in fashioning the summit convexity. Rather does the explanation lie in volume-load relationships on a predominantly water-controlled slope.

Along the summit, fine particles of soil, released during dry periods (thus is the summit slowly lowered) are removed by wash. This however, is at a minimum on the summit and requires an increasing declivity, for the amount of material carried down the slope is cumulative and at first the volume of the water does not increase proportionately, as much is lost by seepage and evaporation. In a storm, however, the initial deficiency is overcome and the amount of water rapidly increases downslope until it is wholly capable of transporting the fine waste on the gentler slopes of the pediment. Transportation is thus the function of sheetwash over the whole swell though slow regrading of the slopes proceeds in relation to base-level changes as already described for the pediment. The rapidity with which the volume of water is able to cope with its load downslope will determine the position of the point or zone of change (straight slope) from a convex to a concave profile and this is borne out by the observation that where a slight concentration of sheetwash occurs in shallow hollows on the sides of the swells the convex summit is of very limited area.

It is important to note that unlike the pediment passes (Howard, 1942, p. 108-9) and domes of arid regions the convex summit is not initiated on the pediment as the mountains diminish in size but after they have disappeared altogether. Structure is, however, the probable cause—homogeneous granitic terrains in the arid regions and a heterogeneous stratified landscape in this region.

The proportion of convex to concave elements varies from profile to profile but the area of the rounded summit does not progressivly become smaller as base-level streams become more closely-spaced. Thus, the proportion of the convex summit to the concave portion of the slope is relatively increased as stream spacing becomes closer. For example, on swells of a maximum gradient of 5 degrees this proportion may be as high as 50 per cent. if streams are only 2,000 feet apart, but in most instances they are three and more times this distance apart so that concave slopes always occupy a larger area.

CONCLUSION.

It has been shown that hillslopes remain steep and retreat parallel to themselves up to a certain stage due to favourable conditions of structure at the crest of scarps and an adequate rate of removal of debris at the base. For the same conditions of climate, and therefore of process, however, would slopes retreat with constant declivity if the influence of structure was eliminated? On shale areas, invaded by the Pleistocene cycle (e.g., between Colenso and Weenen) where no hard cap exists, there is no evidence to suggest that slopes retreat parallel to themselves except in gully headwalls and along stretches kept steep by lateral planation by streams. Elsewhere the rise of the pediment upwards, and the less conspicuous downward growth of the waxing slope cause slopes to flatten quickly.

Slopes on the weak granites of the Valley of a Thousand Hills (outside the area herein discussed) also reflect how they either retain or lose their initial steepness depending on the processes that operate upon them. It is a submaturely dissected region (King, 1942) in which remnants of the End-Tertiary peneplain (2,000-feet surface) are being consumed essentially by the parallel retreat of slopes under the influence of closely-spaced gullies with headwalls of as much as 35 degrees. The angle between the peneplain surface and the headwall is comparatively sharp. In the Valley rain shadow effects are remarkably clear and in certain localities spurs are asymmetrical. They are closely scored by gullies on the one side and are, therefore, steep (more than 30 degrees), while on the other, soils are thinner, thorn trees and euphorbias occur, grasses are poorer, gullies are much more widely spaced, sheetwash is superior to soil creep, for profiles are concave, and slopes are flatter (19 to 24 degrees). Similarly, on upfaulted blocks on the End-Tertiary peneplain where these are thinly capped with Table Mountain sandstone and underlain by granite, the steepest slopes are those on which concentrated waterwork is the chief process (see Plate XIV Fig. 2). On the other hand, where weathering and sheetwash alone are operative slopes are less steep. There is ample evidence of this from profiles drawn in this region.

Thus, omitting the effects of lateral planation by major streams, it would seem that in the sub-humid parts of Natal the retreat of slopes without loss of declivity depends primarily upon the presence of a hard cap-rock, and, if this is absent, then on the work of closely-spaced gully heads which generally corrade laterally at their base. In the interior of Natal, and this applies to much of South Africa, structural control is the primary factor keeping slopes steep. Under these conditions sheetwash is able to remove debris from the foot of the hill at a rate not out of proportion to the supply of weathered waste. On weaker rocks slopes flatten almost as soon as they are generated, but as sheetwash is superior to soil-creep concave slopes are dominant over convex.

In the more humid parts of the province with more than 40 inches of rainfall per annum, for the same conditions of structure, the intensity of sheetwash at the foot of slopes is inadequate to remove debris at a sufficiently low angle, due, among other factors associated with a humid climate, to a greater rate of weathering along the scarp and to retardation of sheetwash by a thicker vegetative cover. Concentrated water flow, however, becomes increasingly important and to closely-spaced headwater gullies and streams must be attributed the retreat, with constant steepness, of the Great Escarpment and the scarp backing the 2,000-feet End Tertiary peneplain.

The retreat of scarps under these humid conditions is naturally more rapid than the retreat of those on which weathering and sheetwash are mainly at work, so that the 4,000-feet peneplain develops westward, as the Great Escarpment retreats, more rapidly than the structural remnants upon it can be consumed and it remains perpetually an imperfectly planed erosion surface.

ACKNOWLEDGMENTS.

The writer is indebted to the Council for Educational, Sociological and Humanistic Research for financial assistance in connection with the study, and to Dr. L. C. King for helpful criticism. Aerial Photograph No. 37627 (Plate XIII) is reproduced under Government Printer's Copyright Authority No. 678 of 21 2 47.

BIBLIOGRAPHY.

DAVIS, W. M.: "Geomorphology of Mountainous Deserts." *16th Intern. Geol. Cong. Report.*, Vol. 2, 1933.
DIXEY, F.: "Erosion Surfaces in Central and Southern Africa." *Trans. Geol. Soc. S.A.*, Vol. XLV, 1942.
FAIR, T. J. D.: "The Geomorphology of the Ladysmith Basin, Natal." *S.A. Geog. Jour.*, Vol. XXVI, 1944.
HOWARD, A. D.: "Pediment Passes and the Pediment Problem." *Jour. Geom.*, Vol. V, Apr., 1942.
JOHNSON, D. W.: Comments in *Ann. Assoc. Amer. Geogrs.*, Vol. XXX, No. 4, 1940.
KING, L. C.: "Notes to Accompany a Topographical Map of Part of The Valley of a Thousand Hills, Natal." *Anns. Natal Museum*, Vol. X, Pt. 2, 1942.
KING, L. C.: "Geomorphology of the Natal Drakensberg." *Trans Geol. Soc.. S.A.*, Vol. XLVII, 1944.
"Landscape Study in Southern Africa." *Pres. Add. Geol. Soc. S.A.*, 1947.
KING, L. C. and FAIR, T. J. D.: "Hillslopes and Dongas." *Trans. Geol. Soc. S.A.*, Vol. XLVII, 1944.
LAWSON, A. C.: "Rainwash Erosion in Humid Regions." *Geol. Soc. Amer. Bull.*, Vol. 43, 1933.
WELLINGTON, J. H.: "Thermal Regions in Natal." *S.A. Geog. Jour.*, Vol. XVII, 1943.
"A Physiographic Regional Classification of South Africa." *S.A. Geog. Jour.*, Vol. XXVIII, 1946.
WOOD, A.: "The Development of Hillside Slopes." *Proc. Geol. Assoc.*, Vol. 53, 1942.

SOUTH AFRICAN NATIVE COLLEGE,
FORT HARE.

Discussion.

Mr. F. A. Steart.

Being familiar with much of the country referred to by Mr. Fair, more particularly in Northern Natal, I have been much interested in studying his excellent and informative paper.

The author points out that thick beds of sandstones alternating with shales, together with the sills of Karroo dolerite now capping many of the hills, impose strong structural control upon landscape.

The effect of the dolerite intrusions seems to have received hardly sufficient prominence. Dolerites, more commonly as sills, extend laterally over great distances; such sills, varying from a few to hundreds of feet in thickness, behave peculiarly in the Natal coal fields. In some places they pass, within a short distance, from one to another sedimentary horizon, perhaps hundreds of feet higher. There are many instances where continuation of an intrusion forming the capping of a hill can be traced almost to the base of the hill. In their reconstructed form, some would appear to be great hollow domes or canopies, each enclosing a mass of sediments, extending in some instances, over a very wide almost circular area. Other cores are elongated, as a result of undulating intrusions, which may have enveloped a mass of sediments many miles in length, though of relatively small width.

In either case, through the protection afforded by the enveloping dolerite, the erosion of the sedimentary core has been delayed. The result of such delayed action may be seen in the splendid isolation of certain hills, or of a long narrow range of hills.

One grand example is Makatees Kop, near Paulpietersburg. Another is Ilanga or Job's Kop in Umsinga Native Reserve, east of Ladysmith. Here, at the base, on its eastern slopes, a great dyke-like intrusion leans against the mass of sediments to form a fine bastion, and rises to build the almost horizontal capping to the mountain.

Bodies of this nature, or great intersecting inclined sills, of which there are many examples, modify slope development in no small degree and their influence should not be underrated in the general scheme of physiographical evolution.

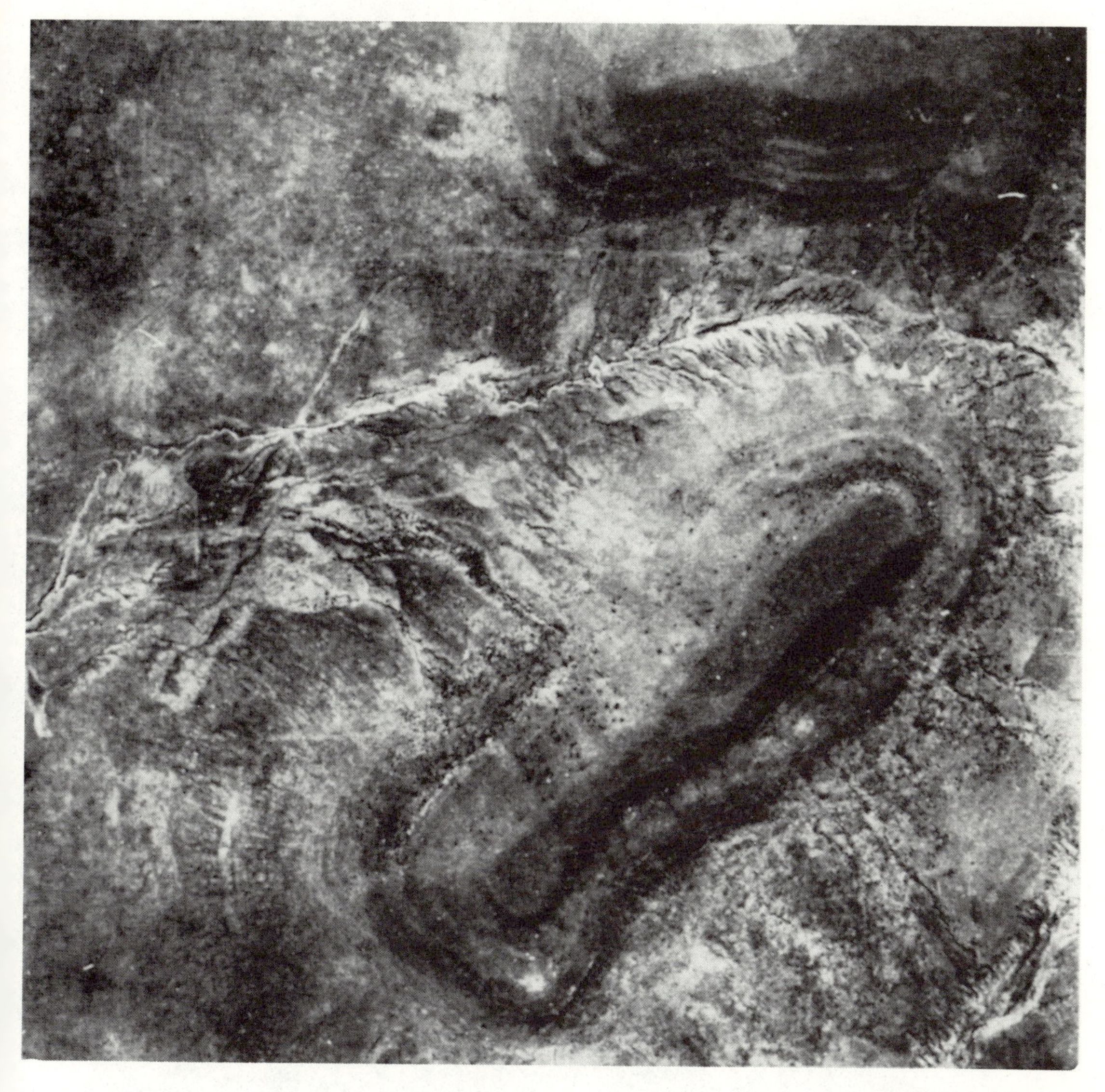

[*Photo: Aircraft Operating Company.*

A pediment developed round the base of a dolerite-capped hill, Northern Natal. Note erosion by dongas.

Fig. 1.
Loskop (see Fig. 2C) a 500-feet high mesa of middle Beaufort sandstone strengthened by a dolerite sill near the crest. Dongas on the talus slope are due to overgrazing. Pediment in the foreground.

Fig. 2.
Inghomunkulu, near Umlaas Road, a granite hill thinly capped with Table Mountain sandstone showing the effects of concentrated water work on the one side and sheetwash on the other.

Reprinted from *Inst. British Geog. Trans.*, **18**, 31–51 (1952)

SOME OBSERVATIONS ON SLOPE DEVELOPMENT IN SOUTH WALES

By R. A. G. SAVIGEAR, B.A.
(*University of Sheffield*)

17

Introduction

THE nature of slope development is complex and has been discussed at length in several books and papers, but the writer knows of no publication containing precise measurements of slope forms derived from field survey. Such measurements would appear to provide a very satisfactory basis for discussing certain aspects of the problem. This paper contains an account of some attempts to measure slope forms in the field and of an attempt to relate those forms to processes that may have had a share in forming and modifying them.

During the course of a departmental field class the writer had the opportunity of examining the abandoned and sub-aerially modified cliffs at the head of Carmarthen Bay. Their nature and origin appeared so clear that it seemed worth while to place on record some accurate measurements of their forms and, with this object in mind, the area was revisited in the late summer of 1950.

At the head of Carmarthen Bay lie the drowned estuaries of the Gwendraeth, Towy and Taf. After this positive movement of base level, cliffing occurred in already sub-aerially modified seaward-facing slopes. A period of deposition followed, and from Pendine a sandy beach (Pendine and Laugharne Sands) grew eastward to, and partly across, the Taf estuary (Figure 1). Behind it a line of dunes accumulated, and behind that again tidal marsh formed and has since been reclaimed from west to east. The cliffs from Pendine eastward have thus progressively lost contact with the sea and have been partially converted to sub-aerially weathered slopes. At Pendine the cliffs that formerly truncated the more gently inclined seaward slopes have been replaced by concave forms, but to the east, towards the Taf estuary, approximately vertical cliffs still terminate the seaward slopes below. Two distinct slope forms may, therefore, be identified; approximately vertical cliffs in the east and concave slopes in the west. Further, between these two extremes, intermediate forms exist representative of certain stages in the sub-aerial modification of nearly vertical slopes. It was this sequence of slope forms that first attracted the writer, and it was these forms which he had determined to subject to careful examination and precise measurement. On closer investigation, however, it became apparent that it would be advantageous to extend the work to include the seaward slopes above the abandoned cliffs, and to other valley and seaward slopes in the immediate vicinity. A more detailed description of the area will illustrate these advantages and serve to outline the general characteristics of its slope forms.

Carmarthen Bay may be said to divide the South Wales Coalfield into two parts, and the valley and seaward slopes on its north side are cut, not in the

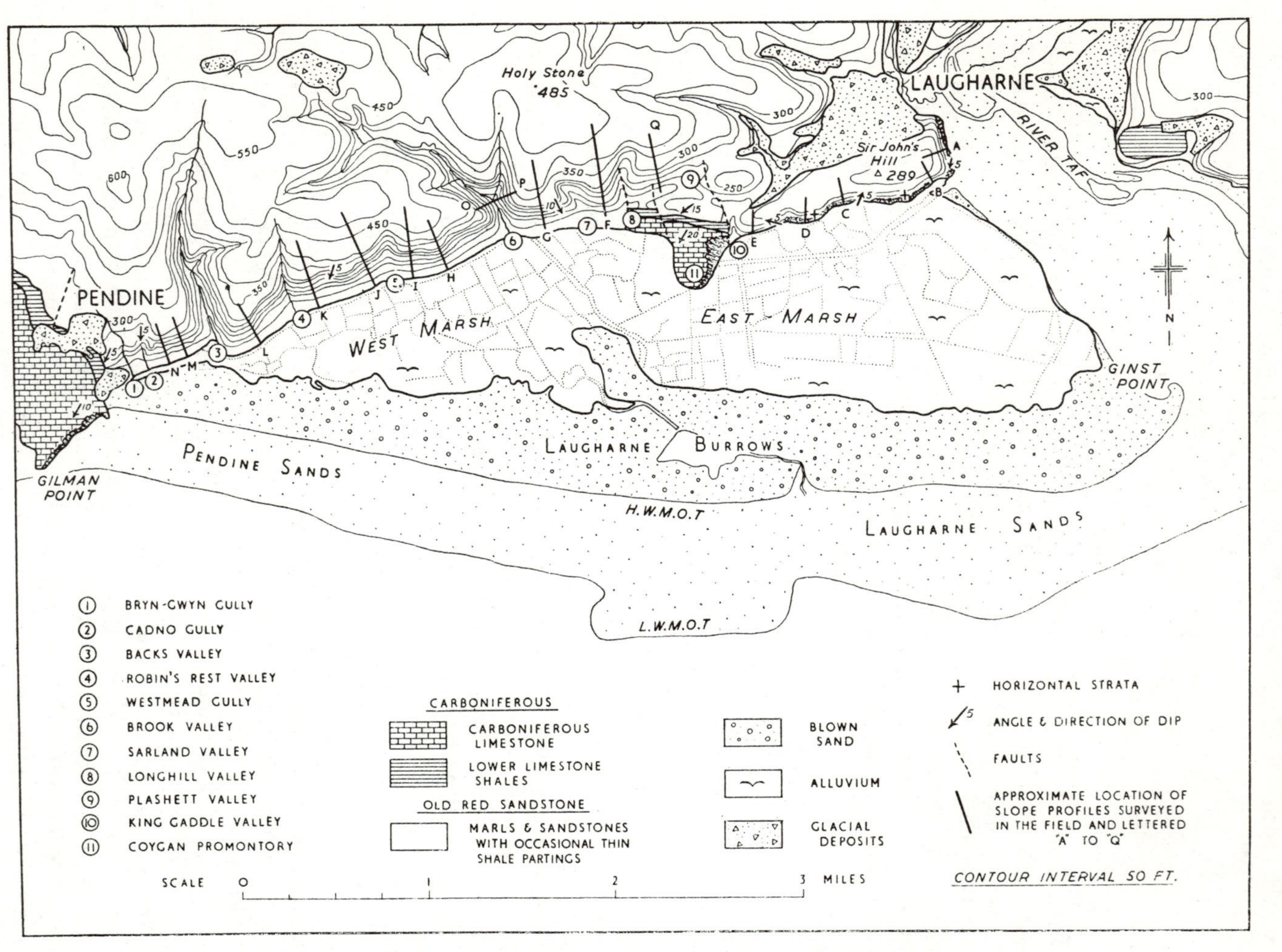

FIGURE 1—The nature of the area between Pendine and the Taf estuary.

Carboniferous, but in the Old Red Sandstone. This consists of varying proportions of sandstone and marl with occasional thin shale partings. Strahan writes '. . . at Pendine, red marls with red sandstone pass up into highly micaceous grey and green sandstone sometimes with thin shale partings'.[1] However, not all the slopes are cut in Old Red Sandstone. Approximately three miles east of Pendine and 1½ miles west of the Taf estuary lies the Coygan Promontory, formed of down-faulted Carboniferous Limestone. It was decided that no measurements should be taken of the slopes on this outlier since it seemed advisable, for purposes of comparison, to keep lithological and structural differences to a minimum. Nevertheless the Coygan Promontory may be said to divide the area into two parts, an eastern and a western.

Immediately north-east of the Coygan Promontory lies the King Gaddle valley. Its western slopes are chiefly in limestone; its eastern in sandstone and marl. From the crest of these eastern slopes at about 200 feet the land surface rises gradually, for approximately three-quarters of a mile, to a maximum of 289 feet on Sir John's Hill. From here it declines east and south in broadly convex, then uniformly inclined, and finally cliffed, slopes to sea-level. In the Taf estuary the sea at high tide still washes the foot of east-facing cliffs. Near the Coygan Promontory, however, the cliffs have been partially replaced by concave slopes. In the 1½ miles between may be seen certain early stages in the subaerial modification of abandoned marine cliffs originally about 100 feet high.

West of the Coygan Promontory, and between it and Pendine, the forms are more complex. However, it is possible to distinguish gully and valley slopes from seaward slopes. The former occur in the three gullies (the Westmead, Cadno and Bryn-Gwyn gullies) and six valleys (the Plashett, Longhill, Sarland, Brook, Robin's Rest and Backs valleys) that dissect the area, and the latter, facing approximately south-east, between them. On inspection the seaward slopes appear to consist of three major elements. These are a lower concavity, which has replaced the former marine cliffs, an intermediate slope of approximately uniform inclination, and an upper convexity. From the upper convexity the land surface rises gradually inland as a uniform or terraced slope to a broadly convex watershed. This lies at approximately 600 feet behind Pendine, but falls eastward to about 350 feet to the north of the Coygan Promontory. The seaward slopes vary in extent and height. Between Pendine and the Backs valley their upper limit is at approximately 300 feet. Between the Backs valley and the Brook valley it rises to between 400 and 450 feet. Farther east, between the Brook valley and the Coygan Promontory, it falls again to between 300 and 350 feet.

Clearly, within the area described, exceptional opportunities exist for the observation and measurement of slope forms representative of certain stages in the modification of cliffs and of seaward and valley slopes.

In the Taf estuary the movement of water at high tide has effectively

[1] A. STRAHAN, 'The geology of the South Wales Coalfield, Part X', *Memoir of the Geological Survey*, No. 229 (1909), 75.

prevented the accumulation of talus and debris. Comparable conditions are presumed to have existed formerly below the south-facing cliffs but, since that period, talus and debris have accumulated and are present in increasing quantities towards the west.' Precise measurements therefore of the forms developed, both in the Taf estuary and at the foot of the seaward slopes, provide an accurate record of certain stages in the modification of the original cliffs which have been modified initially with, and subsequently without, removal of the products of weathering. Further, as the most recent episode of marine cliffing has cut into the base of already sub-aerially modified seaward slopes, and as the growth of Pendine and Laugharne Sands has now left these abandoned by the sea, it is possible in some measure to assess the effect of both corrasion and debris accumulation on the development of these slope forms. In addition the seaward slopes, in varied stages of modification, may be compared with both younger and older valley slopes that have developed contemporaneously in the valleys that dissect the area. It seems important to emphasize also that, although structure and lithology are by no means uniform, the area is not characterized by the juxtaposition of thick and alternating beds of gritstone, sandstone and shale such as is the case in the coalfield to the east. Problems relating to the significance of lithology and structure cannot be ignored but, in this area, they do not obtrude in a consideration of the origin of the slope forms. Nowhere in this area are structural benches of major proportions identifiable.

To obtain measurements of the slope forms that characterize the area, and that now constitute the evidence in the ensuing discussion, it was decided to use a technique developed, and already used successfully, in the southern Pennines.

The attempt to measure slope forms involves the surveyor in a difficult problem since normal methods of topographic survey do not record small, but important, variations in slope form and angle. This problem had been partially solved by making detailed measurements up or down selected slope profiles. The method suggests itself, at least for initial survey, since the evolution of hill slopes is normally considered in profile form, and it was decided that it should be used in this area.

An example of the data obtained from field survey and an illustration of the manner in which they were approximated for the construction of slope profiles is given on the opposite page.

The measurements were made with ranging rods, tape and Abney level. Each profile was surveyed up or down the maximum slope, and the ranging rods were placed at estimated breaks of slope. On smoothly convex or concave forms that could not be treated in this manner, the rods were placed at fixed intervals. The distance between ranging rods was then selected in relation to the radius of curvature of the slope. On long uniform slopes, apparently uninterrupted by any irregularity, test readings were taken to check this estimation and, when this proved to be correct, the distance between two stations was determined by the maximum length of the tape (100 feet).

Profile L — Surveyed Up Slope

	Field Data			*Field Data Approximated*		
Station	*Measured Length feet*	*inches*	*Inclination (up slope and down slope readings in degrees and minutes)*	*Mean of Up Slope and Down Slope Readings (approximated to half a degree)*	*Profile Intercept* (feet)*	*Average Inclination (degrees)*
1	38	11	5.50 5.50	6	38.9	6
2	23	4	7.50 8.00	8		
3	17	11	8.00 7.50	8	67.3	8
4	26	1	7.40 8.00	8		
5	25	5	9.00 8.50	9	25.4	9
6	19	6	12.20 12.10	12½	19.5	12½
7	18	7	15.40 15.40	15½	18.6	15½
8	16	5	18.10 18.00	18	16.4	18

*The writer defines a *profile intercept* as a uniformly inclined portion of a slope profile. It may correspond to a measured length derived from field survey, or it may be obtained by the amalgamation of two or more measured lengths, when these do not vary in average inclination by more than half a degree.

The Observations

(i) *The general forms and average angles of the seaward and valley slopes*

The chief differences in form between the seaward slopes east and west of the Coygan Promontory are clearly illustrated by the slope profiles (Figure 2). The former, represented by profiles A to E,[2] only recently abandoned by the sea, are still partly cliffed. The latter, represented by profiles F to N, last affected by marine processes at an earlier period, possess at their bases well developed concave slopes. The valley slope profiles O, P and Q may be distinguished as those representing younger (O and P) and older (Q) valley slopes. The former are without basal concavities, the latter has one developed, but of shorter radius than those of the concave forms at the foot of the seaward slopes.

[2] Profile E represents a special case since the data for its construction were measured from the spur between the King Gaddle valley and the seaward slopes to the east.

None of these profiles may be said to possess an average angle above 33 degrees, and there are few smoothly convex slopes, although the upper portions of the slopes represented by the profiles F, H, I, M and Q (Figures 2, 3 and 4) appear to approximate to such forms. There is more characteristically a sharp intersection between more and less steeply inclined slope segments,[3] as for example is clearly indicated on the remaining profiles.

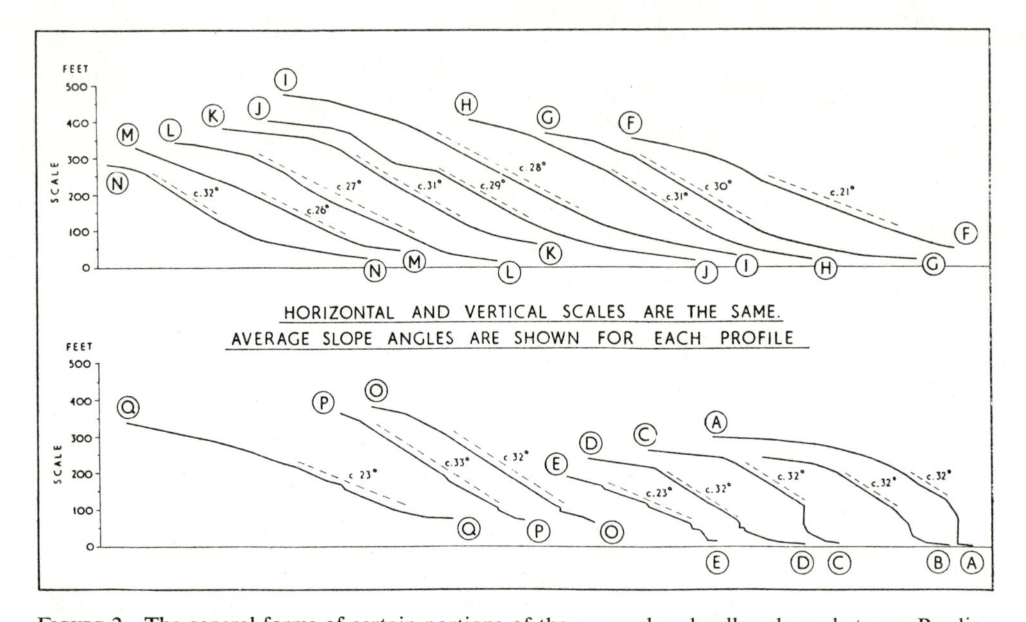

FIGURE 2—The general forms of certain portions of the seaward and valley slopes between Pendine and the Taf estuary, illustrated by slope profiles.
The approximate locations of the slopes, from which the data for the construction of each profile were obtained, are indicated on the map in Figure 1.

Because of the difficulty of obtaining precise measurements of the nature and thickness of the debris cover, only certain generalizations are possible. On the slopes represented by the profiles F to N and Q the debris may be said to increase in thickness progressively downslope, from an average of about one foot near the crest to approximately five feet just above the area where the surface slopes become concave. On the more steeply inclined portions of the slopes, represented by profiles A to D and by O and P, the debris cover is frequently very irregular in thickness, though it would appear to maintain an average of two to three feet. If debris is present on slopes much above 32 or 33 degrees it may be said to form a discontinuous layer and its surface is very uneven.

[3] The writer defines a *slope segment* as a uniformly inclined portion of a slope profile which, composed normally of several *profile intercepts*, differs distinctively in average inclination from those slope segments, or in form from those concave or convex slope elements, which may lie above and/or below it.

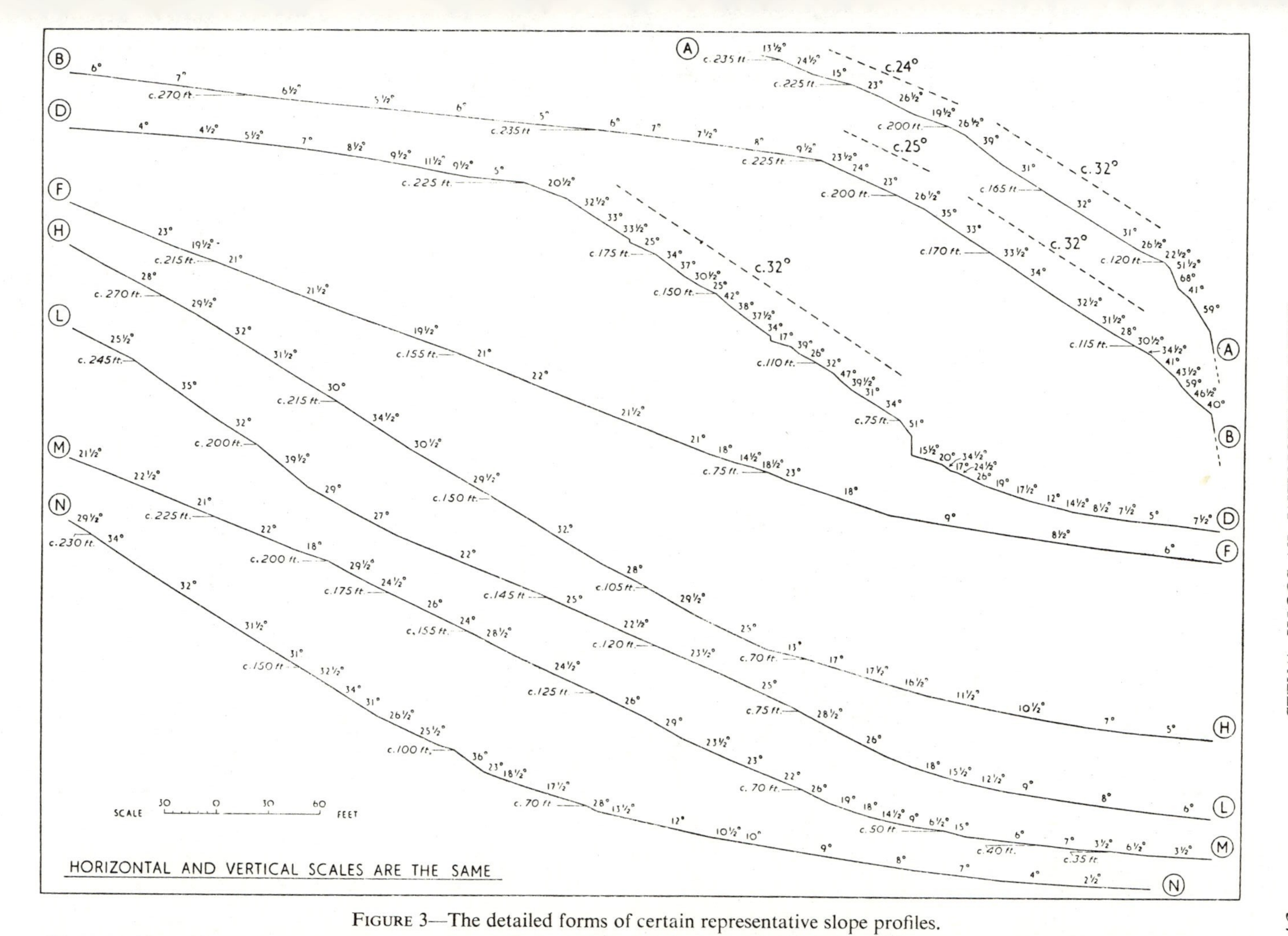

FIGURE 3—The detailed forms of certain representative slope profiles.

These profiles are the enlarged portions of profiles A, B, D, F, H, L, M and N, which appear in Figure 2. Only certain representative angles are shown on each profile. The inclinations of certain slope segments are shown on profiles A, B and D.

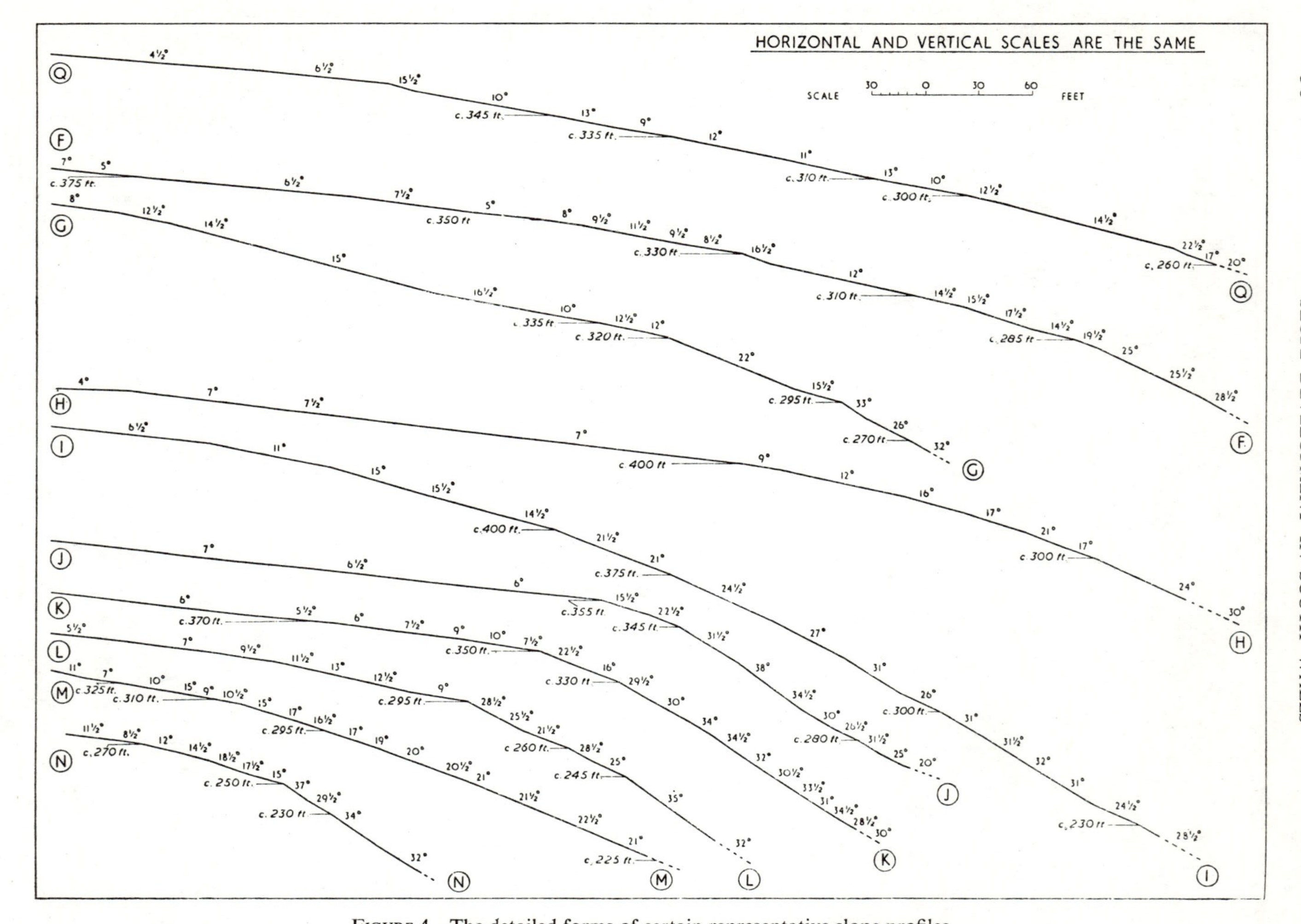

FIGURE 4—The detailed forms of certain representative slope profiles.

These profiles are the enlarged portions of profiles Q, F, G, H, I, J, K, L, M and N, which appear in Figure 2. Only certain representative angles are shown on each profile.

(ii) *The detailed forms of the valley slopes and those of the seaward slopes above the abandoned cliffs*

When the profiles are drawn on a larger scale (Figures 3 and 4) the middle and upper portions are seen to consist of innumerable large and small concavities and convexities. A correct interpretation of the nature and origin of all of these forms is not possible without further investigation. They result from many causes acting individually and collectively, and the task of interpretation is complicated by the possibility that certain features may be the remnants of former valley slopes and terraces, or marine cliffs and abrasion platforms.[4] However the nature of certain characteristic features may be recorded.

Minor features, such as those on profile F above 75 feet, H between 105 and 150 feet, L between 120 and 145 feet and M between 125 and 155 feet (Figure 3), are examples of the irregularities common to these slopes for which there is no obvious explanation. One may suppose that they result from variations in the thickness of the debris cover due to former debris slumping, or spring sapping and gullying. In certain cases they may result from the presence of more coherent sandstones, but such features are never obvious and do not reach large proportions on debris-covered slopes.

Spring sapping and gullying appear to be important processes on these slopes, and are known to have created and to be modifying the slope forms represented by profiles H between 150 and 215 feet, M between 155 and 200 feet, N above 100 feet (Figure 3), G between 295 and 320 feet, J between 280 and 345 feet and L between 245 and 295 feet (Figure 4).

(iii) *The slope forms that have resulted from the modification of the cliffs east of the Coygan Promontory*

In its simplest form the cliff profile, though in detail serrate, consists of one vertical or steeply inclined slope segment which extends from a debris-covered slope above (inclined at about 32 degrees) to the East Marsh, or the abrasion platform in the Taf estuary, below. In certain places, however, the intersection of the cliff face and the debris-covered slope above is bevelled, as illustrated in profiles A, B and D (Figure 3), and from its base one or several more gently inclined slope segments, or a concave slope, may extend to the East Marsh (Figure 5).

[4] The denudation chronology of the region within which this area is located is discussed in the following:

T. N. George, 'Shore line evolution in the Swansea district', *Proceedings of the Swansea Scientific and Field Naturalist Society*, 2 (1938), 23-48.

K. L. Goskar and A. E. Trueman, 'The coastal plateaux of South Wales', *Geological Magazine*, 71 (1934), 468-77.

S. E. Hollingworth, 'The recognition and correlation of high-level erosion surfaces in Britain', *Quarterly Journal of the Geological Society*, 94 (1938), 55-84.

O. T. Jones, 'The Upper Towy drainage system', *Quarterly Journal of the Geological Society*, 80 (1924), 568-609.

A. A. Miller, 'The 600-foot plateau in Pembrokeshire and Carmarthenshire', *Geographical Journal*, 90 (1937), 148-59.

F. J. North, *The evolution of the Bristol Channel* (1929).

Measurement with the Abney level indicates that the inclination of the cliff face varies from 60 to 90 degrees, but that an inclined face is more characteristic than a vertical one, and the average inclination is probably about 80 degrees. The changing relation of the alignment of the cliff face to the dip and strike of the Old Red Sandstone means that the proportions of marl and sandstone with thin shale partings vary from place to place. The detailed form of the face and its average inclination appear to be related to those proportions and to vary with them.

The inclination of the bevel above the cliff face averages 45 to 50 degrees. It is not always present, as is illustrated by profile C (Figure 2), and may be said to occur only if structure and lithology are suitable since a massive and coherent sandstone band outcrops at the crest of the cliffs represented by this profile.

The slopes that lie immediately at the foot of the cliff face (Figure 5) average 30 degrees. However, they may be as low as 15 degrees (profile S) or parts of them as high as 50 to 55 degrees (profiles U, X and Y). The concave slope may be of short (profiles W and X) or long radius (profiles Y and Z).

In the Taf estuary the slopes developed at the cliff foot (profiles S, T and U) are bedrock slopes without a talus cover. They may be contrasted with those below the south-facing cliffs (profiles V, W, X, Y and Z) where a talus or debris[5] cover is present. In both cases, however, the stepped form of the bedrock slope in sandstone and marl is identifiable.

Where only a small quantity of talus has collected this is found on the more gently inclined portion of the bedrock surface between the steeper slopes. With a larger quantity the bedrock surface is entirely obscured and, under these conditions, the surface slope of the talus or debris is either uniformly inclined (profile V and the upper portions of profiles W, X and Y) or concave (the lower portions of profiles W, X and Y and profile Z), according to the quantities of talus and debris present. In the former case the bedrock surface may be said to possess a talus or debris cover of uniform thickness. Where an extensive concave slope occurs there is characteristically a change in composition downslope from talus on the steeper slope segments, at approximately 25 to 32 degrees, to a debris and vegetation cover on the gentler ones, and on the concave slope below (profiles X, Y and Z). The finest debris and soil are found at the toe of slopes such as those represented by profiles Y and Z.

(iv) *The slope forms of the lower portion of the seaward slopes west of the Coygan Promontory*

Between the Coygan Promontory and Pendine only isolated fragments of the modified marine cliffs protrude above the concave debris slopes that extend to the West Marsh. It has not been possible to determine to what extent the

[5] The writer has found it convenient to distinguish between an accumulation of rock waste, that results from the initial weathering of the bedrock, and an accumulation of weathered material, that is produced by the subsequent weathering of the rock waste. The former has been called *talus*, the latter *debris*.

former cliffs have everywhere been modified completely by sub-aerial erosion, and to what extent they have been buried by debris that has moved down from the debris-covered slopes above. In the places, however, where the data for the construction of the profiles were obtained, the former cliffs have been replaced by approximately concave surface slopes. The seaward slopes between the Coygan Promontory and Pendine in profile form (Figure 3) may, therefore, be said to consist of two main elements below the smooth or simulated convexity. These are an intermediate slope segment and a lower approximately concave curve. At the base of the intermediate segment there is an identifiable break of slope and a change in surface angle which accompanies the alteration in slope form. From somewhere about this change in slope angle and form the debris cover may be said to thin progressively up slope, and down slope to thicken considerably over a short distance, until finally it thins towards the West Marsh. In section the debris cover may be described as lenticular thinning both up slope and towards the sea.

Discussion

(i) *The modification of cliffs*

The manner in which a cliff may be modified by sub-aerial weathering has been considered by A. C. Lawson, W. Penck, A. van Burkalow and A. Wood,[6] but only Penck has examined in detail the case in which all talus is removed from the base of the cliff by a transporting agent. The conditions normally considered are those in which all, or most, of the talus accumulates against the cliff face. Such conditions may exist in certain glaciated, or arid, or semi-arid regions, but in a youthful valley developed in the cycle of normal erosion the stream effectively removes the talus from the foot of any river cliff that may develop. An accurate assessment of the manner in which such a cliff is modified by sub-aerial weathering is not, however, usually possible, since the stream is most likely to be actively engaged both in corrading the cliff base and incising its bed. Conditions in the Taf estuary, and between here and the Coygan Promontory, appear, however, to have approximated, for a short period, to those in which the products of weathering are removed, but the base of the cliffs is neither extended nor corraded. The slope forms that have developed under such conditions are those represented by profiles S, T, U and V, and by the slope segment at 30 degrees on profile W (Figure 5). These slope forms appear to correspond to

[6] A. C. Lawson, 'Epigene profiles of the desert', *Bulletin of the Department of Geology, University of California*, 9 (1915), 23-48.
A. Wood, 'The development of hillside slopes', *Proceedings of the Geologists' Association*, 53 (1942), 128-38.
W. Penck, *Die Morphologische Analyse* (1924). See also J. E. Kesseli, 'The development of slopes' (mimeographed, University of California, 1940).
A. van Burkalow, 'Angle of repose and angle of sliding friction: an experimental study', *Bulletin of the Geological Society of America*, 56 (1945), 669-708.

those that Penck reasoned would appear in such circumstances, although his premises were theoretical.[7]

With suitable lithology and structure, several 'steps' of the slope that replaces the cliff may be created instantaneously. This results from the collapse of the cliff which is superseded by a more stable slope at a lower angle. In such circumstances the lower portion of the cliff cannot be removed, since it is supported from below, and this forms the new slope.

Where the lithology is more uniform and the fragmentary destruction, rather than the collapse, of the cliff face occurs, occasional loose blocks or fragments may be found close against the base of the cliff. In these circumstances it may be argued that, whereas the effect of gravity is to remove an initial layer of loosened rock waste instantaneously from the cliff face, the lowest fragment, or block, remains in place for an indefinite period as it is supported from below. Eventually it will be removed, but it will have remained in position long enough to delay preparation for removal of that part of the cliff that lies behind it. It is this part of the cliff that may be said to form the initial part, or step, of a slope that develops at the cliff foot when the second layer of loosened rock waste is removed. With the removal of a third layer from the cliff face (because of the protection of the lowest portion of the cliff face by the block that had remained on the initial step already formed), a second step is produced. For every layer removed from the cliff face, a further step is created.

Whether all, or part, of the slope that *initially* replaces the cliff is created either instantaneously, or fragment by fragment, its form and inclination appear to be determined by the structure and lithology of the rock in which it develops. In these examples (Figure 5) the thickness and cohesion of marl and sandstone, coupled with the distribution of joint and bedding planes, appear to determine the height, inclination and depth of each step. This may be illustrated by comparing the angles of the slope segments and the structure and lithology as shown on profiles S, T and U. However, the form and inclination of the bedrock slope that initially replaces the cliff will reflect the structure and lithology of the rock in which it develops only if talus is prevented from accumulating against the cliff face, a condition illustrated by the surveyed profiles (Figure 5). Such a condition is possible only in certain circumstances, of which one may be the presence of a transporting agent at the slope foot. Given this circumstance, three possible cases are illustrated by these profiles.

With suitable lithology and structure the weathering of the cliff may result in the production of a slope facet[8] less steep than the cliff face, but above the

[7] Penck's premises were, according to Kesseli, (i) 'Uniformity in lithology', (ii) 'Stability of the local base-level of denudation', (iii) 'Rapid rate of removal', and (iv) 'A transporting agent at the foot of the slope able to carry off all the debris shed by the slope'. J. E. KESSELI, op. cit., 2-3.

[8] The writer defines a *slope facet* as a uniformly inclined surface area on a slope which differs distinctively in average inclination from those slope facets, or in form from those concave or convex slope elements, which may lie above and/or below it. It appears as a *slope segment* on a slope profile. See also S. W. WOOLDRIDGE, 'The cycle of erosion and the representation of relief', *Scottish Geographical Magazine*, 48 (1932), 30-6.

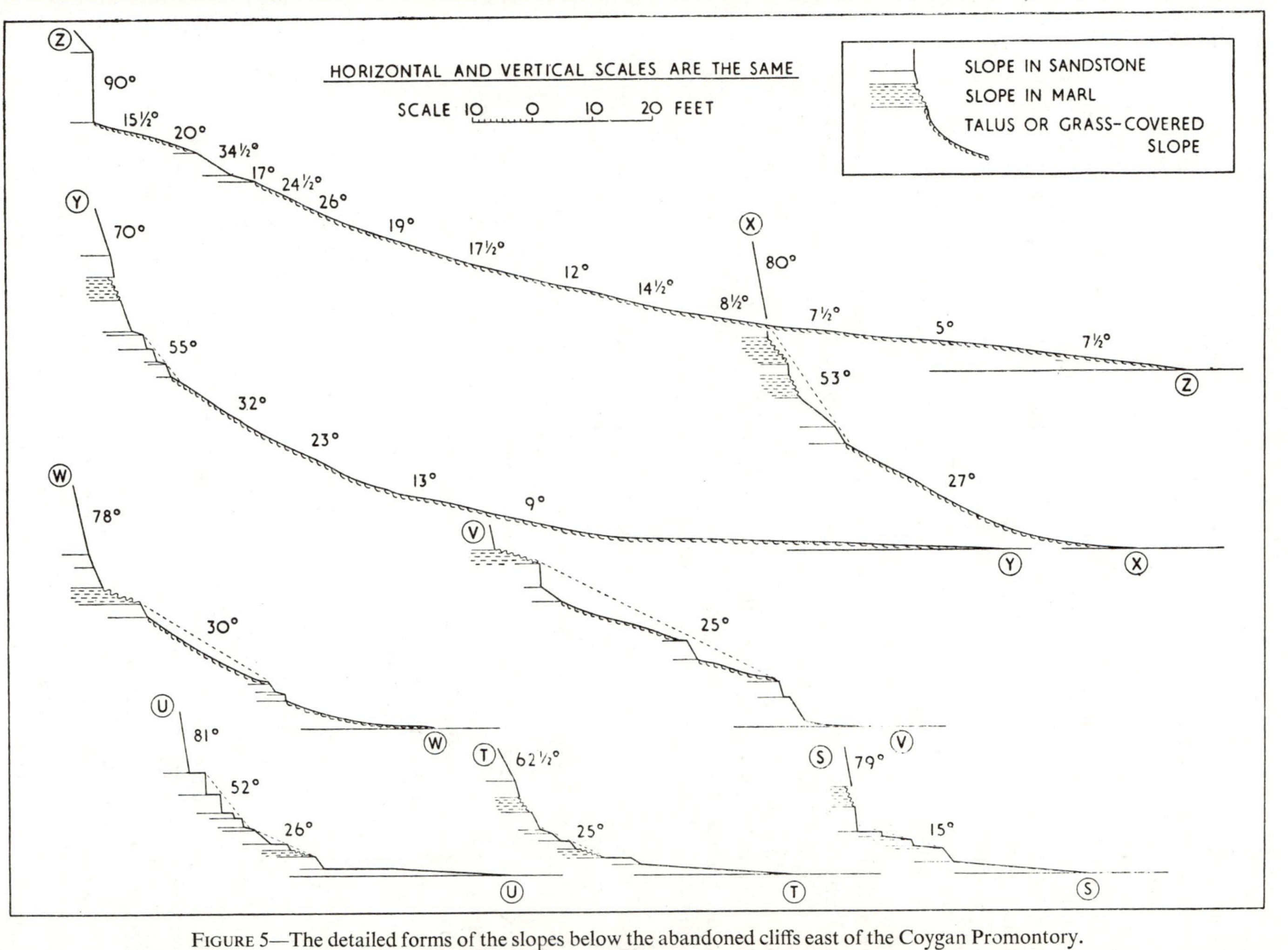

FIGURE 5—The detailed forms of the slopes below the abandoned cliffs east of the Coygan Promontory.

Profiles S, T, and U were surveyed from below the east-facing cliffs in the Taf estuary and the profiles V, X and Z are enlarged portions of the profiles B, C and D which appear in Figure 2. Profile W was surveyed from below the cliffs between profiles B and C, and profile Y from below the cliffs between between profiles C and D. Only certain representative angles are shown on each profile. The inclinations of certain slope segments are shown on all profiles except Z.

angle of repose of loose talus, or of rock waste that weathers from its surface (profiles U, X and Y). Under such conditions no talus or rock waste remains against, or on, the new slope facet. Alternatively, the weathering of the cliff may result in the production of a slope facet approximately at the angle at which rock waste or talus can remain on its surface. In this case a layer of talus collects on the surface of the new slope facet (profiles V and W). If, however, the initial slope that replaces the cliff is much below the angle at which talus will collect, it will normally pile against the cliff face and protect this portion of the cliff face from weathering (profile S), although in the example illustrated special conditions may be said to have prevented this.[9]

In cases one and two the talus does not accumulate against the cliff face, as, for example, Wood[10] implied, and determine the form and inclination of the initial bedrock slope that replaces it. However, the maximum inclination of the bedrock talus or debris-covered slope, that ultimately grows in case one,[11] or initially in case two, is determined by the angle at which rock waste or talus will remain on a bedrock surface. In this area the maximum angle appears to be about 30 to 32 degrees. It is not known whether it approximates to the angle of the slope of repose of loose talus or to the angle of repose, just below that of sliding friction, for rock waste that weathers from the surface of the new slope facet. It might appear to approximate to the former since talus is not piled against the cliff face (Figure 5). However, this suggestion may not be correct. A layer of talus or debris would seem more stable in close contact with a bedrock slope than when loosely piled; further, the impetus possessed by rock waste falling from the cliff face may prevent it coming to rest on a slope of about 30 degrees. In such circumstances loose talus may roll or slide to the base of the slope.[12] Hence the surface angle of the loose talus may be depressed, and the factor which determines the angle at which the bedrock talus or debris-covered slope grows, may be the angle at which rock waste weathered from its surface may remain *in situ* for an indefinite period. This angle may be that just below the angle of sliding friction for the given conditions.[13]

The possibility that loose talus does not pile against the cliff face, when no

[9] This profile has been surveyed from slopes partially covered by water at high tide, and therefore debris is unable to accumulate.

[10] A. Wood, op. cit., 129-30.

[11] If the initial slope facet that develops is steeper than either the angle of the slope of repose of loose talus, or of rock waste that weathers from its surface, it is suggested that it will be *replaced* below by a slope at an angle at which rock waste will remain. *Replacement* may be said to occur rather than decline, since the lowest portion of this slope is most likely to be removed first. The removal of the lowest block, or fragment, for example, forming the base of the slope facets represented at about 50-55 degrees on profiles U, X and Y (Figure 5) would result in the extension of the more gently inclined slopes below at an angle at which talus remains, namely at a maximum inclination of 30-32 degrees.

[12] Penck's third assumption. See footnote 7.

[13] A. van Burkalow, op. cit., 693-700. Note, however, that the talus or debris lies on an inclined surface. In the circumstances postulated, therefore, namely with a transporting agent at the slope foot, there is no extension of a horizontal surface at the base of the slope on which talus or debris can accumulate, and so prevent lowering of the angle of the slope of repose (A. van Burkalow, op. cit., 688-90).

transporting agent is present at the cliff foot, has yet to be examined in the light of field evidence. That this may be possible in certain circumstances is illustrated by profiles X and Y. The cliff face, above the slopes represented by these profiles, would appear to have been modified by sub-aerial weathering since the growth of Laugharne Sands and the East Marsh prevented continuous removal of talus and debris from the slope foot, yet loose talus is not piled against it. In such circumstances there must be a 'rapid rate of removal'[14] of talus from the top of the talus slope.

Where the modification of the cliffs is more advanced, certain slope forms result from the accumulation of talus and debris (Figure 5, lower portions of profiles W, X and Y and profile Z). The surface forms that appear in such circumstances are distinctive. Whereas those that develop with *unimpeded* removal of talus or debris reflect closely those of the bedrock beneath (profiles S, T, U and V, and upper portions of profiles W, X and Y), those which develop when removal of talus and debris is *impeded* show surface forms and inclinations which are, at best, only an approximation to those of the bedrock they cover (Figure 5, lower portions of profiles W, X and Y and profile Z; Figure 3, lower portions of profiles F, H, L, M and N; Figure 6, lower portions of profiles 4, 5 and 6). The general forms and inclinations of the *bedrock*, or the *bedrock debris-covered* slopes reflecting those of the bedrock, may be said to provide evidence of the stage of adjustment of the bedrock to the processes that have modified it. The general forms and inclinations of *debris* slopes reveal no such evidence, and possibly only reflect the approximate composition of the debris.

(ii) *The modification of slopes of approximately 32 degrees and less*

The differences between slope forms that have developed with unimpeded, as distinct from impeded, removal of debris may be emphasized further by a consideration of the valley slopes, and of those portions of the seaward slopes above the present and former cliffs. The seaward slopes above the present cliffs (profiles A to D, Figure 2) and the young valley slopes (profiles O and P) have been, and still are, modified under conditions of unimpeded removal of debris. Such slopes have average angles of about 32 degrees. These may be contrasted with the average angles of the seaward slopes above the former cliffs (profiles F to N) and the older valley slopes (profile Q) which have been subject to impeded removal. The average angles of these slopes vary, but are generally lower than 32 degrees.

A tentative conclusion about the nature of slope development is possible on this evidence: namely, that with *unimpeded removal* of debris, slopes display *parallel retreat*, while with *impeded removal* they *decline*. This conclusion is substantiated by further evidence, upon which may also be based the suggestions that the profiles of slopes modified under unimpeded removal are characterized by the presence of slope segments, which correspond to slope facets on the ground, and which meet in angular intersections. With impeded

[14] See footnote 7.

E

removal slope facets disappear and in their places develop concave or convex slope elements.

A theory of parallel retreat with unimpeded removal of debris may be suggested since there is surely more than chance in the close correspondence of average angles of the younger valley slopes and seaward slopes above the present cliffs. The conditions under which these slopes have developed may be assumed to have varied in many, if not all, characteristics (e.g. lithology, structure, aspect, mode of initiation) except one, namely unimpeded removal of debris from the slope foot. If slopes declined with unimpeded removal such a close correspondence of average angles would be unlikely to exist, rather could they be expected to vary along one slope, and between one slope and the next. That these average angles are approximately 32 degrees appears to be determined by the fact that this is the maximum angle at which slopes can retain a continuous debris cover in this environment.

The evidence for the presence, with unimpeded removal of debris, of the angular intersection of slope facets is provided by the forms of profiles A, B, C and D (Figure 2), and the enlarged portions of profiles A, B and D (Figure 3). If, in such circumstances, slopes declined, the slope facets represented by the slope segments of about 32 and 25 degrees on these profiles could be expected to vary in inclination, as well as in extent, when traced laterally along the seaward slopes and, somewhere along these slopes, grade into each other. The slope facet inclined at 32 degrees, for example, might be found to decrease to, say, 28 degrees when traced towards the Coygan Promontory and grade into the one above, which itself might be found at a lower angle. These slope facets do vary in extent, but they do not appear to grade into each other, nor vary greatly in inclination. These facts, illustrated in profiles A, B, C and D (Figure 2) and A, B and D (Figure 3), were confirmed where readings were taken with the Abney level along the slopes between those surveyed for the construction of slope profiles.

Inseparable from a theory of parallel retreat is the conception that the removal of debris is faster from the steeper slopes. The angular intersection of slope facets may be said to confirm this, no matter how close such facets may be in inclination. When, however, the difference in inclination of intersecting facets is large, a distinction in the character and distribution of processes may be observed between one facet and the next. Where, for example, slope facets of, say, 32 degress and 7 degrees intersect, the processes of spring sapping, gullying, and debris slumping on the former are clearly more competent than those of creep and rainwash on the latter (Figure 6) and the more gently inclined slopes would appear to be consumed from below.

There is less evidence to support the suggestion that slopes decline with impeded removal of debris since it may be argued that, although the average angles of the seaward slopes west of the Coygan Promontory are generally lower than those to the east or in the younger valleys, yet these have only recently been subject to conditions of impeded removal, and the lower angles result from an increase in the thickness of debris down slope. Nevertheless, that slopes do decline with impeded removal may be suggested on general grounds and on the evidence provided by these profiles (Figure 2). On these slopes, although the debris cover does thicken downslope, it is relatively thin and there is a close correspondence between the bedrock and surface forms.

Evidence of the distinctive surface forms that develop with impeded, in contrast to those that develop with unimpeded, removal of debris is provided by the lower portions of profiles F, H, L, M and N (Figure 3). The curved outlines of the lower portions of these profiles may be contrasted with those exhibited by profiles A, B and D (Figure 3) where slope segments may be identified.

In accordance with these observations, and those previously stated, a theory of slope decline may be advanced. It is based partly on the assumptions of Lawson and Wood[15] that a talus or debris cover protects the bedrock from weathering, and partly on the arguments of Lawson and S. Morawetz.[16] It is also assumed that the thicker the debris cover the slower the weathering of the bedrock beneath it.

At the base of seaward slopes abandoned by marine erosion, as for example those between Pendine and the Taf estuary, or at the base of any valley slopes where a flood plain or terrace allows debris to collect, a deposit may be assumed to grow in form comparable to that at the foot of the seaward slopes between the Coygan Promontory and Pendine (profiles F, H, L, M and N, Figure 3, and profiles 5 and 6, Figure 6). Under such conditions, where the debris thickens progressively from the crest of the slope to the lenticular accumulation near the base, protection from weathering increases progressively in the same direction. Where, however, the debris thickens towards the base, not progressively, but rapidly over a short distance, protection from weathering increases in the same ratio. It could therefore be argued that, as long as the predominant processes remain creep and rainwash,[17] and a deposit of this form is present, the upper slopes decline, and a convex bedrock profile develops where the debris thickens rapidly over a short distance against the bedrock slope. If Lawson's or Morawetz's arguments are accepted the upper slopes also become convex. On the one hand, however, a convex bedrock slope is developed as a result of weathering accompanied by the downward creep or wash of the surface debris, on the other, as the result of protection from weathering derived from the lenticular layer of debris that covers it.

The evidence upon which this discussion has been based is too fragmentary for the conclusions to be other than tentative. They must clearly be tested against further measurements of slope forms obtained from field survey. If,

[15] A. C. LAWSON, op. cit., e.g. 24 and 29. A. WOOD, op. cit., 129.

[16] A. C. LAWSON, 'Rainwash in humid regions', *Bulletin of the Geological Society of America*, 43 (1932), 703-24.

S. MORAWETZ, 'Eine Art von Abtragungsvorgang', *Petermanns Mitteilungen*, 78 (1932), 231-33. See also J. E. KESSELI, op. cit., 17-20.

[17] K. BRYAN, 'Gully gravure — a method of slope retreat', *Journal of Geomorphology*, 3 (1940), 89-107.

however, impeded removal results in the decline of slopes, and if impeded removal is the normal condition in the normal cycle, in this cycle slopes will usually decline.

(iii) *Slope nomenclature*

The slope elements called the cliff, the bedrock slope, and the bedrock talus or debris-covered slope, appear to correspond to those referred to by Wood[18] as the 'free face', the 'buried face', and the 'constant slope'. Such terms are to be preferred to those of Penck,[19] since they clearly describe the feature to which they refer. None of the English terms, however, quite corresponds to Penck's 'Haldenhäng', which describes the special condition, with unimpeded removal of debris, when the 'buried face' and the 'constant slope' are parallel. [20] 'Constant slope' is to be preferred to Meyerhoff's term for the corresponding slope element, namely 'gravity slope.'[21] This latter term suggests free removal of debris under gravity alone, and such a condition applies to the 'free face' rather than the 'constant slope'.

In order to distinguish between the concave debris slope that develops with impeded removal (when the forms of the debris slope and the bedrock slope it covers may be quite different) from the concave bedrock debris-covered slope that may be found to develop at the slope foot under certain conditions of unimpeded removal, Wood's term 'waning slope'[22] may perhaps be reserved for the former, and Meyerhoff's term 'wash slope'[23] for the latter.

Summary and Tentative Conclusions

Partly from observations in the field, but mainly from the evidence provided by surveyed slope profiles, certain facts have been recorded concerning the

[18] A. Wood, op. cit., 129-30.

[19] W. Penck, op. cit., chapters 6 and 7.

[20] Kesseli translates 'Haldenhäng' as 'slope of talus material' or 'rubble slope' and writes that it is a slope 'consisting of bedrock and covered only by a veneer of weathered material shed by the cliff' (J. E. Kesseli, op. cit., 3).

[21] H. A. Meyerhoff, 'Migration of erosional surfaces', *Annals of the Association of American Geographers*, 30 (1940), 247-54.

[22] A. Wood, op. cit., 130.

[23] H. A. Meyerhoff, op. cit., 251.

Figure 6—Diagram to illustrate five possible stages in the sub-aerial modification of cliffed and abandoned seaward slopes.

The forms and inclinations of profile 1 are based on the measured and observed characteristics of the seaward slopes and abandoned cliffs east of the Coygan Promontory. Those of profiles 2 to 6 are based partly on field measurements and observations, and partly on the tentative conclusions concerning the nature of slope development that have been derived from them. The slope forms and inclinations below approximately 80 feet on profiles 2 to 6 are based on field observations and measurements. The slope forms and inclinations above approximately 80 feet are based on the assumptions that slopes of 32 degrees and less, when subject to unimpeded removal, are modified by parallel retreat of slope facets, the steeper consuming the gentler; but, when subject to impeded removal, by decline and the development of convex and concave slopes. The convexity at the crest of profile 6 reflects the form of the bedrock; the concavity at the base, however, is a slope developed in debris and obscures the convex bedrock surface that develops beneath. The nature of the predominant processes on slopes of differing inclinations is based on field observations in this area.

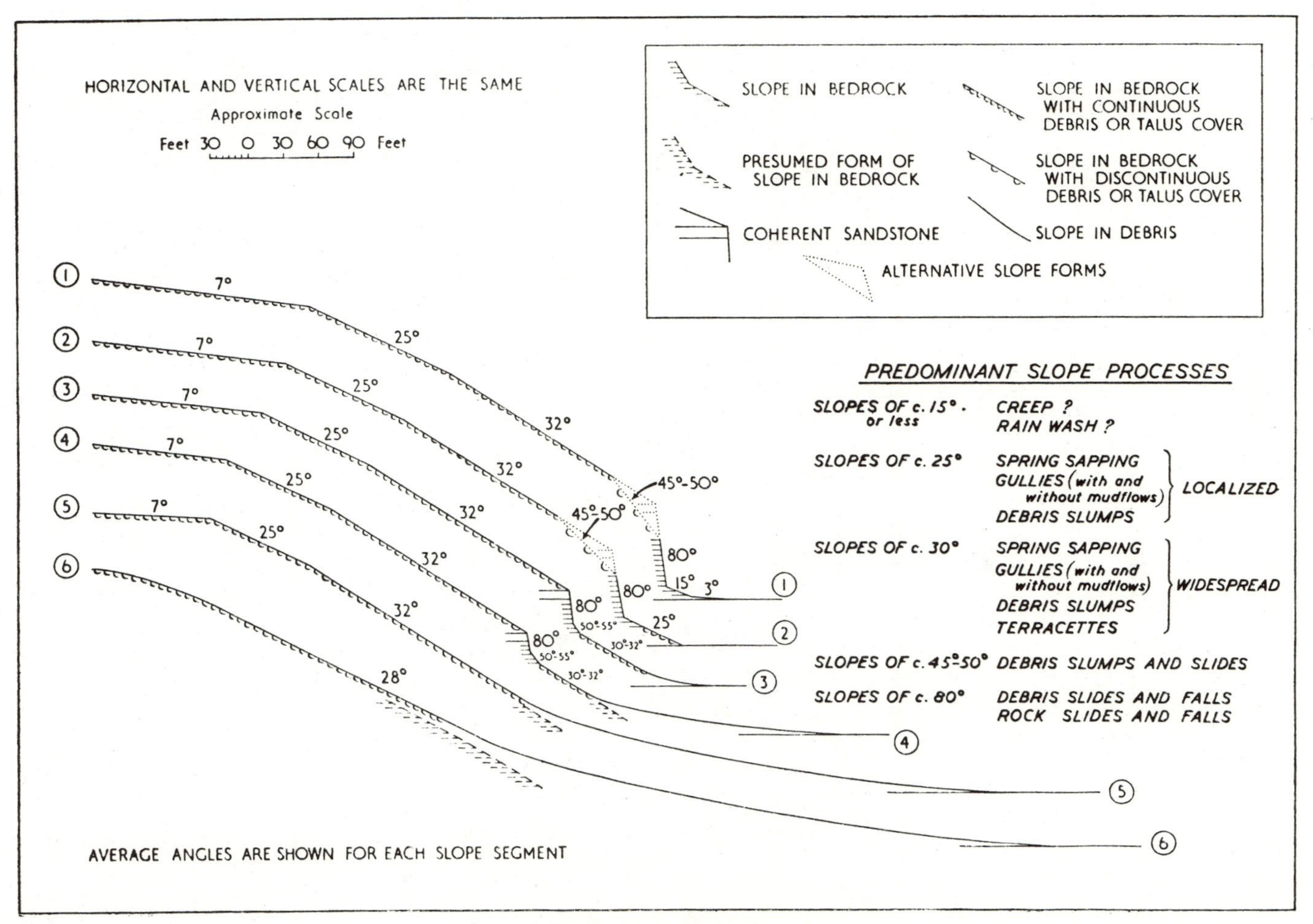

FIGURE 6—For explanation see foot of page 48.

nature of the seaward and valley slopes at the head of Carmarthen Bay in South Wales. In this area the slope forms may be classified as those associated either with younger or older valley slopes, or with cliffed or formerly cliffed seaward slopes.

A smooth upper convexity is more generally characteristic of older valley and formerly cliffed seaward slopes than of younger valley and cliffed seaward slopes, but in all cases it may be simulated, either wholly or in part, by *slope facets*. Below the real, or simulated, convexity there extends, on all slopes, a lengthy slope facet of approximately uniform inclination. On younger valley slopes and on cliffed seaward slopes it averages thirty-two degrees. On older valley slopes, and on formerly cliffed seaward slopes, it is usually below this angle. Twenty-one degrees is the lowest inclination recorded. A lower concavity is developed on older valley and formerly cliffed seaward slopes. Elsewhere it may be partly developed but usually is absent.

On detailed examination the surfaces of all slopes are seen to be formed of innumerable large and small concavities and convexities. Some may be attributed to marine or fluvial erosion at former and higher base-levels, some to the processes of spring sapping, gullying or slumping, some to inexplicable variations in the thickness of the debris cover, and some to differences in the cohesion of sandstones and marls.

A debris cover is discontinuous or absent from slopes with inclinations greater than an average of thirty-two degrees. On slopes of less than this inclination the debris cover thickens progressively down slope from about one foot at the crest to about five feet at the slope foot, or where the basal concavity develops. At the foot of the seaward slopes the basal concavity is in debris. In such circumstances the debris cover may be described as lenticular thinning both up slope and towards the sea.

The abandoned marine cliffs immediately west of the Taf estuary vary in inclination from sixty to ninety degrees, but have an average inclination of about eighty degrees. Where structure and lithology are suitable, namely where coherent sandstones outcrop at the top of the cliff, the cliff face and the seaward slope above the cliff face intersect in a sharp angle; elsewhere this intersection is bevelled. The slope of the bevel averages forty-five to fifty degrees.

Where talus and debris have been prevented from accumulating against the cliff face, lithology and structure appear to have determined the angle of the bedrock slope that is formed as a result, either of the weathering, or of the collapse, of the cliff. The inclination of this slope may vary from fifteen degrees to fifty-five degrees. Where, however, a small quantity of talus or debris is present other factors appear to have operated, and a bedrock slope with a thin layer of talus or debris is formed. The inclination of this slope is nowhere above thirty-two degrees, and averages thirty degrees. Where a large quantity of talus or debris has accumulated a concave surface slope is developed. Any one, two, or all three of these forms may occur at the foot of the cliff, either alone or in combination.

From an analysis of these facts certain tentative conclusions have been drawn concerning the sub-aerial modification of (i) cliffs and (ii) hill slopes of approximately thirty-two degrees and less. In both cases the nature of slope form and of slope evolution is attributed to the relation between the rates of weathering of the bedrock and of transport of the residual debris. Slope forms and slope evolution, it is suggested, may therefore be related to the dominance of either *impeded* or *unimpeded* removal of residual debris (Figure 6).

(i) *Cliffs.* With unimpeded removal cliffs are replaced initially by bedrock slopes whose inclinations are determined by lithology and structure. In theory the inclinations of these slopes may be above, at, or below, the angle at which talus or debris will remain (although in the process of transport) on a bedrock surface. In this area the initial slopes normally develop above, or approximately at, this angle. It is presumed to be just below the angle of sliding friction for rock waste that weathers from a bedrock surface under the given conditions and is approximately thirty to thirty-two degrees. If the slopes that initially replace the cliffs are above thirty-two degrees they are replaced below by bedrock talus or debris-covered slopes at about this angle. In this area therefore under conditions of unimpeded removal bedrock slopes, with a thin layer of talus or debris on their surfaces, grow at an inclination of approximately thirty to thirty-two degrees.

With impeded removal cliffs are replaced by concave debris slopes whose surface forms differ distinctively from those of the bedrock slopes they cover.

(ii) *Slopes of thirty-two degrees and less.* With unimpeded removal of debris, slopes of thirty-two degrees and less are modified by parallel retreat. This conclusion is tentatively suggested since the lengthy intermediate slope facets of both cliffed seaward and younger valley slopes are inclined at approximately the same average angles, namely thirty-two degrees, and because such slopes are more generally characterized by angularly intersecting slope facets.

With impeded removal of debris, slopes of thirty-two degrees and less decline, developing smoothly convex and concave forms. This conclusion is tentatively suggested since the older valley and formerly cliffed seaward slopes are inclined at various angles, usually below thirty-two degrees, and because such slopes are more generally characterized by concave and convex forms. A theory of slope decline is substantiated by the observation that, on such slopes, the debris cover thickens progressively down slope, and is based on the assumption that protection from weathering varies according to the thickness of the debris or talus cover.

It is also argued that a convex bedrock slope may develop when a lenticular deposit of talus or debris has accumulated, and is retained, at a slope foot. A convex bedrock slope may develop therefore as a result of protection from, as well as exposure to, weathering.

ACKNOWLEDGMENT

The author gratefully acknowledges the financial assistance provided from the Research Fund of the University of Sheffield and is greatly indebted to Professor S. W. Wooldridge for his suggestions and encouragement and to Professor D. L. Linton for his help and interest.

[AMERICAN JOURNAL OF SCIENCE, VOL. 253, JANUARY 1955, PP. 44-52]

18

CLIFF RETREAT IN THE SOUTHWESTERN UNITED STATES

DONALDSON KOONS

ABSTRACT. Studies of escarpments developed in nearly horizontal rocks of various characteristics in the southwestern United States show that slope angles vary within a surprisingly limited range, depending upon the angle of repose and angle of sliding friction of the materials involved: that scarp retreat is a discontinuous process consisting of sudden rock falls separated by long periods of stability during which sliderock is gradually removed; that active retreat takes place only when these sliderock accumulations can be removed; and that retreat of the escarpment is by formation of successive parallel cliffs and slopes.

INTRODUCTION

The problem of cliff development and retreat has been the subject of intermittent study by the author since 1941. Several facets of this problem have been particularly puzzling, including the exact mechanism by which retreat occurs, the influence of various rock types on cliff forms, the reasons for the similarity of forms developed in diverse rock types, reasons for variability in location of the boundary between cliff and basal slope, and cause of certain minor slope forms.

The present study is part of a coordinated research program of quantitative analysis of erosional landforms supported by the Office of Naval Research, Geography Branch contract N6 ONR 271, Task Order 30, with Columbia University, Professor A. N. Strahler Project Director.

In this report the general term "scarp" or "escarpment" is defined as the compound form consisting of one or more vertical or nearly vertical cliffs, and one or more slopes either bare rock or sliderock-covered at some angle distinctly less than the cliff angle. The typical escarpment in the Southwest consists of an upper cliff of variable height, comprising 10 to 75 percent of the total height of the escarpment, and a lower slope, comprising 25 to 90 percent of the total height of the escarpment.

Areas selected for study include parts of the Southwest in which escarpments of large size have been developed in a variety of horizontal or nearly horizontal rock types. Representative examples are shown in figures 1-3, and plate 1.

FIELD OBSERVATIONS

Observations made at each locality include: angle of talus slope (talus angle), angle of bare rock slope below the cliff (rock slope angle), height of escarpment, height of cliff, length of slope, character of cliff-making formation, character of slope-making formation, character of talus blocks, thickness of sliderock cover, where visible, and angle of sub-talus slope, where visible.

The last two observations, while of considerable significance, could not be made in every case and hence are not included in the tables. Thickness of sliderock cover varied from 2 or 3 inches, equivalent to one layer of sliderock blocks, to 15 feet, but averaged about 3 to 6 feet, somewhat thinner than anticipated. The angle of sub-talus slopes ranged from 26 to 36°.

Five principal types of cliff-making formations and five principal types of slope-making formations were recognized. The data for these types are summarized in figure 4 and tables 1 and 2. The talus angle is essentially the angle of repose for the material involved; the rock slope angle is essentially the angle of sliding friction (Van Burkalow, 1945).

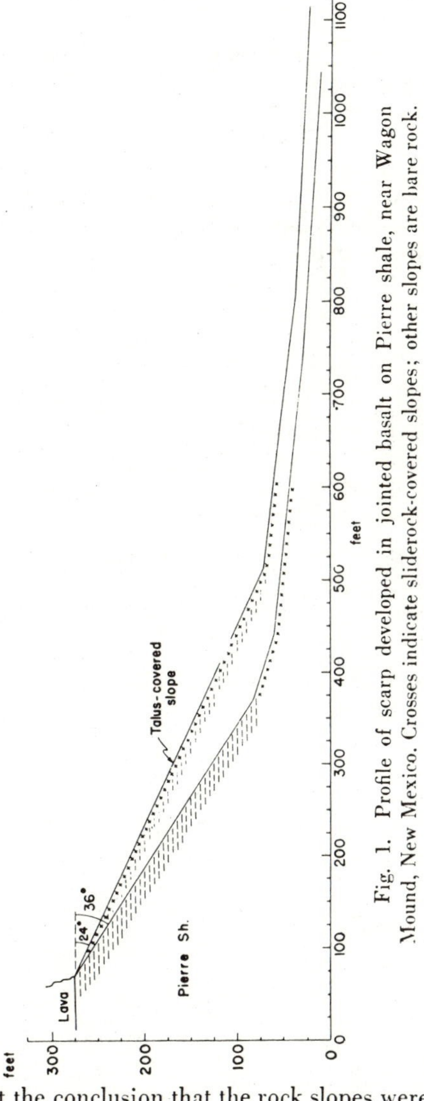

Fig. 1. Profile of scarp developed in jointed basalt on Pierre shale, near Wagon Mound, New Mexico. Crosses indicate sliderock-covered slopes; other slopes are bare rock.

In order to test the conclusion that the rock slopes were actually the slopes of sliding friction, a laboratory model was constructed consisting of an adjustable inclined plane which could be surfaced with different materials, including unfinished sandstone slab and various grades of sandpaper selected for their textural similarity to the exposed rocks of bare slopes. The results of experiments with fragments of different materials, including sandstone, limestone, shale, conglomerate, basalt, and granite, are given in figure 4. The mean sliding angle of all types of material is 35°. Presence of a thin sheet of running water on the inclined plane produced no significant change in the angle of sliding friction. Agreement of field observations and laboratory experiments is very close.

Reference to figures 1-3 and tables 1 and 2 shows that the talus angle is always less than the angle of sliding friction in the areas studied. This contradicts the findings of Van Burkalow (1945, p. 703-705). Disagreement of field and laboratory data is to be expected in this case, since natural sliderock accumulations do not consist of precisely screened material and have been exposed to the action of weathering for a considerable period of time. Whereas creep may reduce a talus angle below the angle of repose, it cannot similarly affect a bedrock slope.

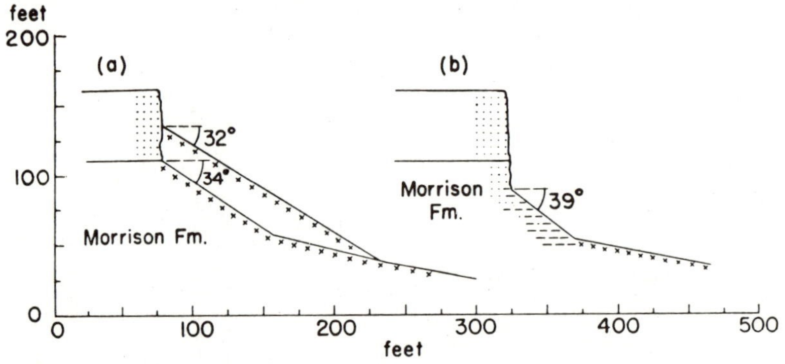

Fig. 2. Profiles of escarpments developed in the Morrison formation, near Rito, New Mexico. Angles and relations of cliff to slope are characteristic.

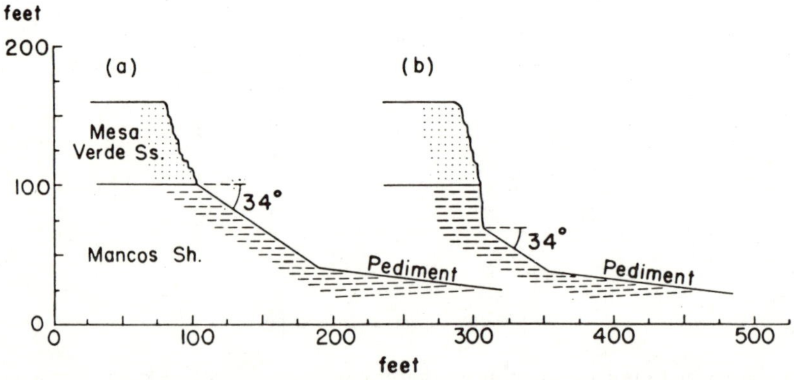

Fig. 3. Profiles of escarpments developed in the Mesa Verde and Mancos formations, near Shiprock, New Mexico. No sliderock cover present, and the position of boundary between cliff and slope varies widely. See also plate 1.

PLATE 1

Southeast corner of Table Mesa, near Shiprock, New Mexico. Mesa Verde formation above, Mancos shale below. Note the wide range in contact between cliff and slope, showing no relation to variations in resistance. The slopes are devoid of sliderock cover, except for the small accumulations of rubble in the center and near left edge of view. Photo by A. N. Strahler.

INFLUENCE OF ROCK TYPE ON SLOPE FORM

Talus and rock slope angles have a high degree of uniformity with one exception. The mean of talus angles measured below a cap rock of columnar jointed basalt is 5 degrees lower than the mean of all others, and the standard deviation is larger (table 1). This results, in part at least, from the relatively higher sphericity of talus blocks derived from jointed basalts and the very wide range of fragment sizes compared to fragments derived from sedimentary materials.

With this exception, the data indicate that rock type has little influence on form of escarpments, though rate of retreat may be directly connected with rock type.

POSITION OF BOUNDARY BETWEEN CLIFF AND SLOPE

A puzzling feature of many escarpments is the variation in position of the boundary between cliff and slope. Frequently the boundary is at the contact of the cliff- and slope-forming formations, rarely above this contact, and often below it, forming a vertical cliff in what is elsewhere a slope-making formation (figs. 2-a and b, 3, pl. 1). This variability is the key to understanding the mechanism of scarp retreat.

MECHANISM OF SCARP RETREAT

In many textbooks the formation of a talus at the base of a cliff in arid or semi-arid regions is described as the result of continuous shedding of fragments from the associated cliff. By implication, scarp retreat is then

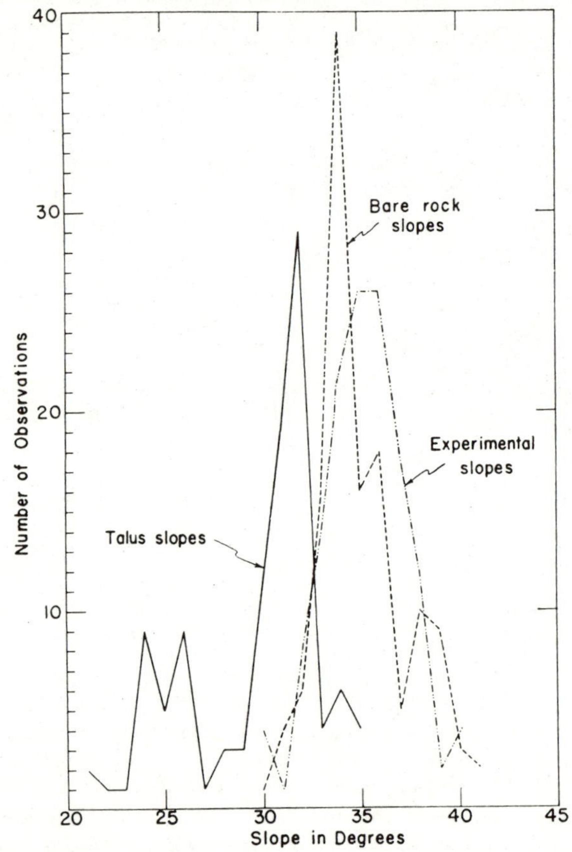

Fig. 4. Distribution of angles measured on talus slopes, bare rock slopes, and experimental apparatus. The secondary concentration of talus angles near 25° reflects the influence of columnar jointed basalt (see text and fig. 1).

visualized as the result of this continuous shedding of material along the whole cliff face. If this were the mechanism of retreat, however, there is no reason why the boundary between cliff and slope should vary in position, unless it reflected variations in the character of the slope- and cliff-making beds.

Detailed examination of cliff faces where variations in boundary between cliff and slope occur disclosed no related variations in rock character. A significant relationship was found to exist between the presence of sliderock on the slope and the state of weathering of the cliff face. Large sliderock accumulations were always associated with obviously fresh unweathered cliff faces, indicating recent large rock falls. Bare rock slopes and extensions of the cliff into slope-making formations were always associated with old, weathered cliff faces. While occasional individual fragments are undoubtedly shed from the cliff, the relationship indicates that large sliderock accumulations result almost wholly from single falls of large volumes of material, and that the shedding of individual weathered blocks is not a significant part of scarp retreat. It follows, then, that retreat must consist of a process which permits periodic large falls of rock from the cliff face, and preserves the characteristic cliff and slope angles through long periods of time. These conditions can be met by assuming the development of a fracture surface in the slope-forming material, at some angle near 45°, extending upward to the cliff-forming beds. This fracture should develop when the cliff extends such a distance into the relatively weak slope-forming beds that the strength of the material is exceeded by the weight of the overlying load. Slipping will then occur, with fracture of the cliff-forming beds along joints.

Obviously, a fracture surface of this type would be difficult to recognize in the field, since sliding would occur very soon after the fracture first developed. However, one such fracture surface, dipping 48° toward the cliff face and with displacement of approximately ½ inch, was observed in a small cliff in the Moenkopi formation near Lees Ferry, Arizona.

The process of retreat may be summarized as follows:

1. Formation of an escarpment by faulting or canyon-cutting, and development of a talus at the base of the cliff at an angle of approximately 32° (fig. 5-a).

2. Removal of the talus by weathering and erosion, developing a bare rock slope at an angle of 34-38° (fig. 5-b).

3. Erosion of bare rock slopes, partly by rill and sheet wash, partly by direct abrasion by sliderock, increasing the proportion of cliff to slope (fig. 5-c).

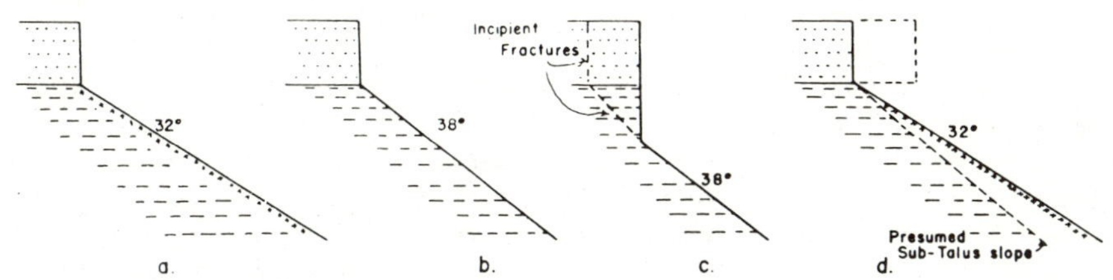

Fig. 5. Idealized profiles showing stages in retreat of an escarpment.

4. Yielding of the slope-making formation along a fracture surface at 40 to 50°, and collapse of cliff-making beds along joint planes, forming a talus over the rock slope at approximately 32° (fig. 5-d).

Further retreat is negligible until this talus is destroyed, when the cycle will be repeated. Rapid retreat of the scarp ceases when bare rock slopes at an angle equal to or greater than the angle of sliding friction can no longer be maintained. During the period of active retreat, successive cliffs and slopes would be approximately parallel to the originals, as indicated by Bakker and Le Heux (1947, p. 963-965). Once this phase of scarp retreat is passed, the conditions which produce parallel retreat are no longer present, and further reduction of the escarpment is accomplished by lowering of slope angles.

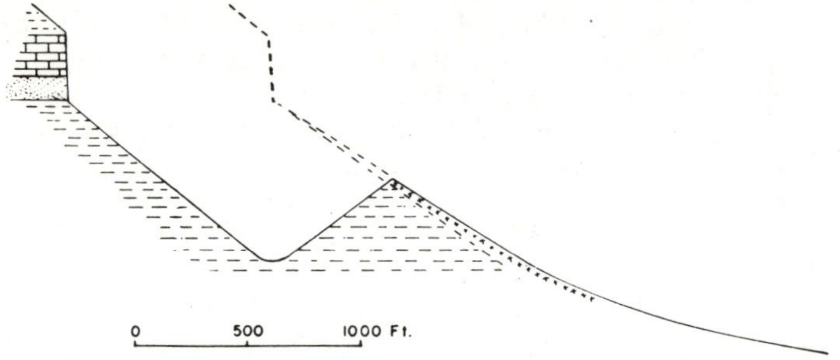

Fig. 6. Diagrammatic section of talus flatiron characteristic of western Grand Canyon. The talus flatiron is roughly triangular in plan.

The process outlined above would produce results which correspond to observed conditions. The preservation of the characteristic angles of cliff and slope, the relatively thin cover of sliderock over smooth rock slopes, and the correlation between large talus accumulations and fresh cliff faces are all explained. Variation in position of boundary between cliff and slope is a necessary consequence of this process. The similarity of scarp forms in different materials is explained as the result of the uniformity of talus angles and angles of sliding friction in different materials. Departures from the average result from special conditions, as in the case of the basalt cliffs mentioned above.

TALUS FLATIRONS

"Talus Flatirons" are minor slope forms having the general appearance of flatirons produced by inclined beds of resistant rocks, but are developed in rocks of relatively low resistance with horizontal or nearly horizontal attitude. They develop only where the cliff-forming beds are especially resistant to weathering, and where the slope-forming beds are relatively thick and weak (fig. 6).

The outer face of the talus flatiron consists of sliderock at the angle of repose characteristic of the particular materials. The inner slope, toward the escarpment, is undisturbed bedrock exposed by weathering and erosion. In western Grand Canyon talus flatirons form a conspicuous part of scarp topography. Here, the chief cliff-formers are the resistant Toroweap and Kaibab formations; the slope-former is the thick, weak Hermit shale. The talus, from 3 to 15 feet thick, forms an inclined resistant protective cover over the weak Hermit shale, and as a result produces forms similar to the true flatiron resulting from differential weathering and erosion of steeply tilted sediments of varying resistance.

The form and history of talus flatirons confirms the conclusion that the sliderock-covered slope is relatively stable and that retreat of the scarp occurs only when the talus cover has been removed and the bare rock slope exposed. Otherwise, such remnants of the slope would not now be isolated from the rest of the escarpment.

TABLE 1
Talus Angles

Type of material	Number of observations	Mean, to nearest .1 degree	Standard deviation, to nearest .1 degree
1. Basaltic lava with columnar jointing; talus blocks variable, from 2″ to 8′	35	27.2	3.1
2. Blocky sandstone, with closely spaced joints; talus blocks 2″ to 24″	32	32.1	2.2
3. Massive cross-bedded sandstone; blocks equidimensional, 2″ to 6′	18	32.4	2.3
4. Massive sandstone, joints 6′ to 20′; cross-bedded, few talus blocks	14	32.2	1.2
5. Limestone; talus blocks blocky to platy, 1″ to 36″	6	32.1	1.2

TABLE 2
Rock Slope Angles

Type of material	Number of observations	Mean to nearest .1 degree	Standard deviation to nearest .1 degree
1. Thin-bedded silty sandstone	43	34.2	2.3
2. Thin-bedded friable siltstone and shale	10	36.1	2.1
3. Thin-bedded clays and bentonites	21	34.9	3.3
4. Thin-bedded siltstone and sandstone	21	35.1	1.3
5. Thin-bedded variable siltstone and sandstone	34	36.3	2.8

SUMMARY

Talus slopes and bare rock slopes fall generally within a very small range of angles in a wide variety of rock types. Scarp retreat in the regions studied is a discontinuous process consisting of sudden rock falls separated by long periods of relative stability. Active retreat occurs only when the sliderock accumulations can be removed and bare rock slopes established at approximately the angle of sliding friction for the materials involved. During this phase of development retreat is essentially by formation of successive parallel cliffs and slopes.

REFERENCES

Bakker, J. P., and Le Heux, J. W. N., 1947, Theory on central rectilinear recession of slopes, I and II: K. Akad. Wetensch. Amsterdam Proc., v. 50, nos. 8 and 9.

Van Burkalow, Anastasia, 1945, Angle of repose and angle of sliding friction, an experimental study: Geol. Soc. America Bull., v. 56, p. 669-707.

DEPARTMENT OF GEOLOGY
COLBY COLLEGE
WATERVILLE, MAINE

[American Journal of Science, Vol. 254, November 1956, P. 693-706]

THE ROLE OF CREEP AND RAINWASH ON THE RETREAT OF BADLAND SLOPES

19

STANLEY A. SCHUMM

ABSTRACT. In badlands near Wall, South Dakota, two distinct types of topography have developed. The broadly rounded interfluves on the Chadron formation are interpreted as due largely to creep, whereas the steep, straight slopes formed on the overlying Brule formation are attributed to dominance of rainwash erosion.

Measurements of erosion depth along stake profiles on badland slopes reveal that erosion is rapid, ranging from 0.8 to 1.5 inches under 32 inches of rainfall during a 25½-month period. Erosion depth increases with slope angle on Brule slopes controlled by rainwash, but is a maximum on the convex divides of the highly permeable Chadron slopes. Pediments adjacent to the badland slopes were lowered 0.9 inch during the same period.

The topographic differences appear to resemble those existing between characteristic humid and arid topographic types, or between the ideal forms postulated by Davis and Penck in their respective cycles of slope retreat. It is suggested that the dominance of creep over rainwash erosion causes the development of broadly rounded divides and convex slopes, whereas predominant rainwash maintains steep, straight, parallel-retreating slopes heading in narrow divides. The contrasting forms observed may be viewed as end members of a continuous series.

INTRODUCTION

In the past the assignment of roles of creep and rainwash in the development of landforms has occasioned some controversy. The importance of creep in forming convex hilltops has been emphasized by Davis (1892), Gilbert (1909), Baulig (1940), Birot (1949, p. 23), and Penck (1953, p. 78), and the importance of rainwash stressed by Fenneman (1908) and Lawson (1932). Both concepts have been summarized by Baulig (1940), Cotton (1952), and Van Burkalow (1945).

Creep has been defined by Bryan (1922, p. 87) as "the slow movement of soil and rock waste down the slope from which these materials have been derived by weathering. Creep is due primarily to gravity but is facilitated by the presence of water, alternate wetting and drying, freezing and thawing, growth and decay of roots, and the work of burrowing animals." Gilbert (1909) generalized the mechanism of creep in the statement that "whatever disturbs the arrangement of particles permitting any motion among them promotes flow or creep."

Rainwash has been defined by Bryan (1922, p. 89) as "the water from rain, after it has fallen on the surface of the ground and before it has been concentrated into definite streams." Rainwash is the "unconcentrated wash" of Fenneman (1908).

During investigations of slopes and drainage systems of badlands in South Dakota, Arizona, and New Jersey (Schumm, 1956), examples were found which may illustrate the contrasting action of creep and rainwash. In Badlands National Monument, South Dakota, two distinct types of topography have developed. Steep, sharp-crested slopes, on which the mean of a sample of maximum slope angles measured 44°, have formed on the Brule formation, whereas the topography developed on the underlying Chadron formation, where exposed by the retreat of the overlying Brule, is composed of broadly rounded interfluves with a mean maximum slope angle of only 33° (pl. 1-A).

PLATE 1

Topography and slope surfaces at Badlands National Monument, South Dakota.

A. Typical topography of the Brule and Chadron formations. Contact between the two formations, just above the center of the photograph, separates the steep slopes developed on the Brule formation from the gentle, rolling topography developed on the Chadron formation.

B. Closeup of loose mosaic of aggregates forming the surface of slopes developed on the Chadron formation.

C. Small residual of the Brule formation showing straight steep slopes despite short slope length. Note pediment surface rising to make a sharp junction with the residual slope.

D. Water poured onto the surface of the Chadron formation follows subsurface channels to the base of the slope where it reappears on the pediment surface, sapping back the base of the slope.

The surface appearance of the slopes developed on each of the two formations is markedly different. In the dry state the Brule surface is hard; rill channels testify to copious runoff, although desiccation cracks are numerous. The dry Chadron surface, on the other hand, is composed of a loose mass of clay aggregates (pl. 1-B) which may be scooped up by hand.

The rapidity of breakdown of the Chadron clay and the looseness of its surface might suggest rapid erosion, and yet the harder, seemingly more resistant Brule has retreated far back from the margin of the Chadron (pl. 1-A). The Chadron thus appears to act as a more resistant rock. An explanation was apparent when a canteen was emptied on both types of slopes. Runoff occurred almost immediately on the Brule slope; whereas the water was completely absorbed by the loose Chadron aggregates. This suggested that the Brule slopes retreat more rapidly under the action of surface runoff, or rainwash, while the Chadron slopes with high infiltration rates are modified more slowly by the action of creep. To check these conclusions a 5-gallon hand pump was used to spray water on the slopes. On slopes developed on the Brule formation runoff usually occurred almost immediately and in all cases before 1 gallon of water was sprayed on an area of about 6 square feet. The water was quickly concentrated into rill channels and flowed away. In contrast 4½ gallons were sprayed onto about 6 square feet of Chadron slope before runoff began. By then the Chadron surface had become a sticky mud on which each aggregate had softened and slumped down until it came in contact with the next lower aggregate on the slope. From plate 1-B it is possible to visualize how the aggregates would move, sliding on the underlying fine-grained surface which apparently becomes sealed when wet.

EROSION MEASURED ON BADLAND SLOPES

Erosion was measured on badland slopes developed in a clay-sand fill at Perth Amboy, New Jersey (Schumm, 1956) by driving wooden stakes into the slopes along profile lines orthogonal to the contours of the slope. The stakes were oriented normal to the surface and driven flush with the slope surface. Erosion depth was measured by the length of stake exposed after each storm. Using similar techniques, stake profiles were established on some slopes in the South Dakota badlands in July 1953. The stakes were 18-inch segments of concrete reinforcing rods, ⅜ inch in diameter.

After the installation of the profiles the areas were revisited 15 months (October 1954) and 25½ months (August 1955) later. Between July 1953 and October 1954 precipitation was about 20 inches but only 12 inches during the second period. The total was thus about 32 inches of rainfall between July 1953 and August 1955 at Badlands National Monument Headquarters (J. H. Fraser, personal communication).

Stake profiles are plotted in figures 1 and 2 with the measured erosion for the two periods indicated above each stake. Profiles A and B on figure 1 are from small residuals developed on the Chadron formation. Profiles A and B of figure 2 are on a larger residual composed of both the Brule and Chadron formations. Stake 4 of profile A (fig. 2) is on the contact between the two

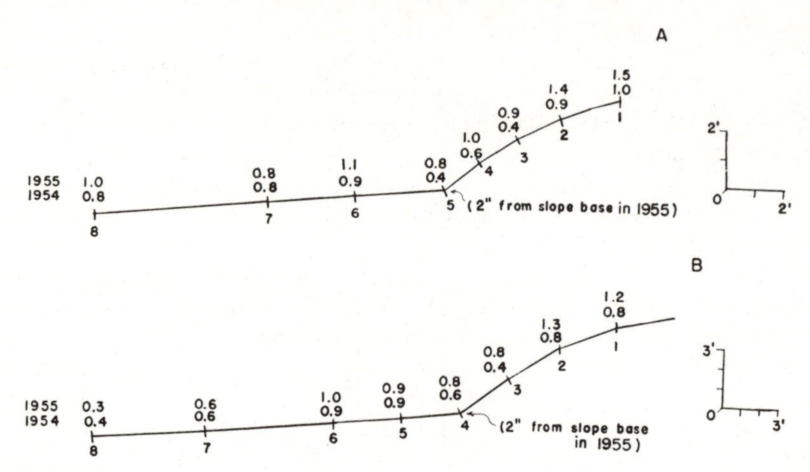

Fig. 1. Profiles showing slopes and pediments developed on two small badland residuals composed of the Chadron formation. Ticks on profiles show position of stakes; numbers above profiles indicate depth of erosion in inches measured in 1954 and 1955. The number assigned to each stake is listed below the profile. During the period of study the slopes of the two profiles have retreated 2 inches from the stakes driven at the junction of pediment and slope (stake 5, profile A; stake 4, profile B).

formations. The last vestige of Brule shale has been removed above profile B (fig. 2) which represents a slope composed entirely of the Chadron formation.

No stake profiles were established wholly on the relatively resistant Brule formation residuals with low infiltration rates, but profiles established on the fill at Perth Amboy illustrate erosion on slopes of similar appearance under the action of rainwash. At Perth Amboy the steep, straight slopes whose mean maximum angle is 49° intersect in narrow-crested interfluves very similar in appearance to those of the Brule. Among the many slope profiles measured at Perth Amboy that were straight from crest to base were several with convex summits. In these cases the parallel-retreating straight slopes had not entirely planed away remnants of the original upland surface. The break in slope profile between the upland and bordering straight slope was rounded by rainwash passing over the break in slope. The three Perth Amboy slope profiles selected for comparison with the Chadron profiles have convex summits and are illustrated in profiles A, B, and C of figure 3. The rainwash erosion on the relatively impermeable, convex Perth Amboy slopes may then be compared with the action of creep on the permeable, convex Chadron slopes.

Comparison of the profiles reveals differences in erosion rates between slopes here attributed predominantly to creep (profiles A and B, fig. 1; profile B, fig. 2) and those on which rainwash is assumed to be dominant (fig. 3 and possibly the combination slope fig. 2-A). On the Perth Amboy profiles (fig. 3) erosion at any point appears to be related to slope inclination. On the steep, straight portions of the slope, erosion is at a maximum and decreases on the convex summit areas. Compared to erosion on the summits, erosion on the straight portions of the slope is essentially uniform. Differences in erosion depth exist on the straight portions, but for the most part are due, as

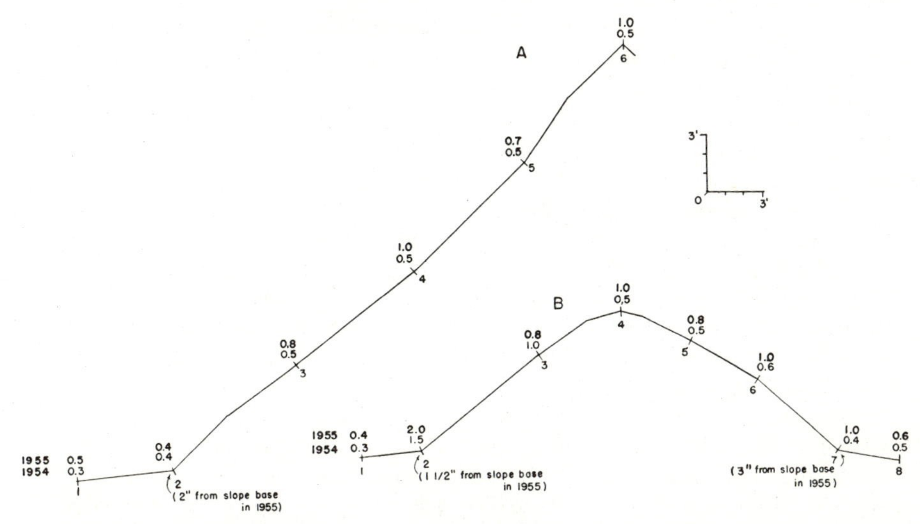

Fig. 2. A. Profile of slope composed of both the Brule and Chadron formations. Stake 4 is driven into the slope at the contact between the two formations. Stake numbers are given below each profile; erosion depth in inches above each profile.

B. Profile across same residual as in A above. Brule formation has been eroded off leaving slopes composed entirely of Chadron formation.

at stake 9, figure 3-B, to a slight concavity or in other cases to a convexity on the slope surface. A statistical comparison of erosion on 16 slopes during the 10-week period and of slope angles measured at the same points in 1949 and 1952 at Perth Amboy cast no doubt upon the hypothesis that the steep, straight portions are retreating parallel under the action of rainwash (Schumm, 1956).

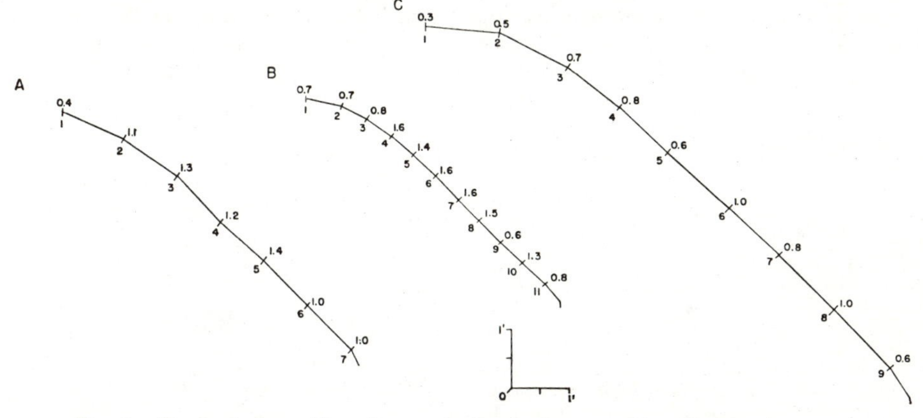

Fig. 3. Perth Amboy, New Jersey, badland slope profiles. Numbers above each profile give measured depth of erosion for a 10-week period in 1952. Ticks on profiles indicate position of stakes.

Erosion on the Chadron residuals of figure 1 shows an opposite relationship. Total erosion depths measured after 15 and 25½ months are greatest on slopes of least inclination. On profile B, figure 2, no systematic relationship is apparent between slope angle and erosion depth, yet a convexity is apparently developing. Note that stake 3 on profile B showed no erosion for the latter 10½-month period, but rather deposition of 0.2 inch. The lack of stakes on the lower steeper parts of this profile precludes the acquisition of data critical for the understanding of mass removal on the slope.

On profile A, figure 2, erosion was uniform over the length of the slope during the first period of erosion and is essentially uniform for the 25½-month period despite the fact that the mean slope below the contact between the Brule and Chadron formations (stake 4) is 39° and that above on the Brule formation is 44°.

The opposite relations as to erosion depths displayed by Chadron profiles in comparison with those of the Perth Amboy badlands (fig. 3) are considered most important to this discussion. Reversed relationships of slope steepness to erosion depth in these two areas can perhaps be reconciled by appealing to dominance of two different geomorphic processes: creep and rainwash.

Runoff or rainwash where observed on badland slopes occurs as surge or subdivided flow (Horton, 1945; Schumm, 1956). Minute obstacles check and deflect the runoff in its downslope course; the movement of the eroded material is therefore not continuous. The runoff neither constantly accelerates nor becomes overloaded toward the slope base on straight slopes in the cases observed. Erosion measured on straight badland slopes is essentially uniform over the length of that slope, although during any one storm the mean erosion depth for the slope as a whole may not be reached or may be exceeded on any part of the slope.

If the slope has a convex summit, due in this case to rounding of the junction between the valley-side slope and original upland slope by runoff, erosion will be less on the gentler slopes, and the steeper portion of the slope will encroach on this convexity until two straight slopes unite to form a sharp crest. In other words, when rainwash acts on a slope any particles A, B, C,

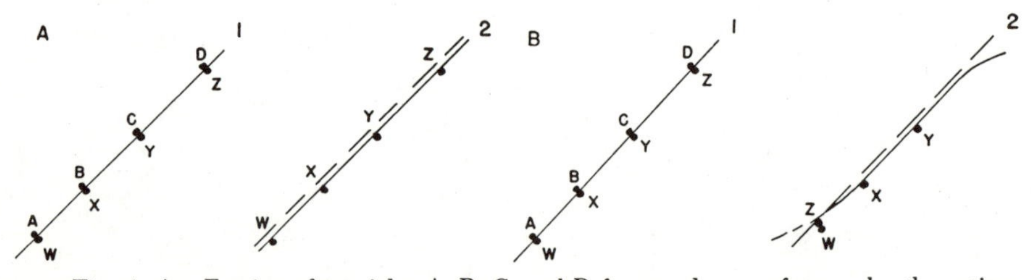

Fig. 4. A. Erosion of particles A, B, C, and D from a slope surface under the action of rainwash. The grains are not necessarily removed in sequence, but each layer of grains is removed before erosion of the next can begin.

B. Removal of particles A, B, C, and D from a slope surface by creep. The grains were removed in sequence, followed by particle Z, causing maximum erosion on the crest of the slope.

and D (fig. 4-A) spaced equally along the slope may be removed from the slope during any one storm or during several storms, but in any case before the underlying particles W, X, Y, and Z are removed. Rainwash is thus an eroding mechanism tending to remove a uniform thickness of material from a straight slope during a storm. It does this because as an eroding agent it attacks the steepest part of the slope with the greatest energy. Thus, when particle A (fig. 4-A) is removed from the slope, in effect, the base level for the particles above is lowered and they are quickly removed. If by chance particle D is removed first, a slight concavity results and erosion is less there until the lower particles on the slope have been removed. This type of action has been documented by repeated measurements along the stake profiles at Perth Amboy (Schumm, 1956).

Creep has been proposed as the dominant process active on the Chadron residuals. The individual aggregates move downslope by swelling from wetting, shrinking from drying and filling of the desiccation cracks from the uphill side, and possibly by rainbeat and sliding and slumping on the sealed, underlying, finer-grained surface. Unlike the action of rainwash, in which a sheet of material moves on the entire slope surface, a particle D (fig. 4-B) engaged in creep will not be removed from the slope at the same time or before a particle A but must traverse the entire slope length behind the particle downslope from it. Thus, the aggregates will be removed from the slope in the order A, B, C, D, Z. Since the movement of the creeping material should be greatest on the steepest slopes, it would seem that erosion would logically be greatest on these slopes. However, it may be that these slopes are principally slopes of transportation. The creeping material from upslope moves over these slopes, protecting the bedrock from rapid erosion.

Perhaps the above explains the partial burial of stake 3 on profile B of figure 2. Furthermore, a stake profile established on a badland slope in Sioux County, Nebraska also illustrates the partial burial of the stakes on steeper slopes (fig. 5). The slope illustrated in figure 5 is steep, bordered at its base by an ephemeral stream, and its surface is composed of highly permeable aggregates derived by weathering of the Chadron formation. After 1 year the upper two stakes of the profile were exposed by the action of creep. Of the

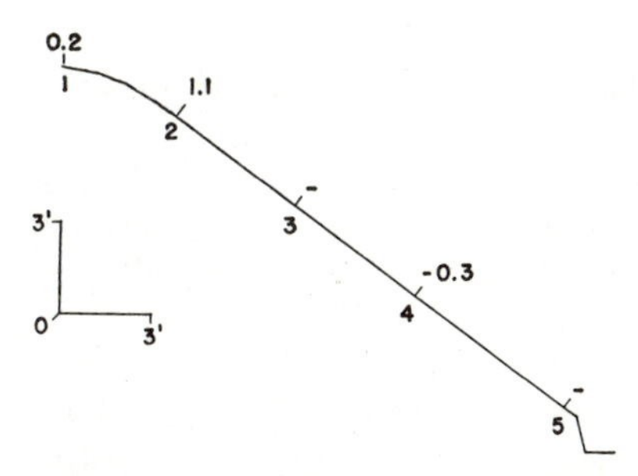

Fig. 5. Profile of badland slope developed in the Chadron formation (Sioux Co., Nebraska). Erosion is greatest at the upper two stakes, while the lower three were buried by downward creeping material from the slope crest. Erosion depth in inches is given above the profile.

lower three stakes on the steepest part of the slope, only stake 4 was found, buried 0.3 inch. Stakes 3 and 5 are assumed to have been buried as well.

If, as observed on slopes modified by creep, wasting is most rapid on the gentler slopes near the summit of a hill, the convexity will broaden and encroach on the less rapidly eroding side slopes, as illustrated by the lowering of the Chadron residuals. A lune-shaped wedge of material is thus removed from the slope as indicated by Lawson (1932) but not, as he thought, by rainwash. Perhaps, also, the supply of material from the flatter summit slopes and the ease of movement on these slopes is increased by the greater amount of rain falling per unit area as compared with the steeper side slopes

EROSION ON MINIATURE PEDIMENTS

Stake profiles were extended across the miniature pediments bordering the badland residuals in figure 1-A and B. In each case the pediment was lowered, presumably by rainwash. Similar reduction of miniature pediment surfaces has been described by Bradley (1940) and Smith (in press).

In this area Smith recognized that while some pediments were being dissected and lowered others were being raised by aggradation. He concludes that the sheet wash is not a major erosional agent but is chiefly one of transportation. He expects slow lowering of the pediment, however, when it is graded to a "slowly falling base level stream."

In profile the pediments are slightly concave-up. Both pediments are graded to a small ephemeral stream. McGee (1897) and Davis (1938) agree that under such conditions the pediment probably will be degrading and probably concave in longitudinal profile.

The hypothesis favored here is that although modified by rainwash the pediment is not formed by it. Both Paige (1912) and Davis (1938) considered that formation of the pediment is by retreat of a steeper slope leaving the pediment behind as a surface of transportation to the nearest drainage channel or interior basin. In the South Dakota badlands the miniature pediments lie at the bases of retreating badland slopes. Slope retreat is accelerated by undercutting of the upper slope which differs on the two lithologic units. Greater runoff on the Brule shale slopes and "the spreading of sheetwash at rill channel mouths undercuts the escarpments" (Smith, in press). On pediments formed at the base of Chadron residuals the slope retreat is aided by basal sapping of the slope by the appearance of subsurface flow at the junction of slope and pediment surface (pl. 1-D).

During 25½ months the bases of the residual slopes retreated on an average of 2.0 inches from the stake originally placed at the junction of pediment and badland slope (figs. 1 and 2). The pediments were lowered on an average of 0.7 inch during the first 15-month period but only 0.2 inch more during the second period for a mean of 0.9 inch for the 25½ months. During the second period deposition occurred at the lower stake of profile B, figure 1. It may be that the greater total rainfall of the first period, including one storm which totaled 3.46 inches during 24 hours, produced much greater runoff which lowered the pediments by sweeping depositional material from the pediment surface.

EROSION IN ARID AND HUMID CLIMATES

The differences observed between the two types of badland topography in Badlands National Monument have been interpreted as due to the different processes acting on the slopes: creep and rainwash. Because the topographic differences appear to resemble those existing between characteristic landforms of humid and arid climates, the explanation of the differences in the badlands may be projected to include differences in humid and arid topographic forms. Figure 6 shows the writer's concept of slope development on the highly permeable Chadron formation contrasted with that on the impermeable Brule formation. The differences are perhaps also illustrative of the distinctions between the classical slope retreat concepts of W. M. Davis and W. Penck. Under the action of rainwash (fig. 6-A) the initial slopes retreat at an essentially constant angle, leaving a pediment at their base adjusted to the angle required to allow removal of the debris brought to the slope base, thereby

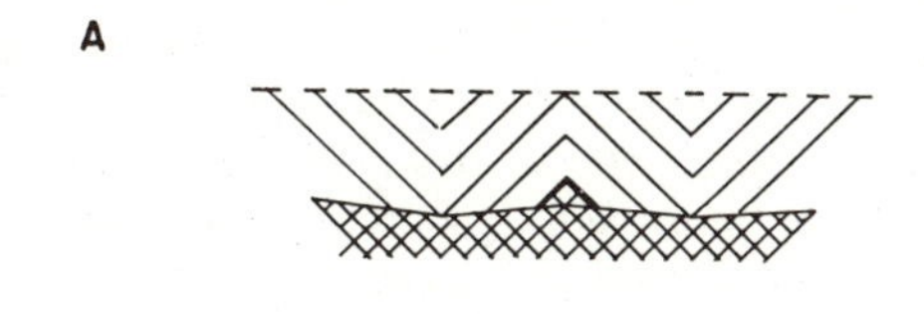

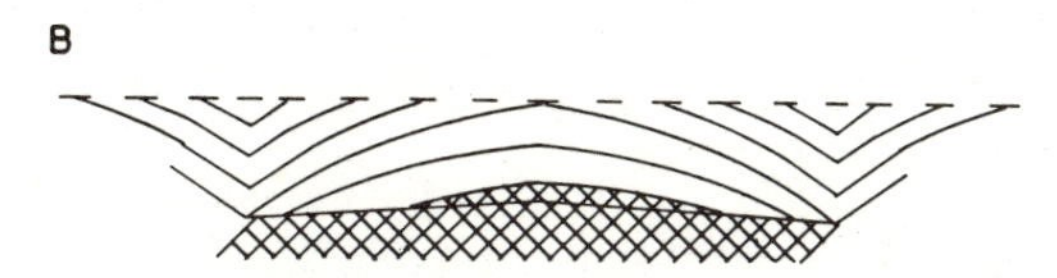

Fig. 6. A. The cycle of slope development on the Brule formation.
B. The cycle of slope development on the Chadron formation.

perpetuating the maintenance of the straight slope. The slopes remain constant until only a small residual remains surrounded by a pediment surface (pl. 1-C).

Slope angles measured on Brule residuals of all sizes are essentially constant in comparison to those measured on Chadron residuals, substantiating parallel slope retreat in this type of landform evolution (Schumm, 1956). Some reduction of slope angle with erosion of the Brule shale slopes may occur, however, due to the alternation of resistant and less resistant layers. The slopes capped by or composed of several resistant layers are somewhat steeper than those where the resistant layers have been removed by a shortening of the slope. Smith (in press) has shown a slight but significant lowering of slope angles on slopes ranging in length from 45 to 3 feet. He attributes this decrease in angle to a lessening of the undercutting of the slope base as it becomes shorter, causing a declining slope retreat even on the Brule formation residuals. This is not substantiated by the basal slope retreat measured on the profiles of figures 1 and 2, which is about 2 inches for all the slopes.

In the development of slopes controlled by creep (fig. 6-B), rounding of the upper slopes occurs early in the cycle, and the radius of curvature increases through much of the later slope development. In this example (fig. 6-B), however, illustrating the development of the Chadron residuals, a pediment also forms at the base of the steeper lower slope. In this area the pediment is apparently formed by basal sapping of the slope by the appearance of subsurface flow at the slope base (pl. 1-D). If this basal sapping did not occur transport would be slowed at the base of the steep slope, and deposition there would probably aid the formation of a concave lower slope analogous to the concave basal slopes of humid climates.

In an effort to obtain some information on the progress of the erosion cycle on the Brule and Chadron formations, a series of slope profiles was measured on remnant buttes exemplifying the results of erosion on slopes of each type.

In figure 7 three profiles were measured across the divide on a residual composed of the Brule formation. The mean slopes of each profile are not markedly different; a small decrease in mean angle is the result of the removal of a nodular layer which causes the convex appearance of the right slope of profile 1. Areas of more impermeable Brule maintain even steeper slopes until late in the cycle (pl. 1-C).

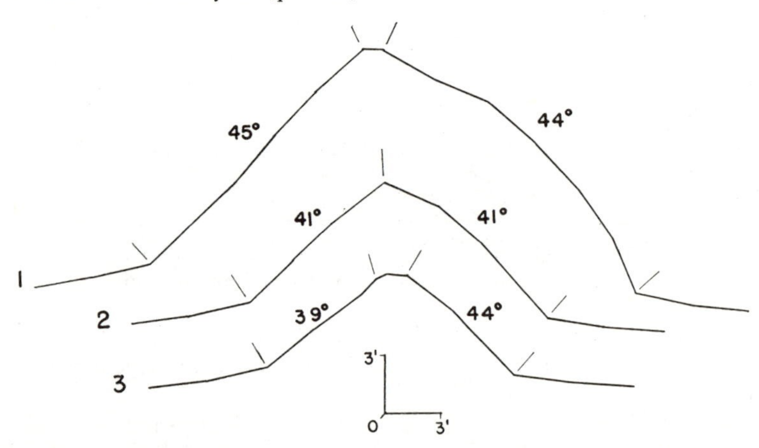

Fig. 7. A series of slope profiles measured on a residual of the Brule formation. Mean slope angles change only slightly with shortening of the slope on this relatively impermeable material.

Figure 8 shows the profiles of figure 2 with two additional profiles added to show the decline of the mean slope angle after the removal of the overlying Brule shale. Profile 1 (fig. 8) is a slope composed of both the Brule and Chadron formations. The dashed line indicates the contact between the overlying Brule and the stratigraphically lower Chadron. With the removal of the Brule from the upper part of the slopes, profiles 2, 3, and 4 show a progressive decline in mean slope angle.

Figure 9 shows a series of profiles across a residual of highly permeable Chadron clay. The texture of the topography in this area is relatively coarse for badlands, as would be expected in areas of permeable soils. The profiles show the marked reduction in slope angle as the erosion of the residual progresses.

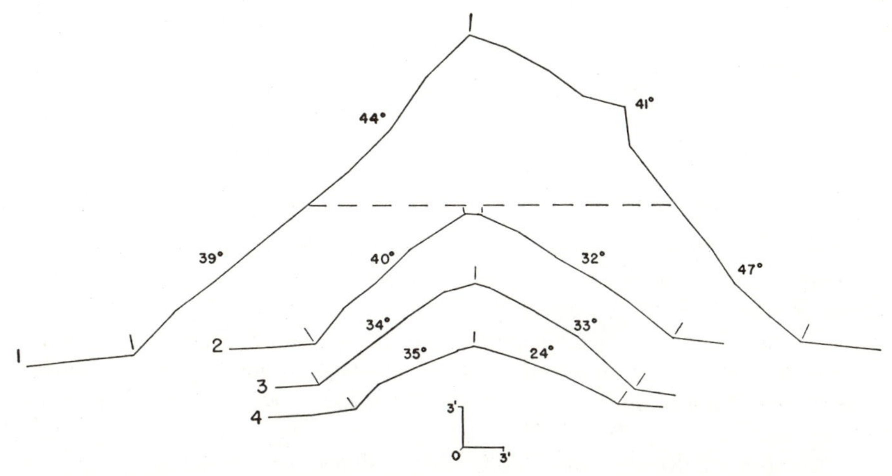

Fig. 8. Series of slope profiles on the residual where the stake profiles of figure 2 are located. The lower three (Chadron) profiles have moderate infiltration rates in comparison to the low rate of figure 7 profiles and the high rate of figure 9 profiles. Dashed line (profile 1) indicates contact between the overlying Brule and the Chadron formation.

The differences between erosion on the Brule and Chadron formations may be more clearly illustrated by comparing the rate of lowering of the residual divides with that of the reduction in width of the residual. In the three profiles of figure 7 (Brule) between profile 1 and 3 the width of the residual was reduced by 50 percent and the height by 56 percent. On the Chadron profiles of figure 8 between profile 2 and 4 the width was reduced only 56 percent while the height was lowered by 72 percent. On figure 9 (Chadron) between profiles 1 and 3 the width was reduced by 63 percent and the height by 81 percent. Therefore, in spite of variations in slope angles caused by lithologic units of differing resistance, the height of the Brule formation residual was lowered only 6 percent more than the width, indicating a small decrease in slope angle, whereas the height of the Chadron residuals is decreased by 28 percent and 18 percent over the width, causing a marked decrease in slope angle. These series of slope profiles seem to substantiate the hypothetical cycle illustrated in figure 6 and indicate that in this badland area the topographic differences are most reasonably explained by the difference in infiltration rates of the slope surfaces.

A further distinction between landforms fashioned by creep and those eroded by rainwash is suggested by a comparison of the rates of erosion on the Perth Amboy, New Jersey badlands and Chadron slopes in Badlands National Monument. The contrasts in mean slope angles and texture may be considered analogous to differences between any two areas of high relative relief fashioned by the two processes. Also, the comparison is justified on the

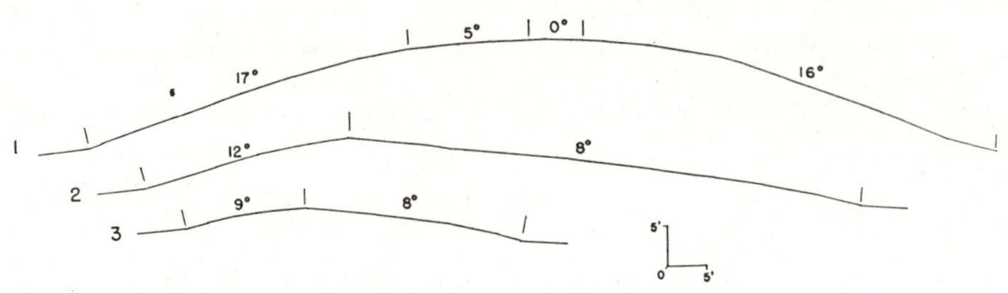

Fig. 9. Series of slope profiles on a badland residual located in a highly permeable area of the Chadron formation.

assumption that the clay-sand fill at Perth Amboy is probably about as easily eroded as any relatively impermeable material, while the Chadron is a rapidly eroding area in which permeability is high.

On the Chadron slopes 32 inches of rain fell, while a mean depth of 1.1 inches of material was removed. At Perth Amboy the mean depth of erosion was 0.9 inch in a period in which about 15 inches of rain fell (U. S. Weather Bureau, 1952).

Comparing directly, the Perth Amboy slopes were reduced an average of 0.2 inch less than the Chadron but with 17 inches less rainfall. With equivalent rainfall over the two areas (32 inches) the Perth Amboy slopes would, at the same ratio, be eroded to an average depth of about 2.0 inches as compared to the 1.1 inches on the Chadron. In other words, slope reduction would proceed at a rate about twice as rapidly by rainwash as by creep.

Influencing the reliability of this estimate are many factors of which the most important is probably the precipitation intensities involved. It is, however, reasonable to suppose that in areas otherwise comparable, rainwash erosion will proceed at rates nearly double that by creep. The hills of Chadron formation lying in front of the retreating Brule escarpment in the South Dakota badlands seem to support, at least qualitatively, the above estimate.

This concept may be of importance in attempting to extend the relationships between the geomorphic and hydrologic characteristics of drainage basins in semiarid regions to more humid areas and also as a partial explanation of the high rates of erosion in semiarid regions.

CONCLUSION

In humid regions rapid rock disintegration produces thick soil horizons protected by good vegetation cover. As Davis (1930) has suggested, creep is probably dominant there. Abundant evidence (Sharpe, 1938) has been cited of its importance. In arid regions with poor vegetation cover and thin soils, rainwash is probably the dominant process. Undoubtedly, both occur in each climatic region, but in humid areas rainwash erosion probably occurs only in exceptional storms; whereas in arid regions creep is important probably only after several consecutive rains have saturated the soil. It may be that rainwash erosion and parallel slope retreat are dominant in the early stages of the erosion cycle in humid regions, when relief is great and slopes are steep

(Birot, 1949, p. 23), but that with the development of thick soil cover and grading of the slopes the action of creep and declining slope retreat become the characteristic process and form.

It is possible that the areas in which creep dominates over rainwash, or vice versa, are end members of a continuous series ranging through all proportions of both processes depending on prevailing vegetation, soils, and climate. Holmes (1955) has suggested that the classic examples of the arid and humid geomorphic cycles, as contrasted in the Penck and Davis concepts of slope retreat, are end members of such a series. This concept appears to be supported by the badland studies.

If creep is responsible for convex interfluves in the mature and later stages of erosion (after any initial surface of low relief is completely consumed), there is no need to resort to Penck's explanation that the recent diastrophic history of a region controls the slope profiles. In Penck's system the straight slopes of the Brule formation residuals would indicate uniform uplift (gleichformige Entwicklung) whereas the convex slopes of the Chadron residuals would indicate increasing rates of uplift (aufsteigende Entwicklung). Both diastrophic histories cannot be true within the same small area. Penck claims that such local exceptions do not invalidate his concepts. Perhaps this is true, but by the same token the diastrophic control is merely an unnecessary complication whose introduction is unjustified in many areas.

ACKNOWLEDGMENTS

This study is part of a quantitative investigation of erosional landforms sponsored by the Geography Branch of the Office of Naval Research as Project No. NR 389-042, under Contract N6 ONR 271, Task Order 30 with Columbia University. The writer wishes to acknowledge the suggestions made by Professor A. N. Strahler of Columbia University and Messrs. M. G Wolman and L. B. Leopold of the U. S. Geological Survey for improvement of the manuscript.

REFERENCES

Baulig, Henri, 1940, Le profil d'équilibre des versants: Annales de géographie, v. 49, p. 81-97. (Reprinted in Essais de Géomorphologie, 1950: Paris, Soc. Edition, Les Belles Lettres.)

Birot, Pierre, 1949, Essai sur quelques problèmes de morphologie générale: Lisbon, Instituto para a Alta Cultura, Centro de Estudos Geograficos.

Bradley, W. H., 1940, Pediments and pedestals in miniature: Jour. Geomorphology, v. 3, p. 244-254.

Bryan, Kirk, 1922, Erosion and sedimentation in the Papago country, Arizona: U. S. Geol. Survey Bull. 730, p. 19-90.

Cotton, C. A., 1952, The erosional grading of convex and concave slopes: Geog. Jour., v. 118, p. 197-204.

Davis, W. M., 1892, The convex profile of badland divides: Science, v. 20, p. 245.

————, 1930, Rock floors in arid and humid regions: Jour. Geology, v. 38, p. 1-27; 136-158.

————, 1938, Sheetfloods and streamfloods: Geol. Soc. America Bull., v. 49, p. 1337-1416.

Fenneman, N. M., 1908, Some features of erosion by unconcentrated wash: Jour. Geology, v. 16, p. 746-754.

Gilbert, G. K., 1909, The convexity of hilltops: Jour. Geology, v. 17, p. 344-350.

Holmes, C. D., 1955, Geomorphic development in humid and arid regions: A synthesis: Am. Jour. Sci., v. 253, p. 377-390.

Horton, R. E., 1945, Erosional development of streams and their drainage basins; hydrophysical approach to quantitative morphology: Geol. Soc. America Bull., v. 56, p. 275-370.

Lawson, A. C., 1932, Rain-wash erosion in humid regions: Geol. Soc. America Bull., v. 43, p. 703-724.

McGee, W. J., 1897, Sheetflood erosion: Geol. Soc. America Bull., v. 8, p. 87-112.

Paige, Sidney, 1912, Rock-cut surfaces in the desert ranges: Jour. Geology, v. 20, p. 442-450.

Penck, Walther, 1953, Morphological analysis of landforms: New York, St. Martin's Press. (Translated by Hella Czech and K. C. Boswell.)

Schumm, S. A., 1956, Evolution of drainage systems and slopes in badlands at Perth Amboy, New Jersey: Geol. Soc. America Bull., v. 67, p. 597-646.

Smith, K. G., Erosional processes and landforms in Badlands National Monument, South Dakota: Geol. Soc. America Bull., in press.

Sharpe, C. F. S., 1938, Landslides and related phenomena; a study of mass-movements of soil and rock: New York, Columbia Univ. Press.

Van Burkalow, Anastasia, 1945, Angle of repose and angle of sliding friction, an experimental study: Geol. Soc. America Bull., v. 56, p. 669-707.

Building 25
Federal Center
Denver, Colorado

Reprinted from *Inst. British Geog. Trans.*, **32**, 31–47 (1963)

CONTRASTS IN THE FORM AND EVOLUTION OF HILL-SIDE SLOPES IN CENTRAL CYPRUS

20

C. E. EVERARD, M.SC.

(*Lecturer in Geography, Queen Mary College, University of London*)

CYPRUS is a land of 'steep slopes, sharp angular forms and a stepped topography'.[1] In this island of great geological contrasts, torrential stream erosion and rapid rejuvenation have wrought a bewildering diversity of landforms, valuable studies of which have been contributed by E. de Vaumas, W. F. Schmidt and N. J. W. Thrower.[2] Of particular interest are the slope profiles of the Central Lowland. Concave slopes dominate this area and there is evidence that they are 'back-wearing', or evolving by parallel retreat with the production of broad pediments. Concavo-convex slopes are also to be found and these, conversely, appear to be evolving by 'down-wearing' or decrease of height and slope-gradient with age. In addition to these contemporary contrasts in form and evolution, there is evidence of a fundamental change in slope form since the Late Pleistocene, at which time concavo-convex slopes were dominant in the Lowland. Man appears to have been an important agent in bringing about this change.

Cyprus may be divided into three principal geomorphological regions (Fig. 1). In the south lies the Troodos massif (6403 feet), a deeply dissected block of basic and ultra-basic rocks, formerly part of the Arabian foreland. Running parallel to the north coast is the slender, wall-like Kyrenia range (3357 feet), upthrust in the Alpine orogeny. Separating the two mountainous regions is the Central Lowland, largely below 800 feet in height. Its rocks and landforms are intimately related to the two flanking highlands, for from them in Tertiary and Quaternary times have come both infilling sediments and dissecting streams.

The geology of the Central Lowland is far from simple, for the terrestrial and marine sediments display rapid facies changes from the margins towards the centre.[3] Geomorphologically, the significant feature is the occurrence of horizontal or gently inclined limestones (Pliocene to Recent in age) at various levels, forming resistant cap-rocks on underlying shales, marls and silts. When dissected these give a distinctive landscape of flat-topped ridges, mesas and buttes. The oldest cap-rock is the Pliocene Nicosia–Athalassa calcareous sandstone. In the south and south-east of the Central Lowland (Fig. 1: Region 3) the cap-rock is *kafkalla*, a solid surface veneer of evaporational or 'secondary' limestone. In the south-western portion of the Lowland (Fig. 1: Regions 1 and 2) the cap-rock is the *kafkalla*-cemented upper layer of a series of great alluvial fans. Deposition of the coarse 'fanglomerate' began in the late Pliocene and continued through the Pleistocene.[4]

Where the cap-rocks have been destroyed one finds extensive plains cut in the underlying argillaceous rocks, or along the north of the Lowland, sandstones and shales of the Kythrean Beds dipping nearly vertically. Limestones outcrop along the margins of the Troodos massif.

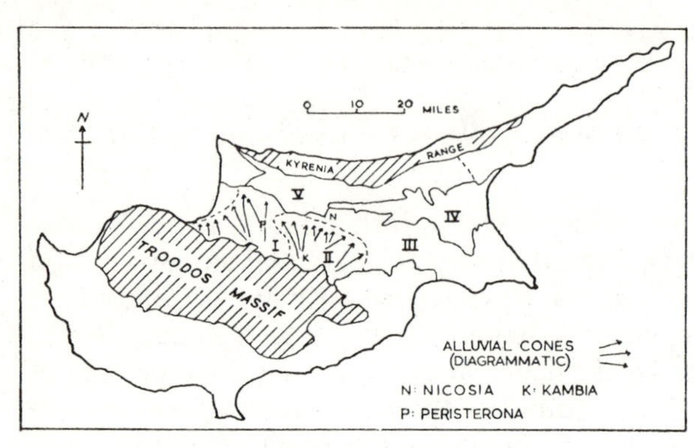

FIGURE 1—Regions and places referred to in the text.

1. Western alluvial fan region (fans furrowed by concavo-convex valleys; little *kafkalla*).
2. Eastern alluvial fan region (flat-topped ridges, mesas and buttes; *kafkalla* widespread).
3. *Kafkalla* plateaux region.
4. Alluvial lowlands and Mesarka.
5. Kythrea belt.

(Based upon D. Christodoulou (Note 1) and E. de Vaumas (Note 9)).

The hill-slope profiles of the Central Lowland are of two main types. The concave profile consists of a gently inclined apron (5 degrees or less) which sweeps gradually upwards to gradients of 25–35 degrees. Above this there is invariably a change of lithology to a hard cap-rock. The profile closely resembles the scarp and pediment form described by L. C. King.[5] It is characteristic in areas of *kafkalla* sheets, *kafkalla*-cemented fanglomerate and gently folded Nicosia–Athalassa limestones.

Secondly, there are the concavo-convex slopes, reminiscent of the classic Davisian slope profile.[6] They are to be found (*a*) on the relatively undissected western fanglomerate beds (for example in the Potami area, 738611);[7] (*b*) in areas of uniform impermeable rocks with limited available relief; (*c*) in areas of permeable rocks, such as massive chalks in the Lapithos beds (for example near Kambia) and gypsum lens in the Kythrean Beds; and (*d*) very occasionally where a *kafkalla* bed extends for some distance down-slope.

Environmental Factors

Before the evolution of these slopes can be discussed it is necessary to refer briefly to some aspects of the environment in which they have been shaped.

The intense erosion characteristic of so much of Cyprus is largely the product of the rapid rejuvenation to which the island has been subject. R. A. M. Wilson[8] and de Vaumas[9] record at least four major post-Pliocene stages of uplift, which have carried Athalassa Beds above 1200 feet.[10] There are no perennial rivers in the Central Lowland at present, and the valleys, free faces and relative relief of today are inherited from the rivers of wetter spells in the past and from the waves of rejuvenation that affected them.

Quaternary climatic fluctuations in the Near East have recently been reviewed by K. W. Butzer[11] who has traced a sequence of Pluvials and Interpluvials which correlate with the waxing and waning of the Pleistocene ice-sheets. Lowland Cyprus in the Würm Pluvial was, according to Butzer[12] covered with 'Subtropical Forest and Scrub', and D. K. Jones and others note that there is historical and biological evidence for pre-Neolithic forest in the Central Lowland.[13] Vegetation has a profound influence upon processes and slope profiles[14] and it is necessary to consider the factors that have brought about the meagre vegetation cover of the present day. The Late Glacial and Post-glacial have been times of increasing desiccation, with temporary reversals of the trend at about 9000 and 5000 B.C.[15] In addition to the effect of this climatic change upon the vegetation, Man has interfered with it by burning and by grazing flocks and, whatever the original natural vegetation may have been, slopes in the Central Lowland have for several thousand years evolved under a plagio-climax vegetation in a Man-dominated environment.

The Post-glacial desiccation has produced a semi-arid climate in the Lowland. In two places annual rainfall averages (1916–50) are under 12 inches, and everywhere below 16 inches. It is important to note that rain comes in heavy, erosive downpours both winter and summer (all but about 2 inches falls in winter). Of sixty records of 'unusually heavy rainfall' (2·8 inches or more in 24 hours), fifty-four are for the months November to January[16]. Mean daily temperature maxima reach 97° F. in July and August, and soil temperatures reach 140° F. Deep sun-cracks weaken semi-consolidated shales and marls, and hard rocks fracture, especially when suddenly cooled in a shower. The high evaporation rates have two important geomorphological consequences.[17] The first is the formation on argillaceous rocks of a dry, surface skin, about half an inch deep, which is minutely cracked and extremely unstable. It is readily dislodged by treading and a heavy rainstorm will strip the whole skin off a hill-side. Secondly the upward movement of soil-water by capillarity and its evaporation at the surface has led to the accumulation over wide areas of the hard, crust-like *kafkalla* or 'secondary limestone', and a softer under-layer of *havara*. The details of its formation have been discussed by D. J. Burdon.[18] The *kafkalla* visible today appears to be subject to erosion, and there is no clear evidence that it is forming to any great extent at the present. Once formed, it restricts evaporation and therefore can only accumulate to a limited thickness. Its rapid accumulation would be favoured by the greater amount of soil water available for evaporation in the early stages of the change from a humid Pleistocene climate to the drier conditions of today.[19]

Present-day vegetation encourages rapid erosion. On the uncultivated areas, almost invariably the hill-sides, grow tufted annuals and dwarf scrub.[20] There is no vegetation mat and so, although the soil has some protection from the direct impact of rain and hail, there is practically no binding influence at ground level. Small bushes and grass tufts often support masses of loose rock on their uphill side, testifying to the continual surface movement of loose debris.

On the cultivated flatter land erosion is still rapid. One can trace a 'cultivation nick' on most slopes at the upper limit of ploughing, which maintains a steeper slope gradient than would otherwise be the case. Unirrigated land is under a cereal/fallow rotation which leaves large areas exposed to the storms of a winter and two summers when in fallow, and liable to erosion by autumn rains even when planted. Burdon has estimated that bare and agricultural land erodes nearly six times as rapidly as heavily forested areas.[21]

Free-range grazing is practised over much of the Lowland. The sheep and goats, particularly the latter which climb higher and on to steeper slopes, dislodge loosened rock and the fragile sun-dried crust, causing a continual downhill movement of material at a far faster rate than would otherwise be the case. These animals have probably been in Cyprus since at least 4000 B.C. There are no early records of numbers but the impression is that, significantly, goats exceeded sheep. Goats prefer a more woody diet than sheep and are potentially more damaging to Mediterranean vegetation. Neither can do much harm to a mature vegetation, but should it be degraded by other means, for example by cutting for fuel, they keep it so.[22] Hence grazing has maintained a *garrigue-botha* scrub on slopes, with all the consequential effects on their erosion.

The Slope Profiles

The chief processes eroding the slopes are rain-drop erosion, sheet-wash and gullying, their potency stemming from the environmental factors noted earlier.

Large rain-drops and hailstones play an important part in moving material down-slope by the force of their impact. Small pieces of loose rock are dislodged and the sun-dried layer is especially vulnerable. A slope of 25 degrees, cut in Kythrean shale, was studied during a heavy thunderstorm. Ninety per cent of the hillside was bare at surface level. The first rain-drops evaporated rapidly on the hot, dry surface but as their intensity increased small (one-tenth inch) pieces of the dry surface skin were knocked downhill, by rolling, jumping or sliding, on both the lee and weather sides of the hill. At the height of the storm there was a continual downward movement of small fragments, and when it had passed a layer of dislodged rock ran along the foot of the slope.

According to D. L. Christodoulou[23] sheet erosion is widespread. Bare, smooth and unfurrowed hill-sides of shale, marl or silt bore witness to its work. A good example was found on an 8-degree slope cut in Lapithos marl. Upon it were numerous pebbles perched on pillars of marl about one inch high, circum-eroded while a layer one inch thick had been stripped off the slope. Where the stones had rested upon it, surface drying had not advanced far and so the marl had retained its cohesion and resistance to erosion. Ungullied vegetated slopes, although very complex in detail, are also severely attacked by sheet erosion, debris being easily moved by unabsorbed, unchannelled water sweeping down-hill.

Closely spaced rills furrow many slopes, often starting well below the steepest facet on concave slopes, but frequently reaching the crest on straighter slopes. The rill profile determines the slope profile, as the minor crests run parallel to the rill floors. Rills may also occur on the lower facet of concavo-convex slopes, heading at the point of inflexion.

Deep gullies trench many concave slopes. Their heads are almost vertical and they are often separated by knife-edge ridges. The gully-floors and ridge-crests are both concave, and determine the hill-slope profile. They are only occasionally water-bearing, and at other times denudation is by the processes described above.

Concave slopes develop wherever a resistant stratum overlies a weaker one. If there is no cap-rock on the latter, the profile is concavo-convex. So persistent is this relationship that the cap-rock appears to be essential for the development of a concave slope. Immediately below the cap-rock the gradient averages 35 degrees and then gradually decreases down-slope in a sweeping concave curve to become 5 degrees or less as it merges with the plain. This curved profile is in contrast to the rectilinear slopes of constant angle which rise abruptly from pediments in other semi-arid and arid areas.[24] J. A. Mabbutt states: 'Such rectilinearity of profile indicates that the slope angle is rarely modified by running water.' The annual rainfall of the Central Lowland is higher than that of most areas from which rectilinear slopes have been described, and the dominance of running water would appear to be a critical factor in the development and maintenance of the smooth concave slopes of Central Cyprus.

The profile is constantly repeated on valley sides, plateau edges and isolated mesas and buttes, in spite of differences in rock type, relative relief and aspect. Isolated hills, not long separated from a larger plateau, are symmetrical and have the same profile as the main mass. Mesas maintain their height and characteristic slope profile until the last vestige of cap-rock has gone. The constancy of this profile, cut in the solid rock and under active erosion, is clear evidence that at the present time slopes are developing by parallel retreat.[25] River incision in the late Pleistocene cut deep valleys through the cap-rocks,[26] and parallel retreat of the scarp-like valley sides away from the streams has left wide valleys and plains, which seem at first sight inexplicable in comparison with the narrow, dry stream beds of today. Rejuvenation was intermittent, and each phase sent a wave of scarp-retreat across the landscape.

The cap-rock facet. Where this is composed of cemented fanglomerate or Nicosia–Athalassa limestone it is flat-topped, with practically no upper convexity (that is, there is no waxing slope). Below this is a vertical face, four to six feet high, which overhangs the lower slopes. If the cap-rock is *kafkalla* the overhang is greater, the edge is thinner and small caves develop beneath in the softer *havara*. The cap-rock retreats by collapse of its undermined portions, thus maintaining a sharp edge with the ridge-top. The fallen blocks gradually disintegrate where they lie so that the upper concave slope is littered with their fragments, but it is not a scree slope in the strict sense of the term as the slope gradient is not necessarily that of the angle of rest of the detritus. On the larger scale, the cap-rock has a protective function. It resists any tendency

E

towards down-wearing or the production of an upper convex facet. The slopes below it are moulded by running water and consequently develop a concave profile. By maintaining the relative relief and preventing the upper facet of the slope from eroding too rapidly, the cap-rock allows this facet to develop and sustain the maximum gradient compatible with the rock type and the environmental conditions. The cap-rock would appear to be an essential factor if the parallel retreat of the slope is to develop and to be maintained.

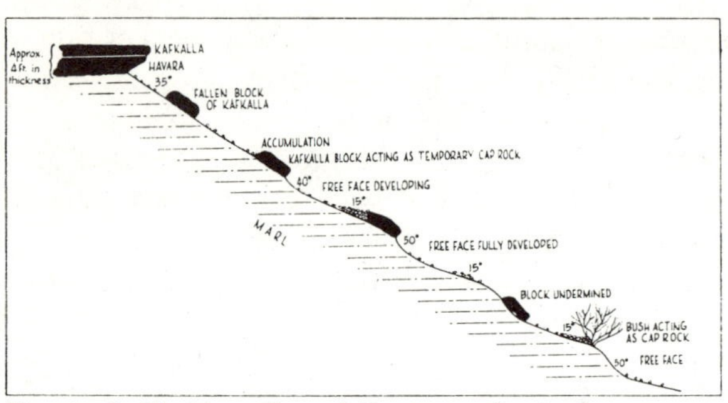

FIGURE 2—Sketch-section of the upper facet of a concave slope capped with *kafkalla* (based upon mesa 107657).

Below the cap-rock the gradient gradually decreases and sub-division of the slope on the basis of this alone would be purely arbitrary, but a threefold division on other grounds is possible. A typical, semi-vegetated slope cut in argillaceous rock is considered first.

The *upper facet* is inclined at gradients of 25–35 degrees (Fig. 2). Large slabs of cap-rock rest upon it, sloping at the facet gradient but normally quite stable, although rocks the size of a half-brick will roll down if dislodged. One might expect the softer rock of the concave slope to be eroded only when the top was exposed by removal of the cap-rock but, in fact, erosion of the concave slope overtakes that of the cap-rock and undermines it. The erosion of the topmost portion of the concave slope depends upon that of the slope immediately below it. As the latter weathers and erodes the topmost facet becomes steepened and eventually unstable. A *kafkalla* cap may overhang three or four feet, showing the extent to which erosion can proceed with the cap in place. In winter, water percolation may aid weathering and rock decay at this point, but the overhang is too uniform and continuous to be the result of spring-sapping.

Apparently smooth in appearance from a distance, the upper facet really consists of a number of steps and treads two or three feet high, and of a similar length, scattered haphazardly over the hill-side. The steps are small free-faces, usually cut in the solid rock with gradients of up to 50 degrees. Above is a cap consisting of a large fallen slab of limestone or a grass tuft or bush, usually supporting a layer of broken and fallen rock on the uphill side (Fig. 2). Initially the gradient above and below the cap is that of the average slope at that point (usually about 30 degrees), but the bare rock below it is soon eroded by rain beat, treading and sheet-wash to a steeper gradient. The little free-face then retreats parallel to itself, gradually undermining the bush or rock slab. The cap eventually collapses, the upheld debris moves downhill a short distance, and the process begins again. This, repeated over the whole facet, is the basis of its parallel retreat. Treading by goats increases the rate of downhill movement and a heavy storm causing extensive sheet-wash will bring about a wholesale reorganization of the minor cap-rocks. The whole facet is thus subject to bedrock erosion, for the waste material is only locally and temporarily at rest.

Where this facet is of alternating hard and soft beds the principle remains the same—for example, limestone beds form vertical faces, the intervening shale beds retreat parallel to themselves and leave the limestone beds overhanging and unsupported.

The *middle facet* of 15–20 degree gradients has a 25 to 50 per cent vegetation cover, and is usually littered with rock fragments averaging eight inches to a foot in diameter. These blocks, in contrast to those on the higher facet, are partly buried by silt on their uphill side. As we are here lower down the hill there is more eroded material in transit, and the gentler gradient aids the lodgement of it against obstacles. On the down-slope side further erosion creates free-faces, but these are less steep than those recorded higher up. Rock fragments remain at rest longer on this gentle gradient and weathering *in situ*, especially on the damp under-surface, occurs.

The *lowest facet* has gradients of 5 degrees and less and forms a long sweeping slope which merges with the plain. In many cases under 25 per cent of the surface is vegetation-covered. Where it lies below vegetated slopes this facet is only slightly gullied, and sheet-wash seems to be the main erosive agent. It was on one of these facets that the example of sheet-wash, described earlier, was seen. A retreating scarp, according to King, leaves at its foot a gently sloping, concave surface of transport, the pediment.[27] This lower facet fulfils the first two conditions, but it is difficult to demonstrate that it is a surface of transport. Sections are rare, but there was usually a layer of alluvium, a few inches thick, which in a sudden rain-storm would be swept up and carried towards the nearest valley axis. The lowest slope facet here described is therefore taken as the upper part of the ped: ıent. In no case was a sharp change of gradient found.

In the case of mesa 107657 there was evidence that the lowest facet is partly erosional and partly depositional. A wave of rejuvenation has reached the north-west side and two very narrow, vertically sided, trough-ended gullies or *dongas*[28] are eating into the facet. The sections revealed five feet of fine-grained colluvium, with bands of pebbles and stones derived from the cap-rock, resting on marl which showed spheroidal weathering. The ground is here slightly undulating and the colluvium had collected in a shallow valley whose rock floor was about 15 feet below the flanking ridges, which were cut in solid rock. W. G. V. Balchin and N. Pye[29] record that on pediments in the Sonoran and Mohare deserts 'dissections of up to fifteen to twenty feet in depth were by no means uncommon...' But for the sections the depression could easily be taken as having a solid rock floor. The smoothness of many Central Lowland pediments may therefore be the result of infilling of small shallow valleys and when traced parallel to the hill-face they may consist of alternate erosional and depositional zones, almost indistinguishable in the absence of sections.

Where there is no vegetation on the hill-side the profile is similar to that described above, but differs in detail as the processes are different. The upper facet may attain a gradient of 40 degrees just below the cap-rock, and is usually remarkably smooth and eroded by sheet-wash. At a varying distance down-slope, but approximately one-third of the total height below the crest, rills develop and become the main erosive agent on this, the middle facet. This rill zone is characteristic of all bare slopes, but is absent on vegetated slopes. A small mesa (480580) was noted in the former Akhyritou Reservoir on which one quadrant was vegetated while the rest were bare. The former has no rills, the latter many. The rills develop as the volume of water moving down-slope increases and is concentrated by small irregularities. The slope as a whole retreats parallel to itself as King has shown.[30] As gradients decline to 8 to 5 degrees the rills die out and may pass into very flat deposition cones, particularly in the Akhyritou area. As the rill water ceases to be concentrated, sheet-wash again becomes dominant.[31] This zone of deposition was always narrow, and sometimes absent.

Proximity to an intermittent stream that can carry off sediment seems to determine whether one has a true erosional pediment at the foot of a slope or whether it becomes a series of coalescing fans. The latter are concave towards the valley axis, and difficult to distinguish from pediments.

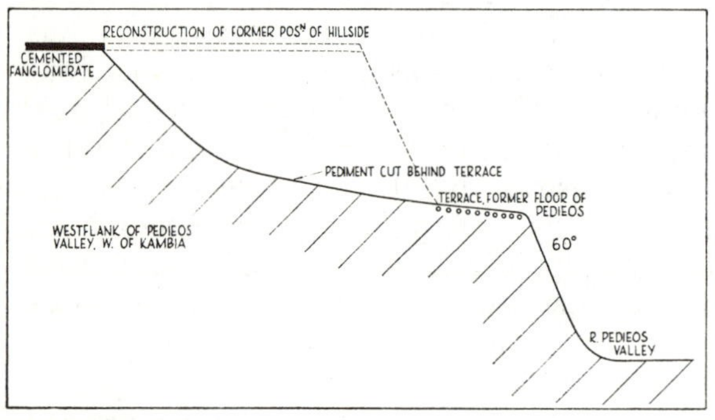

FIGURE 3—Evolution of the Pedieos valley west of Kambia (925504).

Two final points concerning the evolution of the concave slopes can be illustrated by reference to the Pedieos valley, near Kambia (925504) (Fig. 3). The river Pedieos is here flowing in a steep-sided gorge about 100 feet deep incised into one of its terraces. The latter is backed by a concave rear bluff, 150 feet high, capped by fanglomerate. The terrace gravels form a cap-rock on the gorge side, which is almost vertical in places. It is at present weathering back to a gentler gradient and shows no sign of parallel retreat. This initial weathering to maximum gradients of about 35 degrees must have occurred on all slopes now developing by back-wearing. The bluff at the rear of the terrace was presumably also at one time nearly vertical, but has now retreated so far from its original position as to leave a broad pediment behind the terrace, demonstrating that back-wearing slopes evolve independently of base-level, now well below the level of the pediment.

In certain instances back-wearing persists long after the cap-rock has been destroyed. In the Arg Petrati valley (953514) near Kambia the fanglomerate-capped valley sides slope at 40 degrees in places and are almost straight for heights of 80 to 90 feet. These gradients are maintained on knife-edge ridges that rise from the valley floor, and have long lost their cap-rock. One, 20 feet high, had a slope of 38 degrees and another, seven feet high, had a slope of 37 degrees, both with straight slopes and sharp summit convexities. Only when reduced to about four feet in height are there signs of a decrease in gradient and a broadening of the summit.

This tendency was investigated further in a series of gullies cut in nearly vertical Lower Kythrean shales two miles north of Nicosia. The maximum slope gradient was measured on small ridges of different relative relief:

Relative relief	*Gradient*	
1 foot	7½°	
2 feet	10°	
3 feet	13°	
5 feet	24°	Above this size all slopes are about 25° in their upper parts, declining to 20° in centre of gully
8 feet	26°	

The ridge tops were sharply convex. Other examples were noted of the persistence of parallel retreat once it has been established, even though the cap-rock had been destroyed.

Concavo-convex profiles were noted in four areas, and were studied in detail in two of these. In the Ovgos valley (880705) such slopes are developed on the marly facies of the Lapithos Formation. Broad convexities are found with maximum gradients of 11 degrees. Similar slopes are found in the minor valleys of the western fanglomerate fans, where surface dissection has not progressed far enough to breach the fanglomerate and so there is no cap-rock. The rivers are therefore cutting into a fairly uniform material, as the fanglomerate here seems to be thicker than in the mesa country farther east and also less cemented by *kafkalla*.

The Pedieos river enters its west–east course north of Nicosia and the river runs in a small gorge about 16 feet deep which becomes the stream bed in time of flood. Above this are two, and sometimes three, terraces cut in fine-grained argillaceous rock. The lowest bluff is the gorge-side, which is straight and has gradients of 55–60 degrees, maintained largely by undercutting. The other bluffs are concavo-convex where cut in uniform rock, such as silty alluvial fill or Kythrean Shale. The bluff behind the first terrace averaged 25 degrees on the former and 35 degrees on the latter, but in both cases it was almost a straight slope displaying only slight rounding at the top and bottom. The bluff at the rear of the second terrace, on the other hand, has a well-developed concavo-convex profile, with a maximum gradient of under 20 degrees and occasionally as low as 10 degrees. We have here evidence of a decrease in gradient with age and this was found to be generally the case where slopes are cut in rocks of uniform lithology without a cap-rock.

Other examples of concavo-convex slopes, in which the convexity dominates, were found on the massive chalks of the Lapithos Formation. The Lapithos chalk in the Kambia area is thinly bedded and has a slight northerly dip. The summits are broadly convex, with gradients of 12 degrees just below the summit, maximum gradients of 25 degrees on the flanks, decreasing to 10 degrees and less towards the valley floors. The summit and flanks are underlain by solid rock, and covered by a thin veneer of platy debris. Unfortunately all the valley floors are cultivated, and this has produced a sharp break of slope at the edge of cultivation. In some cases the gradient changes abruptly from 15 degrees to 6 degrees at this point. The valley floors are filled with alluvium, but it was possible in one case to trace the solid rock almost to the valley centre. The alluvium was about five feet thick, and the solid floor beneath appeared to be concave.

Evidence that these concavo-convex slopes decline with age was found near by, where the chalk rises above the fanglomerate surface. The latter has acted as a fixed local base-level for the chalk slopes for much of the Pleistocene. The concavo-convex chalk slope passes smoothly into the fanglomerate surface and its maximum gradient is about 8 degrees. The convexity on the chalk becomes very broad as the slopes decrease in gradient.

A number of processes are at work on these slopes including, of course, solution of the chalk but it was noted that there was no residual insoluble material on the ridge tops. It is probably removed by sheet-wash. The slopes can be divided into a series of stepped outcrops, two or three feet high, and sheet-wash carrying loosened rock downhill seemed as potent a medium as any in the removal of material loosened by solution and by wetting and drying. The small-scale changes closely resemble those on the concave slopes, but solution and the absence of a cap-rock are all-important in producing the different profile.

Former Slope-profiles

The influence of Pleistocene climatic fluctuations upon slope development has been discussed by C. A. Cotton and F. Ahnert[32] among others, and in the Central Lowland of Cyprus there is clear evidence that slope evolution in areas of concave slopes is, at the present time, very different from that in the Pluvials of the Late Pleistocene. On many valley sides the simple concavity is modified by a series of separate triangular cuestas of up to 50 feet in height, the scarped apices of which face inwards towards the ridge (Fig. 4). They resemble tilted, inverted flat-irons and have been so named. D. A. Osmund[33] first noted them (terming them steps) in the Ayios Ioannis area (877570), but they are in fact very widespread wherever *kafkalla* formation is possible. A typical profile, starting at the top of a ridge, is: horizontal cap-rock; concave slope, at first 30 degrees and decreasing downhill to perhaps 15 degrees; a rise up an infacing concave scarp, of maximum gradient 30 degrees; inclined cap-rock on the top of the flat-iron; and a back-slope of 8 to 5 degrees, which merges with the plain. The apex of the cuesta and part of the back-slope invariably have a capping of either *kafkalla*, or *kafkalla*-cemented fanglomerate detritus identical with that capping the adjacent ridges and obviously derived from them. The flat-irons are dissected relics of earlier hill-slopes, and as they run parallel to the present ridges (Fig. 5) it follows that the ridge- and valley-pattern of this earlier period was almost identical to that of today. H. Brammer has noted similar (but bigger) features in the Haute Volta, Ghana.[34] Here sloping-topped buttes with iron-pan caps parallel the hills, or are 'concentrically disposed around inselbergs'. The back-slopes of the flat-irons are concave, and so, therefore, were the lower slopes of the former profile. Furthermore, as the flat-irons lie to the side of the present ridges, the former ridges were much broader. It is believed that above the

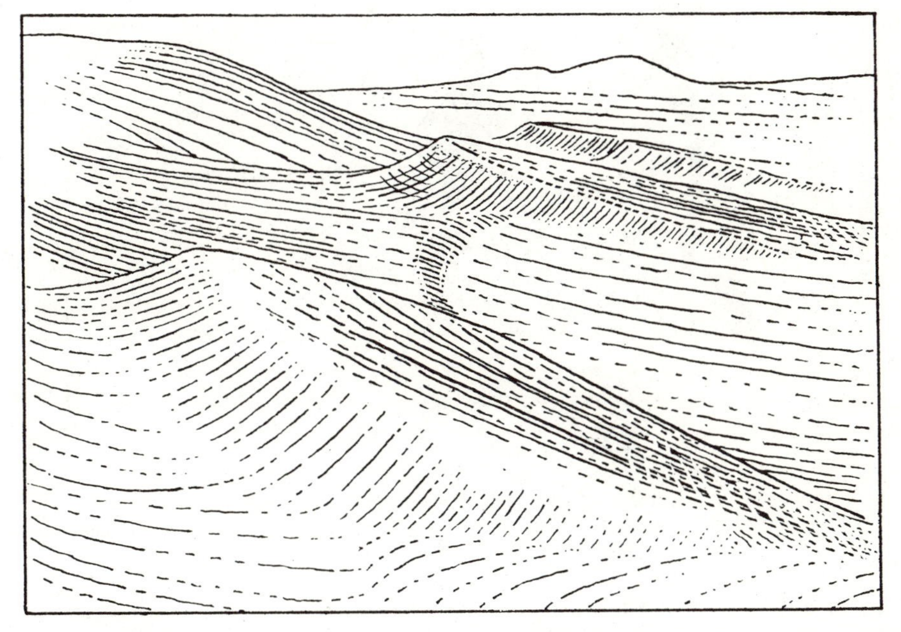

FIGURE 4—Flat-irons flanking a hill-slope near Meniko (841617). (Drawn from a photograph.)

concavity the slopes became convex, and were part of broad, rounded ridges, in contrast to the scarp-edged ridges of today. Near Ayios Ioannis, for example, and in a small valley (983640) west of Lakatamia, convex upper slopes mantled with *kafkalla*-cemented fanglomerate have survived. These areas are in every other respect typical of dissected fanglomerate country and the formation of the convex slopes is not regarded as the result of special circumstances. The mantling of the flat-iron back-slopes with fanglomerate detritus could only be accomplished by some form of creep. The mass wasting processes at work on the slopes today do not produce a uniform layer of waste.

The large number of flat-irons traceable on the ground show that the concavo-convex profile was formerly dominant in areas now characterized by parallel-retreating concave slopes.

The development of the concavo-convex form and its subsequent destruction are closely related to fluctuations of base-level, climate and vegetation in the late Quaternary. The sequence of events can be fairly readily reconstructed from the field evidence but the timing of these events is, for the present, uncertain.

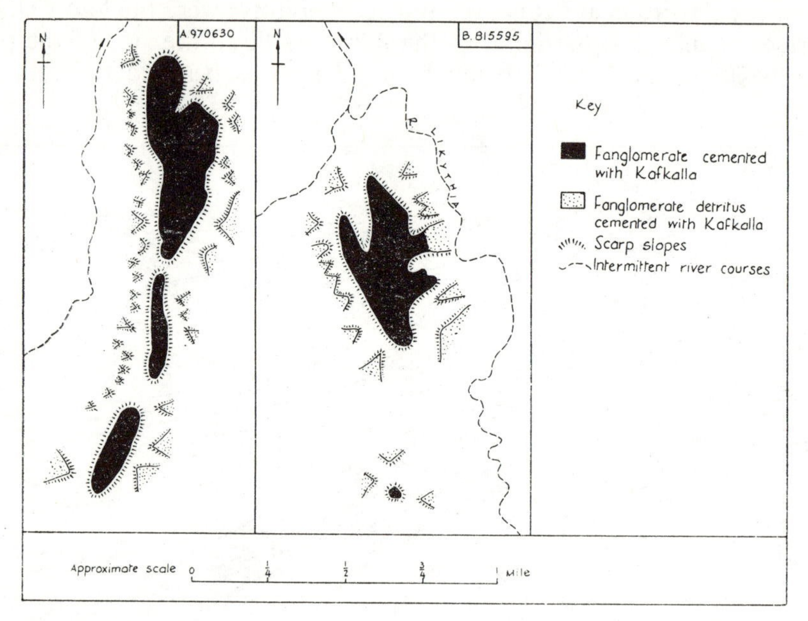

FIGURE 5—Sketch-maps of two areas in the alluvial fan region illustrating the relation between the flat-topped ridges and the in-facing 'flat-irons' rising from their flanks. Grid references are given for the map centres.

The deposition of the alluvial fans was probably completed in the Early Pleistocene and they were subsequently dissected by streams, aided by an increase of precipitation during Pluvial periods[35] and also by negative changes of base-level. At a time when the drainage pattern and relative relief were much as at present the valley slopes were mantled with fanglomerate debris, with the formation of rounded ridges. In the areas not covered by alluvial fans the argillaceous Tertiary rocks carried a scrub forest vegetation and in these areas soil-creep was the dominant mass-wasting process and concavo-convex slopes were formed. The climate then became drier and intense evaporation led to the accumulation of *kafkalla* at the surface, which fossilized these slopes (Fig. 6a). Renewed erosion gave rise to gullies that broke through the protective *kafkalla* layer on the hill-sides, cutting bottle-necked valleys, whose heads coalesced to isolate the flat-irons. In some narrow valleys the flat-irons plunge

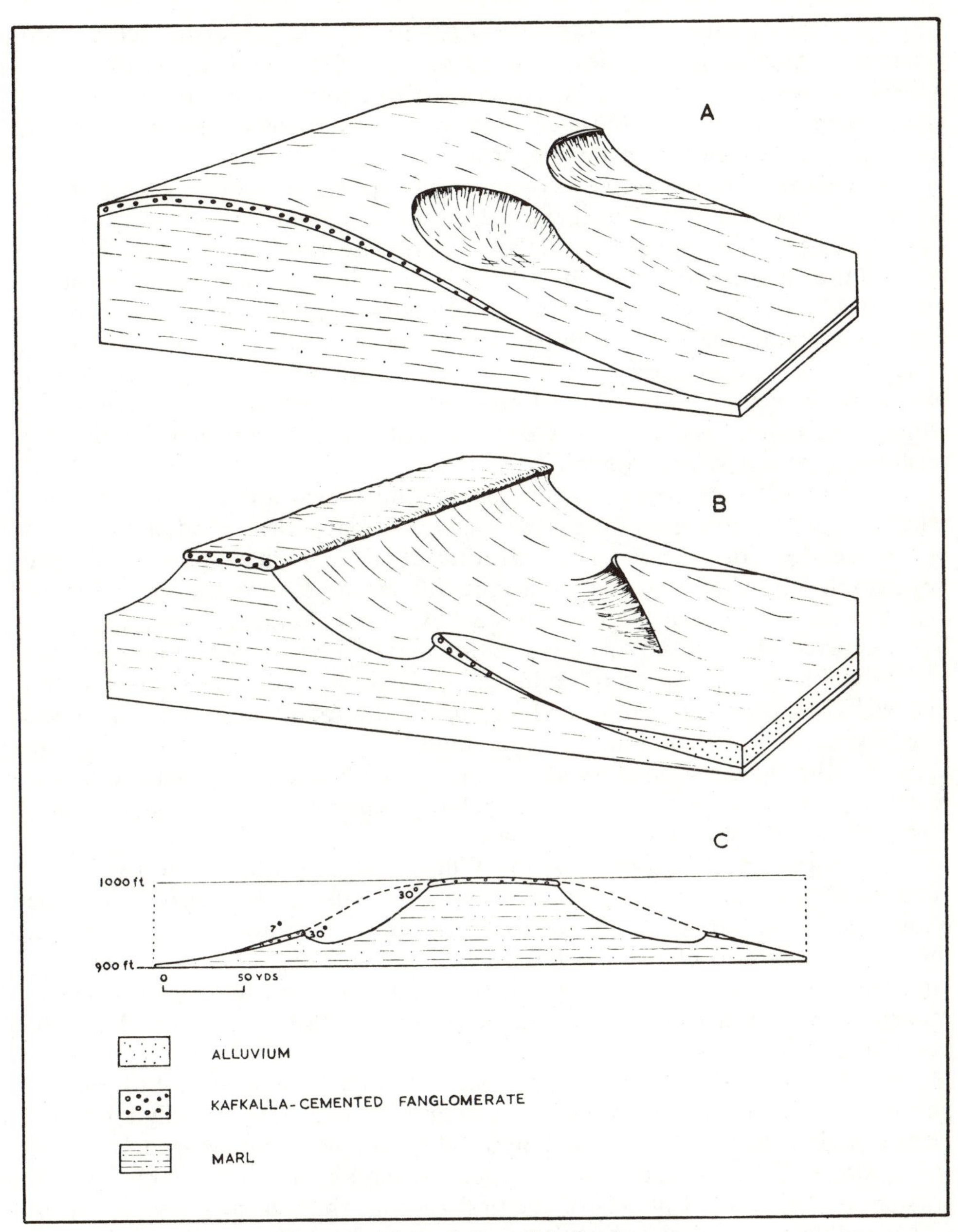

FIGURE 6—The evolution of the flat-irons.

A. Gullies cutting into the flanks of concavo-convex ridges, which have been mantled with *kafkalla*-cemented fanglomerate detritus.
B. The gully-heads have joined leaving isolated, undissected remnants of the former slopes (the flat-irons) rising above the new hill-side. This rapid erosion has aggraded the valley floor.
C. Sketch cross-profile of mesa 975625.

beneath the valley-floor, because intense erosion by the gullies has led to considerable aggradation (Fig. 6b). This fill is usually little, if at all, dissected. The hill-slopes produced by the destruction of the convex facet are the concave slopes seen today, which have evolved from the united gulley-heads in the manner described earlier in this paper.

The isolation of the flat-irons by gullying appears to be of very recent date, for the following reasons. They are closely related to a relief pattern very similar to that of the present and cannot, therefore, pre-date it by a long period, bearing in mind the intensity of erosion in Cyprus. The undissected nature of the valley-fill derived from the erosion of the gullies points to the recent nature of its deposition and the general sharpness and angularity of the ridge and mesa tops, which have been developed from the coalescence of the gully-heads, could not have survived denudation during a Pluvial, the mass-wasting processes of which would have blurred the outlines. One is led to the conclusion that the gullying post-dates the last Würm Pluvial.

At the time of the concavo-convex slopes the pattern and dimensions of the relief were similar to those of the present, and this places the period of formation of the rounded ridges in one of the Pluvials of the Late Pleistocene. A climatic desiccation following a Pluvial would provide the necessary conditions for the accretion of *kafkalla* on the concavo-convex slopes. *Kafkalla* formation must have occurred in the post-Würm desiccation and in the Riss/Würm Interpluvial, and probably in the earlier Interpluvials as well. The concavo-convex slopes may well have been initiated in the Riss Pluvial and fossilized by *kafkalla* in the Riss/Würm Interpluvial. Later in the Interpluvial erosion may have set in again but the increased precipitation of the Würm Pluvial, accompanied by the reappearance of a more extensive vegetation cover[37] would probably have re-established the concavo-convex slopes.

It is suggested, therefore, that the Würm Pluvial, with its forest or scrub vegetation, saw the final shaping of this now destroyed slope profile and the mantling of it with fanglomerate debris in the area of the alluvial fans and that the increasingly arid climate of Post-glacial times gave rise to the cementing *kafkalla*. The latter does not form a continuous cover on the rapidly eroding modern slopes, and one must postulate a period of slope stability for it to accumulate.

The above tentative chronology assigns to the Late Glacial–Post-glacial period both the formation of the *kafkalla* and its breaching by gullying. Early in this period its rapid accumulation would have been encouraged by fairly stable slopes (it does not form a surface sheet under contemporary active erosion) and the initial phases of the desiccation. The crucial phase is that in which gullying was initiated and the protective limestone mantle breached. Once broken its destruction proceeds rapidly, as can be observed today. There is no evidence of a recent rejuvenation that could have initiated the gullies, as in many cases the valley floors are not incised. A more probable cause was the depletion of the vegetation as rainfall totals fell, accompanied by an increase in rainfall intensity, the sudden rapid runoff locally cutting into the *kafkalla*. Butzer notes temporary increases of rainfall *c*. 9000 B.C. and *c*. 5000 B.C., which may have induced more rapid erosion, rather than an increase in the vegetation cover.[38] But probably the most potent factor has been Man. By clearing the vegetation and particularly by introducing grazing flocks he has caused small patches of the underlying shales and silts exposed by local denudation of the *kafkalla* mantle, to erode rapidly and undermine the *kafkalla*, stripping it off whole hill-sides. As an accelerator of erosion he has had no equal, and he is, in fact, responsible for much of the minor morphology of the Central Lowland.

The degree of erosion here envisaged for the past few thousand years is less surprising when it is realized that in the eastern Lowland (Mesarka) deposition raises the level of the plain about two feet per century.[39]

Flat-irons are also to be found in areas where fanglomerates do not occur. Their origin is, of course, the same, but there is no gravel detritus to indicate so clearly how they were formed. Flat-irons are absent from the plateau edge overlooking the Akhyritou Reservoir, which has all the conditions necessary for their formation. This has some relevance in the present context, for these slopes are probably sea-cliffs, as this part of the central Lowland was an arm of the sea in late Quaternary times.[40] At the time when the concavo-convex slopes were under formation, the sea was probably cliffing this plateau edge and the absence of the flat-irons is therefore to be expected.

Osmund has suggested that the valley-side *kafkalla* was a travertine deposit.[41] This is an interesting possibility, but does not account for the concavo-convex form, and presupposes a calcareous layer on the ridges but not on the valley-sides.

Conclusion

This survey of slope profiles in the Central Lowland of Cyprus has revealed how complex are the factors that influence the evolution of hill-side slopes, and how important it is to consider the whole environment, including Man where relevant, within which the slopes are being shaped. The conclusions may be conveniently presented under three headings:

1. *Rock type*. In this small area of essentially uniform environment rock-type and structure have a profound influence upon slope-profiles. The parallel retreat of slopes and extensive pedimentation, resembling that described by King, is admirably displayed in places where nearly horizontal cap-rocks exist, and persists in such areas on the underlying softer beds when the cap-rock is destroyed until the residuals are of very small size. In these last stages the slope gradients become gentler with the passage of time. The parallel retreat of slopes is undoubtedly aided by a large relative relief and rapid removal of debris, but the cap-rock structure appears to be the essential requirement.

In areas of soft, uniform rock (such as marls and shales), where there is no cap-rock, and on permeable rocks, concavo-convex profiles predominate, and hills are lowered by down-wearing. *Kafkalla* accumulation upon soft, impervious rocks produces a cap-rock that completely re-orientates landscape evolution, leading to the back-wearing of slopes and pedimentation. This is well illustrated in the river Koladhoes valley (for example near Aradiou 905570).

2. *Climatic fluctuations*. The profiles described are characteristic of the present environment and appear to be adjusted to the prevailing conditions of climate and human interference. Evidence has been presented to show that under wetter conditions in the Pleistocene slope-profiles and mass-wasting processes were unlike those now prevailing. Small relics of these earlier concavo-convex slopes have survived but are rapidly being denuded. Even so, it is clear that a full explanation of the present morphology requires consideration of climatic fluctuations in the Quaternary, a point already emphasized by Cotton and Ahnert.

3. *The influence of Man*. The importance of considering the whole environment has already been emphasized, and Man is a vital factor in the environment. At the present time grazing by sheep and goats maintains a poor vegetation cover on the hill-slopes that encourages rapid erosion. The few remaining trees of the Lowland are in constant danger of destruction by fuel-seeking villagers or hungry flocks.

On the land of gentler gradient the system of fallowing encourages further rapid erosion in a climate of torrential rain-storms.

Man's influence, it is suggested, goes beyond that of maintaining the *status quo*. His activities, particularly those concerned with the virtual destruction of the vegetation, appear to have greatly accelerated erosion in the Central Lowland in the past few thousand years and to have brought about the change to concave, back-wearing slopes more rapidly than would otherwise have been the case. In this he has, of course, been aided by the climate and by the waves of rejuvenation that are still eating their way into the landscape. To say that much of the detailed geomorphology of the Central Lowland is attributable to Man may seem an unwarranted assertion until one takes into account the abundant evidence of the rapidity of erosion.

ACKNOWLEDGMENTS

The writer wishes to record his indebtedness to Dr. D. Christodoulou, for his hospitality and lively interest, and to Mr. Fuad Sami, Dr. F. Ingham, Mr. J. Leafe, Dr. P. Loizides, Mr. C. Soteriades, Mr. P. Pantelides, Mr. G. Grivas and many others for the facilities they so generously made available.

The author acknowledges a grant from the Central Research Fund, University of London, for the field-work, and from the Department of Geography, Queen Mary College, towards the cost of the illustrations.

NOTES

[1] D. Christodoulou, *The evolution of the rural land use in Cyprus*, World Land Use Survey, Monograph 2 (1959), 9.

[2] E. de Vaumas, 'The principal geomorphological regions of Cyprus', *Annual Report of the Geological Survey of Cyprus* (1959), 39–43; W. F. Schmidt, 'Der morphogentische Werdegang der

Insel Cypern', *Erdkunde* 13 (1959), 179–201; N. J. W. THROWER, 'Cyprus, a landform study', Map Supplement No. 1., *Annals of the Association of American Geographers*, 50 (1960).

[3] F. R. S. HENSON, R. V. BROWNE and J. MCGINTY, 'A synopsis of the stratigraphy and geological history of Cyprus', *Quarterly Journal of the Geological Society of London*, 105 (1949), 1–41.

[4] R. A. M. WILSON, *The geology of the Xeros-Troodos area*, Geological Survey Department, Cyprus, Memoir No. 1 (1959), 135; L. M. BEAR, *The geology and mineral resources of the Akaki-Lythrodondha area*, Geological Survey Department, Cyprus, Memoir No. 3 (1960), 140.

[5] L. C. KING, *South African Scenery* (1951), 45.

[6] W. M. DAVIS, 'Rock floors in arid and humid climates', *Journal of Geology*, 38 (1930), 1–27.

[7] Grid references are taken from the G.S.G.S. 1 : 50,000 map of Cyprus.

[8] R. A. M. WILSON, op. cit., 14.

[9] E. DE VAUMAS, op. cit., 40.

[10] See also W. F. SCHMIDT, op. cit.

[11] K. W. BUTZER, 'Quaternary stratigraphy and climate in the Near East', *Bonner Geographische Abhandlungen*, 24 (1958).

[12] K. W. BUTZER, op. cit., 140.

[13] D. K. JONES, L. F. H. MERTON, M. E. D. POORE and D. R. HARRIS, *Report on Pasture Research, Survey and Development in Cyprus* (Nicosia, 1958), 52.

[14] L. C. KING, 'Canons of landscape evolution', *Bulletin of the Geological Society of America*, 64 (1953), 721–62.

[15] K. W. BUTZER, op. cit., 142.

[16] *The Rainfall of Cyprus*, Cyprus Meteorological Office Technical Note No. 2 (1959), 39.

[17] P. A. LOIZIDES, 'The cereal-fallow rotation in Cyprus', *Proceedings of the First Commonwealth Conference on Tropical and Sub-tropical Soils*, Commonwealth Bureau of Soil Science, Technical Communication No. 46 (1948), 211.

[18] D. J. BURDON, *The underground water resources of Cyprus* (Nicosia, 1953).

[19] R. A. M. WILSON, op. cit., 46.

[20] D. K. JONES and others, op. cit.

[21] D. J. BURDON, 'The relationship between erosion of soil and silting of reservoirs in Cyprus', *Journal of the Institute of Water Engineers*, 5 (1951), 676.

[22] I am indebted to my colleague, Mr. D. R. Harris, for comments upon this section. Reference should also be made to: D. R. HARRIS, 'The distribution and ancestry of the domestic goat', *Proceedings of the Linnean Society of London*, 173rd Session, 1960–61 (1962), 79–91.

[23] D. L. CHRISTODOULOU, op. cit. 41.

[24] W. G. V. BALCHIN and N. PYE, 'Piedmont profiles in the arid cycle', *Proceedings of the Geologists' Association*, 66 (1956), 167–81; J. A. MABBUTT, 'Pediment land forms in Little Namaqualand, South Africa', *Geographical Journal*, 121 (1955), 77–83.

[25] L. C. KING, op. cit. (1953), 748 (Canon 6).

[26] W. F. SCHMIDT (note 2).

[27] L. C. KING, op. cit. (1953), 748 (Canon 16).

[28] Ibid., 734.

[29] W. G. V. BALCHIN and N. PYE, op. cit., 172.

[30] L. C. KING, op. cit. (1953), 733.

[31] Ibid.

[32] C. A. COTTON, 'Alternating Pleistocene morphogenetic systems,' *Geological Magazine*, 95 (1958), 123–36; F. AHNERT, 'The influence of Pleistocene climates upon the morphology of cuesta scarps on the Colorado plateau', *Annals of the Association of American Geographers*, 50 (1960), 139–56.

[33] D. A. OSMUND, *Report on some Cyprus soils*, Colonial Office (1954).

[34] H. BRAMMER, 'A note on former pediment remnants in Haute Volta', *Geographical Journal*, 122 (1956), 526–7.

[35] W. F. SCHMIDT, op. cit.

[36] L. C. KING, op. cit. (1953), 735; H. BAULIG, 'Peneplains and pediplains', *Bulletin of the Geological Society of America*, 68 (1957), 913–30.

[37] K. W. BUTZER, op. cit., 140; D. K. JONES and others, op. cit., 52.

[38] K. W. BUTZER, op. cit., 142.

[39] D. CHRISTODOULOU, op. cit., 16.

[40] D. J. BURDON, op. cit., 10; E. DE VAUMAS, op. cit., 41.

[41] D. A. OSMUND, op. cit., 2.

Reprinted from *Nach. Akad. Wissen. Gottingen, Math.-Physik. Klasse,* **15**, 229–237 (1963)

Importance of Soil Erosion for the Evolution of Slopes in Poland

21

By *Alfred Jahn*

Instytut Geograficzny Uniwersytetu Wrocławskiego

Vorgelegt von H. Mortensen in der Sitzung vom 9. November 1962

In Poland, the problem of soil erosion has for many years been chiefly dealt with by soil scientists whose achievements, though duly appreciated, were not made sufficient use of by the geographers and geomorphologists of our country.

In investigating soil erosion, the following methods were applied:

1. Measurement of (a) the size of erosional forms (gullies, ravines, etc.) in fields, and of (b) the quantity of material deposited at the slope base.
2. Detailed nivelation (levelling) of chosen fields at regular intervals, i. e. every ten or twelve years.
3. Study of soil profiles (thickness of layers with organic material, humus).
4. Regular measurements of the effects produced by erosion in experimental fields by means of special containers where the washed off material is being collected.

This paper contains some remarks on the mechanism of slope processes. These remarks of the writer chiefly based on the Polish publications on soil science. Unfortunately in these publications the morphological aspect of the problem is not properly taken into account, except in the paper of Ziemnicki, Mazur [11]. What remains to be done, is to examine the original observational data, which were hardly ever published, and the writer regards this as a separate task in his further investigations.

According to a synthetic work by A. Reniger [7] four well-defined regions were distinguished in the area of Poland, the morphological surface of which shows a different degree of resistance to erosive action. Loess areas are most liable to soil erosion; they are represented, on the one hand, by the loess upland in the Vistula basin: Lublin Plateau and Małopolska (Little Poland), and on the other hand, by the loess areas of Silesia and the Subcarpathian depression. The wash extention of the erosive action of rain-water is shown by the fact that in some areas of the loess upland, the average lowering of slope ranges from 6 to 16 mm, and in some places it amounts to 50 mm yearly. In such areas the rise of valley bottom by accumulation reaches some 33 mm yearly (Bac [1]).

The thickness of the soil providing a basis for evaluation of the intensity of erosion and accumulation on slopes, two examples will be described, one referring to loose rock slopes (loess), the other—to compact rock slopes (marl, limestone)

I. Soil profiles were examined in the Lublin Plateau.

A typical profile is represented on figures 1 and 2. On top, above slope edge, there is a full soil profile with accumulated humus horizon (A). On the upper, steep part of the slope (over 10% grading more than 5°), the soil is so severely eroded that unweathered loess can be seen at the surface. The lower slope parts and its base are covered with deposited (rain-washed) soil. The slope has a concave-convex profile.

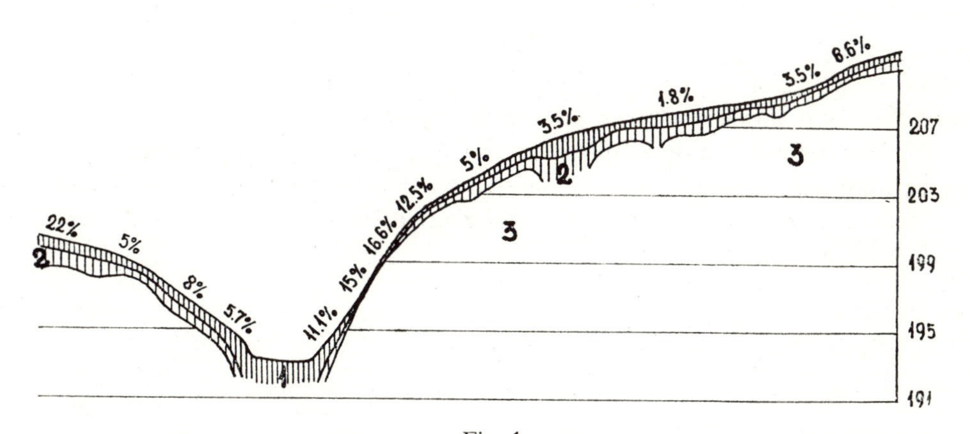

Fig. 1

Loess slopes. A section across the valley near Werbkowice, the Lublin Upland, eastern Poland. (After Dobrzański and Ziemnicki [4])

1. Humus horizon, 2. transitional layer, 3. Loess.

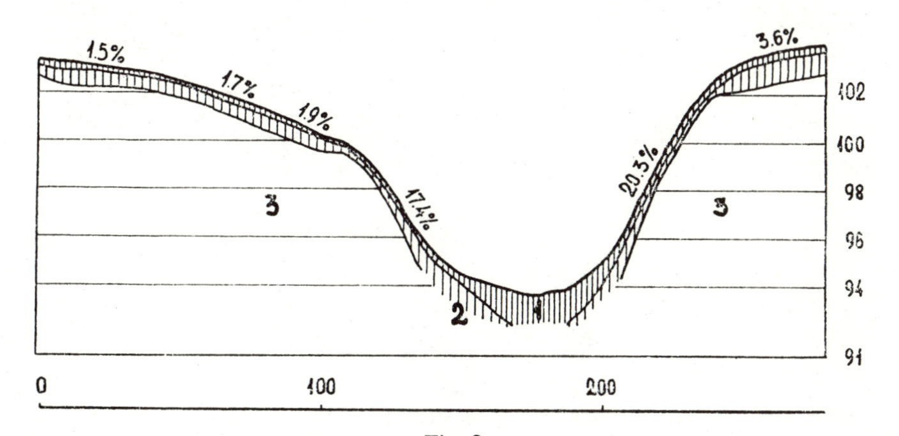

Fig. 2

Loess slopes. A section across the valley in Elizówka near Lublin. (After Ziemnicki [12])

1. Humus horizon, 2. transitional layer, 3. loess.

The same changes in the features of the soil profile of loess, can be detected on the map (Fig. 3). The carefully mapped slope of a small valley near Sławin, in the Lublin area, provides a good example.

It exhibits the following regularity. First, the upper part of the slope is degraded while the lower represents an area of deposition—accumulation. Secondly, within the forms extending down the slope, i. e. convexities and concavities, erosion is limited to the convex parts of the slope, and accumulation occurs in the concave parts, in trough-like valleys.

Material washed off from the upper part of the slope is deposited in its lower part. This is the rule. Such a transportation of material occurs also in individual slope sections. Each concave section terminates in a thickened humus soil layer. This is deposited humus. It must be added that, as a result of transportation, some of the physical properties of the material underwent a change. It has been shown that deposited loess in the profile at Werbkowice (Fig. 1) is less porous and thus more impervious than loess in the upper part of slope.

II. The second example illustrates the action of erosive processes in places where cretaceous rocks form the substratum of the Lublin Plateau. These are calcareous marls, i. e. rocks hardly permitting percolation of water, and therefore producing a soil of the "rendzina" type.

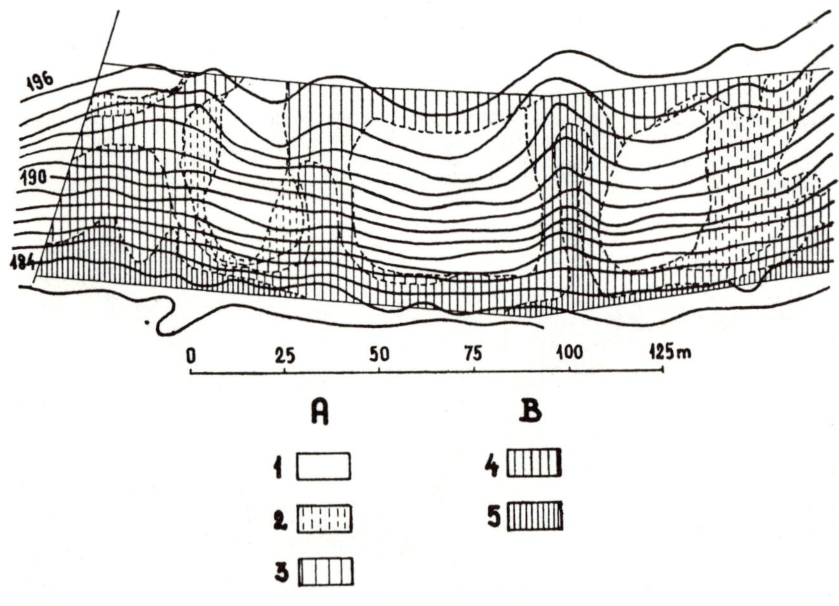

Fig. 3

Effects of erosive processes on a loess slope in the village Sławin near Lublin. (After Ziemnicki [12])

A: Eroded soil: 1. intense erosion, 2. medium erosion, 3. poor erosion, B. Deposited soil: 4. thin layer, 5. thick layer.

Here, too, a complete soil-profile can be observed above the upper slope margin (Fig. 4). Degradation is conspicuous where the upper part grades more than 5°. The field surface is, as a rule, covered with debris from parent rock. The soil with horizon A and that with deposited material are found in all the concavities irrespective of the location of cavities on the slope. Hence in the spring, when the fields have just been ploughed, such slopes present a pattern of dark patches alternating with light ones.

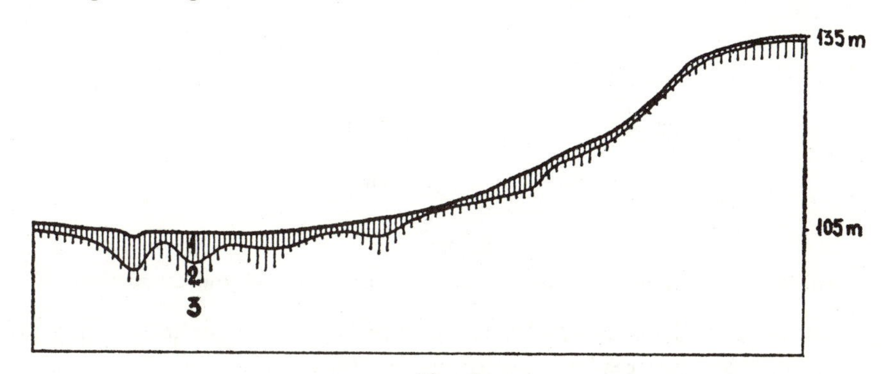

Fig. 4
Slopes of marly rocks in the village Nowosiółki, east of Lublin.
(After Dobrzański and Borowiec [5])
1. Humus layer, 2. transitional layer, 3. marl.

Slope processes of the two types described above—those in loess and those in marls—obviously differ in character. The first type is characterized by the interdependece of slope processes throughtout the whole profile of the slope. Degradation at the top is combined with deposition at the bottom. The slope has a clearly marked section of maximum steepness. In general, the slope is rather short, shows a characteristic convex-concave shape, and is comparatively well-levelled, i.e. the profile is practically free of disturbances. Compact rocks (marls) have, on the contrary, a long slope composed of various sections which are separated from one another by small concavities. Soil material stripped off the summit is being deposited in every concavity along the slope. On marls, there is no sharp contrast which is so characteristic of loess slopes between the upper and the lower part of the slope.

These data should be considered now in the light of the concept which the present writer termed "denudational balance" (Jahn [6]) and which Tricart [10] called "bilan morphogenetique".

This balance is represented by the equation

$$A = S \pm M$$

where A stands for the positive element of the balance: the accumulation of soil material on the slope, both by the production of waste and by sedimentation; S denotes erosive processes (down wash); and M—the gravitational movement of soil masses.

Thus *S* and *M* constitute the negative element of the balance, i. e. the losses in soil material.

On loess slopes, the positive balance[1] is marked right above the edge (accumulation of waste) and right below the concavity (accumulation of deposits). The whole central slope portion is remarkably negative. On marly and limestone slopes, the sections of positive balance alternate with those of the negative one (Fig. 5).

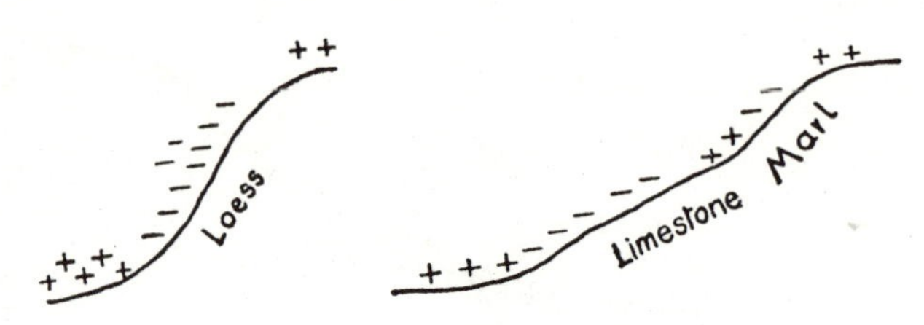

Fig. 5

Denudational balance on loess- and marly slopes; loss of mass (—) and accumulation (+).

The writer believes that a marly-linestone slope is not as liable to levelling as a loess slope. There are two reasons to account for that:

1. in resistant rock, processes of degradation are laborious and therefore retarded, which is self-evident;
2. certain structural factors, i. e. joints, are responsible for the development of local cavities resembling sinkholes, along which the slope splits into separate sections.

A second, and probably more important inference drawn by the author concerns the intensity of erosion on convex and concave slope forms. As convex slope portions show degraded soil profiles, and the concave ones are lined with accumulated soil, the action of erosion may be assumed to affect the convex forms more intensely than the concave ones.

However, a direct measurement of the intensity of erosion in loess areas near Puławy in the Lublin Plateau, taken in spring (A. Reniger [8]) provided evidence to the contrary.

The intensity of soil erosion was determined on the basis of the size and quantity of erosion rills. Wherever a concave niche appears on the slope, erosional rills are found to be far more numerous here than they are on the convexity.

The number expressing (in millimetres) the lowering of the slope in the concave part is more than the double (exactly 2,5 times as much) of that referring to the convex part (Fig. 6). The process is easily comprehensible, since the extent of degradation (down wash) depends upon the amount of water which is undoubtedly more abundant in the lower, concave part of the slope than in the upper, convex one.

---

[1] With reference to my work published in 1954, I wish to notify some terminological modifications: a negative denudational balance stands for a section of decay (decrement), while a positive denudational balance denotes a section of accumulation or increased production of weathering.

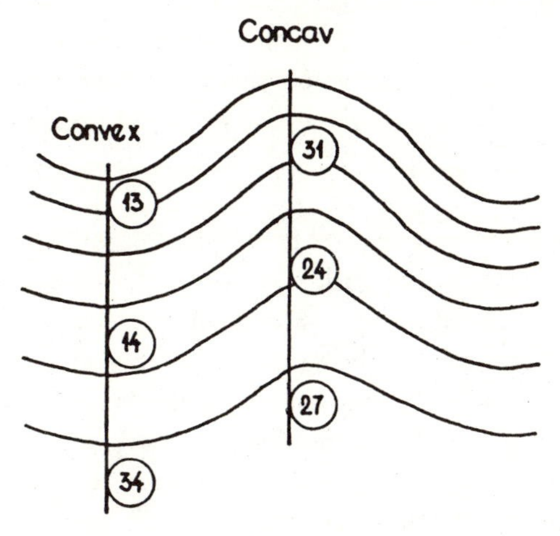

Fig. 6
Intensity of erosion, as determined by the size of rills and ravines, on convex and concave parts of a loess slope. Numerical values enclosed in rings indicate the extent of surface lowering (in millimetres) after the action of spring melt water. (Computed according to materials collected by A. Reniger [8])

This seemingly indicates a contradiction: on the one hand, measurements made according to the method applied in investigating the dynamics of the process (depth of rills, ravines, erosive action of rivulets) point to degradation of the slope concavity; on the other hand, measurements following the static method (soil profile, sediments) indicate agradation (accumulation) in the concave section of the slope. Nevertheless the contradiction is only apparent, because the concave section of the slope, which is actually a surface of predominant accumulation, is, at the same time, a section of predominant line degradation (gully erosion). The latter process appears to be less important in the general development of the slope than the surface processes (Fig. 7).

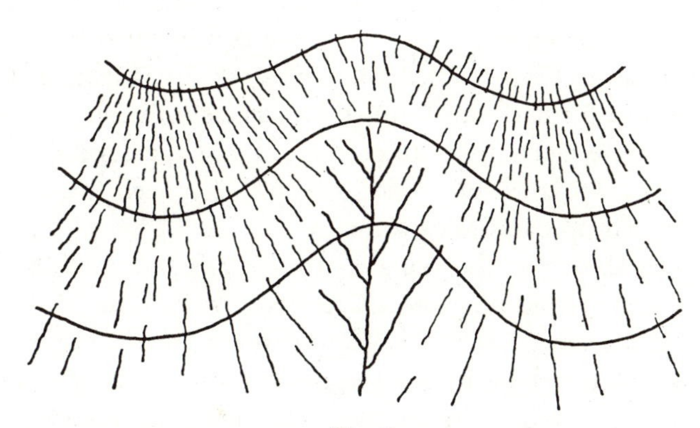

Fig. 7
A diagram illustrating the action of diffused water upon convex parts of a slope, and that of concentrated rills upon its concave parts.

How are these phenomena to be accounted for?

Processes which have been called here surface processes (agradation, degradation) actually result from the action of small rivulets which are formed after every major rain-fall. It is the action of the so-called unconcentrated or diffused run-off, a process which is both general and common. The line or gullying action, on the other hand, belongs to a category of processes which are rather catastrophic and sporadic. Every now and again, following major spring thaws or heavy summer rain-falls, large ravines or gullies are formed on loess slopes. The examples referred to above seem to justify the conjecture that the sum-total of downwash action exceeds that of sporadic gullying degradation.

It is essential to bear in mind that the slopes discussed in this paper are areas under cultivation. Ploughing has a double function here: in the first place, it systematically shifts the soil layer down slope; in the second place, it obliterates erosional rills, and levels the slope surface.

Both actions clearly display the characteristics of a surface (areal) process, not those of a line process.

The conclusions deduced from the evidence cited above should be formulated with caution.

1. As observed on slopes subject to soil erosion, the general tendency to alteration is to transform convex and straight slopes into concave ones.

2. Soil erosion is most conspicuous in the upper part of the slope just below the convex summit. Degradation proceeds here most rapidly and, at the same time, it is most widerspread. This is the zone of a remarkably negative denudational balance. Such places are exposed to the action of diffused run-off erosion which is a surface-, not a line processes. This process is largely facilitated and accelerated by ploughing. A retreat of the zone of intense degradation is an essential condition for the development of a slope concavity.

3. The slope base and the downward part of the slope concavity are those places where material accumulates and which exhibit the rills produced by concentrated run-off. It must be said, that these rills, cutting the slope transversely do not affect its long-profile. These rills which subsequently change into small valleys divide a formerly uniform slope surface into clearly marked ribs.

Considering all these facts, it should be stressed that the development of a slope profile results primarily from the action of fine, unconcentrated run-off, for it is the latter alone that operates as a surface agent.

H. Baulig's [2] theory of the slope concavity being formed by concentrated run-off, has not gained confirmation. Other factors seem to be involved in the process which are described by P. Birot [3] in his theory of slope formation. Surface modelling of the slope leads to accumulation of material at its base. In pervious rocks, e. g. loess, the transition from the zone of degradation to that accumulation is rather sudden. The slope concavity is sharply marked. In impervious rocks, the distance between degradation and accumulation zones is much longer (slope of cretaceous marls). The resulting vast zone might be called "transportation sections" ("pentes de transport" to follow R. Souchez's [9] terminology). In a profile of this type, the concavity is smaller.

Thus a transformation of a convex profile into a concave one is due to retreat (and not incision) of the steep section of the slope, as shown in Fig. 8. As the retreat proceeds, this section becomes shorter and shorter and its position on the slope becomes higher and higher. Its inclination is reduced only negligibly.

A slope retreating under the influence of erosive action is characterized by a convex-concave profile which is made of three sections (Fig. 8 sections *A*, *B*, *C*).

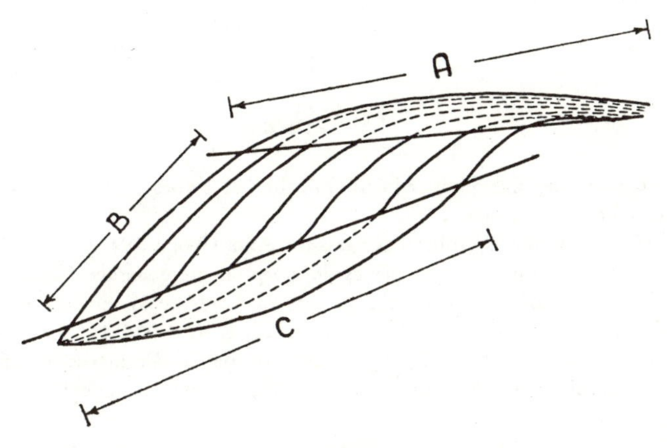

Fig. 8

A diagram illustrating successive stages of transformation of a convex slope into a concave one. For explanations see the text.

A. The convex section, a zone of scanty erosive action of water. Inclination is slight, removal of material—minimal, denudational balance—positive.

B. The rectilinear section, a zone of action of diffused water which passes into concentrated rills. Considerable inclination; large amounts of material transported outside; a negative denudational balance. Alterations of position (backwards and upwards) essentially determine the form of the slope.

C. The concave section, a zone of action of concentrated water which acts linearly (rudiments of gullies). Slight inclination; a positive denudational balance. The amount of material transported outside is smaller than in section *B*.

The middle section alone, i. e. section *B*, appears to be active and dynamic. The retreat of the two adjacent sections, *A* and *C*, has been made possible—notwithstanding their positive, as a rule, denudational balance—only because section *B* has been retreating. This is a temporal change: as soon as section *B* is reduced to 0 (i. e. disappears altogether), the remaining sections will fulfil their proper functions in accordance with their positive denudational balance. Accumulation (sedimentation and weathering) may then occur, though this is not necessarily equivalent to surface raising.

The present remarks concerning slopes situated in areas which have been brought under cultivation, are based on changes observed in soil thickness. The essential fact is that slope retreat occurs not through incision, but through parallel surface degradation of the middle section, i.e. one which exhibits maximum inclination.

The author is still unable to resolve the question to what extent this tendency to surface retreat has been caused by the agency of ploughing. Thus the mechanism of slope evolution, as presented here, remains subject to verification in genuinely natural conditions, so as to establish its full validity.

Literature

[1] Bac, St.: Conserning the changes in micro-relief of slopes in loess-areas. Roczniki Nauk Rolniczych i Lešnich. Poznań 1928.

[2] Baulig, H.: Le profil d'equilibre des versants. Ann. Geogr. 1940.

[3] Birot, P.: Essai sur quelques problèmes de morphologie générale. Lisbonne 1949.

[4] Dobrzański, B., Ziemnicki, S.: Project of land-use system on eroded chernosyems at Werbkowice. Annales Univer. M. C. S. Lublin, S. E. v. VI. 1951.

[5] Dobrzański, B., Borowiec, J.: Water erosion in Cretaceus Rendzina terrains as observed in the soils of the Anti-erosion field at Nowosiołki. Annales Univer. M. C. S. Lublin, S. E. v. XIII. 1958.

[6] Jahn, A.: Balance de denudation du versant. Czasopismo Geogr. Wrocław 1954.

[7] Reniger, A.: Intensity and range of the potential soil-erosion in Poland—an attempt of appraisal. Studies on the soil-erosion in Poland. Warszawa 1950.

[8] Reniger, A.: Soil erosion during the period of rainfall and downflow of spring waters as depending upon climatic conditions. Roczniki Nauk Rolniczych i Leśnych. Poznań 1959.

[9] Souchez, R.: Theorie d'une évolution des versants. Bull. Soc. Belge de Géographie. 1961.

[10] Tricart, J.: L'evolution des versants. L'Information Géographique. 1957.

[11] Ziemnicki, S., Mazur, Z.: An outcrop of slope as exposure of soil erosion. Annales Univer. M. C. S. Lublin, S. E. v. X. 1955.

[12] Ziemnicki, S.: Anti-erosion measures. Wiad. IMUZ. T. 1. Warszawa 1960.

Process

V

Editors' Comments on Papers 22 Through 29

Before hillslope evolution and the development of characteristic and limiting or threshold angles can be fully understood, the erosion processes responsible for landscape modification must also be understood. There is still disagreement as to the relative importance of the various processes, especially in different climatic areas, although creep and overland flow have received most attention. Mass movement phenomena, which tend to be sporadic and local in occurrence, will not be considered in this volume, except for the example of major slope modification reported by Rapp.

The first four papers in this part all deal with creep. The first three use both field and experimental techniques because the slow rate at which creep operates makes field observation difficult. Davison's short article, written almost ninety years ago, is exemplary for its methodology and reasoning, and for its use of controlled experimentation. His anticipation of modern concerns with lunar erosion processes is remarkable. One cannot help but regret that, during the following sixty years, geomorphology failed to take the road to which Davison and Gilbert pointed.

Young's discussion of techniques available for obtaining field measurements of erosion processes has proved to be of great value; the term "Young pit" is now well established in the geomorphologist's vocabulary. Kirkby's study uses a wide range of techniques, including those suggested by Young, to assess erosion rates in an area of Scotland. The comparison of a mathematical model of hillslope processes with actual measurement of the processes in the field and laboratory is noteworthy. Kirkby's results also enable one to put the significance of hillslope erosion in perspective. He concludes that, in his study area, the dominant denudational agent is neither creep nor slopewash, but stream bank erosion.

Schumm, like Kirkby, notes the difficulty of carrying out field studies of geomorphic processes, but is able to conclude that the rate at which creep proceeds is significantly related to slope sine. The papers by Young, Kirkby, and Schumm amplify the remark by Chorley (Paper 1) that even the slowest geomorphic processes are susceptible to measurement. They show that, for a proper understanding of slope form and evolution, the characteristics and behavior of the regolith, even on a microscopic scale, must be examined.

The preceding articles have described investigations of more or less continuous processes; the next two in this part point out the importance in at least some areas of infrequent, catastrophic occurrences. Rapp's description of the debris slides in Ulvadal reminds us that episodic events are capable of modifying slopes and that they may, in fact, be a dominant, though infrequent, process. However, this conclusion should be compared with his discovery that in the Karkevagge, solution was by far the most important agent of denudation (Rapp, 1960).

Terzaghi's contribution suggests the manner in which valley-side slopes may initially develop by failure to a stable angle of inclination. The techniques and approach of soil and rock mechanics clearly have a great contribution to make to slope studies; it is encouraging that more and more geomorphologists are becoming aware of the engineering literature.

Finally, two papers from the agricultural engineering literature are included. Research personnel of the U. S. Department of Agriculture have been extremely active in the study of land erosion and its control, and many data have been collected at agricultural experiment stations. Results of this research are presented in numerous official reports, and in such journals as the *Transactions of the American Society of Agricultural Engineers.* This profession has been deeply involved with studies of erosion processes, especially rilling, sheetwash, and rainsplash on low-angle agricultural lands, and it has adopted a field of study which geomorphologists have largely disregarded. So much of the earth's surface, especially that part which is used by man, is characterized by low-angle slopes, and it is unfortunate that geomorphologists have passed up the chance of making a significant practical contribution in this area.

Reprinted from *Quarter. J. Geol. Soc. London,* **44**, 232–238, 825–826 (1888)

22

19. *Note on the* Movement *of* Scree-Material. By Charles Davison, Esq., M.A., Mathematical Master at King Edward's High School, Birmingham. (Read February 29, 1888.)

(Communicated by Prof. T. G. Bonney, D.Sc., LL.D., F.R.S., F.G.S.)

The slope of screes being, as a rule, not much under the greatest angle at which it is possible for their component material to rest, it follows that a very slight force is in general required to put the surface-stones in motion*. This is evident, too, from the number of stones dislodged when a block falls down from above; also from the difficulty experienced by persons in trying to cross them. "Every movement," says Scoresby, describing the descent of a mountain covered with loose stones in Spitzbergen, "was a work of deliberation. The stones were so sharp that they cut our boots and pained our feet, and so loose that they gave way almost at every step, and frequently threw us backward with force against the hill. We were careful to advance abreast of each other, for any individual being below us would have been in danger of being overwhelmed with the stones, which we unintentionally dislodged in showers"†.

The instability of scree-material being so great, the causes of its motion are consequently numerous. Many have at various times been pointed out, more especially in considering the origin of different accumulations of angular débris, such as the limestone-breccias of Gibraltar or the stone-rivers of the Falkland Islands, the main difficulty in these cases being, however, to account for the transport of the material over surfaces inclined at a small angle. References to these well-known discussions are perhaps hardly necessary, it being the object of this paper to call attention to one other cause of movement which, at least in the present application of it, seems to have passed unnoticed.

While sitting near a shale-heap some time ago on a dry warm summer day, I was surprised by the fall close beside me of several blocks of shale, followed by a number of smaller pieces dislodged by their movement. And, again, in a slate-quarry, I have noticed fragments of slate on a waste-heap tumbling down and carrying others along with them in their course. In neither case, so far as I could see, was there any visible reason for the disturbance. All around being still, the movement could only be attributed to the expansion of the stones by the sun's heat during the day‡.

* Mr. Ruskin has given a useful list of the angles of screes observed by him in different parts of Switzerland in his 'Modern Painters,' vol. iv. p. 317. In making the experiments of the first kind afterwards described, I found that when the bricks rested at angles nearly equal to the angle of friction, the tremors due to carts passing at a distance of 8 or 9 yards were sufficient to shake them down. It was for this reason that I afterwards made the experiments with angles so low as 20°.

† Arctic Regions, vol. i. p. 129.

‡ Does not the fact that so many stones on the surface of screes are *just on the point* of slipping show that the cause of movement is not, as a rule, paroxysmal, but continuously acting and gradual in its effects?

The creeping of the lead on the roof of Bristol Cathedral through a distance of 18 inches in less than two years is well known, and arose, as Canon Moseley proved*, from the alternate expansions and contractions of the lead during changes of temperature taking place mainly downwards, being assisted by gravity in that direction.

Let AB represent a bar of lead or other substance resting on an inclined plane. When the temperature rises the bar expands; but, since it requires a less force to push a body down, than up, an inclined plane, the part, AC, pushed down the plane, is longer than the part, BC, pushed up it. Again, when the temperature falls the bar contracts, and the part, BD, pulled down the plane, is longer than the part, AD, pulled up it. In both cases, then, the descent of one end is greater than the rise of the other, and therefore the bar, as a whole, descends.

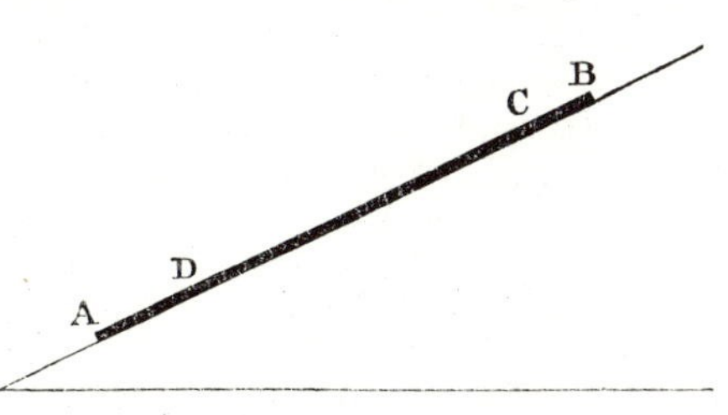

Let a feet be the length of the bar, μ the coefficient of friction between it and the plane, α the inclination of the plane, and e the coefficient of expansion of the bar for a rise of 1°; then, if the temperature rise τ°, and fall subsequently by the same amount, the total descent of the bar, Canon Moseley shows, will be

$$\frac{a\,e\,\tau\,\tan\alpha}{\mu}$$

feet †. Hence the descent is greater the longer the bar, the greater the coefficient of expansion and the range of temperature, the higher the slope of the plane, and the less the coefficient of friction.

Canon Moseley also tested his theory by a simple experiment. A sheet of lead, 9 feet long and $\frac{1}{8}$ of an inch thick, was placed on a flat wooden surface inclined at an angle of 18° 32′. The average daily movement, from the 16th of February until the 28th of June, was ·1745 inch. The movement was found to be greatest on those days when there were cold winds or passing clouds, there being then many changes of temperature during the day ‡.

It appeared to me that, in the same way, all stones free to move on the surface of screes must be slowly creeping downwards, and that this might be the explanation of the fall of the stones in the two instances given above. But it yet seemed desirable to make further experiments with slabs of stone, instead of with lead-sheeting: first, because the granular surfaces of rocks might offer effective

* "The Descent of Glaciers," Roy. Soc. Proc. (1855), vol. vii. pp. 333–342.

† *Ibid.* pp. 334, 335.

‡ "On the Descent of a Solid Body on an Inclined Plane when subjected to Alternations of Temperature," Phil. Mag. (1869), 4th series, vol. xxxviii. pp. 99–118.

resistance to a motion so minute; and, secondly, because, on account of their greater thickness, the changes of temperature might not sufficiently penetrate the stones in the short periods during which such changes sometimes take place. The experiments were of two kinds—the first qualitative, the second quantitative. I will now give an example of each.

Experiment 1.—The object of this experiment was to discover if any downward motion took place in a stone lying upon an inclined plane, and, if so, whether the motion were to be attributed to the alternate expansions and contractions of the stone during changes of temperature*.

A brick (B) was fixed with its upper surface inclined at an angle of 20° towards the south-west. On this was laid another brick (A) whose movements, if any, were to be observed by means of a level, resting at one end on the upper edge of the brick A, and at the other on a support made of similar brick. If the brick A did not move relatively to B, the vertical components of the expansion or contraction of the supports at either end of the level would be the same, and the level would indicate no change of position. If, however, the brick A did move in the way anticipated, the nature and manner of the movement should be in conformity with the theory given by Moseley.

The level was read frequently during the day, generally every half-hour. The temperature was also observed at the same times. As the experiment was intended to show the nature only of the movement, I did not attempt to determine the value in angular measurement of each division of the level. The following conclusions were brought out very clearly by this experiment:—

1. The upper end of the brick rises slightly with a rise of temperature, but descends beyond its first position with a corresponding fall of temperature.
2. The movements accompany, or take place a short time after, the changes of temperature.
3. The total downward movement is least on those days on which the sky is clouded and the range of temperature small.

A similar experiment might have been made to determine the motion of the lower end of the brick, by resting the level on a projection near the end and on a brick support near it. But the above experiment, several times repeated, seemed to me to show conclusively that stones resting freely on an inclined plane will gradually creep down the plane owing to the expansions and contractions of the stones, due to changes of temperature, taking place mainly downwards under the influence of gravity.

* [Since this paper was read, it has been pointed out to me that "in Canon Moseley's experiments ...... the lead which moved had an expansibility different from that of the wood on which it rested," and that if the two bricks in Experiment 1 "had expanded equally no motion could have resulted." The movement in this, and the following, experiment must therefore be attributed to the unequal heating of the upper and lower stones. The distinction is important, though it does not affect the subsequent results.—*Note added* March 24, 1888.]

Experiment 2.—In this experiment, the object was to determine the actual distance which a stone of given length lying on a given slope will descend in a given time.

Two slabs of a fine-grained sandstone (called York stone by the mason) were cut, each 3 feet long, 5 inches broad, and 2 inches thick. One face and one side of each were smoothed. One of the stones was fixed, so that its upper surface sloped at an angle of 17° towards the south. The other was placed on this, the smoothed faces being in contact, and the ends of both stones in line at starting. Fine scratches were cut in the same straight line on the smooth sides of the stones, in the middle and at either end, their subsequent displacement determining the amount of the movement. The whole was well protected by a light wooden framework covered with wire-netting. It was exposed to sunshine for about five hours a day at the commencement of the experiment, diminishing to about two hours a day at the end. Readings were, as a rule, taken once a week, at the same hour in the afternoon, and, except at these times, no artificial shadow was ever allowed to fall upon the stone.

The experiment began on May 5, 1887, and ended on September 22. In this time the total descent was $5\frac{5}{6}$ mm.*, *i. e.* an average of ·00164 inch per day, or ·599 inch per year. The period of the experiment was naturally divided into three intervals, the first from May 5 to June 9, the second from June 10 to August 25, and the last from August 26 to September 22. During the first interval there were 14 days on which rain fell, the sky was usually overcast and on one day only was fairly free from cloud: the average daily range of temperature was 12°·0 F., and the average daily descent of the stone ·00187 inch. The second interval was remarkable for its prolonged summer weather, rain fell but seldom, and the sky was cloudless for days together: the average daily range of temperature for the first eight weeks of this interval was 18°·3 F., and the average daily descent for the whole interval ·00119 inch. During the last interval there were 8 days on which rain fell, and on half the days the stone was frequently shaded by passing clouds: the average daily range of temperature was 14°·7 F., and the average daily descent ·00258 inch. From this experiment we may conclude that:—

1. The descent is greatest on those days on which there is bright sunshine intercepted frequently by passing clouds.

2. Rain slightly increases the rate of descent†, probably, though perhaps not entirely, by diminishing the friction between the stones.

Assuming the average rate of descent throughout the year to be that above given, namely, ·00164 inch per day, the upper stone will, by the creeping movement alone, have advanced far enough to fall over the lower one after a period of about $29\frac{1}{2}$ years. This is, of

* This distance being correct to within one sixth of a millimetre, it follows that the error in the average daily descent is less than ·00004 inch.

† This follows from a comparison of the movements during the first and second intervals.

R 2

course, a slow movement, almost imperceptibly slow, but, as it will be seen, far from unimportant.

Apart from the resistance to motion offered by vegetation and earthy matter, the stability of scree-material depends largely on the form and lie of the stones and the slope of the surface, these conditions being themselves connected. If the stones be nearly cubical in form, they will be found to lie at all angles, often with their edges leaning on surfaces sloping in opposite directions. The stability is naturally great in such a case, unless the stones be small. On the other hand, if the stones be flat-shaped, as in most slaty screes, they rest with their flat surfaces on the edges and faces of those below, inclining outwards and downwards at angles more or less approaching that of the scree-talus. Large blocks are more stable than small ones, not only on account of their greater weight, but also because they are generally imbedded amongst a number of smaller stones, and thus can hardly be regarded as surface-stones. Still, it is far from unusual to meet with blocks five feet or more in length lying quite on the surface of screes and in a position suitable to creeping.

A good example of a case in which the conditions are favourable to movement, and especially to creeping, occurs near the top of Hindscarth, a mountain in Cumberland, 2385 feet in height. On the west side of this mountain, not far below the cairn, are several sheets of loose fragments of well-cleaved slate. The sheets may be a foot or more in depth, and are inclined at an angle of about 20°. The pieces of slate are of all sizes up to a foot or a foot and a half in length, and are generally very thin, the largest not being more than about an inch in thickness, and even this amount is unusual. They lie with their flat surfaces on the bare hill-side or resting on those of other stones, nearly all inclined at the same angle as the slope. They are mostly long-shaped, and, with few exceptions, both large and small stones *lie with their longer axes pointing down the slope*, showing that, during motion, they have placed themselves in the position of least resistance.

Such conditions are I suppose unusual; but, as a general rule, the majority of the stones on the screes with which I am acquainted slope outwards and downwards, and are therefore in a position for creeping. The effects of creeping, moreover, are not confined to the mere descent of the stones. The movement of a stone in this way may withdraw its support from others resting on it. It may easily be imagined also that the stones may be of such a form and so arranged that a very slight movement in one may cause both it and some of those in contact with it to topple over; and, once in motion, they will drag many others along with them before they finally come to rest.

Again, the entire surface of screes is exposed to every change of temperature, and, throughout their whole extent, every stone that is free to move will make, with every change of temperature, a small slip downwards. The importance of this will be best shown by an example.

Let us consider a scree-talus one mile in horizontal length: let the average height measured along a line of slope be 1000 feet, and the average thickness of the surface-stones 6 inches. Let us suppose that the stones on one half only of the scree-surface are in a position to creep, and that, on the average, every stone on this half creeps downward one-thousandth of an inch a day. Then, the total amount of movement is equivalent to 1320×1000 cubic feet moving through one thousandth of an inch, or 110 cubic feet through one foot, every day.

This is of course a mere approximation; but it will serve to indicate the order of magnitude of the movement contemplated. If we had similar numerical estimates for the other moving agencies, we should be able roughly to compare them in efficiency. The importance of each cause varies, however, with the climate and so many other conditions that it would be difficult to assert at any time and of any screes that one cause is more efficient than another. In all probability, creeping is very far from being the most important cause of movement, yet it is possible to imagine conditions under which it might in time be most effective, as in the stone-rivers of the Falkland Islands, and at least one case where it may be almost the only agent at work.

I refer here to the conditions which probably obtain on the surface of the moon, where, as is well known, there are no seas, no appreciable trace of an atmosphere, and where the most diligent telescopic search for many years has failed to detect the slightest sign of present volcanic action. Deprived of the most potent agents of geological change, there remain the effects which can be produced by sudden alterations of temperature. The change from the intense heat of the lunar day to the cold of the lunar night or shade being untempered by any intervening atmosphere, the strain that results from the sudden cooling will be amply sufficient to break up the surface of any known rock. In this way, screes must accumulate on the mountain-sides, and the surface blocks that are free to move will creep gradually downwards, perhaps more rapidly than they would do on the earth, owing to the absence of vegetation and disintegrated rock upon the moon and the great range of temperature, long though its period be, to which its surface-rocks are exposed. Unbalanced by volcanic action, the actual rate of degradation may even be greater than on the earth; but, whether this be the case or not, there can be little doubt that in great and sudden alterations of temperature there exists a very important source of change upon the surface of an otherwise dead world.

Discussion.

The President expressed some surprise that the observations had not been made before, and congratulated Mr. Davison on the neatness of his demonstration.

Prof. Bonney agreed with the Author that changes of temperature do cause many stones to change their level, and considered the investigation a valuable one. It had often occurred to him that there were more minute movements in loose materials than was generally recognized, and that frequently a stratification was produced in this way, as could frequently be seen in talus from excavations.

Rev. EDWIN HILL agreed that we were indebted to Mr. Davison. He suggested that besides surface-motion an internal rearrangement of the scree-stones was produced by such movements.

The PRESIDENT called attention to observations by Newbold upon musical sand-hills, which showed that the sliding-down of the sand whereby a musical note was produced was due to the sun's heat.

* * * * * * *

47. SECOND NOTE *on the* MOVEMENT *of* SCREE-MATERIAL. By CHARLES DAVISON, M.A., Mathematical Master at King Edward's High School, Birmingham. (Read June 6, 1888.)

(Communicated by Prof. T. G. BONNEY, D.Sc., F.R.S., F.G.S.)

[Abridged.]

THE first results of the experiment described in this note, namely, those relating to the period from May 5 to September 22, 1887, have already been recorded in a paper read before the Geological Society on February 29, 1888 *.

After a brief interval the experiment was continued under the same conditions as before, from October 4, 1887, to May 5, 1888, with the object of comparing the rates of descent in the winter and summer halves of the year, and also of determining the effects on creeping of rain and snow.

Allowing a distance of ½ mm. for the interval of 12 days during which the experiment was suspended, the total descent during the year was $13\frac{1}{6}$ mm. (*i. e.* a little more than half an inch), the mean rate of descent being therefore ·00140 inch per day.

Comparison of the Rates of Descent during the Winter and Summer Months.—Dividing the year of the experiment into winter, from October 4, 1887, to April 3. 1888, and summer, from May 5 to October 4, 1887, and April 3 to May 5, 1888, we have:—

	Average daily range of temperature †.	Total descent in mm.	Rate of descent in inches per day.
Summer (184 days)	14°·4 F.	8	·00171.
Winter (182 days)......	8°·0	$5\frac{1}{6}$	·00112.

Had the creeping movement been proportional to the range of temperature, the average daily descent during the winter, compared with that during the summer, would have been rather less, namely ·00095 inch per day. Not only, however, is the heat of the sun more intense in summer than in winter, and consequently the effects produced by passing clouds so mnch the greater, but also for about three months of winter the experimental stone was entirely shielded from the sun by surrounding houses. Clearly, then, other causes must have operated in producing the comparatively rapid rate of descent during the winter months.

Influence of Snow.—The heavy snow-storms which visited many parts of England during the last winter were represented at Birmingham by very meagre falls. Except between February 14 and March 28, the snow seldom lay upon the ground, and when, on several occasions between these dates, it did lie for a short time, the

* Quart. Journ. Geol. Soc. for May 1888, p. 232.

† Excluding 20 days from August 6–25, and 8 days from February 15–22.

snow was nearly always driven by the wind from the experimental stone before the middle of the day, or melted by the increasing heat of the sun.

During the 12 weeks from November 23 to February 14, the average daily range of temperature was 7°·5 F., and the average daily descent only ·00078 inch. From February 15 to April 3, a period of 7 weeks, the average daily range of temperature (during all but the first 8 days) was 8·2 F., and the average daily descent (during the whole time) ·00147 inch, nearly twice as great as in the preceding period. I believe that this difference was chiefly, though not entirely, due to the influence of snow.

By contact with the snow the upper stone is more thoroughly and quickly cooled than by contact with the air. Moreover, the lower stone is directly cooled only by the air, and as the movement depends on the difference of the temperatures to which the stones are at any time subjected, the effect of short and repeated contact with a covering of snow is evidently to increase the rate of descent.

On the other hand, snow, when it lies thickly and for long periods, prevents the stones from fully participating in the range of temperature to which they would otherwise be subjected, and the effects of mere creeping are then reduced to a minimum.

Reprinted from *Nature*, **188**, 120–122 (1960)

23

SOIL MOVEMENT BY DENUDATIONAL PROCESSES ON SLOPES

By Dr. ANTHONY YOUNG
Department of Agriculture, Zomba, Nyasaland

TWO aspects of geomorphology which have recently received increased attention are the rate at which changes in landforms take place, and the study of erosional and denudational processes. Estimates of the rate of landform evolution have been made from indirect geological evidence[1-3], or by relating measured river loads[4-6] or rates of sedimentation[7-9] to the catchment areas from which they are derived. These results combine the material removed in bank and bed erosion with that supplied to the rivers by denudational processes. Direct measurements of the rapidity of denudational processes on hill slopes have related mainly to slope-wash, undertaken in connexion with investigations of

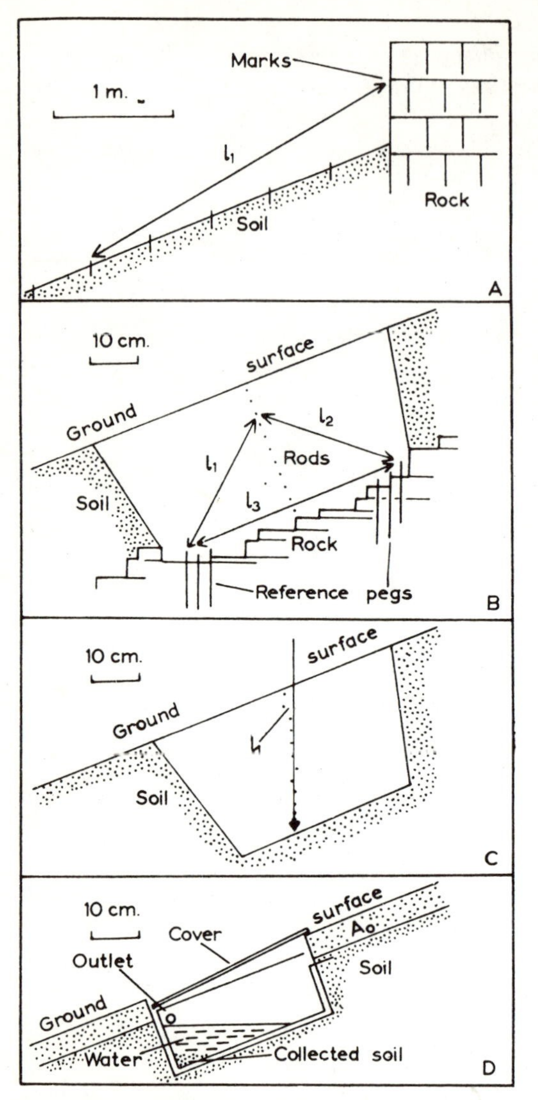

Fig. 1. Techniques for field measurement of the rate of soil-creep and slope wash. l_1, l_2, l_3 are measured lengths

accelerated soil erosion. This article describes techniques evolved for measurement of the rate of soil-creep under natural conditions, and reports interim results; experiments reproducing this process are described, and the relative effectiveness of soil-creep and slope-wash compared.

In the initial attempt at measuring soil-creep, metal pegs were inserted into the ground surface at short distances downslope of a bedrock outcrop (Fig. 1*A*). Three marks were engraved on the rock, and distances from each of these to the pegs measured at intervals of time. Down-slope movements are directly obtained by this method, but their interpretation is uncertain; if the surface soil moves more rapidly than that beneath, it will tilt the pegs, exaggerating movements of their upper ends. Measurements during 3½ years indicate that, on grassed slopes of 30°, down-slope movement close to the ground-surface is of the order of 1 mm. They also show clearly that on slopes steeper than 35°, with a scarred or incomplete turf cover or with terracettes, rate of movement increases rapidly with increase in angle. All results refer to valley slopes in the basin of the Upper Derwent, Southern Pennines, underlain by sandstones with interbedded shales, belonging to the Millstone Grit Series.

To obtain movements of soil in depth, it is necessary to bury markers and afterwards re-excavate, since a vertical soil-face left exposed to the air creates artificial conditions of moisture, temperature and biological activity. Two methods of making such measurements have been employed. In the first (Fig. 1*B*) a pit was dug through the soil to the rock beneath, and six pegs driven firmly into the rock; when filling the pit, these were protected by covers. A line of metal rods (diameter 2 mm., length 10 cm.) was inserted into one side of the pit, the ends of the rods being flush with the undisturbed soil. Measurements from pegs to rods fixed the positions of the latter in two dimensions. In the second method (Fig. 1*C*), a vertical line of fine rods, 1-mm. diameter, was inserted flush with the pit side, and their position measured with respect to a plumb-line suspended close to their ends; on re-measurement, the deepest rods, close to bed-rock, were assumed to have remained stationary. This method is of comparatively high accuracy, but gives only the horizontal component of movement. Nine pits were established, and measured on re-excavation at intervals of 6–12 months over 2 years. Disturbances of rods during burial and excavation were distinguishable as random movements, contrasted with systematic movement of adjacent rods due to soil movement. Results indicate that, on grassed and wooded slopes of 20°–30°, down-slope movement of the upper 10 cm. of soil (excluding the organic horizons) is of the order of 0·25 mm./yr. The unit for measuring volumetric down-slope movement of soil is the volume of soil moved annually across a plane perpendicular to the ground-surface and parallel with the contour of the slope, for unit horizontal distance along the plane, expressed in $cm.^3/cm./yr.$ The interim results show a movement of the order of 0·6 $cm.^3/cm./yr.$ They also show that movements in the organic A_0 horizon, both under grass and woodland, were very much greater than in the immediately subjacent mineral soil, amounting to 0·5–2·0 mm./yr. It is hoped to obtain more precise results by re-measuring after the pits have been left undisturbed for 3–5 years.

A list of the causes of soil-creep has been given by Sharpe[10]. The majority of these are essentially causes of disturbance in the relative positions of soil particles; down-slope movement is produced by the constant factor of gravity acting on re-adjustments to these positions. Quantitatively, one of the greatest causes of disturbance is the swelling and shrinkage of colloidal components of the soil on wetting and drying. Experiments to reproduce soil-creep by this means have been carried out. Blocks of soil were placed in glass-sided troughs, and pins inserted into the sides of the soil to record movements; the troughs were placed at an angle of 30°, and the soil surface wetted and dried artificially. On wetting, expansion took place perpendicular to the surface; on drying, gravity tends to cause contraction in a vertical direction, but due to cohesion the observed contraction was in a direction intermediate between vertical and perpendicular to the surface. Net down-slope movement resulted in the manner shown in Fig. 2*A*. Frequently cracks perpendicular to the surface developed, following which each block of soil acted as a separate unit: during expansion a plane of no movement was found up-slope of the centre of the block, and during contraction a corresponding plane down-slope of the centre, resulting in a net down-slope movement of each block (Fig. 2*B*). Difficulty was encountered in preventing movement from occurring mainly at the junction of the soil with the base of the trough, but in some cases differential down-slope movement of the upper layer of soil relative to the lower took place.

Creep due to wetting and drying was further reproduced under semi-natural conditions. An undisturbed block of soil 100 × 20 × 20 cm., with overlying turf, was placed in a glass-sided trough, which was tilted to 25° and exposed to natural weather conditions in Sheffield. Movements were recorded monthly for 2 years. Expansion and contraction of the soil corresponded closely with periods of rainfall and drought. The soil acted as a single block, with movement taking place mainly, but not entirely, along the base, at a rate of 5·0 mm./yr. In a second trough filled to 30 cm. depth with disturbed and re-packed soil, the rate of movement decreased regularly from the surface downwards.

These results suggested an indirect means of measuring soil-creep. At three sites on a slope, on the 26° steepest part and at 18° and 7° on the concavity, variations in moisture content of the soil at three depths were obtained by monthly sampling during 2 years. By means of shrinkage tests, the percentage expansion and contraction of the soil for unit change in moisture content at each of these nine points was obtained. Combining these results and interpolating between the sampled depths gave

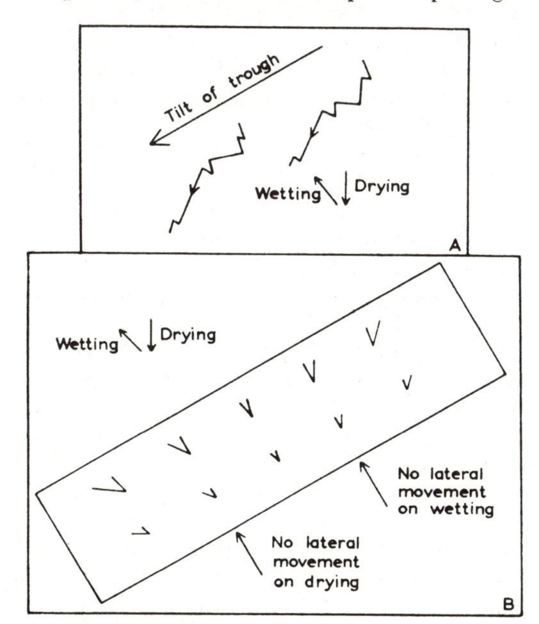

Fig. 2. Movements in a tilted block of soil subjected to alternate wetting and drying. *A*, trace of movements during five periods of drying separated by four of wetting; *B*, diagrammatic representation of movement in a single cohesive block of soil during one cycle of drying and wetting

Table 1

Slope angle	Depth (cm.)	Linear shrinkage for 1 per cent moisture loss* (per cent)	Moisture change + and − (per cent/yr.)	Change in linear dimensions (per cent/yr.)	Down-slope movement (cm./yr.)	Volumetric down-slope movement (cm.³/cm./yr.)
26°	0–5	0·08	37	3	0·86	1·07
	5–10	0·06	20	1	0·41	
	20–30	0·02	11	0·2	0·08	
18°	0–5	0·06	34	2	0·58	0·83
	5–10	0·06	15	1	0·33	
	20–30	0·03	9	0·3	0·08	
7°	0–5	0·10	46	5	0·61	0·79
	5–10	0·14	24	3	0·36	
	20–30	0·04	13	0·5	0·05	

* Prior to residual shrinkage, and corrected for gravel content of soil.

annual expansion and contraction of the soil on the slope. It was assumed that down-slope movement = sin (slope angle) × expansion and contraction perpendicular to the ground-surface. The results are summarized in Table 1. They show a movement of 0·5–1·0 mm./yr. close to the ground-surface, and 0·05–0·1 mm./yr. below 20 cm. depth. The volume of soil movement is 0·8–1·0 cm.³/cm./yr. The most striking feature is that movement on the 7°-concavity was only slightly less than that of the 26°-slope; this is attributable to the greater fineness of the soil and larger variation in moisture content on the concavity, and probably also to the presence there of expanding-lattice clay minerals.

The rate of the movement of the soil by slope-wash was measured on sites close to those used for soil-creep. Tins were placed in the soil with projecting lips on their up-slope sides (length 30 cm.) inserted just below the boundary of the A_0 horizon and the upper surface of the mineral soil (Fig. 1*D*). Soil was collected from the tins at 6–18 month intervals during 3 years. The mean of 49 results was 5·1 gm./yr. of soil, equivalent to a down-slope movement of 0·08 cm.³/cm./yr.

The relative ineffectiveness of slope wash is explicable from the presence of an organic cover, the A_{00} horizon of grass roots or leaf litter and the A_0 humus horizon. Besides their direct protective effect, these have a large capacity for absorbing moisture during storms, afterwards releasing it slowly to the soil beneath. Samples of the humus horizon taken during storms showed it to be capable of holding moisture contents of 75–90 per cent by weight, while retaining a firm structure.

These interim results indicate that on the slopes investigated the rate of down-slope soil movement by soil-creep is of the order of ten times as rapid as that by slope-wash.

This work was assisted by a grant from the Research Fund of the University of Sheffield.

[1] King, L. C., *Trans. Geol. Soc. S. Africa*, **43**, 153 (1940).
[2] King, L. C., *Bull. Geol. Soc. Amer.*, **67**, 121 (1956).
[3] Linton, D. L., *Adv. Sci.*, **14**, 58 (1957).
[4] Cavaille, A., *Rev. Géomorph. Dynam.*, **4**, 57 (1953).
[5] Wundt, W., *Erdk.*, **6**, 40 (1952).
[6] Corbel, J., *Rev. Géomorph. Dynam.*, **8**, 97 (1957).
[7] Steiner, A., *Geog. Helv.*, **8**, 226 (1953).
[8] Journaux, A., Union Géog. Int., 1r. Rap. Com. Ét. Versants, 133 (Amsterdam, 1956).
[9] Young, A., *Proc. Yorks. Geol. Soc.*, **31**, 149 (1958).
[10] Sharpe, C. F. S., "Landslides and Related Phenomena" (Columbia Univ. Press, New York, 1938).

24

THE JOURNAL OF GEOLOGY

July 1967

MEASUREMENT AND THEORY OF SOIL CREEP[1]

M. J. KIRKBY
Department of Geography, University of Cambridge

ABSTRACT

Although soil creep is widely thought to be an important agent in hillside development, its rate has rarely been measured. The role of creep therefore has been a subject for speculation, unsupported by evidence. In this paper, a theoretical model of soil creep is proposed, using soil mechanics principles. This model was tested in a laboratory study, which provided moderate support for the theory. Creep was also measured in two ways at each of fifty sites in Kirkcudbrightshire, southwest Scotland, over a period of 2 years. Soil moisture was shown to be significantly related to changes in soil moisture, and was significantly downhill, averaging 2.1 cm^3/cm. year for slopes of which the median angle was 13°. The high variability between individual measurements, at least part of which is inherent in the process of soil creep, disguised the existence of possible correlations with slope, aspect, or soil type. Estimates of the relative magnitudes of soil creep due to various agents indicated that, for southwest Scotland, soil moisture, freeze-thaw, burrowing animals, and temperature changes were the most important, in that order. Measurements of soil wash showed that its average rate was not more than 0.09 cm^3 cm. year, which is significantly less than the rate of soil creep. Thus the sum of creep and wash together, if acting at their present rates, would only have lowered the drainage basin by an average of one-third of an inch in the last 10,000 years.

INTRODUCTION

So few measurements of soil creep have been made that its role in slope evolution is largely a matter of speculation. All measurements of creep contribute to our knowledge of the relative rates of mass wasting processes, and thus to our knowledge of the manner in which hill slopes develop. The essential need is for field measurements, but our knowledge of the process and the agents which cause it can be only partially studied under field conditions, where changes in the soil are not controlled and where measurements cannot be made so accurately as in the laboratory. In order to control conditions more closely, a block of soil was brought into the laboratory for more detailed study of soil behavior during wetting and drying, and the results were compared with a theory of soil creep which has been developed.

Soil creep has been defined as being imperceptible except to measurements of long duration (Sharpe, 1938, p. 21). Creep may be considered as either "seasonal" or "continuous" (Terzaghi, 1950, p. 84-85). Continuous creep in a slow landslide, in which the downslope component of the weight of the soil causes a slow, continuous yield. This state of slow yielding represents a delicate balance of forces, and it is argued here that this balance could not be exactly maintained over the whole of a hillside. Instead, the inhomogeneities of soil thickness and composition, and of water content, would almost certainly give rise to equilibrium without movement in some parts of the hillside, and to recognizable landslides in other parts. Therefore, on any hillside without recognizable landslide scars, continuous creep cannot be more than locally active. Seasonal creep,

[1] Manuscript received January 10, 1966; revised October 18, 1966.

on the other hand, is caused by cyclic changes in the surface layers of the soil, and the changes generally decrease in magnitude with depth. For example, changes in soil moisture beneath a level soil surface cause vertical expansion and contraction of the soil (Ward, 1953). Where the soil is on a slope, the expansion and contraction will tend to be normal to the surface, but the downhill component of the weight of the soil encourages both expansion and contraction to take place on the downhill side of the normal to the surface. In this way, each point of the soil moves downhill in a zigzag path, with no general failure of the soil, as is required for continuous creep to occur. The forces set up by these cyclic movements and the resulting downhill movement will be present over an entire hillside, without any likelihood of causing a landslide. The hillsides considered below are without landslide scars, and it can therefore be assumed that any soil creep measured will be, except perhaps locally, "seasonal" in character and not "continuous."

THE THEORY OF SOIL CREEP

Davison (1889) considered that the expansion of the soil, during cycles of freeze and thaw, should be exactly normal to the surface, but that the subsequent contraction should be exactly vertical (fig. 1, *a*). This model of the downhill movement assumes a soil cohesion which is great enough to completely prevent displacement parallel to the surface during expansion, but which is zero during the subsequent contraction of the soil. His practical experiments showed that, in a soil where ice needles were not present, the expansion was almost normal to the surface, but that cohesion caused the contraction movement to take a line intermediate in direction between the normal and the vertical (fig. 1, *b*). The result of this zigzag movement will be to produce a net displacement in the soil, parallel to its surface. This displacement tends to decrease with increasing depth, so that the net result appears similar to that which would be caused by a laminar shear, although the actual motion is more complex. This resultant relative displacement is referred to below as the "net shear." This theory can be used to predict the velocity profile which is to be expected in the ground, due to soil creep. The rate of net apparent soil shear is assumed to be proportional to the frequency of freezing and thawing times an expansion coefficient for the soil; consequently the rate is greatest at the surface and declines approximately exponentially with depth (fig. 1, *c*).

Davison's analysis leaves out of consideration the forces which are tending to move the soil downhill. The only force present which is able to do this is the weight of the soil overburden. At the surface, there is no overburden and so the soil cannot suffer a net shear, even though the amount of movement normal to the surface is at a maximum. At great depth, the soil does not expand or contract, so that there can be no net shear component, however, great the overburden (provided that it is not great enough to initiate continuous creep). In between, at some finite depth will be a zone of maximum net shear rate, so that the velocity profile is qualitatively as shown in figure 1, *d*.

This analysis can be put into mathematical terms in order to show the physical quantities involved more clearly. It must, however, be emphasized that the theory represents only the average behavior of the soil. The actual movement is made up of the making and breaking of contact among individual soil aggregates, with sizes comparable to the mean distance moved, so that large random variations in rate are to be expected. In order to simplify the theory, the following assumptions must be made: first, that the action of gravity is solely through the weight of overburden above any point in the soil, although it is recognized that there will be random variations about mean values in both direction and value of the overburden pressure; second, that the cyclic forces are alternately exactly opposed in direction, and that in each case the force increases until the soil yields; and third, that the direction of movement, at the failure, is in the direction of maximum stress. This last assumption is equivalent to one of minimum work.

Figure 2, *a*, shows a soil surface inclined at an angle θ. At a point P, at vertical depth z below the surface, suppose that a cyclic stress Q_1 acts, in a direction inclined to the normal to the surface at an angle β. Suppose that, in yielding, the soil element around P, of thickness δz, expands by an amount δl_1 in the direction of the yield. Then the direction of the yield differs from the direction of Q_1 by an angle γ_1, in the downhill direction. This angle can be seen (fig. 2, *b*) to be given by:

$$\sin \gamma_1 = \frac{\rho_1 g z}{f_1} \sin(\beta + \theta)\,, \quad (1a)$$

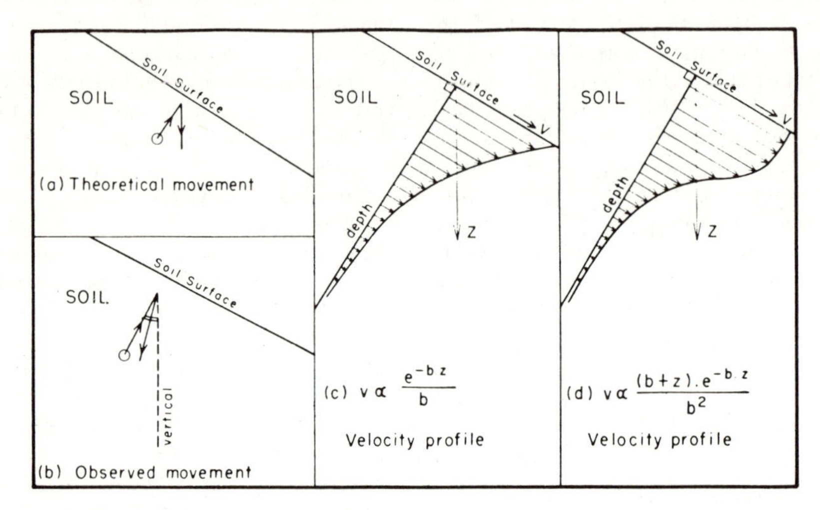

FIG. 1.—*a*, Theoretical path of a soil particle (after Davison); *b*, actual path of a soil particle (after Davison); *c*, predicted velocity profile from Davison theory; *d*, predicted velocity profile from present theory.

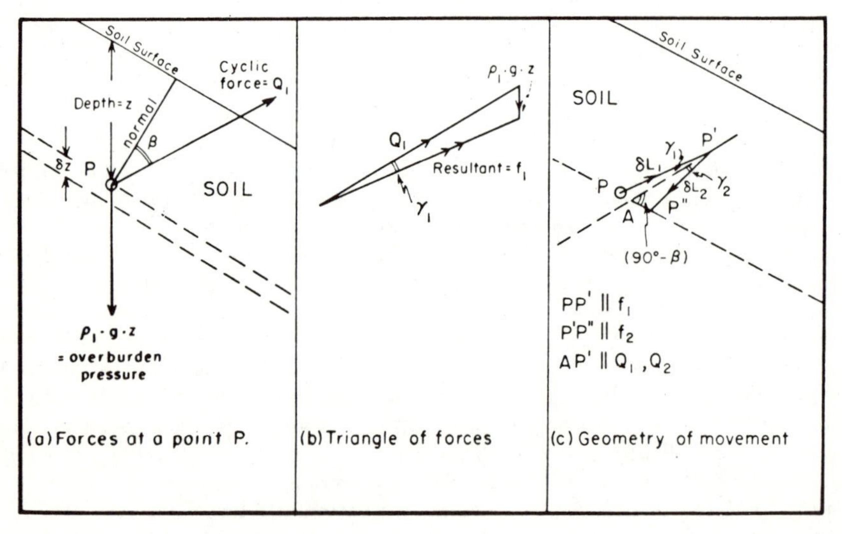

FIG. 2.—*a*, The forces acting at a point in the soil; *b*, the triangle of forces for the point; *c*, the geometry of the movement, if no net consolidation is to occur.

where f_1 is the yield strength of the soil in the direction of Q_1, and ρ_1 is the density of the soil during the expansion. In the contraction part of the cycle, with suffixes $_2$, similarly,

$$\sin \gamma_2 = \frac{\rho_2 g z}{f_2} \sin(\beta + \theta) \quad (1b)$$

the sign of γ_2 being taken so that it is again positive for a deviation in the downhill direction. For soil creep to persist over many cycles, there can be no long-term consolidation of the soil; and δl_1, δl_2 must be related (fig. 2, *c*) so that no net consolidation occurs. Suppose the movement, within the element of thickness δz, to consist of a movement $PP'P''$. Then if $P'A$ is in the direction of Q_1 and Q_2, the angle $P'AP'' = (90° - \beta)$ and the creep components PA, AP'', are given by:

$$PA = P'A \frac{\sin \gamma_1}{\cos(\beta - \gamma_1)};$$
$$AP'' = P'A \frac{\sin \gamma_2}{\cos(\beta + \gamma_2)}. \quad (2)$$

Adding these equations, and writing $P'A = \delta l$, the total net creep, δC, in a direction parallel to the surface, contributed by the element, is given by:

$$\delta C = \delta l \frac{\cos \beta \sin(\gamma_1 + \gamma_2)}{\cos(\beta - \gamma_1)\cos(\beta + \gamma_2)}. \quad (3)$$

If it is assumed that the slope is not subject to continuous creep, then $Q_1, Q_2 \gg \rho_1 g z$, and γ_1, γ_2 are small. Substituting for γ_1, γ_2 in (1), the equation then simplifies to:

$$\delta C = \delta l \cdot gz\left(\frac{\rho_1}{f_1} + \frac{\rho_2}{f_2}\right) \sin \theta \times \left(1 + \frac{\tan \beta}{\tan \theta}\right). \quad (4)$$

Dividing through the δz, and letting δz tend to zero, gives the net shear rate, dC/dz, at depth z. Integrating this expression for the shear rate, from a great depth up to depth z, gives the total net movement parallel to the surface in one cycle at depth z:

$$C(z) = g \sin \theta \int_z^\infty z \frac{dl}{dz}\left(\frac{\rho_1}{f_1} + \frac{\rho_2}{f_2}\right) \times \left(1 + \frac{\tan \beta}{\tan \theta}\right) \cdot dz. \quad (5)$$

In the most general case, all of the terms inside the integral sign may be functions of z, and may vary from cycle to cycle, but some approximations and simplifications can be made for more usual circumstances. First, most geomorphic agents act more or less uniformly over the surface of a hillside, so that the expansions they cause are normal to the surface, and $\beta = 0$. This leaves $\sin \theta$ as the only slope-dependent term in the expression, so that the rate of movement should, other things being equal, be proportional to the sine of the slope angle. Second, the term $[(\rho_1/f_1) + (\rho_2/f_2)]$, the mean ratio of density to yield strength, will be relatively conservative, probably not varying by more than an order of magnitude throughout the soil profile. It may be replaced by a pseudo-constant, k, which is a property of the soil alone. Third, the term dl/dz is the product of the coefficient of linear expansion of the soil (for the relevant agent) and the accumulated change of the geomorphic agent (e.g., frost frequency or accumulated change of soil-water content). Now the coefficient of linear expansion is another order-of-magnitude pseudo-constant, s, of the soil; while the accumulated moisture change, M, is partly a function of the soil but is much more a function of climatic conditions over the year. As a first approximation, it may be assumed to die away exponentially with depth. As a mathematical equation:

$$\frac{dl}{dz} \sim a \cdot M \cdot e^{-bz}, \quad (6)$$

where b is another "constant" of the soil. Substituting (6) in equation (5), and summing over a year:

$$C(z) \sim a \cdot k \cdot g \cdot \sin \theta \cdot M\left(\frac{b + z}{b^2}\right)e^{-bz}. \quad (7)$$

Using the same approximations, Davison's theory would give a velocity profile:

$$C(z) \sim \frac{a \cdot M}{b} \cdot e^{-bz} \tan \theta. \quad (8)$$

If $C(z)$ is to be measured in a horizontal direction, rather than a downslope one, then each of these expressions should be multiplied by cos θ. On slopes of up to 30°, however, this has only a minor effect on the result. The difference in form between equations (7) and (8) is the same as has already been qualitatively deduced, and is shown in figures 1, *c*, and 1, *d*, respectively.

Culling (1963, p. 130–134) has developed a theory of soil creep based on the operation of random forces on individual soil particles. This physical picture may represent the action of many small forces in the soil, such as weathering and the activities of small animals, much more accurately than the cyclic assumptions of the previous paragraphs. Culling shows that the result of random movements of particles is a diffusion of particles from regions of high particle concentration (i.e., density) to regions of lower concentration. He assumes that the slope surface itself represents a density gradient, but this assumption is an oversimplification. The only way in which a density stratification is likely to come about naturally is as a result of consolidation. On a uniform slope, the lines of equal density will, if they exist, be parallel to the soil surface. Therefore, the diffusion will have a net outward drift in a direction normal to the surface. This drift, under the action of diffusion agents, will be balanced by continued further consolidation if equilibrium is maintained. The consolidation will occur under the action of gravity but will be prevented from occurring vertically by the cohesion of the soil in exactly the same way as in the contraction during cyclic seasonal creep, discussed above.

Creep under random forces, although physically very different from cyclic creep, can be described mathematically by the same equations, with β, the angle of deviation of the cyclic force from the normal, equal to zero; and with γ_1, the angle of deviation of the expansion from the direction of the cyclic force, also equal to zero. The displacement, δl, must be interpreted as the mean free path of the diffusing particles, and the number of cycles as the number of displacements suffered in unit time. Thus all forms of soil creep considered may be described, to a first approximation, by the mathematical theory proposed here, although different geomorphic agents will be responsible for different values of the constants in the equations, and although the present theory deals with mean movements of the soil considered as a uniform medium, whereas other theories, especially that of Culling (1963), consider the soil as an aggregate of interacting particles. Whether the present theory is accepted or previous theories are preferred, it can be seen that creep processes should always be at a rate which increases as the sine of the slope angle, to a good order of approximation.

A LABORATORY EXPERIMENT

A block of soil was brought into the laboratory for study to measure soil movements with greater accuracy than is possible in the field, and to limit the causal agents to changes in soil moisture. This study allows a much closer comparison with theory, both qualitatively and quantitatively, than is practicable in the field.

The block of soil was taken from the surface, over Chalky Boulder Clay, near Cambridge, England. It was transferred, with a minimum of disturbance, to a wooden box, open at the top, and with one glass side. The block measured 80 cm. × 30 cm. (deep) × 27 cm. (wide), and its faces were cut to fit the box as closely as possible, except at one end where the upper corner of the block was cut away (fig. 3). The whole box was mounted on a slope of 21°, so that the cut end of the block was almost vertical, at the lower end of the box. Sliding of the block at its base was prevented by a grid of 5-cm. nails projecting into the soil through the base of the box. Before the glass side of the box was finally secured, twenty-two wires, 1.5 mm. in diameter, were inserted horizontally across the width of the block, so that their ends could be seen when the glass was replaced. Five of the wires were 7 cm. long; the remainder, 25 cm. long. On the glass, opposite each wire, a pair of rectangular axes was scratched parallel to the sides of the box. The position of each wire was then referred to its own axes and measured with a traveling microscope to an estimated accuracy of ± 0.05 mm. Some readings were lost owing to obscuring of the inside of the glass by soil washed down in front of the wires.

The experiment was set up indoors, and the block was alternatively allowed to dry by evaporation at room temperature and watered from above, in units of 8–16 mm. of water, allowing between 6 and 24 hours for the soil to reach equilibrium between units of water. The experiment was run for 5 months, during which the block was wetted four times and dried three times.

Figure 3 shows a diagram of the box and block of soil, orientated so that the edges of the paper are horizontal and vertical. The lettered points, A–V, show the positions of the wires, each being located at the dot. The zigzags beside each wire position show its movement, magnified twenty times relative to the rest of the figure, during the wetting and drying cycles. Wire V fell out early in the experiment, and wires F, J, N, and Q showed negligible movement. Wire U, while still attached to the block, was moving relatively fast, on a separating flake of soil.

The movement of each wire consisted of an expansion on each wetting, broadly outward from a center, and a contraction on drying toward a center. Over several cycles each wire moved in a zigzag pattern with a steady downward trend (with the exception of S). Since the block of soil was of limited length, there were marked end effects, and the block had a tendency to expand outward in all directions, as it was not confined at either end. With this limitation, the movement of the wires is in qualitative agreement with the theory of soil creep outlined above, since the zigzags open out downward and away from the vertical. Furthermore, the amount of movement in expansion and contraction is greatest near the exposed faces of the block. The wires appeared to move with the soil throughout, and there is no evidence of movement of the mass of the soil relative to the wires.

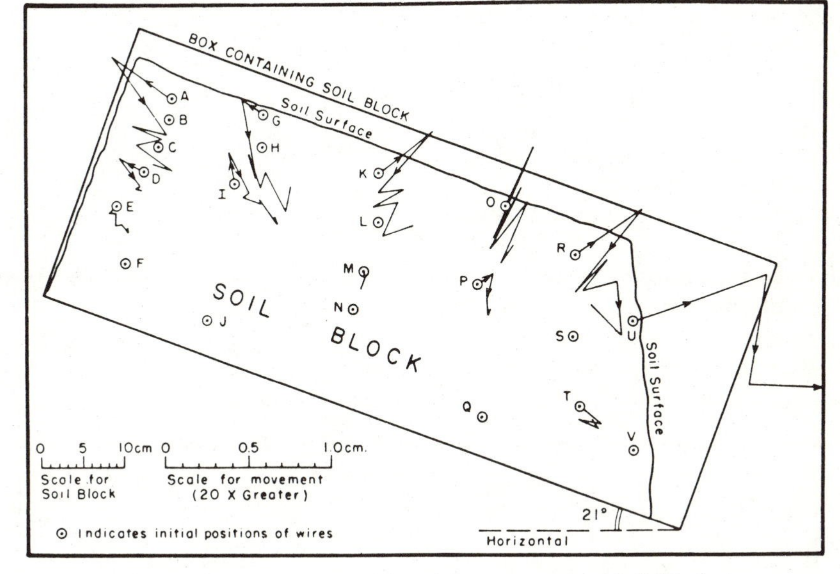

FIG. 3.—The laboratory soil block and the movements of individual wires

The motion of the wires can be analyzed into a cyclic component which is back and forth in a constant direction, representing the direction of the cyclic force in figure 2; and a non-cyclic component at right angles to this mean direction, representing the net soil creep. (Since wetting and drying may not be equal over a small number of cycles, the creep must be taken at right angles to the cyclic force, to eliminate any component of it.) The internal consistency of the movements of the wires was tested in two ways: for the consistency of direction of movement of each wire on consecutive wetting and consecutive drying periods; and for the consistency of relative magnitude of movements between wires in consecutive periods. These tests indicated that more than 50 per cent of the directions lay within 30° of the mean for that wire and phase (i.e., wetting or drying); and more than half of the magnitudes lay within a factor of two times the normal for the wire and period. These figures do not show unanimity but random variations about a uniform trend. A wide range of variation is to be expected, as has been pointed out above.

The mean expansion movement (for all wires) in each wetting period is linearly related to the amount of water added in that period (fig. 4), which shows that the assumption of a constant coefficient of expansion of the soil is a reasonable one. Isolines of equal expansion movement can be drawn at right angles to the mean direction of movement of each wire, showing that the motion is a bulk expansion of the soil mass (fig. 5). Drying could not be controlled as well as wetting, but was prolonged and therefore more or less complete. This is supported by the close relationship between

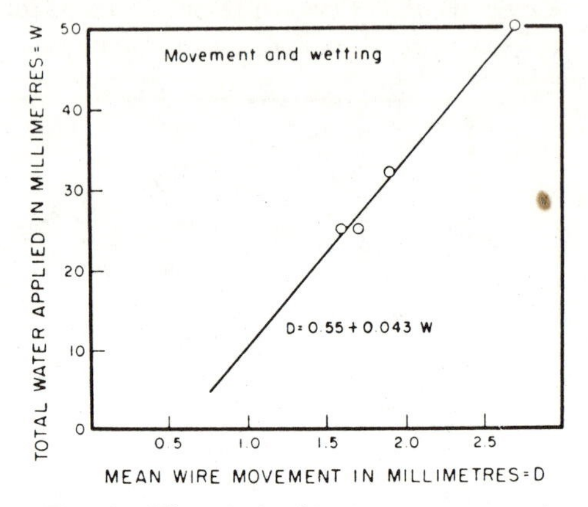

FIG. 4.—The relationship between mean wire movement and the quantity of water added in laboratory experiment

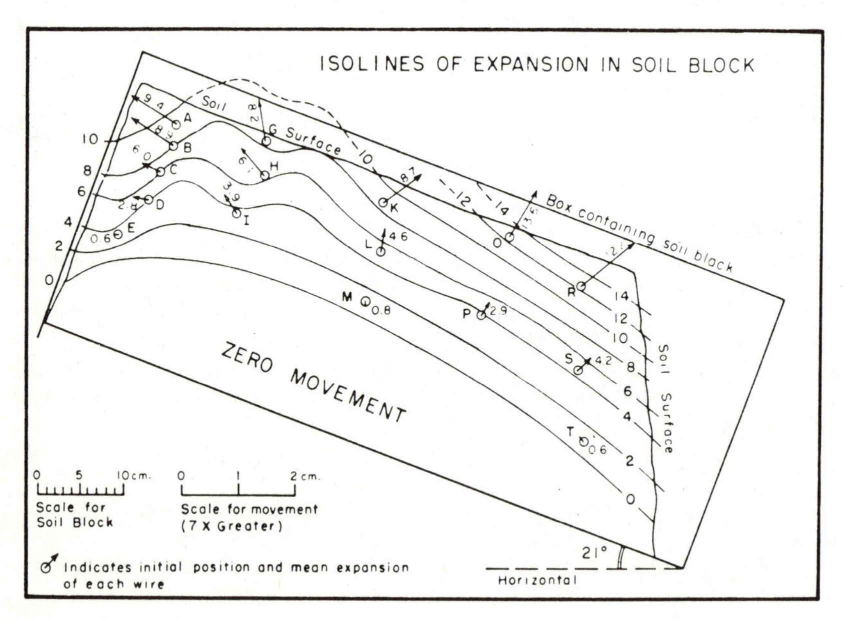

FIG. 5.—The cyclic expansion component in the laboratory soil block, showing individual wire expansions and isolines of expansion.

total wetting movement and total drying movement for the average of all wires.

A comparison can be made between the measured values of creep in this experiment and the theory previously discussed. Equation (5) above may be used in the form:

$$C(z) = k \cdot g \int_z^\infty z \cdot \frac{dl}{ds} \sin \theta \cdot ds , \quad (9)$$

and numerically integrated along orthogonals to the isolines of figure 5 to obtain a value of creep which is proportional to that required by the theory above, where θ is understood to be the inclination of the isolines to the horizontal, and s is the length measured along the orthogonals. Figure 6 shows the relationship between the values calculated by equation (9) and the measured values. There is a significant (95 per cent) correlation between the two sets of data. Owing to the random influences present, this is considered to be moderately good evidence for the validity of the theory suggested in this paper, but further experiments of this type would be needed to test the theory conclusively.

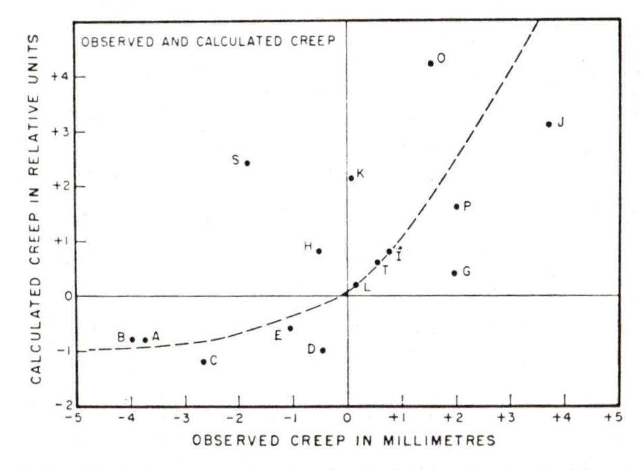

FIG. 6.—The relationship between measured and calculated soil creep in laboratory experiment

FIELD STUDY AREA: THE WATER OF DEUGH DRAINAGE BASIN

The area chosen for field measurements is in southwest Scotland, in the drainage basin of the Water of Deugh, in Ayrshire and Kirkcudbrightshire (fig. 7). The highest point in the basin is the summit of Cairnsmore of Carsphairn, 2,613 feet above sea level, and the outlet of the basin is at an elevation of 800 feet. The area of the basin is 28 square miles of rolling mountain grassland, which is almost exclusively used for sheep grazing. Southwest Scotland was glaciated during the Pleistocene period, and the higher ground was probably subject to a late-glacial re-advance (Charlesworth, 1926, p. 17) at about 8,000 B.C. The valley bottoms are filled with till from the last main glaciation. Small patches of till from the late-glacial stage are also found. The high rainfall and low evaporation rates produce conditions suitable for the formation of peat, which is widespread on gently sloping hill crests, and near the valley axes. Drainage ditches, cut and maintained for at least 100 years, have increased the drainage density from its natural value of 3.0 miles per square mile to about 60 miles per square mile. The Deugh and its tributaries flow on a floodplain, which is incised in the till except in the upper parts of the basin. The bluffs bordering this floodplain are locally steep, at 35°–40° angles, and are up to 40 feet high. They appear to degrade, over a period of several hundred years (Kirkby, 1963, p. 279–286) to slope angles of about 15°–20°, if the base of the bluffs is not attacked by the river before this can occur.

Bedrock is rarely exposed in the basin,

but lies close to the surface on the steepest slopes of most valley sides. Around the Cairsnsmore of Carsphairn, there is an area of granite, but most of the basin is in Silurian metamorphosed shales and flagstones, which are very tightly folded, with a regional northeast southwest strike, cut by many dikes of Caledonian age (Pringle, 1948, p. 1–10).

Approximate climatic data have been collected from a number of sources, and referred to the Deugh basin, at an elevation of 1,100 feet. A rain gauge has been maintained at the Mair of Deugh since 1933 (Meteorological Office, 1933–60). Evapotranspiration has been averaged from two independent sources: from standard evaporation tanks at Amlaird Filters, Fenwick, Kilmarnock, Ayrshire (Meteorological Office, 1936–40 and 1941–50), corrected for grassland on the basis of Penman's measurements (Penman, 1948, p. 138–139); and from data for actual evapo-transpiration at Clatteringshaws Loch from 1937–53 and 1954–60, collected by the Galloway Hydro-Electric board (Briggs, 1961, personal communication). Temperatures for Eskdalemuir (Meteorological Office, 1953) have been corrected for the observed difference between Eskdalemuir and the Mair of Deugh, while temperature records were kept at the latter site during 1962–63. The results of these estimates are in table 1.

Slopes in the Deugh drainage basin are characteristically convexo-concave. The up-

TABLE 1

CLIMATE OF THE WATER OF DEUGH (1,100 FEET)

Month	Average Monthly Rainfall (Inches)	Average Monthly Evapo-transpiration (Inches)	Monthly Average of Mean Daily Temperatures (°C)
January	7.23	−0.01	1.3
February	4.73	+0.01	1.5
March	3.77	0.24	2.9
April	4.11	0.80	4.9
May	3.95	1.72	7.9
June	4.18	1.82	10.8
July	5.62	1.58	12.7
August	4.66	1.59	12.2
September	6.84	0.88	9.9
October	7.72	0.42	6.8
November	7.55	0.23	3.5
December	8.00	0.02	2.0
Year	68.40	9.30	6.4

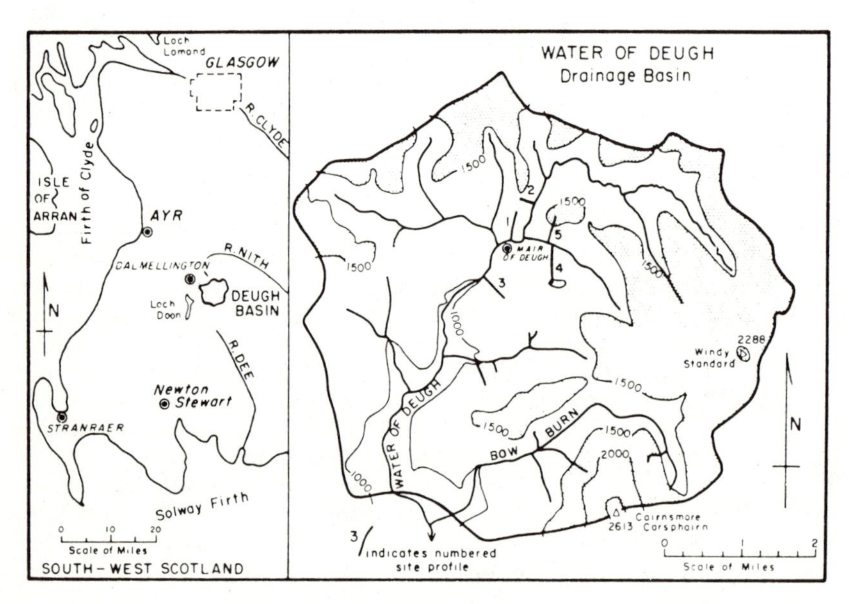

FIG. 7.—The Water of Deugh drainage basin, and its location in southwest Scotland

per convexity is generally peat covered, but the peat thins out downslope to a shallow, humus-rich soil overlying bedrock. The concave section of the slope is generally covered in till, which may be up to 50 feet thick along the valley axes, but which thins up the hillsides. The more gentle slopes on the concavity are also peat covered.

Five slope profiles were selected, each from hill crest to valley bottom, and facing in different directions. On each of these slope profiles, ten sites were selected, on a variety of slopes and soil types (fig. 8). At each of the fifty sites, a set of instruments was installed. For the purposes of this study, the soils were divided into four broad categories, as follows:

1. "Peat": Acid organic soils more than 12 inches deep with little mineral soil.
2. "Rock": Humus rich soil with high mineral soil content, less than 12 inches deep.
3. "Till": Acid clay soils, with an organic layer less than 12 inches deep, but developed to a considerable total depth on a clayey till.
4. "Bluff": Similar to "till" but with improved drainage, less developed profiles, and on steep slopes (formed by slumping into the river).

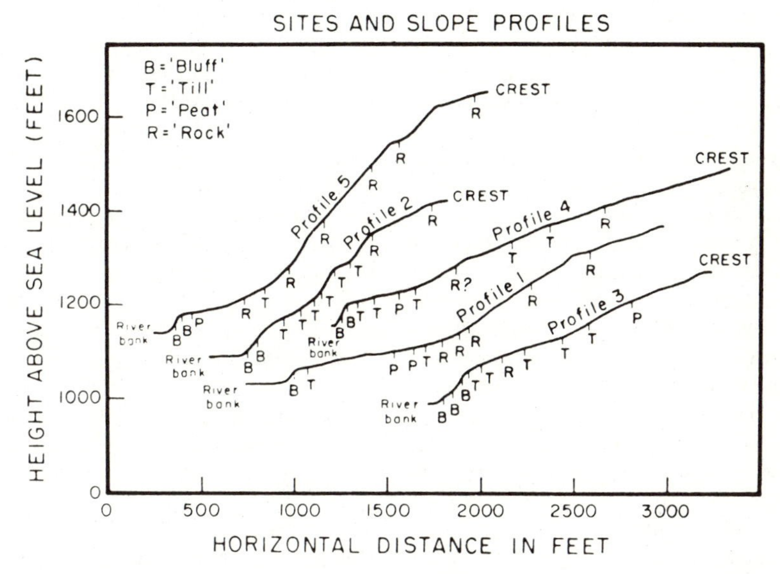

FIG. 8.—Surveyed slope profiles, showing the positions of the measurement sites

FIELD METHODS

Two techniques were used to measure soil creep in the field. The most direct method was the digging of "Young pits" and measurement of creep differences between wires in the walls of the pits, at relatively long intervals of time. The second method was the insertion of T-shaped pegs into the soil and measuring the tilt of the pegs with a sensitive spirit level placed on the cross-piece of the T.

The Young pit was first used by A. Young (1960, p. 121, fig. 1, *c*). The following description is of the pits used for this study. A pit was dug as deep as possible into the soil, with a vertical face running parallel to the line of steepest slope for a horizontal distance of about 18 inches. Into this face were inserted a number of stiff metal wires, 2-4 inches long, each placed at right angles to the face, with its tip just showing.

When the wires had been inserted, a plumb line was hung close to the face, and ideally the wires all lay close to it, so that errors of measurement would be minimized. The horizontal distances between the tips of the wires and the plumb line were then measured, parallel to the cut face, with a pair of calipers. Practical difficulties probably limited the accuracy of measurement to about ±0.5 mm. When the measurements had been completed, the face was covered with a sheet of paper, and the hole refilled, taking great care not to disturb the measurement face. After a period of at least 6 mo. the hole was re-excavated, again with great care, and the measurements repeated.

All of the Young pits in this study were in a sheep-grazing area, where marker stakes are liable to removal by sheep, which rub their bodies against them. This rubbing not only tends to remove the stake in a matter of days but also subjects the neighborhood of the site to more than average trampling. Despite this difficulty of marking sites, thirty-three out of forty sites were refound at the end of a year. When sites are remeasured, the difference between the new and original figures is a direct measurement of the net differential shear between the wires. Absolute values for downslope translation can only be assured if the bottom wire is secured in bedrock, and this is not normally practicable. For the pits in this study, the depth of the bottom wire varied between 8 and 15 inches.

The advantages of this method are immediately apparent. The measurement is direct and therefore unequivocal. Provided that the digging is done with sufficient care, the measurement site is subject to no more outside disturbance than the surrounding ground. However this proviso can never be guaranteed; there is always some risk of disturbance. Further, there is the likelihood that the sides of the hole allow more water to permeate than elsewhere, perhaps giving rise to an atypical response. Because of the risk of disturbance every time a measurement is made, the reliability of the over-all result declines as the frequency of measurement increases. Furthermore, the method cannot give results of much meaning in the very open soil close to the surface, where the wires are not securely held by the soil and almost all methods break down. Nevertheless, for directness and simplicity the Young pit is one of the most suitable methods for long-period readings. Its only practical rival to date appears to be the "test pillar" method (Rudberg, 1958, p. 115–116). In this method, an auger hole is made approximately 1 inch in diameter. A thin-walled tube fitting the hole is inserted, and then a column of short, separate, wooden or plastic cylinders is pressed down inside the tube. The tube is then withdrawn, leaving the column of cylinders in the hole. This method overcomes some of the objections to Young pits (inhomogeneous water flow; invalid readings close to the surface; disturbance of the soil) and is very simple to install provided the ground is not too stony. It is, however, still less sensitive, as there must be sufficient space around the cylinders to withdraw the tube, so that there will be a minimum error of about ±3 mm. in the initial position. The use of radioactive tracer particles is theoretically an almost ideal method for measuring soil creep, but high cost and the need for safety precautions have so far prevented their actual use in the field.

A second measuring device, called a T-peg, was developed for this study. A T-peg consists of a steel rod 9–15 inches long and $\frac{1}{4}$ inch square in cross section, welded to a cross-piece (fig. 9). The cross-piece is formed from a steel bar, $\frac{1}{8} \times 1$ inch in section and 11 inches long, bent into a long U, whose limbs are 1 inch apart. The midpoint of one limb of the U is welded to the upright square rod, so that the long U, on its side, forms the cross-piece of the T. On top of the upper, unwelded limb of the U, two brass V's are mounted, so that a cylinder may be laid across both of them, parallel to the limbs of the U. An adjusting screw connects the two limbs of the U so that the angle between the upper limb of the U and the rod can be varied. The rod is inserted vertically in the soil to a depth of 6–12 inches (3 inches above ground in each case). As the rod shears with the soil, the tilt in the rod is measured with a specially constructed spirit level mounted on a cylinder. The cylinder is true to 0.0001 inch, and the spirit level is graduated in 10 seconds of arc divisions, so that angular movements of 4 seconds, or relative movements of 0.005 mm. over a 12-inch depth of soil, can be detected.

In use, a pair of T-pegs, 9 and 15 inches long and inserted to 6 and 12 inches depths, respectively, were mounted side by side at each measurement site to measure the mean soil shear over two separate depths. The rods were placed as near as possible to the vertical, so that the weight of the cross-piece had as little disturbing effect as possible. The main advantage of this method was the ability to take frequent and sensitive measurements without disturbing the instrument in the ground. However, the exposure of part of the instrument above the ground surface made it liable to a wider variety of disturbances than the Young pit, of which kicking by animals was the worst and most obvious. Kicking often caused rotation of the cross-piece so that it was no longer aligned downslope, but smaller disturbances could not be detected with any certainty. The likelihood of disturbance increased with time, so that the method could only be used for relatively short periods of measurement. A second disadvantage was the indirect nature of the measurement. Conversion of the angular reading involved the unproven assumption that the rod averaged the shear in the thickness of soil through which it passed. In fact, a T-peg placed in the laboratory block of soil described above showed that, for the soil used the T-peg tilt was too low by a factor of about five times, showing that the soil moved appreciably relative to the T-peg. Hence the T-peg could only be relied upon to give a pattern of relative movement to compare short periods with one another, and not for a long-term average rate of movement. The T-peg will also cause some local disturbance of the soil during insertion, but this disturbance should be slight for a $\frac{1}{4}$-inch rod. It can be seen that the Young pit and T-peg methods are complementary. The former gives direct measurements of long-term average relative displacements, and the latter shows the angular movements over short periods, so that these movements can be compared with other, simultaneous, physical changes in the soil.

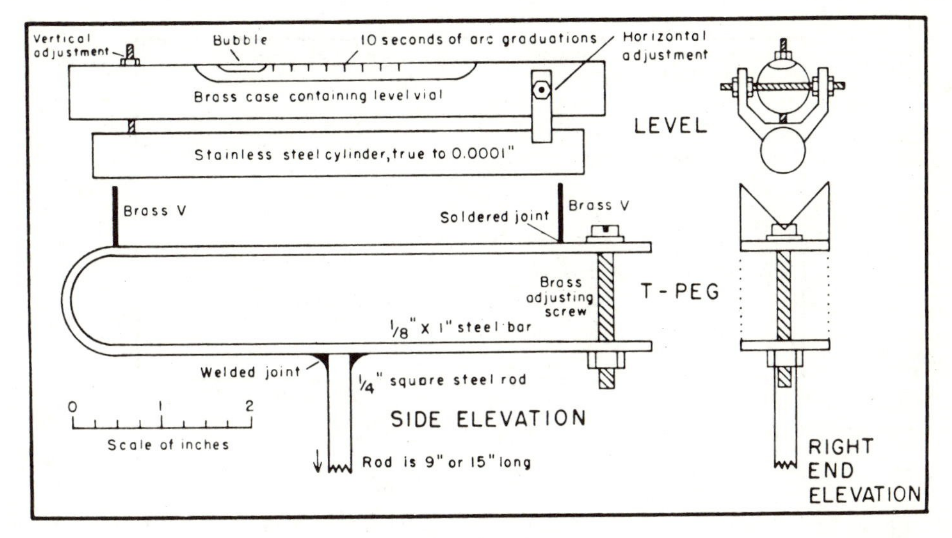

FIG. 9.—Diagram of a T-peg

Additional installations were made in the field to measure slope wash and soil moisture. Slope wash was collected in a trough 12 inches wide, with an upper lip which extended into the soil on the upslope side at a depth of about 1 inch below the surface (Young, 1960). This method always tends to overestimate the volume of soil moved, as there is a cut face immediately above the trough which is liable to erode at an unnaturally high rate. It will, however, give the order of magnitude for slope-wash transport.

Nylon resistance moisture meters were installed in an attempt to measure changes in soil moisture, but those used proved to be insensitive over the most important range of soil moisture change. Samples of soil were taken to measure soil moisture, but the samples were very small, as no proper drying equipment was available. As a result, the variability among samples in space was at least as great as the changes over time, and no valid conclusions could be drawn. In the absence of direct measurements, changes in soil moisture were inferred from daily rainfall data, as is described below. Daily rainfall figures were available for one point in the basin, and from October, 1962, until May, 1963, daily maximum and minimum temperatures were recorded. In January, 1963, near the end of the longest cold period in Britain since 1947, some measurements of soil temperatures were made with a thermistor probe. Even during this very cold winter, frozen ground rarely extended to depths greater than 4 inches. Therefore, in the area of study, where winters are relatively mild and snowfall is heavy, freeze-thaw probably is not so important as moisture changes in causing soil creep. Finally, some measurements of soil strength at all sites were made with a Vicksburg penetrometer.

The laboratory experiment described above showed that moisture changes can cause significant amounts of soil creep. Therefore, if daily rainfalls could be related to measured movements of the soil, this relationship would represent the causal chain from rainfall to soil moisture to soil movement. In making a correlation, a function of daily rainfalls was chosen which would behave roughly in the same way as the soil moisture. Figure 10 shows the structure of such a model. In the study area, rainfall is at least twice the evaporation for every month of the year, so that drainage is more important than evaporation in causing changes in soil moisture. The model, therefore, is a model of drainage, and has been made linear for simplicity. The model is of the water content of a surface layer of soil, say the top 12 inches. Zero water content corresponds to "field capacity," that is, the water content to which the soil would drain after several days without rainfall or evaporation (Veihmeyer and Hendrickson, 1931, p. 181). Field capacity is an approximate rather than a real constant.

Immediately after a short period of rain-

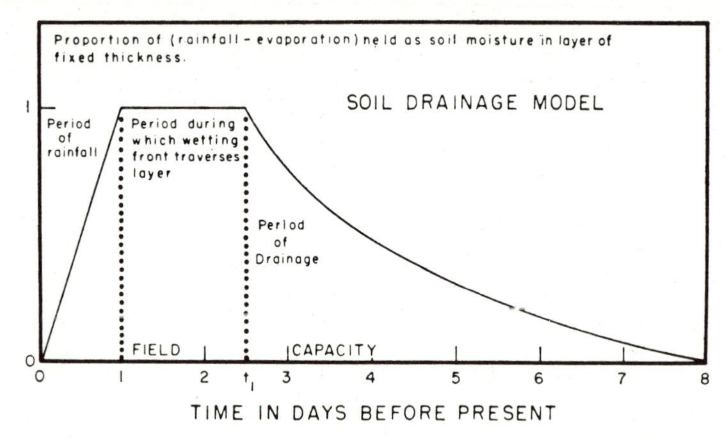

FIG. 10.—Model of soil moisture drainage, in terms of rainfall

fall, all of the rain which fell will, if no runoff occurred (the usual case), be within the soil layer considered. For a time, t_1, the wetting front will be traversing the layer, and no water will leave the layer. At the end of the transit time, t_1, the wetting front will have reached the bottom of the layer, and water will begin to drain from the layer. The rate of loss through draining will slow down and tend to zero as field capacity is approached (Youngs, 1958). For simplicity the soil moisture content during this drainage period will be assumed to decrease exponentially; that is, a proportion, p, of the water will remain in the layer after a period of 1 day, for some constant p less than 1.0. This simple model is broadly representative of the changes of soil moisture during drainage, both in its tendency to drain to field capacity and in the existence of a transit period, and the model is completely defined by the two constants, t_1 and p. Various values of these constants were tried for best fit between observed rainfall and movement. Although a correlation of this type is valid, direct moisture data would have been more valuable, had they been available. The correlation is likely to mask some of the secondary causes of variability, in particular that with slope angle.

At each of the fifty instrument sites, a Young pit, a pair of T-pegs, a soil-wash trough, and two nylon resistance moisture meters were installed and measurements taken regularly between July, 1961, and May, 1963. During this period each site was visited a total of fifteen times, and appropriate measurements made at each visit. The T-peg tilts were measured at every visit, and other variables at intervals of between 2 weeks and 1 year.

RESULTS OF FIELD MEASUREMENTS

Of a total of about 1,100 individual T-peg measurements made, about 550 were spoiled by gross disturbance, and were rejected. The remaining readings were arbitrarily divided into two classes, those for period of less than, and those for more than 2 weeks. As in the laboratory experiment, there should be a cyclic and non-cyclic component of the motion. For shorter periods, the cyclic component, corresponding to expansion and contraction of the soil normal to its surface, would be dominant; while, for longer periods, the net creep would become increasingly important. If a T-peg is vertical and responds in an ideal manner to movement of the soil, then a fractional expansion of q will give a rotation proportional to $q \cdot \sin \theta$, where θ is the slope angle. The measured tilts for the short periods were compared with rainfall functions representing soil moisture, of the type discussed above, and the best fit was with the time constant, $t_1 = 2$ days, and the decay constant, $p = 0.5$. Thus the estimate of soil moisture content, m is:

$$m = r_1 + r_2 + 0.5\, r_3 + 0.25\, r_4 + 0.125\, r_5 + 0.0625\, r_6 + \dots \qquad (10)$$

where r_i is the (rainfall-evaporation) i days ago. In making this correlation, the tilts from the 6-and 12-inch pegs were pooled on the basis of the best fit relation $T_{12} = 0.75\, T_6$, where T_6 and T_{12} are the tilts of the 6- and 12-inch pegs, respectively. The combined tilts are then plotted against the estimated moisture, and a significant correlation at a 95 per cent level obtained:

$$T = 1.75\, m - 0.33\,, \qquad (11)$$

where T is either T_{12} or $0.75\, T_6$, measured in minutes of arc, and m is measured in inches of water. Thus relationship takes no account of either slope angle or soil type.

If each of the four soil categories described above is assumed to have a constant coefficient of expansion as water content changes, then a second formulation of the relation between moisture change and movement can be attempted, taking slope angle into account. Since theoretical considerations demand a slope term proportional to the sine of the angle, the second formulation is (fig. 11):

$$T = m \cdot k \cdot \sin \theta\,, \qquad (12)$$

where k is a constant of the soil type. In fact, this form gives a correlation which is as significant, but no more so than that of equation (11). It is however, to be preferred on theoretical grounds. The best-fit values for k in equation (12) were: for "peat," 2.8; for "till," 4.3; and for both "rock" and "bluff," 1.4. The tilt component of equation (12) is considered to represent the short-term, cyclic part of the movement.

If the measured tilts for long periods are now reduced by the amount of cyclic movement estimated from (12), a measure of the long-term component of the tilt is obtained which should correspond to the net creep of the soil. Comparison of the number of positive (downhill) and negative values of this long-term component (194 positive against 133 negative) shows a preponderance of downhill movement which is significant at the 99 per cent level; and the subtraction of the cyclic component significantly (97.5 per cent level) increases the number of downhill values. This is considered to be good evidence that a measurable amount of soil creep exists.

If the T-peg responds in an ideal manner to a uniform net shear q in the soil, then the peg will tilt through and angle proportional to $q \cdot \sin\theta \cos\theta$. If the tilts for 6-inch pegs and 12-inch pegs are compared for long periods, the best-fit relation is $T = T_6 = T_{12}$, and this has been used for combining the tilts for long periods. The net creep should also be proportional to the accumulated moisture increase over the period of measurement, M. Thus it seems best to formulate the relationship between long-term tilt component and moisture as:

$$T = M \cdot K \cdot \sin\theta \cdot \cos\theta\,, \qquad (13)$$

where K is a constant for the soil type. This formulation can only be justified on theoretical grounds, as the data are extremely scattered. If an equation of the form of (13) is assumed, however, it is possible to calculate that the average value of the constant K for all soils is 0.16.

Of the fifty Young pits installed in 1961, forty were examined in 1962; the remaining ten were examined in 1963, and twenty of the others were re-examined. A total of fifty-two measurements were obtained, the remainder being lost through inability to locate the measurement face exactly or through accidental damage. Two methods were used to analyze the results: first, the volume of soil moved between measurements was summed for each site; second, the mean shear for each 1-inch depth interval was calculated and the values pooled for each soil class. The second method relies on the accuracy of the bottom wire's position less than does the first method. In each case,

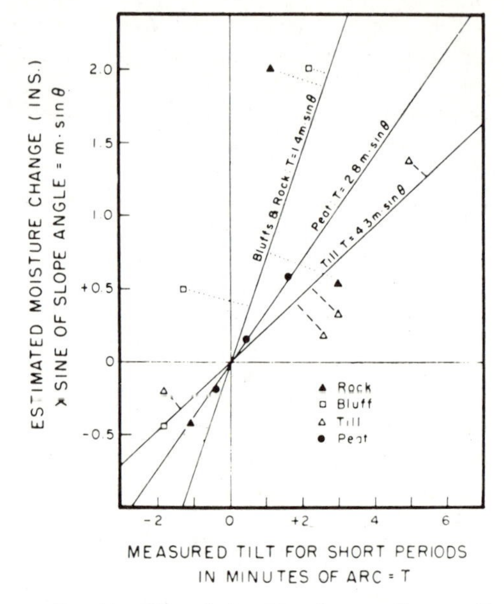

FIG. 11.—The relationship of T-peg tilts to soil moisture content, slope angle, and soil type, for short periods (less than 2 weeks).

the large scatter of the data concealed any possible correlation with slope angle.

Using the volume method of analysis, a total of thirty-four pits showed downhill transport and eighteen uphill transport. This division is "probably significant" at the 95 per cent level. Comparing different periods of time on the assumption that creep is proportional to accumulated moisture increase within each period, the over-all median creep rate was 2.40 cm^3/cm. year, corresponding to the median slope angle of 13° for the measurement sites.

Using the mean shear method of analysis, an over-all transport rate of 1.80 cm^3/cm.

year was obtained, and it is possible to build up velocity profiles for the movement. Figure 12 shows the velocity profiles obtained for each of the four soil types, with the reliability of each. Unfortunately, no reliable measurements could be made for the top 5 cm. of soil, where the texture was too open to hold a wire securely, so that no conclusion can be drawn about the total form of the velocity profile. It has been seen above that theoretical considerations suggest a reduction in the rate of net shear near the surface. If such a reduction exists, then it must occur in the top 5 cm. of soil. It can be stated, however, that the movement was essentially confined to the top 20 cm. of soil. In the peat soils, there was some indication that the surface soil was moving more slowly than the soil layers immediately below it, though the evidence is far from conclusive. Such a velocity profile might be expected if the surface vegetation mat was stronger and more rigid than the peat immediately below it. No similar effect could be observed for other soils.

Averaging results for the two methods of analysis, and assuming that the rate of creep movement is proportional to the sine of the slope angle, the relative rates of creep in different soils can be calculated.

$$S = A \cdot M \cdot \sin \theta , \qquad (14)$$

where S is the mass transport in $cm^3/cm.$ year, and A is constant for each soil type. The values of A derived from this assumed equation are: "Rock" $A = 8.5$; "Till" $A = 4.1$; "Peat" $A = 20.3$; and for the average of all soils, $A = 9.6$. These constants should be of the order of the expansion coefficient divided by the soil strength, while the constant, k, of equation (12) should be of the order of the expansion coefficients alone. Combining these, the relative behaviors of the soils appear to be roughly as in table 2. These values appear to be qualitatively reasonable, the clayey "till" apparently having the highest strength and expansion coefficient, and the "peat" expanding more than the "rock" soil.

Measurements of soil creep at a point include large random variations. The distribution of annual transport, measured in Young pits, may give some idea of the magnitude of variation to be found. However, these re-

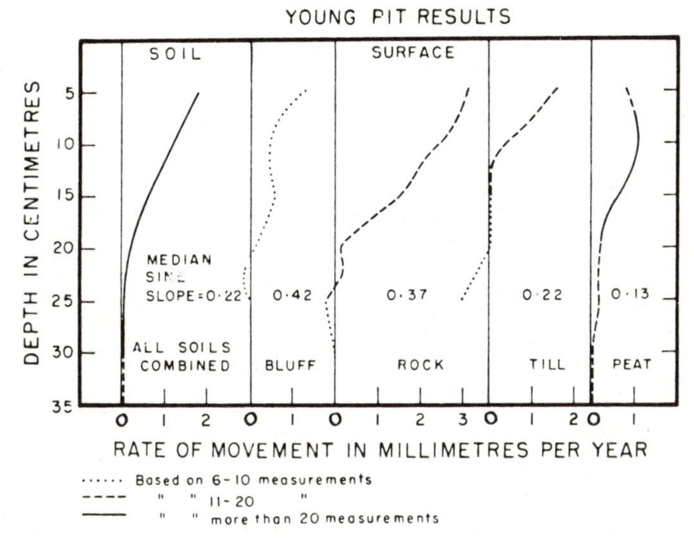

FIG. 12.—Velocity profiles from Young pit measurements. *a*, All soils combined; *b*, "bluff" soils; *c*, "rock" soils; *d*, "till" soils; *e*, "peat" soils.

sults contain experimental errors which may be relatively large as well as some systematic differences in factors not considered in the analysis. The actual distribution of the measurements at the fifty-two pits is approximately normal, with a mean of 2.1 cm^3/cm. year, and a standard deviation of 5.5 cm^3/cm. year. The standard deviation of the mean is 0.39 cm^3/cm. year, so that the mean is significantly positive. To get individual readings that are significantly positive, and so, perhaps, to get valid correlations with such features as slope and soil type, the measurements must be continued for a period of the order of 20 years, assuming that the measured values constitute a random sample. A more practical method of getting results is to duplicate measurements at the same site. For example, a period of 4 years would be as good, if a group of five installations was made around the same point.

Comparing T-peg and Young pit results, the mean annual tilt from the velocity distribution of figure 8 is 40 min. of arc, whereas the corresponding value for the T-pegs is only 3 min. This exaggeration is even greater than that measured in the laboratory trial of the T-peg (5 times), again showing that the T-peg is valueless for long-period measurements of creep rates, but remains valuable for correlations with causal factors affecting the soil.

RELATIVE IMPORTANCE OF AGENTS CAUSING SOIL CREEP

In this paper, it has been assumed that soil moisture changes have been the most important agent causing creep in the Deugh drainage basin. Soil freezing measurements described above and measured air temperatures allow some estimates to be made about the relative effectiveness of the meteorological agents: soil water, soil freezing, and temperature expansion. The relative magnitude of soil creep is estimated as:

Relative magnitude = depth at which movement is halved × amplitude of change × coefficient of expansion of soil with respect to changing agent × number of cycles per year.

TABLE 2

RELATIVE SOIL PROPERTIES FROM CREEP MEASUREMENTS

Soil Type	Relative Expansion Coefficient, k	Relative Strength
Peat	2	1
Till	3	6
Rock	1	1

TABLE 3

RELATIVE EFFECT OF METEOROLOGICAL AGENTS IN CAUSING CREEP IN THE DEUGH BASIN

	Soil Water	Soil Freezing	Temperature Expansion
Depth (inches)	6*	1/2*	4†
Amplitude	5% water*	1 freeze	7°C.‡
Coefficient of expansion	0.08%/%§	2%/freeze‖	0.001%/°C.#
Number of cycles	80%*	51*	365
Relative magnitude	192	51	11

Source: * Based on measurements in Deugh basin; † Geiger (1950, chap. iii); ‡ Meteorological Office (1952); § Young (1958); ‖ Davison (1889); # Coefficient of linear expansion of rock.

Appropriate values for the Deugh area are thought to be as in table 3. These figures are considered to be the best available estimates for the Water of Deugh drainage basin, but no generality should be applied to them.

Besides these meteorological agents, a number of other processes probably cause creep. Most of them are biological in character. Roots and burrowing animals make holes: if it is assumed that holes are opened equally upslope and downslope and that they are eventually closed by material moving from upslope, a maximum rate of creep, corresponding to optimum populations, can be estimated. Earthworms also contribute to soil movement by ejecting soft material as casts at the surface, which tends to be spread out by rain, more downslope than upslope. Darwin (1882, p. 262–273) has estimated that worms may, in this way, be responsible for up to 0.25 cm³/cm. year of soil movement under ideal conditions. Trampling by hoofed animals was thought to be a possible cause of local movement, but a calculation showed that the normal pressure exerted by a sheep's foot was only 17 lb./sq. in., whereas the average pressure required to shear soils in the Deugh area, measured with a Vicksburg penetrometer, was about 200 lb/sq. inch. It was concluded that only where an animal was frightened, or walked in an unusually unstable place, would it move an appreciable amount of soil. Estimated rates of soil creep caused by grass, worms, and rabbits are set out in table 4.

TABLE 4

MAXIMUM RATES OF CREEP DUE TO GRASS, WORMS, AND RABBITS

Agent	Length of Hole per Square Meter of Surface (*m*)	Mean Hole Diameter (mm.)	Mean Life of Hole (Years)	Maximum Movement Rate (cm³/cm. year)
Grass* (*Poa pratensis*)	8.6×10^4	0.215	3	0.0034
Worms†	1.5×10^4	5	1	0.15
Rabbits	0.7	100.	2	0.10

Source: * Dittmer (1938, p. 655); † Darwin (1882, p. 262–273).

These rates are all considered to be maxima, and are therefore unlikely to dominate the sum of meteorological causes under normal conditions. It is therefore considered reasonable to divide the measured rate of soil creep in the proportions 192:51:11 calculated for the meteorological agents above, giving rates of 1.58 cm³/cm. year for soil moisture, 0.42 cm³/cm. year for freeze-thaw and 0.10 cm³/cm. year for soil temperature changes. Comparison of these values with those shown in table 4 suggests that, in the Deugh basin, the most important agents in causing soil creep, in descending order of importance, are soil moisture, freeze-thaw, worms and other burrowing animals, soil temperature changes, and plant roots.

RELATIVE RATES OF SLOPE PROCESSES IN THE WATER OF DEUGH DRAINAGE BASIN

In assessing the contribution of each slope process to the total erosion of the drainage basin, it is convenient to assume that material is transported by each process into the natural stream, and the stream is then able to export all the material. This is some simplification, but it is sufficiently good to allow orders of magnitude to be compared. At the lower end of the basin, a small dam collects most of the coarse material from 21 sq. miles of the basin. Gravel behind the dam is periodically flushed out, so that comparison of the material collected with the material in the till allows an estimate to be made of total erosion of the basin. Subtraction of the amount of erosion due to creep and other slope processes gives an estimate of the amount of erosion caused during valley incision and widening by the river. The rate of flood-plain widening can also be estimated from a calculation of the total volume of material exported from the basin during the incision of the inner valley, at some time since the end of the main glaciation.

The measured rate of soil creep is 2.1 cm³/cm. year, as has been shown above. The associated measurements of soil wash showed an average rate of 0.089 cm³/cm. year for a total of seventy-one measurements, and it should be remembered that this is likely to be an overestimate. No correlation could be found between rate and slope, or distance of overland flow, but this may be due to the large spread in the data. About 55 per cent of the wash (by volume) was of organic material. If these rates are summed over 6 miles/sq. mile of natural channel banks, their total contributions to the degradation of the basin are 0.031 inches per 1,000 years for soil creep, and 0.0014 inches per 1,000 years for soil wash.

Fifteen gullies, not associated with drainage ditches, were found over the area of the basin. If these represent the total erosion of this sort in the last 15,000 years, they represent a mean erosion of 0.0071 inches per 1,000 years. A substantial proportion of this material probably has not yet reached the river but is spread out in alluvial fans on the lower parts of the slopes below the gullies. These gullies, although contributing very little to the total erosion of the basin, are, however, the most apparent evidence of postglacial erosion on the hill slopes where they occur.

The gravel behind the dam accumulated at the rate of 250 cu. yd. per year between 1937 and 1960 (Briggs, 1961, personal communication). This represents a much larger volume of material than that contributed by all the processes discussed so far. Therefore, most of this material probably came from the erosion of the bluffs as the flood plain was widened by the river. The bluffs are composed of heterogeneous material. The coarsest materials may never reach the dam, and the finest may wash over it. In comparing the grain size distributions for material behind the dam and material in the bluffs, a concentration ratio can be calculated for each size class, equal to the percentage in dam basin deposit divided by the percentage in bluff material. Differential transport and trapping would give this ratio low values at both fine and coarse sizes, with a maximum at an intermediate size. The total amount eroded from the bluffs can then be calculated as 250 cu. yd. per year times the maximum concentration ratio (4.4), assuming that all of the corresponding fraction (1–2 inches) is transported and trapped. The figure of 1,090 cu. yd. per year, calculated in this way, compares well with the figure of 1,320 cu. yd. per year calculated from the total quantity of material removed since the Pleistocene period. The former figure is equivalent to a bluff recession of 0.09 inches per year, or a mean basin degradation of 0.59 inches per 1,000 years. Comparison of this figure with those for creep, wash, and gullying show that about 93 per cent of the total degradation today takes place in the widening of the flood plain by river action (table 5).

TABLE 5

RELATIVE RATES OF EROSION OF THE DEUGH BASIN

Process	Rate (Inches per 1,000 Years)
Soil creep	0.0310
Soil wash	0.0014
Gullying	0.0071
Flood-plain widening	0.5900
Total	0.6300

If, as seems probable, conditions in the past have been sufficiently like those prevailing today for the rates of processes measured in the Deugh basin not to have been materially different during the last 8,000 years, then creep and wash have had no perceptible effect on the landscape during the postglacial period. Thus, the only marked changes would involve an increase in the number of small gullies and a growth of the flood plain or inner valley, both headward and laterally. If, however, the late glacial period—from 13,000 to 8,000 B.C., say—was characterized by intense periglacial activity, greatly accelerated slope movement may have stripped the till from the upper hill slopes and concentrated it in its present position, closer to the valley axes.

CONCLUSIONS

Soil creep can be measured in the field, if a sufficient number of installations are made or sufficient time is available. The measurements on which this paper are based were continued for a period of 2 years, during which the spread of the data was too great to show the effect of soil type, aspect, or slope angle on the measurements. A mean rate of 2.1 cm^3/cm. year was obtained, and this value was significantly (99 per cent level) greater than zero. It is suggested that future measurements, if they are to have any validity for a point, should be extended for at least twenty installations times years, a group of installations around a point being the most practicable method. The T-pegs appear to be most useful for demonstrating the mechanism of creep, and to establish correlations of movements with moisture changes in the field.

Estimates of relative movement rates in the Water of Deugh area suggest that soil moisture, soil freezing, burrowing animals, and soil temperature changes are the most important agents causing soil creep, in that order. Measurements suggested that the magnitude of soil was, on average, one-twentieth that of soil creep. These relative magnitudes cannot be considered valid for areas with climates unlike that of the Deugh area. Extrapolation of measured rates of slope processes suggests that gullying and flood-plain incision are the only appreciable changes in the landscape since the end of late glacial cold conditions.

REFERENCES CITED

Charlesworth, J. K., 1926, The glacial geology of the Southern Uplands of Scotland, west of Annandale and Upper Clydesdale: Roy. Soc. Edinburgh Trans., v. 55, p. 1-23.

Culling, W. E. H., 1963, Soil creep and the development of hillside slopes: Jour. Geology, v. 71, p. 127-162.

Darwin, C., 1882, The formation of vegetable mould through the action of worms, with observations on their habits: Edinburgh, John Murray, 298 p.

Davison, C., 1889, On the creeping of the soil cap through the action of frost: Geol. Mag., v. 6, p. 255.

Dittmer, H. J., 1938, A quantitative study of the subterranean members of three field grasses: Am. Jour. Botany, v. 25, p. 654-657.

Geiger, R., 1950, The climate near the ground (2d ed.): Cambridge, Mass., Harvard Univ. Press, 494 p.

Kirkby, M. J., 1963, A study of the rates of erosion and mass movements on slopes, with special reference to Galloway: Unpub. thesis, Cambridge Univ., 411 p.

Meteorological Office, 1933-60, British rainfall: London, H.M.S.O.

——— 1952, Climatological atlas of Great Britain: London, H.M.S.O.

——— 1953, Averages of temperature for Great Britain and Northern Ireland, 1921-1950: London, H.M.S.O.

Pringle, J., 1948, British regional geology. The south of Scotland (2d ed.): Edinburgh, H.M.S.O., 87 p.

Penman, H. L., 1948, Natural evaporation from open water, bara soil and grass: Roy. Soc. [London] Proc., ser. A, v. 193, p. 120-145.

Rudberg, S., 1958, Some observations concerning mass movement on slopes in Sweden: Geol. Foren. Stockholm Forh., v. 80, no. 1, p. 114-125.

Sharpe, C. F. S., 1938, Landslides and related phenomena: New York, Columbia Univ. Press, 137 p.

Terzaghi, K., 1950, Mechanism of landslides, application of geology to engineering practice: Geol. Soc. America, Berkley volume, p. 83-123.

Veihmeyer, F. J., and Hendrickson, A. H., 1931, The moisture equivalent as a measure of the field capacity of soils; Soil Sci., v. 32, p. 181-193.

Ward, W. H., 1953, Soil movement and weather: Internat. Conf. on Soil Mechanics and Foundation Engineering, 3d, Switzerland, Proc., v. 1, sess. 4, p. 477-482.

Young, A., 1958, Some considerations of slope form and development, regolith and denudation process: Unpub. thesis, Sheffield Univ.

——— 1960: Soil movements by denudational processes on slopes: Nature, v. 188, p. 120-122.

Youngs, E. G., 1958, Redistribution of moisture in porous materials after infiltration: Soil Sci., v. 86, p. 117-125.

25

Reprinted from Science, February 3, 1967, Vol. 155, No. 3762, pages 560-561

Rates of Surficial Rock Creep on Hillslopes in Western Colorado

Abstract. *The average rate of downslope movement of rock fragments on shale hillslopes is directly proportional to the sine of the slope angle or that component of the gravitational force which acts parallel to the hillslope. The rates of surficial rock creep range from a few millimeters per year on a 3-degree slope to almost 70 millimeters per year on a 40-degree slope, but these rates vary with natural variations in soil characteristics and microclimate, as well as with accidental disturbances.*

Information on rates of landform erosion and evolution has been increasing steadily during recent years; however, it is rare that the data are adequate to establish a statistical relation between the rate at which erosion progresses on hillslopes and the factors determining these rates (*1*). A 7-year record of the downslope creep of rock fragments on eight hillslopes has been obtained as part of a study of the erosion and hydrology of landforms developed on the Mancos shale of Late Cretaceous age in western Colorado (*2, 3*), and these data illustrate the relation that exists between rates of rock creep and hillslope inclination, as well as the great variability that occurs among data collected in this manner.

Thin platy fragments of sandstone weather out of the Mancos shale and occur scattered over the shale hillslopes. One hundred and ten of these rock fragments, which ranged in thickness from 3 to 6 mm, and from 25 to 75 mm in maximum dimension, were marked with a small dot of aluminum paint. These rocks were then placed along transects normal to the hillslope contours, and the position of each rock was established with reference to metal stakes driven into the slopes. Hillslope angles were measured over a 30.5-cm (1-foot) length of slope extending downslope from each marker. The downslope movement of each rock was obtained by repeated measurements of its progressively increasing distance from a reference stake, which remained fixed in bedrock. The positions of the stakes did not change during the study, and therefore the bedrock was stable. The marked rocks were placed on the transects in 1958 and measurements of rock movement were made yearly in the spring and autumn between September 1958 and September 1962. A final measurement was made in September 1965.

The hillslopes on which the rocks were placed are located in two small drainage basins about 8 km northeast of Montrose in western Colorado. These basins drain to the west from a divide which forms the western edge of Bostwick Park (U.S. Geological Survey, Red Rock Canyon, 7½ minutes topographic map).

The Montrose study area has been described in some detail in previous publications (*3*). Annual precipitation at the Montrose (No. 2) weather station is 23.1 cm (9.1 inches). During the 7-year period of investigation 172.7 cm (68 inches) of precipitation fell at this station (*4*). Maximum relief within the drainage basins is about 76 m. The slopes are sparsely vegetated with shadscale, saltbush, and forbs. All hillslopes have a convex summit, and where they are not graded directly to a channel, they have a lower concave segment which is separated from the summit convexity by a straight segment.

The Mancos shale, which underlies all of the hillslopes, is a saline marine shale. Frost action in the weathered shale mantle is the major factor causing downslope creep of the weathered shale lithosol and the rocks. During the winter the surface of the slopes is loosened and heaved by the formation of granular ice crystals in the upper few inches of the lithosol. The loosened surface is subsequently compacted by summer rainbeat. This annual cycle of heaving and compaction is an effective mechanism causing episodic creep of the soil surface (*3*). Some disturbance of the slopes also occurs when sheep move through the area enroute to seasonal pastures.

Of the original 110 rocks placed on the slopes in 1958, 30 could not be located in 1965. Some of these were lost by burial, others were displaced from the profile by the trampling of livestock, humans, or other animals, and some undoubtedly were lost because their identifying paint mark was obliterated. Of the remaining 80 rocks, eight were eliminated from this analysis either because they showed an unusually large amount of movement during one measurement period, which suggested a disturbance other than that caused by the natural processes that generate creep, or because they showed a negative or upslope movement, which was considered adequate proof of disturbance.

The average rates of rock creep (velocity in millimeters per year), as measured for the 7-year period from September 1958 to September 1965, for each of the remaining 72 rocks are plotted against the sine of the angle of slope inclination in Fig. 1.

The rate of rock creep is directly proportional to the sine of hillslope inclination (s): velocity = 102 s −0.7, or essentially velocity = 100 s. The coefficient of correlation is .77, indicating that 59 percent of the total variation is explained by slope inclination alone. The coefficient of correlation is significant at well above the .001 level. The sine of the hillslope angle is used because it is proportional to that part of the total gravitational force which acts along the surface of the slope (*5*).

Earlier I had concluded that the relation between slope angle and marker movement was exponential (*3*). That relation was derived from an analysis of the data obtained during the period 1958 to 1961. I now consider it incorrect, because I calculated the relation by averaging the rates of movement of various types of markers (painted rock fragments, metal washers, and wooden blocks) which behaved differently. Raindrop impact accelerated the rate of movement of the wooden blocks, whereas many of the washers were buried and therefore moved more slowly than the rocks. No attempt was made at that time to consider each measurement separately, as the objective then was to demonstrate the episodic or seasonal rates of movement (*3*).

The plotted points of Fig. 1 scatter widely about the regression line, but when the data are averaged for 0.10 increments of the sine of hillslope inclination, these points fall near the regression line. When only these average values are considered, hillslope inclination explains 99 percent of the variation of the rate of rock creep. However, the scatter is not surprising in view of the accidents that can befall a given marker as it moves down the slope. For example, its movement can be retarded by the growth of annual vegetation or by being tipped into the desiccation cracks, which are common on the barren slopes. When trapped in the soil cracks, the markers assume the rate of creep of the soil; whereas the markers which rest on the surface move individually in direct response to surface heaving and compaction. Therefore, although the rate

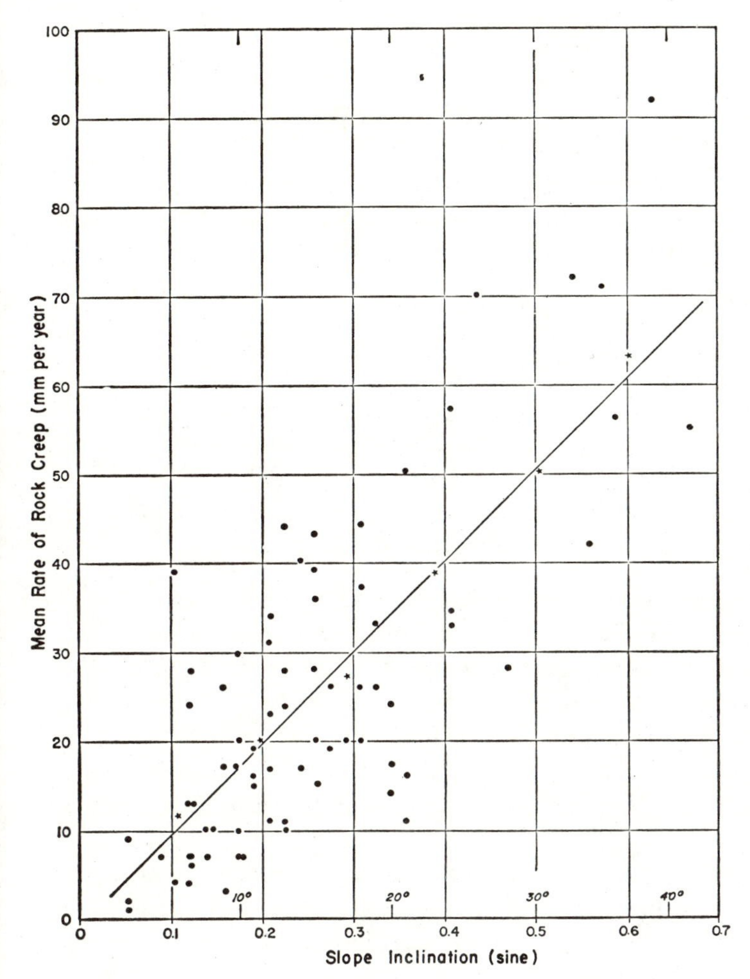

Fig. 1. Relation between the sine of slope inclination and the rate of movement of rock fragments on Mancos shale hillslopes in millimeters per year. Average rates of rock movement for 0.10 increments of the sine of slope inclination are shown as stars. Standard error is 11.8 mm. Correlation coefficient is .77.

of marker movement reflects the movement of the soil, the surface markers undoubtedly move more rapidly than the soil itself.

Notes were taken on circumstances which could have affected the movement of each marker, but there are no consistent relations between the nearness of vegetation or the orientation and the size of each marker on marker movement. All markers were located on slopes which faced to the south; thus a significant microclimatic variable was eliminated. The variability of the rates of rock creep, therefore, was assumed to be the result of variations in soil properties and microclimate at each marker location. For example, if at any given location on the hillslopes the soil was more susceptible to swelling, or if that part of the hillslope remained moister and was, as a result, more susceptible to frost action, then a marker at that location should move more rapidly than its neighbors.

To summarize, the measured rates of surficial rock creep on Mancos shale hillslopes are directly proportional to the component of gravitational force acting parallel to the hillslope. However, on slopes of identical inclination, a great range in the rate of movement occurs. This must be expected in any investigation of mass movement under natural conditions because of the variability of soil characteristics and microclimate, as well as the several accidental factors that can influence the rate of downslope movement. Perhaps this variability will always be greatest on poorly vegetated hillslopes; nevertheless, the scatter of the data indicate that if studies of rates of erosional processes are to be fruitful, not only must a large sample be obtained but also a large range of the independent variable must be included in the sample. For example, if only ten measurements of rock creep were obtained or if the markers were placed on slopes having only a 10-degree range in inclination, it is probable that no significant trend could be established from the data plotted in Fig. 1.

Finally, these rates of surficial rock creep, when compared with measurements of mass movement phenomena made elsewhere, indicate that the rates of rock creep occurring on the Mancos shale hillslopes in a region of low rainfall are more rapid than the very slow rates of soil creep measured in humid regions, but they are less rapid than the rates of talus creep and solifluction measured in the cold high latitudes (*5*).

S. A. SCHUMM

U.S. Geological Survey,
Federal Center, Denver, Colorado

References and Notes

1. S. A. Schumm, *J. Geol. Educ.* **14**, 98 (1966).
2. G. C. Lusby, G. T. Turner, J. R. Thompson, V. H. Reid, *Paper 1532-B* (1963).
3. S. A. Schumm and G. C. Lusby., *J. Geophys. Res.* **68**, 3655 (1963); *Z. Geomorphol. Suppl.* **5**, 215 (1964).
4. U.S. Weather Bureau, *Climatological Data, Colorado 63–70* (1959–1965).
5. A. N. Strahler, *Bull. Geol. Soc. Amer.* **67**, 571 (1956).
6. A. Rapp, *Geogr. Ann.* **42**, 67 (1960); A. Young, *Nature* **188**, 120 (1960); A. L. Washburn, *Bull. Geol. Soc. Amer. Special Paper* **63**, 292 (1962).
7. Publication authorized by the director, U.S. Geological Survey. We thank F. C. Ames and N. J. King, U.S. Geological Survey, who reviewed this paper; and R. N. Forbes, Jr., for preparing the figure.

18 November 1966

Reprinted from *Nach. Akad. Wissen. Gottingen, Math.-Physik. Klasse,* **13**, 195–210 (1963)

26

The debris slides at Ulvådal, western Norway An example of catastrophic slope processes in Scandinavia

By *Anders Rapp*

Department of Geography, University of Uppsala

Vorgelegt von H. Mortensen in der Sitzung vom 9. November 1962

"... each landslide is an opportunity to understand a little better the makeup of the earth and the history of its surface."

(E. B. Eckel in *Landslides and Engineering Practice,* p. 1).

This article is a preliminary report of a geomorphic case study, concerning certain slides and associated processes in a valley of the Scandinavian mountains. The slides at Ulvådal occurred during heavy rains on June 26, 1960. They were observed and reported by an eye-witness, Mr. B. Seyffarth of Oslo. The present author and Mr. Seyffarth examined the locality in July, 1961, together with Mr. E. Fridén and Mr. G. Larsson of Uppsala. The details of the erosion and deposition forms of the slide complex will be examined and described in another connection by G. Larsson.

As regards continuity, the sculpturing of mountain slopes is brought about by two different groups of processes, here simply called (a) continuous and (b) sporadic[1]. Among the latter, the largest cases can be called catastrophic. The continuous processes are those working on a slope every year (weathering, chemical solution and transportation, solifluction, talus creep, etc.). The catastrophic processes are extremely large, momentary and sporadic phenomena, such as large rockfalls, mudflows, slides, etc[2]. Both these types of processes must be considered in an analysis of slope development. The first-mentioned type can be examined, for instance, by recording the annual development within selected test areas in the field (cf. Schumm [1956], Rougerie [1960], Starkel [1960], Jahn [1961], Rapp [1961], etc.). Such investigations were recommended by the I. G. U. Commission on the Study of Slopes (cf. e. g. Macar and Birot [1955], p. 14 f.) as a main topic in the programme of this Commission. On the other hand,

[1] Cf. further discussion on this problem in Wolman and Miller (1960); Tricart (1961), and Rapp (1960).

[2] To most people the term "catastrophic" describes an event which in some way destroys human life or property (H. Poser, discussion in Göttingen). Here we think only of a catastrophic break in the continuous slope development.

the catastrophic events, as they are more rare and present more individual features can be examined and described as "type cases" whenever and wherever they have occurred. Examples of such detailed investigations of landslides and their continued development are the studies made at Mackenröder Spitze near Göttingen (Ackermann [1953], Mortensen and Hövermann [1956], Mortensen [1960]). Other investigations of landslides and their probable role in a cycle of landscape development are those by Wentworth (1943) and White (1949), both concerning steep slopes in tropical forest in Hawaii.

Careful analysis of fresh landslides and other mass-movements is a necessary basis for knowledge of some of the laws which control the sculpturing of slopes. In most cases of rapid mass-movements there are certainly both individual and general features to observe. Therefore a helpful starting-point for case studies like this could be the question: "What events and forms in this actual case are general and what are local or individual?" The purpose of such case studies must be to make constructive comparisons as a basis for general conclusions, and not to describe "unique" or "curious" events.

Terminology

Many types of landslides and similar rapid mass-movements can occur on slopes covered by loose deposits. The processes are often difficult to classify as in nearly every case they were not observed while in movement and as they are often transitional to each other. For these reasons the classification and terminology are not yet definitely established.

Among monographs on slides with classifications of slide types we may mention those by Heim (1933), Sharpe (1938) and Varnes (1958). Here we mainly refer to the terminology and classification in the last-mentioned paper, which is partly a further development from the two earlier ones.

The classification by Varnes is based on (a) the type of material involved; and (b) the type of movement. The main types of movement are three: fall, slide and flow (op. cit., p. 21 and Pl. 1). There are, according to this terminology, four types of mass-movements which are related to the Ulvådal slides, viz. "debris slides", "debris avalanches", "debris flows" and "mudflows". The first-mentioned type is classified as slide, the last three as flows. All these types form a gradational series from dry to wet masses and from sliding to flowing movement.

The Ulvådal slides seem to be either "debris slides" or "debris avalanches" if we use the terms from Sharpe and Varnes. But the distinctions between these two types are rather vague ("With increase in water content or with increasing velocity, debris slides grade into the flowing movement of debris avalanches" [Varnes, p. 29]). Furthermore, in the opinion of the present author, it is necessary to drop the term "debris avalanche" in this connection and use the word "avalanche" only for mass-movements of snow, either pure or mixed with avalanche debris (cf. Rapp [1961], p. 125). A third reason for calling the Ulvådal phenomena debris slides is that the main movement there was probably a sliding, not a flowing one, even if parts of the masses were occasionally moving by flowing (see further below).

It would seem to be a constructive task for future investigation to check if there is any need in the terminology for a special type of mass-movement in between "debris slides" and "debris flows" or "mudflows". If so, we could perhaps distinguish between "dry" and "wet" debris slides.

In this article the author prefers to include debris flows in the term "mudflow", in agreement with S. Sharpe (1938), R. Sharp (1942, p. 122) and Fryxell and Horberg (1943), for the following reasons:

1. The "mud" fraction is probably essential not only in flows of purely fine-grained earth but also in flows of bouldery material (debris flows). This may later be washed clean by rain (Sharp [1942], p. 226).
2. The term mudflow is short and easy to use, e.g. in combinations like mudflow levées, mudflow lobes, etc. This is not the case with the term "debris flow".

The first argument, however, is open to question.

Another way to clear up this terminology could be to keep the terms "mudflow" and "debris flow" as two sub-types, grouped under a new-coined, neutral, collective term.

The slide area

The Ulvådal valley is a southern branch of the great Romsdal valley in western Norway (Lat. 62° N., Long. 8° E., cf. Fig. 1). Like the whole landscape in this area it is strongly marked by glacial erosion and the Ulvådal valley itself is a glacial trough.

Table 1.

Precipitation (Verma, Lesjaskog) and air temperature (Aursjöen) measured at the weather stations nearest Ulvådal on June 20—27, 1960. Prec. = diurnal precipitation in mm. Temp. = air temperature in ° C, measured at 13ʰ. Source: personal communication from "Det Norske Meteorologiske Institutt", Oslo.

	Station	June							
		20	21	22	23	24	25	26	27
Prec.	*Verma*, alt. 265 m, 12 km NW. from Ulvådal	1	—	—	—	—	12	—	1
Prec.	*Lesjaskog III*, alt. 624 m, 20 km E. from Ulvådal	—	—	—	—	—	—	19	8
Temp.	*Aursjöen*, alt. 869 m, 40 km N. from Ulvådal	6.9	8.4	18.9	17.7	21.2	19.2	17.2	6.6

The annual precipitation within the area is not known, but can be estimated at about 1000 mm, to judge from the records at the nearest weather stations. Ålesund, on the coast, has about 1200 mm/year and Verma, in the Romsdal valley, at an altitude of 265 m, has about 750 mm (Norsk Meteorologisk Årbok).

The slide area is situated on the south-facing slopes of Mt. Ulvådalstind and Mt. Kabbetind, which reach an altitude of about 1400 m. Their upper parts consist of undulating hills of slightly boulder-strewn, gneiss rocks.

The upper parts of this valley side are mainly rocky slopes, partly walls with a height of 100—150 m. Below the walls are talus slopes of coarse boulders, forming a 1—2 m thick mantle upon fine-grained bottom moraine.

From an altitude of approximately 1000 m down to Lake Ulvådalsvatn (about 850 m a.s.l.) or the River Ulvån, the slope is covered by a thick layer of silty-sandy till (cf. Table 2). Some gully cuts show a thickness of 3—8 m in this till, which rests upon the glacially smoothed, gneiss surface. Most of the fresh slide tracks were formed in this part of the slope. Before the slides occurred the slope was covered by a forest of mountain birches *(Betula pubescens* subsp. *tortuosa)*, except for some small mires close to the lake or the river. The forest limit runs at about 1000 m a.s.l.

A particularly interesting feature is a system of old ice-lake shore-lines (or possibly: lateral drainage channels), which are visible along most of the valley side in the slide area at about 1000 m a.s.l. They were created at the deglaciation, probably about 8000—9000 years ago. As they have remained intact until now, they indicate that this slope has not been greatly affected by mass-movements during the previous part of the Holocene period. Similar systems of shore lines or lateral drainage channels from the end of the Würm glaciation occur in many valleys of the Scandinavian mountains and indicate the stability of many till-covered slopes.

Reconstruction of the events at the slide catastrophe

Some eye-witnesses observed the slides at Ulvådal when they occurred. One of these, Mr. B. Seyffarth, has given the present author much valuable information. The following reconstruction is partly based on Mr. Seyffarth's report, partly on the field studies made in 1961.

A thunderstorm on the evening of June 25, 1960, increased the water content in the ground, but no slides were released on this day. Another thunderstorm lasting 2—3 hours on the evening of June 26, 1960, brought such intense rains that expanding sheet-slides were released along the paths of occasional runnels.

The events can be reconstructed thus.

Phase 1 (slides)

Seyffarth (1960) gives a short description of this phase: "Suddenly we observe a new slide. A mass of earth, boulders, trees and water moves down the slope and a new slide track is formed . . . The river is filled with a porridge of

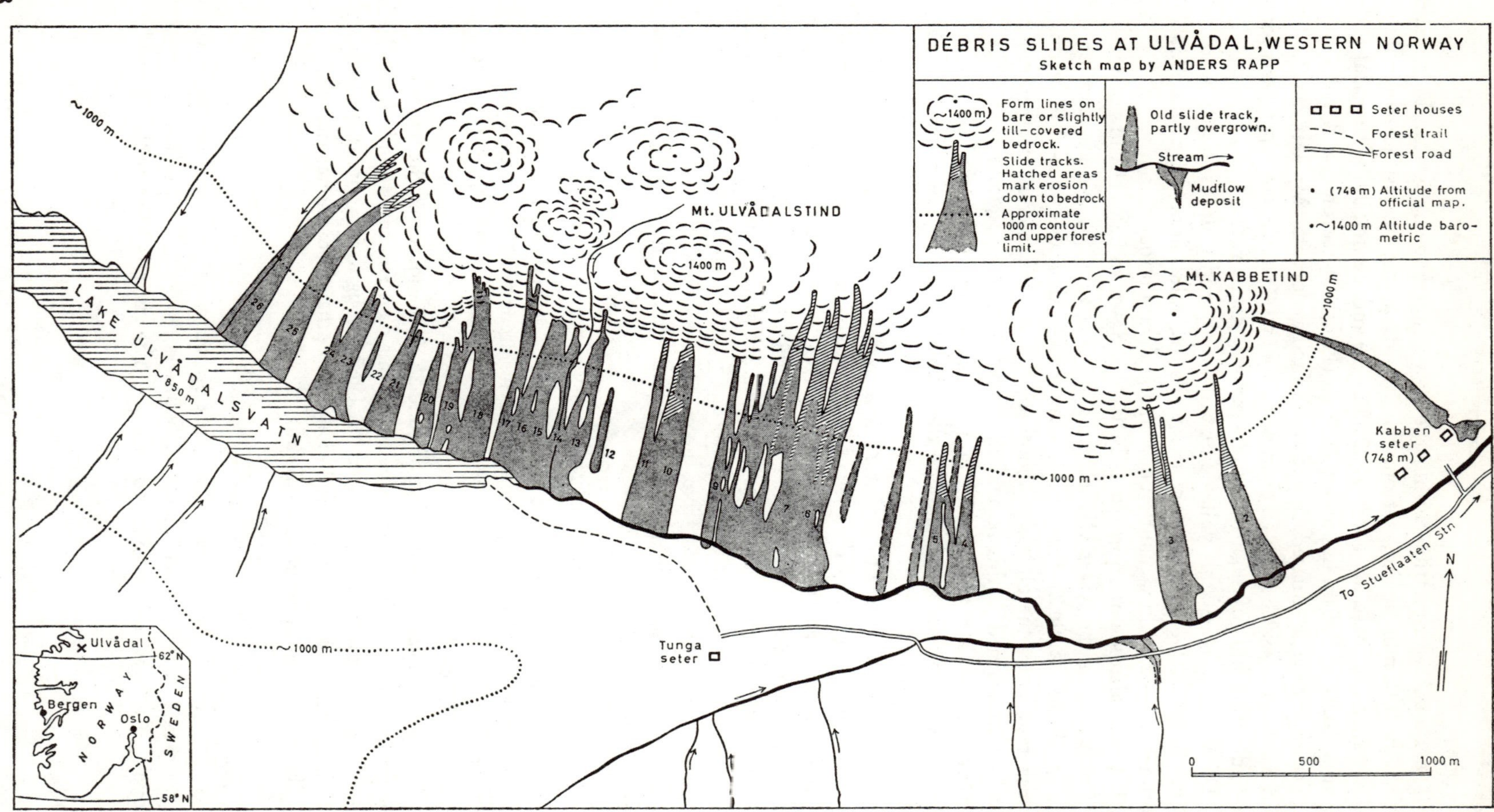

Figure 1
Sketch map of debris slides at Ulvådal, western Norway. Slide track no. 1 was revised by G. Larsson.

earth which flows downstream, mixed with a crowd of naked birch stems, twisting and whirling . . . New slides are coming down. It looks like a wave of water that squeezes earth and trees out of the ground and back again. The trees fall down immediately [Author's note: they were tilted *backwards,* cf. Fig. 3 : B]. Then they are pushed together with the earth and boulders on the way downslope, so they reach the river naked, without twigs and bark. Water sprays out in small cascades from the moving earth." The slides at the lake entered the water with such force that they caused large waves breaking on the opposite shore.

A reconstruction of the supposed slide movement is shown in Fig. 3.

The main moving mass was probably a large frontal lobe, gliding and rolling, heavily laden and lubricated by the surface water from the new slide track behind it, growing by incorporating frontal slabs. The removal of the superficial ground layer in a widening path along the whole slide course was probably caused in this way. In my opinion the almost equal depth of erosion along the whole path proves that the main type of movement was that of a *slide,* not of a flow. In the experience of the present writer mudflows (debris flows) usually move in relatively narrow paths and *often only transport* earth, not actively erode their substratum, due to their viscosity and low friction on the ground.

The mobilization of small slabs and folds in front of the main slide lobe is indicated by several facts: (a) the backward tilting of the birches, typical of the Ulvådal slides, according to Seyffarth, (b) the folded turf layers at the slide margins near the lake in some places (cf. Fig. 7), (c) the comparison with small slumps showing a proximal, primary, slide mass and a distal, secondary, pushed and folded slab (cf. Rapp [1961], Fig. 51, 52).

The supposed mechanism is probably comparable in some respects with the *progressive sliding,* described from large, deep, rotational slides, e.g. in the very detailed studies of slides on the Göta river in Sweden (e.g. Caldenius and Lundström [1958], p. 47).

This kind of movement may also help us to understand the reasons for the widening of most of the slides, their great range even on gentle slopes and their power to tear up the trees, roots and all without breaking the stems. A careful examination of the trees removed (wounds to stems and root systems, orientation of stems, etc.) in similar slide deposits would certainly be of great help in the reconstruction of the types of movement.

Some of the slides were locally narrower downslope, probably those that passed through old ravines, where the slide mass could not reach so far to the sides (e.g. No. 1 in Fig. 1).

Phase 2 (mudflows)

In those slides that passed through pre-existing ravines a part of the slide mass was probably left behind the main lobe. It stayed for a while, damming the brook, and then broke down into mudflows, forming levées and lobes upon the slide track after the main slide had gone.

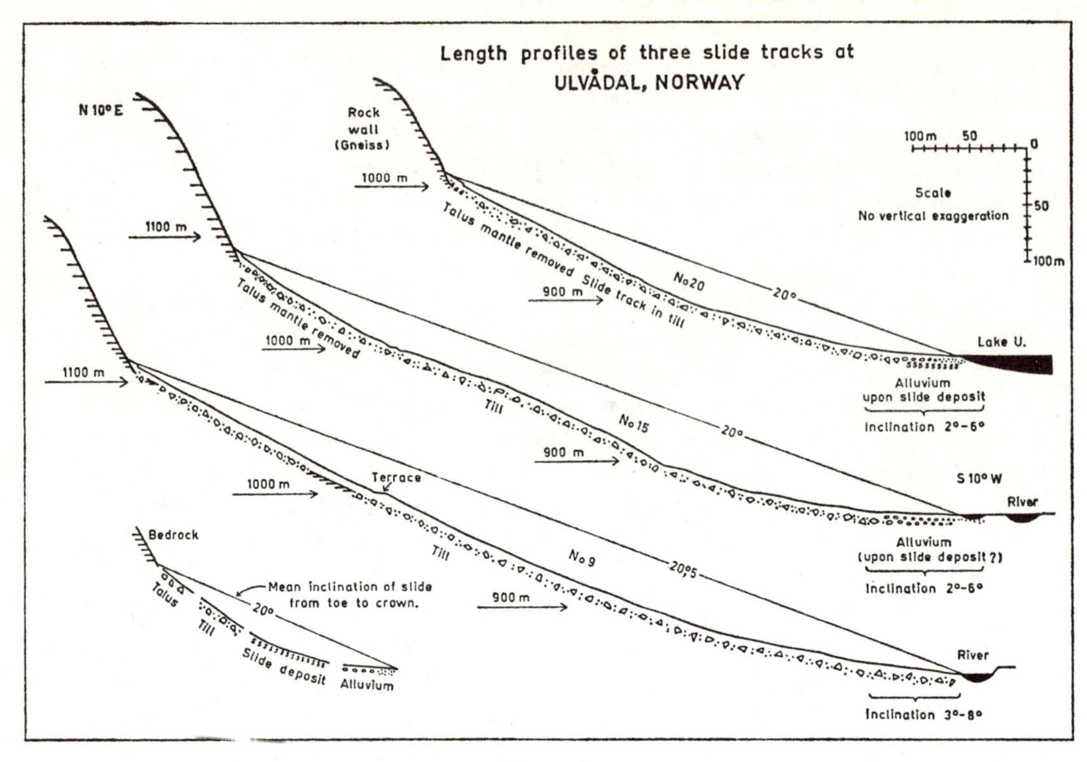

Figure 2
Length profiles of slide tracks nos. 9, 15, and 20.

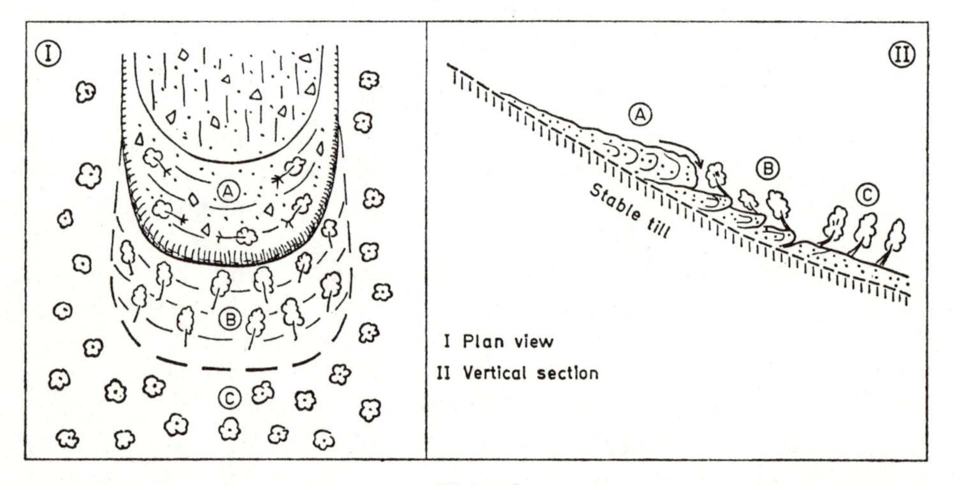

Figure 3
Sketch showing the supposed mechanism of slide movement in the expanding sheet-slides at Ulvådal. A = The main slide lobe, rolling and gliding. B = Frontal folds and thrusts removed by shearing in front of the main lobe. Birches are tilted backwards. C = Intact slope and trees.

Phase 3 (rill and gully erosion)

When protecting vegetation was removed by the slides, the till was open to erosion by the abundant supply of running water at the end of the thunderstorm. Slope wash, rill and gully erosion followed and the lower parts of the slide tracks were partly covered by alluvial deposits in sheets upon the slide masses (Fig. 6, 7, 8).

The slide tracks and their erosion and deposition forms

The sketch map (Fig. 1) shows the general form and position of the slide tracks. The map is based on an air photograph taken by "*Wideröe's flyveselskap*" on September 13, 1961. It is unfortunately the only photograph which covers almost the whole slide area, so stereoscopic mapping was not possible. Fig. 1 was drawn as an enlargement of the air picture after scale corrections taken from ground measurements of three slide tracks. The main topographical features of Mt. Ulvådalstind were drawn from air photographs under a mirror stereoscope.

On the map there are 26 fresh slide tracks marked by numbers. Outside the restricted, five-kilometre-long area of intense sliding there are only two similar slides further west in the Ulvådal valley and none on the north-facing side. There are no obvious differences in height, inclination, soils or vegetation between the slide area and the surrounding valley sides. Hence the limited area of slide tracks probably reflects a local maximum of convective rains in contrast to the wide areas affected by slides due to cyclonic rains (cf. Tricart [1958], Fig. 1; 1961, map; Rapp [1961] p. 161; the former was a rain catastrophe in the western Alps affecting the slopes within an area of at least 250 km in length [Durance to Lötschental] and the latter a similar catastrophe affecting an area of at least 50 km in width [Narvik to Abisko] in northern Scandinavia).

Unfortunately no precipitation records were made in Ulvådal. The small amounts of rain measured on the actual days at the nearest weather stations (Table 1) illustrate the local occurrence of the rains. However, there were intense thunderstorms, causing local slides and floods, at several places in western Norway on June 25—26, 1960. Examples of such cases were reported from Romsdal, Geiranger and Sogn (personal communication by Dr. F. Jörstad, Oslo).

The average length of the slides was estimated as 600 m, the width as 50 m and the thickness of the removed earth layer as 0.4 m. This makes a total slide mass of about 300 000 cubic metres, stripping off 75 hectares of land.

The complex erosion and deposition forms within the slide tracks show that mainly three different types of exogene processes cooperated in the catastrophe, viz. *slides*, *mudflows* (debris flows), and *rill erosion.*

a. Slide forms

The main erosion forms of the slides are the slide tracks. Nearly all of them are narrow in the beginning, widening downslope to 50—100 m width or more (cf. Fig. 1; nos. 2, 3, 25, 26, among others). This form is different from the sheet-

Figure 4
Debris slides nos. 9—19. Lake Ulvådalsvatn to the left. Photograph: A. R. 5. 7. 1961.

Figure 5*a* and *b*

Comparison of photographs taken before (*a*) and after (*b*) the slides at Ulvådal. Note that most of the large gullies existed before the slides occurred (cf. Fig. 6). Fig. 5*a* by H. Seyffarth in December, 1959. Fig. 5*b* by A. Rapp on 5. 7. 1961.

Figure 6

Slide tracks nos. 14—17. Note removal of talus mantle above the four ice-lake shore lines; dissection by deep gullies and small rills; deposits of light alluvial fans at the shore of the lake. Photograph: A. R. 6. 7. 1961.

Figure 7

Detail of the lowest part of slide track no. 20. From foreground towards center: light alluvium, dark slide masses with humus and twigs, intact mire, folded ridge of turf along margin of slide track no. 19. Photograph: A. R. 7. 7. 1961.

Figure 8
Detail of slide track no. 20, showing small rills due to slope wash. View upslope. Photograph: A. R. 5. 7. 1961.

Figure 9
Vertical gully cut through ancient ice-lake shore line (coarse material above note-book). Below is a fine-grained, compact till. From slide track no. 9. Photograph: A. R. 6. 7. 1961.

slides examined by the present writer in Kärkevagge, Lappland (Rapp [1961], p. 152), which generally started as broad slabs of moving earth, but mostly went down into ravines and continued as narrow mudflows etc. This is probably common on slopes which are steeper and more dissected than those at Ulvådal (cf. illustrations in Cholley [1956], Fig. 3a; Derruau [1956], Pl. V; Jäckli [1957], Fig. 40; Louis [1961], Bild 9; Tricart [1961], Fig. 7). The best example of a "narrowing" slide in Fig. 1 is in the lower part of track no. 1.

The superficial, podzolic layer of the earth was torn off all the way down to the lake or the river, except in slides nos. 1, 12 and 22, which did not extend so far.

The slide scars begin on four different sites.

1. Small slabs of till removed from sloping shelves on the rockwall (Fig. 4). The newly exposed bedrock has a light gray colour, in contrast to the other dark-gray rock surfaces. The light scars on the wall were at first erroneously interpreted as rockfall scars (Rapp [1961], Fig. 60). They mark the paths of occasional runnels during the thunderstorm in 1960.

2. Sheet-slides released from the talus mantle (Fig. 6). Some slides started on talus slopes, the debris of which rested upon bottom-moraine. E. g. in slides nos. 16—18 the talus layer was about 1—2 m thick, corresponding to a postglacial weathering and retreat probably less than 1 m, in the steep rockwall above.

These are interesting examples of the catastrophic type of talus removal, similar to cases described earlier from Scandinavia (e. g. Rudberg [1950], p. 144; Rapp [1961], p. 162).

No sliding occurred where the largest brook descended the talus slope (tracks nos. 13—14), probably because this watercourse was already adapted to a heavy water supply, which had earlier removed any possible fine-grained till.

3. Expanding sheet-slides formed by the thin cover of till removed from bedrock slopes inclined at 15—30° and situated above the forest limit (e. g. tracks nos. 6—8, 25, 26). The till mantle there has a thickness of only half a metre or less and is covered by heather vegetation (*Empetrum*, mosses, scattered *Betula nana* and *Salix* scrubs).

The newly exposed bedrock is glacially smoothed and practically unaffected by weathering in postglacial time. The slides did not start in the paths of the permanent watercourses, but between them, as is seen on the western part of Mt. Ulvådalstind (Fig. 1).

The large areas of newly exposed bedrock above the forest limit, for instance, at the beginning of tracks nos. 6—8, is an indication that the sliding of 1960 was one of the greatest in postglacial time at this locality.

4. It is important to note that some few, narrow slides hat started in the birch forest (e. g. nos. 12, 22), in spite of the stabilizing influence of the tree roots.

The main slide planes were probably the bottom of the podzolic layer and the top of the fine-grained substratum till. The podzol profile is very distinct on the till slope covered by birch forest. At the sides of the slide tracks the average thickness of the podzol profile was 50—60 cm, with a rust-coloured B-layer about 30—50 cm thick, resting upon gray, fine-grained and compact substratum till (for grainsizes, cf. Table 2).

Table 2.

Grain size composition in samples from the slopes at Ulvådal. The figures give the fractions in per cent of weight. Pebbles are excluded. The fraction 32—2 mm is put together in one column. Sample no. 5 was taken and analyzed by E. Fridén.

Sample	Percent of sample remaining on sieve						
	32—2 mm	>1 mm	>0.5 mm	>0.25 mm	>0.125 mm	>0.062 mm	<0.062 mm
Podzol, in situ							
1. B-horizon, 880 m alt.	36.2	8.1	8.6	10.2	12.1	8.7	16.1
2. B-horizon, 1000 m alt.	10.4	4.6	7.8	25.0	39.2	10.3	2.6
Till below slide track							
3. 1 m depth, 880 m alt.	23.4	5.5	5.6	6.8	12.1	12.7	33.9
4. 1 m depth, 1000 m alt.	2.4	0.9	1.3	3.3	10.8	15.1	65.9
5. 2 m depth	43.9	7.2	7.8	8.4	9.5	7.5	15.7
Mudflow							
6. Fines in mud-flow lobe, slide 19	29.4	8.3	9.2	10.7	14.2	9.2	18.9
Alluvium							
7. Distal part of alluvial fan	14.1	7.6	10.4	20.5	25.1	12.7	9.3

Several facts indicate that the main slide mass was the podzolic layer. This was obvious along the margins of the slide tracks. But also in the central parts the tracks were partly covered by a thin layer of intact rusty soil. Furthermore the birch stems removed were generally not cut off or broken but had been torn out together with the network of roots about half a metre in thickness. The ice-lake terraces on the upper part of the slope were not entirely obliterated by the slides, a fact which also indicates the thinness of the slide sheets (Fig. 9).

The slides described as soil avalanches by Wentworth (1943) in Hawaii were similar to the Ulvådal slides in many respects: they were thin (only 30—90 cm thick layer) and broad (about 10—50 m), removed the forest vegetation, in many cases slid directly down in the main stream and were "due to frictional failure on a surface close to and essentially parallel to the topographic surface" (op. cit. p. 58). But they were different in some respects too: they came from steeper slopes (40—50°) and slid all the way upon bedrock (op. cit. Pl. 2).

The slide deposits were mainly of two kinds: a) irregular ridges of boulders, birch stems and earth on the margin of the lower part of some slides (e.g. western margin of no. 19) and b) outspread slide or flow masses consisting of a mixture of boulders and fines, humus and twigs. These masses were coloured dark by the humus and were probably to a great extent hidden under water-transported alluvium, deposited after the main slides had passed (cf. below and Fig. 7).

The earth masses of type b generally showed flow structures with a preferred orientation of the twigs, indicating that, just before the deposition, it had been flowing like water-saturated porridge. The report and photographs by Seyffarth show that during the catastrophe a large quantity of slide material and trees filled the outlet of the lake as well as the river and was transported downstream. The lake was also dammed and raised its level 2—3 m. The damming persisted in 1962.

b) Mudflow forms

Levées and lobes from bouldery mudflows (debris flows) occurred on many slide tracks, particularly on the lower parts of those with gullies. Two large mudflow lobes were examined on track no. 19. They were about 10 m wide, 3—4 m high at the front and consisted of boulders 0.2—1 m in size, mixed with fine-grained earth, particularly in the core of the lobes.

The mudflows probably moved down immediately after the main slides had passed. Small mudflow levées, probably from 1961, were observed on track no. 16. They were about 1 m wide.

Only two large mudflow deposits were observed in the vicinity. One of them was at "Kaffevika" on the south side of Lake Ulvådalsvatn, the other almost opposite to slide no. 3 in Fig. 1. They followed steep brooks in a narrow path and accumulated bouldery levées and lobes in the valley bottom, typical of mudflows but in contrast to the slides. The mudflow forms observed were similar to those shown in Figs. 2—4 in Sharp (1942).

c) Erosion and deposition by running water

The traces of rill and gully erosion were distinct and common. In all slide tracks we found a network of small, anastomosing rills, marked by strings of washed stones (Fig. 8). The pattern resembled small stone stripes at intervals of 1/2 to 1 m. Similar features could later be misinterpreted as stone stripes formed by frost action. The orientation of stones is probably different in "frost-caused" stripes.

In many slide tracks were large gullies, 3—8 m deep and 10—40 m wide, in many cases reaching down to the bedrock (Fig. 6). *Most of these large gullies existed before 1960, as is clear from photographs taken in earlier years* (Fig. 5 a, b). They were, however, widened in connection with the slides.

The fine sediments from slides and slope wash were transported by the river to the sea, where they made the water of the Romsdal fjord muddy for many miles (Seyffarth [1960]). But the coarse material (sand, gravel, pebbles) transported by running water on the slide tracks, was deposited as alluvial fans in the lower parts of the tracks. These fans were in many cases the predominant visible accumulation form, probably in part deposited upon dark slide masses (Figs. 6, 7).

As long as the tracks remain naked, without protecting vegetation, the slope wash and gully erosion will probably continue in connection with heavy rains and rapid snow-melting.

The continued sculpturing of the naked slopes will be an interesting object of study from many points of view. a) It may serve as an analogy of the development immediately after the deglaciation, when we may suppose intense slope processes due to lack of vegetation shelter. b) The colonization of plants on the slide tracks at Ulvådal will also furnish interesting parallels to the growth of vegetation on the new ground, exposed after the melting of the inland ice. c) The open tracks will possibly develop by increasing dissection and gullying into y-shaped ravines, like those found in many other areas. If so, the slides have been a link in a cyclic development from smooth, simple trough-valley sides towards dissected ravine topography (Fig. 6).

Discussion and concluding remarks

The complex slides at Ulvådal in 1960 are a good example of the important role played by sporadic mass-wasting in valley widening and slope development. They also give an idea of the geomorphic effect of heavy, local rains in high-latitude mountains. Even if such processes occur at very long intervals—perhaps is this event a local maximum of mass-wasting for the post-glacial period—they have a *predominant and enduring effect upon the morphology and stratigraphy of the slope deposits.*

This and similar events learn, that stratigraphic horizons of alluvium and slide-masses can be formed after only one extreme weather situation. Hence one must be careful with interpretations of similar deposits as indicating a climatic change towards more humid climate, or forest cutting, etc.

And they are not unique. There are several almost overgrown tracks from old slides also at Ulvådal, e.g. between the new tracks nos. 5 and 6 (Fig. 1). Other examples of slide catastrophes which can be traced both in historical tradition and in the stratigraphy of slopes are not rare in the precipitous valleys of western Norway. "Slide-years are often recalled in local history and tradition, and they have sometimes been of a disastrous character. Thus in 1742 a third of the cultivated area of the Opedal farms was covered by material from higher sites. The last slide-year was 1940, when landslides occurred at different localities on the night of 25/26 November" (Låg and Oland [1960], p. 71). This quotation concerns the Hardanger area.

The cooperation of three different slope processes (slides, mudflows, rill erosion) is probably a typical feature in connection with heavy rains on mountain slopes and is reflected both in the erosion and the accumulation forms. The slide types at Ulvådal are also known from the northern Scandinavian mountains, as is shown by similar events, e.g. at Mt. Silkentjakk and Mt. Ullersjöklumpen in Jämtland (Ängeby [1947], p. 102), Lake Ajaure, Lappland (Rudberg [1950]), Kaisepakte, Lappland (Rapp [1961], p. 162) and Ladtjojaure, Lappland.

The Ulvådal slides of 1960, together with the similar cases mentioned above prove that natural birch forest is only a relative but no absolute stabilizer of mountain slopes in Scandinavia. In the period immediately after the deglaciation, when the till-covered slopes were not yet overgrown by forest (the "pre-forest" period), various slides, mudflows, gullying and other exogene processes were probably very active. It is curious that the slopes of Ulvådal were not greatly affected by slides at that time, for instance, to such a degree as to obliterate the ice-lake shore lines.

Why were the slides delayed until the present time and why were the conditions especially favourable for their release in June 1960? No definite answer can be given here to these questions, only some reasonable hypotheses.

1. *The rains* of June 1960 were probably more intense than those occurring at the locality in the "pre-forest" time.

2. The development of a thick *podzol profile* in the till perhaps increased the instability of the earth on the slope. A thick podzol profile with high pore volume and water-bearing capacity which rests upon a fine-grained, compact till is probably a favourable combination to sliding of this type. The instability will probably increase with the growing development of the podzol profile, that is, increase with time. The relation between slide planes and podzol profiles is worthy of a closer examination in this and similar cases.

But the possible influence of the podzol profile was only a contributary factor, as is indicated by the many slide tracks above the forest limit. There the smooth bedrock surfaces functioned as slide planes. This shows that the rains were of such a great intensity that they could release slides without "help" from podzol profiles, layers of frozen ground etc.

3. The eye-witnesses observed lumps of *frozen ground* in the slide masses that had passed across the outlet of Lake Ulvådalsvatn. It was therefore thought that remaining layers of frozen ground could have served as slide planes, to a great extent facilitating the lateral and forward expansion of the slides. In the author's opinion this was not the case. The lumps of frozen ground probably came from peat hummocks in the mires near the lake. In similar undisturbed hummocks between the slide tracks there was frozen peat below a depth of 40 cm on July 4, 1961. No other frozen ground could be traced by digging at that time, either in the mires, or in the birch forest and above the forest limit. Actually the month of June 1960 had been unusually warm, without any freezing temperatures at all in the records from Aursjöen, the nearest temperature station (cf. Table 1).

The other three cases of similar slides in Lappland, mentioned above also occurred during heavy, local rains at the end of June or beginning of July[3], that is to say, shortly after the thawing of the active layer of the ground. P. J. Williams (1959, p. 489) in a study of solifluction concluded that: "A considerable part of the loss of strength at thaw in frost-heaved soil is due to the reduction in the cohesion component following the separation of particles by frost heave." Probably the active layer of the ground is most sensitive not only to solifluction but also to sheet-slides as in Ulvådal released by water saturation shortly after thawing.

A combination of intense freezing of the ground in winter and heavy rains immediately after thawing in early summer is perhaps the most favourable situation for release of "expanding sheet-slides".

In the case of Ulvådal this hypothesis is in good agreement with the conditions during the preceding winter of 1959—1960, characterized by intense freezing of the ground due to unusually poor snow cover (B. Seyffarth, oral communication; cf. also Fig. 5 a).

Acknowledgements

The writer has received much valuable help, particularly from B. Seyffarth of Oslo, who first observed and described the slides and later was our host in his family's hut at Ulvådal. He also assisted in the field work. Much valuable information and other help was also received from F. Jörstad of Oslo, G. Östrem of Stockholm, E. Fridén and G. Larsson of Uppsala and last but not least, from many colleagues at the meeting of the Slope Commission in Göttingen, August, 1962.

N. Tomkinson checked the English text. W. Tiit drew Fig. 1 and E. Ludwigsen Figs. 2 and 3. Financial support was received from Liljewalch's fund, University of Uppsala.

To all persons and institutions mentioned the author wishes to express his sincere thanks.

Zusammenfassung:

Die Erdrutsche im Ulvådal, West-Norwegen
Ein Beispiel katastrophaler Hangentwicklungsprozesse in Skandinavien

Der Artikel ist ein vorläufiger Bericht über die geomorphologischen Wirkungen der Erdrutsche im Ulvådal (Norwegen) am 26. Juni 1960 (Fig. 1). Sie wurden von B. Seyffarth, Oslo, direkt beobachtet und beschrieben und werden von G. Larsson, Uppsala, weiter untersucht.

[3] The dates of these slides were: Lake Ladtjojaure, June 24, 1922 (oral communication by P. Niia); Lake Ajaure, June 25, 1947 (Rudberg [1950], p. 140); Kaisepakte, Abisko, July 8, 1956 (Rapp [1961], p. 162).

Die Rutsche im Ulvådal werden vom Verfasser als „expandierende Schollenrutsche" bezeichnet, und zwar aus folgenden Gründen: 1. Sie hatten überwiegend gleitende, nicht strömende Bewegung (Fig. 3). Es handelt sich also um Rutsche. 2. Die abgerutschte Schicht ist nur ca. 1/2 m mächtig und umfaßt hauptsächlich die Podsolhorizonte mit Bäumen und Wurzeln. 3. Die Breite der Rutsche wurde hangab größer, bis zu 100 m oder mehr (Fig. 1).

Die Erosions- und Akkumulationsformen deuten ein Zusammenwirken von drei verschiedenen Hanprozessen an: Rutsche, Murgänge und Wassererosion. Innerhalb kürzester Zeit haben sich hier durch extreme Starkregen sehr große morphologische Veränderungen vollzogen, die vielleicht ein lokales Maximum der gesamten Postglazialzeit bezeichnen. Möglicherweise wird durch die Rutsche eine weitgehende Runsenbildung und Zerschneidung des Hanges eingeleitet.

Folgende Faktoren trugen vermutlich zur Auslösung der Vorgänge bei: 1. Die Heftigkeit des Regens. 2. Gute Gleitflächen im Kontakt zwischen dem porösen Podsolboden und der unterliegenden, feinkörnigen Moräne. 3. Der labile Zustand der Bodenlagen kurz nach dem jährlichen Auftauen.

Ähnliche ältere Rutsche sind von mehreren Stellen in Skandinavien bekannt; sie waren vielleicht in den Hochgebirgstälern kurz nach dem Abschmelzen des Inlandeises eine verbreitete Erscheinung.

References

Ackermann, E.: Der aktive Bergrutsch südlich der Mackenröder Spitze in geologischer Sicht. Nachr. d. Akad. d. Wissensch. in Göttingen. Math.-Phys. Klasse. 5, 1953. Göttingen 1953.

Caldenius, C. and Lundström, R.: The landslide at Surte on the river Göta älv. Sveriges Geol. Unders., Ca 27. Stockholm 1956.

Cholley, A. and others: Relief form atlas. Inst. Geograph. Nationale. Paris 1956.

Derruau, M.: Précis de Géomorphologie. Masson et Cie. Paris 1956.

Eckel, E. B.: Introduction. Landslides and Engineering Practice. Highway Res. Board, Spec. Report 29. Washington 1958.

Fryxell, F. and Horberg, L.: Alpine mudflows in Grand Teton National Park, Wyoming. Bull. Geol. Soc. Am. 54:1. New York 1943.

Heim, A.: Bergsturz und Menschenleben. Vierteljahrsschrift d. Nat.forsch. Gesellschaft Zürich, 77, 1932.

Jahn, A.: Quantitative analysis of some periglacial processes in Spitsbergen. Nauka o Ziemi II, ser. B 5. Warszawa 1961.

Jäckli, H.: Gegenwartsgeologie des bündnerischen Rheingebietes. Beiträge zur Geol. der Schweiz, geotechn. Serie 36. Bern 1957.

Louis, H.: Allgemeine Geomorphologie. 2. Aufl. Berlin 1961.

Låg, J. and Oland, K.: Natural and man-made changes in the cultivated soils in Ullensvang. Chapter 6 in "Vestlandet. Geographical studies". Bergen 1960.

Macar, P. and Birot, P.: Commission on the evolution of slopes. I.G.U. Newsletter, VI. 1955.

Mortensen, H.: Neues über den Bergrutsch südlich der Mackenröder Spitze ... Ztschr. f. Geomorph. Suppl. 1. Berlin 1960.

Mortensen, H. and Hövermann, J.: Der Bergrutsch an der Mackenröder Spitze bei Göttingen. Prem. rapp. de la comm. pour l'étude des versants. Amsterdam 1956.

Norsk Meteorologisk Årbok. Oslo.

Rapp, A.: Literature on slope denudation in Finland, Iceland, Norway, Spitsbergen and Sweden. Ztschr. für Geomorph., Suppl. 1. Berlin 1960.

—: Recent development of mountain slopes in Kärkevagge and surroundings, northern Scandinavia. Geograf. Annaler, XLII 2—3. Stockholm 1961.

Rougerie, C.: Le façonnement actuel des modelés en Côte d'Ivoire forestière. Mém. de l'Inst. Franç. d'Afrique Noire, 58. Dakar 1960.

Rudberg, S.: Ett par fall av skred och ravinbildning i Västerbottens fjälltrakter. Geol. Fören. Förhandl. 72:2. Stockholm 1950.

Schumm, S. A.: The role of creep and rainwash on the retreat of badland slopes. Am. Journ. of Science, 254. 1956.

Seyffarth, B.: Fjellsiden som raknet. Article in the Norwegian newspaper Aftenposten 5.11. 1960. Oslo 1960.

Sharp, R. P.: Mudflow levées. Journ. of Geomorph. V:3. New York 1942.

Sharpe, C. F. S.: Landslides and related phenomena. New York 1938.

Starkel, L.: Rozwoj rzezby Karpat Fliszowych w Holocenie. English summary: The development of the Flysch Carpathians relief during the Holocene. Polska Akad. Nauk, Inst. Geogr., Prace Geogr. 22, Warszawa 1960.

Tricart, J.: La crue de la Mi-Juin 1957 sur le Guil, l'Ubaye et la Cerveyrette. Rév. de Géographie Alpine. XLVI. Grenoble 1958.

—: Mécanismes normaux et phénomènes catastrophiques dans l'évolution des versants du bassin du Guil (Htes-Alpes, France). Ztschr. für Geomorph. 5, 4. Berlin 1961.

Varnes, D. J.: Landslide types and processes. Chapter 3 in "Landslides and Engineering Practice. 1958 (See further: Eckel, E. B.).

Wentworth, C.: Soil avalanches on Oahu, Hawaii. Bull. Geol. Soc. Am. 54:1. New York 1943.

White, S.: Processes of erosion on steep slopes of Oahu, Hawaii. Am. Journ. of Science, 247. New Haven 1949.

Williams, P. J.: An investigation into processes occurring in solifluction. Am. Journ. of Science, 257:7, 1959.

Wolman, M. G. and Miller, J. P.: Magnitude and frequency of forces in geomorphic processes. Journ. of Geology, 68:1. Chicago 1960.

Ångeby, O.: Landformerna i nordvästra Jämtland. Lunds Univ. Geogr. Inst. Avhand. XII. Lund 1947.

Air photographs nos. K 11: 1228, I 12—13: 1228 taken by Wideröe's Flyveselskap A/S. Oslo.

Reprinted from *Geotechnique*, **12**, 251–263 269–270 (1962)

STABILITY OF STEEP SLOPES ON HARD UNWEATHERED ROCK

27

by

PROFESSOR KARL TERZAGHI, Hon.M.A.S.C.E., M.I.C.E.

SYNOPSIS

This Paper contains a discussion of the factors which determine the degree of stability of steep slopes on hard unweathered rock. These factors include the angle of shearing resistance of the jointed rock, the effective cohesion, and the seepage pressures exerted by the water percolating through the joints of the rock. The Paper also deals with concealed sources of instability which may exist beneath slopes in deep valleys located between high mountains.

The angle of shearing resistance can be estimated on the basis of the results of a joint survey. The effective cohesion cannot be determined by any of the presently available procedures of rock exploration, but its influence on the stability of slopes on jointed rock is commonly much less important than that of the angle of shearing resistance.

The most unpredictable factor determining slope stability is the hydrostatic pressure in the water flowing out of a reservoir or a leaking pressure tunnel through the joints towards a slope, because this pressure depends on the details of the pattern of seepage which is commonly very erratic. Hence, if the slope is located next to and downstream from the abutment of a storage dam, sound engineering requires the elimination of these pressures by radical drainage. If the slope is located near a pressure tunnel, a reliably watertight lining should be installed in those parts of the tunnel from which leakage could have a significant influence on the stability of a slope.

In some very deep valleys located between high mountains, the rocks underlying the slopes of the valley have been displaced and damaged by movement along deep-seated surfaces of sliding. Hence no dam should be built in any deep valley unless a geological survey supplemented by borings has demonstrated conclusively that such movements have not taken place.

Cet exposé a pour objet une discussion des facteurs qui déterminent le degré de stabilité des talus raides sur de la roche non decomposée. Ces facteurs comprennent l'angle de résistance au cisaillement de l'ensemble de la roche, la cohésion efficace, et les pressions de filtration exercées par l'eau filtrant à travers des joints de la roche. L'Exposé a trait aussi aux causes d'instabilité cachées qui peuvent exister sous des talus dans des vallées profondes situées entre de hautes montagnes.

L'angle de résistance au cisaillement peut être évalué à partir des résultats d'une levée détaillée des joints. La cohésion efficace ne peut être évaluée par l'une des méthodes disponibles actuelles d'exploration de la roche, mais son influence sur la stabilité des talus sur de la roche fissurée est d'une facon générale bien moins importante que celle de l'angle de résistance au cisaillement.

Le facteur le plus imprévisible déterminant la stabilité d'un talus est la pression hydrostatique dans l'eau coulant d'un réservoir ou d'un tunnel à pression ayant des fuites à travers les joints vers un talus, parce que cette pression dépend des détails du genre de filtration qui sont communément très imprévisibles. Par suite, si le talus est situé près et en aval de la culée d'un barrage d'accumulation, les principes éprouvés de construction nécessitent l'élimination des pressions par un drainage de base. Si le talus est situé près d'un tunnel de pression, un revêtement dont l'étanchéité est sûre doit être installé dans ces parties du tunnel d'où la filtration pourrait avoir une influence considérable sur la stabilité du talus.

Dans certaines vallées très profondes situées entre de hautes montagnes, les rochers formant le soubassement des pentes de la vallée ont été déplacés et ont été endommagés à la suite de mouvements le long de surfaces de glissement à assise profonde. Par suite, aucun barrage ne devrait être construit dans d'importe quelle vallée profonde avant qu'un relèvement géologique complété par des forages n'ait démontré d'une façon convaincante que de tels mouvements n'ont pas eu lieu.

INTRODUCTION

The failure of the Malpasset Dam and various catastrophic rock slides on slopes above the portals of pressure tunnels have aroused the concern of public building authorities. Consequently they have become more and more reluctant to issue permits to construct on rock unless the applicants demonstrate by means of stability computations that the proposed structure will not cause a rock failure. Yet, in connexion with foundations on rock as well as on soil, natural conditions may preclude the possibility of securing all the data required for predicting the performance of the real foundation material by analytical or any other methods. If a stability computation is required under these conditions, it is necessarily based on assumptions which have little in common with reality. Such computations do more harm than good because they divert the designer's attention from the inevitable but important gaps in his knowledge of the factors which determine the stability of slopes on hard, unweathered rock. This Paper deals with these gaps and with the means at the engineers' disposal for minimizing their consequences.

SIGNIFICANT PROPERTIES OF UNWEATHERED ROCK

The term "hard unweathered rock" is arbitrarily applied to a chemically intact rock with an unconfined compressive strength q_u of more than 5,000 p.s.i. The term as used herein is further limited to those rocks which do not undergo a perceptible volume change upon drying and subsequent saturation. Hence the majority of the rocks classified as shale are beyond the scope of this Paper.

The term "significant properties" will be limited to those properties which have a significant influence on the stability of slopes on the rock.

If an unweathered rock is also mechanically intact, the critical height H_c of a vertical slope on the weakest rock ($q_u = 5{,}000$ p.s.i. unit weight $w_r = 170$ lb/cu. ft) would be very roughly:

$$H_c = \frac{q_u}{w_r} = 4{,}200 \text{ ft} \qquad (1)$$

For intact hard rocks such as granite, H_c is several times greater. Yet no vertical slopes with such height exist, and many gentler slopes with a much smaller height than H_c have failed. This fact indicates that the critical height of slopes on unweathered rock is determined by the mechanical defects of the rock such as joints and faults, and not by the strength of the rock itself.

Joints subdivide the rock into individual blocks which almost fit each other, and the cohesive bond across the joints is equal to zero. The joints may be continuous or discontinuous. A body of rock—for instance the body of rock underlying a slope within a distance approximately equal to the height of the slope—is said to contain continuous joints if it is possible to construct sections across the body which nowhere cut across intact rock. These sections may be approximately plane, irregularly warped, or stepped-up like similar sections across a body of dry brick masonry.

Continuous joints are present in most rocks even at great depth. If a continuous joint is encountered in a tunnel its presence is indicated by a seep or the emergence of a spring. Since perfectly dry tunnels are very rare it can be concluded that continuous joints are almost universally present. For reasons to be discussed under the following heading, the spacing between continuous joints in the proximity of steep slopes is much smaller than beyond the range of influence of the presence of the slope on the stress conditions in the rock. If the continuous joints form a three-dimensional network they transform the rock into a cohesionless aggregate of blocks comparable to dry masonry. With decreasing spacing of continuous joints the influence of these joints on the stability of slopes increases.

In addition to continuous joints, every body of rock contains more or less discontinuous joints with very different degrees of discontinuity. A section following a discontinuous joint cannot enter an adjacent joint without cutting across intact rock. Those portions of a section which are located in intact rock will be referred to as *gaps*. The ratio between the area of the joints located in a section and the total area of the section represents the *effective joint area* of the rock formation along the section. A shear failure along the section is resisted by both pressure-conditioned shearing resistance and the cohesion of the rock located in the gaps between joints. This cohesion will be referred to as *effective cohesion*, c_i. If:

c = cohesion of intact rock,

A = total area of the section through the rock, and

A_g = total area of the gaps within the section,

$$c_i = c\frac{A_g}{A} \qquad (2)$$

The relationship expressed by this equation is of merely theoretical interest because, first, it is impracticable to determine the value A_g for a given section through the rock, and second, for any given rock formation the value of c_i for sections approximately parallel to any given plane may have any value greater than zero.

On account of the decisive influence of the joint pattern on the stability of slopes on unweathered rock it is essential to get as much information concerning the characteristic features of the joint system as our means for rock exploration permit. The first step is a joint survey, involving the determination of dip and strike of a great number of joints exposed on outcrops, in open cuts, in tunnels, and in boreholes. The results of the survey are represented in polar diagrams, using one of several conventional procedures. These methods as well as the sources of error involved in the joint survey and the means for reducing them are discussed in a Paper by Ruth D. Terzaghi (as yet unpublished).

The results of joint surveys for any one area generally show that many joints in the area are approximately parallel to one of two or more intersecting planes. The joints approximately parallel to one of these planes constitute a *set*. If the orientation of the sets does not change significantly over a large area, the sets are said to constitute the *regional joint pattern*.

In some massive (non-stratified) bodies of rock one or more distinct sets of joints may be present whereas in others the joints have practically random orientation. In stratified sedimentary rocks one set of joints is commonly parallel to the bedding and the joints of this set are likely to be continuous over large areas. These will be called *bedding joints*. The bedding joints are commonly associated with two or more sets of much less continuous joints intersecting the bedding joints. These will be called *cross-joints*.

In many areas, the joints of one set are wider and more continuous than the others. Some of the open joints are locally filled with products of weathering. Furthermore, within the same rock formation the average spacing between joints may vary between wide limits. Owing to variations in these details, joint systems with identical patterns may have very different effects on the shear characteristics of jointed rock. Yet, at the very best, these details are known at only a few spots or along a few lines. Hence from an engineering point of view even the results of a conscientious joint survey still leave a wide margin for interpretation.

In the proximity of steep slopes the shear characteristics of jointed rock are further complicated and modified by the effects of the intense shearing stresses produced by the removal of the lateral support of the rock adjacent to the slope. These stresses break many of the cohesive bonds connecting the blocks between joints, thus reducing the average effective cohesion of the jointed rock. In the vicinity of the slope the stresses are also likely to produce a local joint system, superimposed on the regional one. Within the depth of frost penetration and of significant daily and seasonal temperature variations the effective cohesion of the rock may be completely eliminated. These relatively superficial but very important modifications of the regional joint system will be discussed under the following heading.

ROCK CONDITIONS IN PROXIMITY OF STEEP SLOPES

The condition of the rock in the proximity of steep slopes depends chiefly on the type of rock and the geological history of the slope. The overwhelming majority of steep slopes on rock were formed by stream, glacier, or wave erosion. Exceptions are fault scarps, the slopes forming the outer boundaries of flows of extrusive igneous rock, and slopes produced by open excavations. The height of slopes produced by river and wave erosion and that of most fault scarps has increased very slowly from zero to the present value. In order to visualize the successive stages in the development of such slopes we may consider the slopes of a river valley which was carved out of jointed hard rock by rapid erosion.

The deepening of the valley is accomplished almost exclusively by the action of the sediment-laden water on the bottom of the river. In the early stages of erosion, the width of the bottom remains practically unaltered. At this stage most jointed rocks possess considerable effective cohesion c_i (equation 2) and, as a consequence, they can form vertical slopes. Hence, during its initial stage, erosion can produce valleys with vertical sides. Canyons and buried inner gorges with vertical walls are formed in this stage. The maximum height H_0 which the walls of such valleys can attain is determined by the effective cohesion c_i and the pattern of jointing of the rock.

The seat of most of the cohesion of the jointed rock is located in the gaps which interrupt the continuity of joints. As the height of the side walls of the erosion valley increases, the shearing stresses in the rock adjacent to the walls increase. As a consequence more and more of the gaps between joints are eliminated by splitting, whereby the cohesion c_i decreases. More and more of the joints become free to open; the distribution of the normal stresses on the walls of the joints becomes more and more non-uniform and many of the blocks located between joints fail on account of local stress concentration. The surfaces of failure constitute a local joint system superimposed on the regional one. Hence, as the height of the slope increases beyond H_0, a wedge-shaped portion of rock adjacent to the slope, such as *a b c* in Fig. 3(b), will be changed into a practically cohesionless aggregate of angular and more or less irregular blocks which fit each other. Assuming that no weathering takes place, the sides of the valley will now rise at the steepest slope angle compatible with the joint pattern and the orientation of this pattern with reference to the slope. Depending on these factors, this angle may range between about 30° and almost 90°. As the depth of the valley increases, the slope recedes, at a constant slope angle, commonly by a process of ravelling, whereby the blocks located next to the slope drop out of the slope, one by one or several at a time. If the blocks between joints are very large this process assumes the character of intermittent rock falls.

On slopes underlain by stratified rocks with bedding planes dipping towards the valley the blocks drop out by sliding along bedding planes. If the spacing between cross-joints is very large the base of the rock fragments which drop out one at a time may cover many thousands of square feet.

In massive rocks such as granite or gneiss-granite with poorly developed schistosity, the spacing of joints of the regional system may locally be very great. Wherever this condition prevails a set of joints known as *sheeting* is likely to develop. Sheeting joints are roughly parallel to the erosion surface, and rarely occur at a distance in excess of about 50 ft from it. The spacing between these joints increases from a few inches next to the exposed surface to many feet at the inner boundary of the sheeted zone. The sheeting joints are likely to be open over a large portion of their total area. The slabs of intact rock located between them are subdivided by cross-joints commonly spaced from 10 to several tens of feet. If a slope on sheeted rock is undercut, the slabs located between cross-joints may drop out as shown in Fig. 1, or the slabs may fail by buckling. Behind the newly exposed surface, joint formation by sheeting is resumed. A typical case of sheeting and its engineering consequences has been described elsewhere (Terzaghi, 1962*b*).

Fig. 1. Sheeted granite, Yosemite Valley, California (Matthes, 1930)

The physical causes of sheeting are still a matter of controversy. However, it can be considered fairly certain that its development is associated with unloading (Billings, 1954, pp. 121–122; Kieslinger, 1960).

If the rate of down-cutting decreases, the slope angle of the valley walls is gradually reduced by weathering and the removal of the products of weathering by skin creep. The rate at which these processes proceed and the nature of the processes themselves depend on climatic conditions. In all but arctic and arid regions, weathering involves both a mechanical break-up of the blocks between joints and chemical changes which weaken the blocks, whereby factor of safety with respect to a shear failure in the remaining unweathered rock steadily increases. However, the stability of rock slopes in this advanced stage of development is beyond the scope of the present Paper.

Exceptional conditions prevail in valleys produced by river erosion and subsequently invaded by glacial ice which steepened their slopes. The fiords on the west coast of Norway are examples of such valleys. The precipitous slopes of many of these fiords rise to a height of several thousand feet. They acquired their present shape as a result of subglacial erosion. When the ice melted, those few slopes which were not stable without lateral support failed. Those which stood up were exposed for the first time to all the agencies which have weakened the rock adjacent to the slopes of normal river valleys throughout the entire period of down-cutting. Some of the ice-sculptured slopes bordering the fiords remained stable even after long-continued exposure to these agencies. Other slopes failed after varying periods of exposure.

Two types of delayed failure may be distinguished: the relatively superficial *rock falls* and the more deep-seated *rock slides*. Rock falls involve the intermittent detachment and fall of one or more blocks of rock owing chiefly to the weakening effect of frost wedging and important seasonal temperature changes. Rock slides, on the other hand, involve a mass of rock most of which is located below the depth of frost penetration or of notable temperature variations. They are attributable solely to the development of joints under the influence of shearing stresses acting in the rock after the removal of the lateral support once provided by glacial ice. Like rock falls, these slides have continued to take place throughout a long period following the disappearance of the ice. During that time, the shearing stresses in the rock have remained constant. Therefore we must conclude that the gradual development of the local joint system responsible for the slides represents the combined effect of the increase of the shearing stresses and of a slow creep-deformation of the rock acted upon by the stresses.

The factors which determine the stability of slopes formed by open excavation in unweathered rock are essentially identical with those which determine the stability of the slopes of stream valleys during the stage of rapid down-cutting and of ice-sculptured slopes after disappearance of the ice.

CONDITIONS FOR THE STABILITY OF SLOPES ON UNWEATHERED ROCK

Let s be the shearing resistance at a given point P of a potential surface of sliding in a porous and saturated material,

c_i = its cohesion,

ϕ = its angle of shearing resistance,

ϕ_f = angle of friction along the walls of a joint,

ϕ_c' = critical slope angle of jointed rock with effective cohesion, equal to the slope angle of a plane through the foot and the upper edge of the steepest stable slope which can be produced by excavation,

ϕ_c = critical slope angle of jointed rock without cohesion, equal to the slope angle of the steepest stable slope on such rock,

p = the unit pressure at point P,

u = the hydrostatic pressure in the water located next to point P.

According to a well-established empirical law:

$$s = c_i + (p - u) \tan \phi \quad . \quad . \quad . \quad . \quad . \quad . \quad . \quad . \quad . \quad (3)$$

All intact as well as jointed rocks with effective cohesion have the mechanical properties of brittle materials. Failure of slopes on brittle materials starts at the point where the shearing stress becomes equal to s (equation 3). As soon as failure occurs at that point, the cohesion of the rock at that point becomes equal to zero whereupon the stresses in the surrounding rock increase and the rock fails. Thus the failure spreads by chain action and the process continues until the surface of failure extends to the surface of the rock. This process is known as *progressive failure*.

In order to apply equation (3) to problems of rock mechanics, the influence of the joint pattern on the shearing resistance of jointed rock must be considered. If the rock has a random pattern of jointing, equation (3) is valid for any section through the rock. Therefore the rock performs like a stiff clay without joints, or an impure sand with considerable cohesion. If a slope on such material is undercut, the slope fails progressively by shear along a roughly concave surface of failure through the foot of the slope.

If an excavation is made in any of the previously mentioned brittle materials, such as a rock with a random joint pattern, the profile of the steepest stable slope which can be established is S-shaped, like the profile through root and tongue of a clay slide. For a given shape of the profile, the critical slope angle ϕ_c' decreases with increasing height of the slope, but it cannot become smaller than ϕ in equation (3).

Regardless of the character of the joint pattern in cohesive rocks, the real surface of sliding follows the joints wherever possible. The value of ϕ in equation (3) depends on the type and degree of interlock between the blocks on either side of the surface of sliding. In rocks with a random pattern of jointing, type and degree of interlock are independent of the orientation of the surface of sliding. Therefore ϕ in equation (3) has the same value for every surface. On the other hand if the joint system consists of sets of joints which are more or less parallel to one of several intersecting planes, like those in many stratified rocks, degree and type of interlock and the corresponding values of ϕ in equation (3) are very different for different potential surfaces of sliding, depending on the orientation of these surfaces with reference to that of the sets of joints. Yet the general relationship expressed by equation (3) retains its validity.

The only essential difference between a slope on brittle cohesive soil and a rock weakened by joints resides in the means at our disposal for determining the values c_i and ϕ in equation (3). For brittle soils both values can be obtained by means of simple laboratory tests. On the other hand, the c_i-value of jointed rock cannot be determined by any of the presently available methods for rock investigations. The value of ϕ for rocks with a random pattern of jointing can only be estimated on the basis of what we know about the ϕ-value of cohesionless aggregates in general and what we learned from case records as shown under the following heading. The value of ϕ of rocks with a well-defined joint pattern is a function of the orientation of the potential surface of sliding with reference to the joint system. Hence for any one type of rock it can have very different values.

If c_i and u in equation (3) are equal to zero the critical slope angle ϕ_c of rocks with a joint pattern of any kind is independent of the height of the slope. If the foot of the slope is undercut the material immediately underlying the slope descends to the foot and causes the slope to recede parallel to its original position, like the critical slope on a deposit of cohesionless sand.

According to equation (2) the cohesion c_i even of a moderately jointed rock is very much smaller than the cohesion c of the same rock in an intact state. In the proximity of slopes it is further reduced by the development of a local joint system superimposed on the regional one, and on account of the brittleness of cohesive jointed rock the effective cohesion of the jointed rock further decreases as the state of failure spreads. These facts lead to the conclusion that the influence of cohesion on the stability of slopes on jointed rock is relatively unimportant. Furthermore, even if it were important, we could not determine its value by any practicable means. Therefore in all the following discussions of the influence of the joint pattern on the stability of slopes it will be assumed that $c_i = 0$. The concept of effective cohesion will be retained merely for the purpose of assessing the relative importance of the effect of cohesion on the critical slope for rocks with different joint patterns.

The following sections deal with the influence of the joint pattern on the critical slope angle on the assumption that both c_i and u in equation (3) are equal to zero. The influence of the cleft-water pressure u on the stability of slopes will be discussed in a later section of the Paper.

SLOPES ON UNSTRATIFIED JOINTED ROCK

The orientation of joints in massive rocks such as granites and some limestones and dolomites may be almost random or there may be several rather well-defined sets. If the members of at least one set of joints are fairly continuous, the mechanical properties of the rock are similar to those of a stratified rock. Therefore only massive rocks with a random joint pattern will be considered in this section. Rocks of this category are divided by joints into irregular blocks which fit each other and are locally interconnected. In the next paragraph it will be shown that this macro-structure of rock with a random pattern of jointing is a large-scale model of the micro-structure of intact rock. Therefore the shear characteristics of these two materials are similar.

Crystalline rocks such as granite or marble consist of more or less irregularly shaped crystalline particles which fit each other like the blocks between the joints in a rock with a random joint pattern. They are separated from each other by very narrow, slit-shaped voids and each particle is connected with its neighbours at a few points only (Terzaghi, 1945). The tests by Roš and Eichinger (1928), like the older ones by von Karman, were carried out on such rocks. They showed that the angle of shearing resistance of marble, ϕ in equation (3), decreases with increasing normal stress p on the surface of sliding from about 40° at 100 kg/sq. cm. to about 25° at 1,000 kg/sq. cm. This decrease is probably due to the fact that the increase of the normal stress is associated with an increase of the number of failures across grains, which reduces the resistance against sliding due to interlock between grains.

The value of ϕ for intact rocks under pressures less than 100 kg/sq. cm. is not known. However, this gap in our knowledge is at least partly filled by the results of shear tests on aggregates of crushed rock. For crushed aggregate with a porosity of 26% under a load of 9 kg/sq. cm., Silvestri (1961) obtained a value of ϕ of 65°. The value of ϕ of any aggregate increases with decreasing porosity, everything else being equal, and the porosity of jointed rock hardly exceeds a few per cent. Furthermore, the average normal pressure on potential surfaces of sliding in the proximity of steep slopes even with a height of 1,000 ft does not even approach a value of 9 kg/sq. cm., because these surfaces are always located close to the slope, with the exceptions described under the heading "Deep-seated rock slides". Therefore the ϕ-value of jointed rock might be expected to be at least somewhat higher than 65°. This expectation is confirmed by observations made by McDonald (1913). Before deciding what slope angles should be assigned to the sides of the open cuts for the Panama Canal, McDonald measured the slope angles in a great number of rock cuts and canyons in the United States and Central America. The summary of his findings includes the following statement concerning slopes on sound, hard rock such as granite or quartzite. If the jointing and fissuring is "increased to the maximum of that encountered in nature in such rocks" a slope rising at an angle of 71° (3 vertical on 1 horizontal) will be stable. Therefore it can be concluded that the critical slope angle for slopes underlain by hard massive rocks with a random joint pattern is about 70°, provided the walls of the joints are not acted upon by seepage pressures.

SLOPES ON STRATIFIED SEDIMENTARY ROCK

Stratified sedimentary rocks consist of layers with a thickness averaging between a few inches and many feet. These are commonly separated from each other by thin films of material with a composition different from that of the rest of the rock. The bedding planes are almost invariably surfaces of minimum shearing resistance. Therefore in this Paper they are called bedding joints. They are likely to be continuous over large areas.

The cross-joints are generally nearly perpendicular to the bedding joints. They are commonly staggered at these joints. The cohesive bond along the walls of the cross-joints is equal to zero. The intersections between the cross-joints and the bedding planes may be more or less parallel to one of two or more directions, or, less commonly, the intersections may have a nearly random orientation.

Because of the almost universal presence of bedding and cross-joints, stratified sedimentary rock with no effective cohesion ($c_i = 0$) has the mechanical properties of a body of dry masonry composed of layers of more or less prismatic blocks which fit each other. The boundaries between the individual layers of blocks constituting the masonry correspond to the bedding planes of the rock. The cohesion across the joints between all the blocks of each layer is zero, and most of the joints between the blocks of two adjacent layers are staggered at the boundaries between layers. The stability of a slope on a rock with the mechanical properties of such a body of masonry depends primarily on the orientation of the bedding planes with reference to the slope. This relationship is illustrated by Fig. 2(a) and (c) showing the position of the critical slope with reference to the bedding planes. In each one of the fictitious rock formations represented by the figures, the cross-joints are assumed to be perpendicular to the bedding joints, they are staggered at the bedding joints, the angle ϕ_f of friction along the walls of all the joints is assumed to be 30°, and the height of the slope is smaller than H_c, equation (1).

If the bedding planes are horizontal no slide can occur and the critical slope is vertical, $\phi_c = 90°$. If the bedding planes are inclined, the critical slope angle depends on the orientation of the bedding planes with reference to the slope and on the orientation of the cross-joints.

In Fig. 2(a), the bedding planes dip into the slope at an angle α. Line A–A indicates a section through the foot of the slope at right angles to the bedding planes, hence rising at an angle $90°-\alpha$ to the horizontal. If $90°-\alpha$ is equal to or smaller than the angle of friction ϕ_f along the walls of the joints, no failure could occur along planes A–A even if the cross-joints were not staggered. If the cross-joints are parallel to A–A, but staggered, the position of the critical slope depends on the average value of the ratio C/D between the average length of the offset C between cross-joints and the average spacing D between bedding joints, shown in the diagram Fig. 2(b). For any value of α smaller than $90-\phi_f$ the critical slope angle is equal to that of the line BB in the Figure. At any given value of α the critical slope angle increases with increasing values of the ratio C/D and at a given value of C/D it decreases with decreasing values of $90°-\alpha$, until $90°-\alpha = \phi_f = 30°$. At this point the critical slope angle abruptly increases to 90°, because the slope angle of the cross-joints becomes smaller than the angle of friction ϕ_f along the joints. However, as $90°-\alpha$ further decreases and α approaches 90°

the danger of a failure by buckling of the layers located between bedding joints increases. Cohesion along the bedding joints increases the critical slope angle for any value of α smaller than $90° - \phi_f$. If the strata are steeper cohesion practically eliminates the possibility of a failure of the exposed strata by buckling.

If the bedding planes dip towards the valley at an angle smaller than the angle of friction $\phi_f = 30°$, the critical slope is 90°. For values of α greater than 30° the critical slope angle ϕ is equal to α.

If the slippage along bedding joints is resisted by effective cohesion c_i in addition to friction the steepest stable slope is no longer plane. Up to a certain height H it will be vertical as shown in Fig. 2(c) and above it the slope will rise at an angle α.

Let H be the height of the vertical part of the slope,

w = unit weight of the rock,

α = slope angle of the bedding joints,

$\phi_f = 30°$ = angle of friction between walls of the bedding joints, and

c_i = effective cohesion between walls of the bedding joints.

The force which tends to produce a slip along a bedding joint through the foot of the slope is $Hw \cos \alpha \sin \alpha$ per unit of area of the bedding joint as shown in Fig. 2(d) and the force which resists the slip is $Hw \cos^2 \alpha \tan \phi_f + c_i$. Hence the vertical slope shown in Fig. 2(d) will not be stable unless:

$$H \lesseqgtr \frac{c_i}{w \cos \alpha (\sin \alpha - \cos \alpha \tan \phi_f)}$$

An increase of the height of the vertical slope would be immediately followed by a slide along the bedding plane B–B through the foot of the slope.

If the strike of the bedding planes is oriented at right angles to the slope, the critical slope angle depends on the joint pattern.

If the cross-joints have a random orientation the critical slope angle cannot be smaller than that of a massive rock with a random joint pattern. This angle is about 70°. As the angle increases the critical slope angle ϕ_c is likely to decrease from 90° for small values of α to 70° for $\alpha = 90°$. However, a moderate amount of cohesion along the bedding joints increases it to values up to 90°, because the cohesion interferes very effectively with the rotational displacement of the blocks between cross-joints. If the cross-joints are oriented more or less parallel to two planes intersecting at right angles, the critical slope angle depends on the slope angle β of those cross-joints which dip towards the valley.

In the preceding discussions it was shown that the degree of stability of a slope depends primarily on the joint pattern and its orientation with reference to the slope. It also depends to a lesser extent on the effective cohesion c_i. The joint pattern and, as a consequence, the critical slope angle ϕ_c can be determined, at least approximately, on the basis of the results of a joint survey. If a slope rises at an angle of less than ϕ_c it can be considered certain that its factor of safety G_s with respect to sliding is greater than unity, provided u in equation (3) is equal to zero. If the slope rises at an angle greater than ϕ_c it owes part of its stability to effective cohesion c_i and the value of c_i cannot be determined by any of the presently available means. Hence the factor of safety G_s of such a slope may be close to unity, but its real value cannot be ascertained.

SLOPES ON FAULTED ROCK

Faults are fractures along which conspicuous displacement has occurred. Major faults along which displacements of many hundreds of feet have taken place are generally detected during the geological survey of the area. However, minor faults with displacements measured

1*

in a few inches or feet are much more common and these minor faults cannot be located until they are encountered in an excavation, or their existence is disclosed by the occurrence of a slide.

If the strike of the slope of an artificial rock excavation intersects the strike of a fault at an acute angle, the wedge-shaped body of rock located between slope and fault may drop out. The factor of safety with respect to such a failure is determined by the dip of the fault plane, the orientation of the fault with reference to the slope and by the resistance against sliding

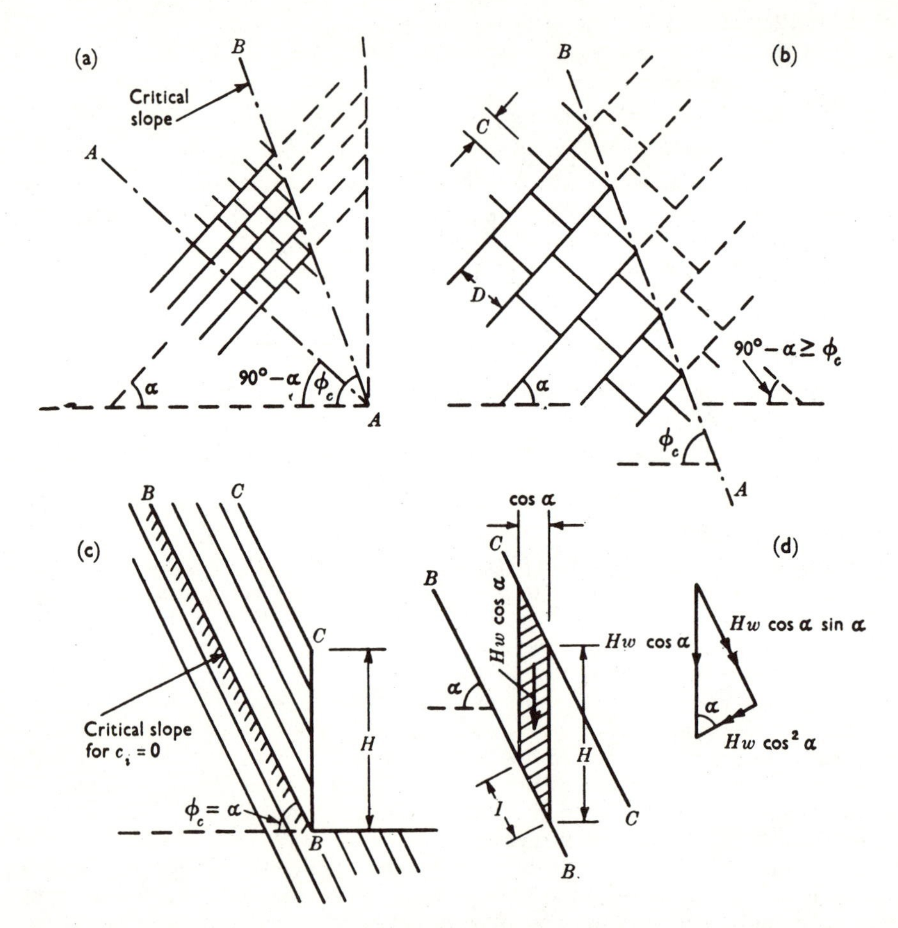

Fig. 2. Diagram illustrating factors which determine inclination of critical slope on stratified rock

along the fault plane. Dip and orientation of the fault can be ascertained as soon as the fault is encountered. The resistance to sliding along the fault plane depends on the condition of the rock along the fault. The walls of many faults are separated from each other, at least locally, by a zone of crushed rock which may be chemically intact or exhibit varying degrees of weathering or hydrothermal alteration. The shear characteristics of these materials may change from place to place and it is impracticable to ascertain their weighted average, because they are accessible at only a few points. Therefore the influence of the presence of such faults on the stability conditions of a slope can only be roughly estimated.

EFFECT OF RAINFALL AND SNOWMELT ON STABILITY

A portion of the precipitation on the higher parts of a topographically dissected area enters the joints in the underlying rock and emerges in the proximity of the foot of the slopes in the form of springs. If the secondary permeability* of the rock were uniform the water-table would assume the shape indicated in Fig. 3(a) and vary between a lowest position (dash line) at the end of the dry season and a highest position (unbroken line) at the end of the wet or melting season. The volume of the continuous open joints in which the water travels through the rock is very small compared to the volume of the rock located between these joints.

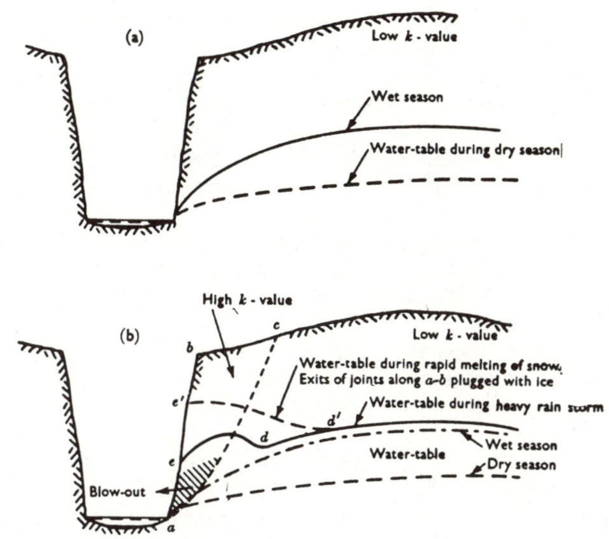

Fig. 3 (a). Location of lowest and highest water-table in jointed rock with uniform but low coefficient of permeability; (b) water-table in the same rock during heavy rainstorms or snow melt if permeability in space *a b c* is much higher than elsewhere. Line *e d* represents the water-table at times when the joints drain freely, and *e′ d′*, when they are plugged with ice

Therefore the vertical distance between the lowest and the highest position of the water-table is measured in tens of feet and not in feet, as in unconsolidated pervious sediments.

In reality the water-table is not as well defined as shown in Fig. 3(a) because the secondary permeability of jointed rock commonly varies more or less erratically from place to place and the water may rise in two adjacent observation wells to very different levels. Yet if the water-table has at least the general characteristics of that shown in Fig. 3(a) the seepage exerted by the flowing ground-water could not possibly be important enough to have a significant influence on the stability of steep slopes.

* The term "secondary permeability" of a rock formation refers to the permeability which results from the presence of open and continuous cracks and fissures in the rock. It depends on the width and spacing of these passages. In contrast to the secondary permeability, the primary permeability has its seat in the voids of the intact rock located between the fissures. Compared to the secondary permeability, the primary permeability is commonly so small that a tunnel driven through intact rock below the water-table appears to be dry.

On account of the low secondary permeability of the jointed rock only a small portion of the rainfall descends through the joints. Part of it is temporarily retained in the voids of the weathered top layer and the remainder flows as surface run-off toward the edge of the slope. However, before it reaches the edge it crosses the upper surface *b c* of the wedge-shaped body of rock *a b c* in Fig. 3(b). On account of the shearing stresses prevailing in this body, the joints may be much wider and more numerous than those in the rock located farther from the slope. Because of this circumstance, the quantity of water which can enter the joint system in this body, per unit of area of its top surface *b c*, is much greater than the corresponding quantity which enters the rock elsewhere. Consequently the water-table in that body may rise temporarily to a position such as that indicated by the unbroken line *d e* in Fig. 3(b).

The water which occupies the joints exerts onto the walls of the joints a pressure u equal to the unit weight of the water times the height to which it would rise in an observation well terminating in the joint. This pressure corresponds to the pore-water pressure in soil mechanics and will be called the *cleft-water pressure*. Like the pore-water pressure, it reduces the frictional resistance along the walls of the joints (see equation (3)) and if a joint is very steep it tends to displace the rock between joint and slope towards the slope.

The cleft-water pressure is zero at the water-table and it increases in a downward direction. Hence if a slope fails on account of cleft-water pressures, the failure will start at the foot of the slope within the shaded portion of the area *a b c* in Fig. 3, whereby the rock will be displaced by the water pressure in a horizontal direction. As a result of the initial failure, the rock located above the seat of the failure is deprived of support and it will descend owing to its own weight.

The influence of the cleft-water pressures on the stability of steep rock slopes is well illustrated by the rock slide statistics prepared by the Norwegian Geotechnical Institute (Bjerrum and Jorstad, 1957). Within the area covered by the observations, the winters are severe, the snowfall abundant, and the heaviest rainfalls occur during the autumn months. In order to find out whether there is a relationship between the climatic conditions and the frequency of rock slides and rock falls, the investigators constructed the diagram which is reproduced in Fig. 4. It represents the seasonal variations of rock falls and rock slides for the period 1951–1955.

The diagram shows that the slide frequency was greatest in April, during the time of the snow melt, and in October within the period of greatest rainfall. However, most of the major slides have taken place in April, because at that time of the year the exits of the joints are still plugged with ice while the snowmelt is feeding large quantities of water into the joints of the rock within the wedge-shaped body of rock *a b c* in Fig. 3(b). Owing to this condition, the water-table adjacent to the slope is raised from position *e d* into position *e′ d′*.

Under exceptional conditions, destructive cleft-water pressures can also develop behind slopes without the assistance of ice. The most notable example is the rock slide which occurred on 13 September, 1936, at the south end of Loen Lake near the west coast of southern Norway. Fig. 5 is a vertical section through the slope at the site of this slide. The slope is located on granitic gneiss with poorly developed foliation and rises at an average angle of 50° to a height of about 3,600 ft above lake level. The middle portion of the slope with a height of about 1,600 ft is nearly vertical. The planes of foliation dip at an angle of about 65° towards the lake as shown in the Figure. The rock behind the vertical face is weakened by joints, probably sheeting joints, oriented parallel to the face, but the rock between the joints is practically impervious. During a heavy rainstorm the rock between the vertical face and one of the vertical joints, with a volume of about 1·3 million cu. yd, dropped from the cliff, descended into the lake and produced a wave with a height of 230 ft which destroyed a village and claimed many victims.

If, during a series of exceptionally wet years, the least stable slopes are "cleaned off" by rock falls and slides, many decades pass before the deterioration of the remaining slopes has advanced far enough to cause a slope failure. The consequences of this condition are illustrated by Fig. 6 showing the variation of the number of rock slides per decade for the period 1650–1900. Between 1720 and 1760 the frequency of rock slides was more than ten times greater than between 1760 and 1810. The period 1720–1760 was exceptionally cold and wet and during that period all the glaciers advanced temporarily.

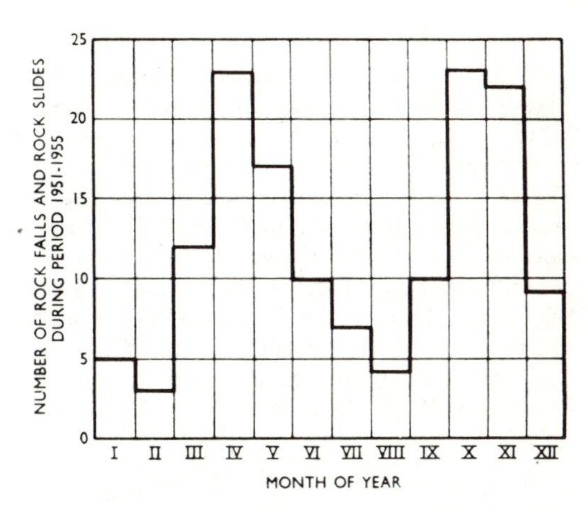

Fig. 4. Seasonal variation of frequency of rock falls and slides on slopes in Norwegian fiords (Bjerrum and Jørstad, 1957)

Cleft-water pressures also impair the stability of slopes in deep open-pit mines in jointed rock. The slope angles for the sides of the pits must be selected before the mining operations start. If the bottom of the pit were located above the water-table, the slopes would be safe if they were made to rise at the critical slope angle ϕ_c (critical slope angle for $c_i = 0$). This angle is determined by the joint pattern and the orientation of the faults with reference to the slope and can, in many instances, be evaluated with rough approximation on the basis of the results of a joint and fault survey. However, in most cases the bottom of the pit is located at a great depth below the water-table and the influence of the cleft-water pressures on the degree of stability of the slopes is commonly underestimated. Therefore slope failures in open pit mines due to cleft-water pressures are very common (see, for instance, Wilson, 1959).

Forecasts of the intensity and distribution of the cleft-water pressures caused by the flow of ground-water towards the foot of slopes on jointed rock are, at the very best, unreliable. Therefore the mining engineer has to choose between risking the occurrence of a slide and eliminating the cleft-water pressures by adequate drainage of the rock adjacent to the slope while excavation proceeds.

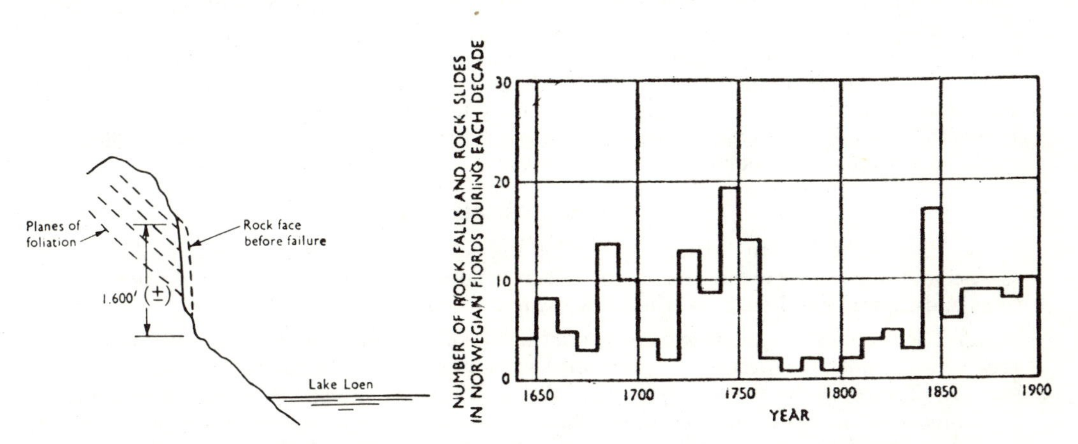

Fig. 5. Rock slide on Loen Lake, Norway (Bugge, 1937)

Fig. 6. Secular variations of frequency of rock falls and rock slides during period 1640–1900 (Bjerrum and Jørstad, 1957)

* * * * * * *

[A discussion of seepage from reservoirs and pressure tunnels was deleted.]

DEEP-SEATED ROCK SLIDES

If a canyon is cut to moderate depth in an area of gently rolling topography, the shearing stresses in the rock remain low except in the immediate vicinity of the canyon walls. On the other hand, in very high mountains dissected by deep valleys, the rock underlying the slopes of the valley may be the seat of high shearing stresses. If erosion is accelerated, a canyon like that shown in Fig. 10 may be cut into the valley floor. Outside the zone indicated by shading, the percentage increase of the shearing stresses in the rock is very small and if the canyon were located in a gently rolling region it would be inconsequential. However, in the case illustrated by Fig. 10, the shearing stresses are already close to the shearing resistance of the rock. Hence as the depth of the canyon increases, a state may be reached in which further downcutting initiates a slide along a deep-seated surface of sliding, such as *a–b* in Fig. 10.

There is increasing geological evidence that rock slides along such surfaces have occurred (Ampferer, 1939, 1940; Stini, 1941, 1942*a*, 1942*b*, 1952). Various geomorphological features

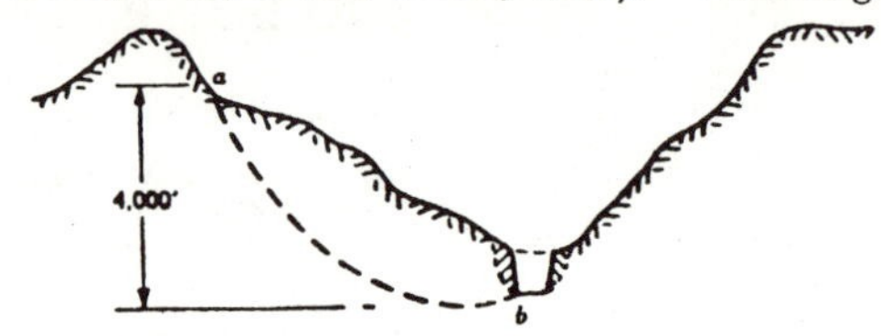

Fig. 10. Deep-seated rock slides beneath slope of deep valley between high mountains

in the Alps, formerly ascribed to ice action, are now believed to be the surface manifestations of deep-seated rock slides. At several dam sites in exceptionally narrow sections of deep valleys in the Austrian Alps, it was found that the rock forming one of the slopes had advanced towards the valley in a more or less horizontal direction, whereby the rock involved in the movement remained relatively intact. At one of these sites ground moraine was encountered in a boring at a depth of about 300 ft below the surface of what appeared to be rock in situ. (Stini, 1952).

Practically nothing is known concerning the mechanics of these deep-seated, large-scale rock slides. It is not known whether the slides took place rapidly or slowly, and it is doubtful whether they are preceded by important creep deformation of the rocks located within the shear zone. However, it is known that the rock located above the surface of sliding has been damaged at least to a moderate extent. Existing joints have opened and new ones have been formed. Hence the compressibility and secondary permeability of the rocks has increased. Furthermore, in the immediate proximity of the surface of sliding, the rock is completely broken or crushed. Hence a site for a high concrete dam should not be considered suitable unless there is positive evidence that the underlying rock has never been subject to displacement by a deep-seated rock slide.

CONCLUSIONS

(1) The results of computations concerning the degree of stability of slopes on jointed rock can be very misleading unless they are supplemented by a detailed statement of the observations and experimental data on which the computations were based.

(2) The foremost requirement for evaluating the degree of stability of a slope on jointed rock which is not acted upon by significant cleft-water pressures consists in a thorough joint survey. The results of the survey furnish the basis for estimating the critical slope angle ϕ_c of the rock on the assumption that the effective cohesion c_i and the cleft-water pressures are equal to zero. Depending on the joint pattern the value of ϕ_c may range between 30° and 90°. Every slope rising at an angle of ϕ_s of less than ϕ_c has a factor of safety with respect to sliding of more than unity and the value of this factor increases with increasing difference between

ϕ_c and ϕ_s. On the other hand, if a slope rises at an angle of more than ϕ_c it owes part of its stability to effective cohesion c_i. Since the value of c_i cannot be determined by any of the presently available means it should be assumed that the factor of safety of such slopes with respect to sliding is equal or very close to unity.

(3) The most unpredictable among the factors determining the stability of slopes on rock are the cleft-water pressures due to leakage from artificial sources. Therefore sound engineering requires either that such leakage be prevented or that the flow of seepage be intercepted at a safe distance upstream from the slope.

(4) Failure of slopes on hard rock due to leakage from pressure tunnels can be reliably prevented by providing the tunnel with an adequately designed lining between the portal and the point at which the depth of the overburden H_c, measured at right angles to the ground surface, becomes greater than one-half of the hydrostatic head H_w of the water in the tunnel.

(5) Harmful effects of leakage out of reservoirs towards the slopes downstream from the abutments can most effectively be eliminated by drainage.

(6) In high mountains, the rocks exposed on slopes above the bottom of the very deep valleys may have been displaced and damaged as a result of a deep-seated rock slide in the geological past. Hence, if a dam site is located in an exceptionally narrow section of a deep valley, an elaborate site investigation should not be started until conclusive evidence is secured that the rocks underlying the dam site were never involved in such a rock slide.

REFERENCES

AMPFERER, O., 1939. "Uber einige Formen der Bergzerreissung" ("Some types of mountain splitting"). *Sitzgsb. Akad. d. Wiss. Wien, Math. Nat. Kl.*

AMPFERER, O., 1940. "Zum weiteren Ausbau der Lehre von den Bergzerreissungen ("Further contributions to our knowledge of mountain splitting"). *Sitzgber. Akad. Wiss. Wien*, Abt. I, 149 : 51–70.

BILLINGS, MARLAND, P., 1954. "*Structural geology*," 514 pp., *Prentice-Hall, Inc.*; especially pp. 121–123.

BJERRUM, L. *and* F. JØRSTAD, 1957. "Rockfalls in Norway." *Norwegian Geotech. Inst.*

BUGGE, A., 1937. "Fellskred fra Topografisk og geologisk synspunkt." ("Rock falls from a topographical and geological viewpoint"). (With summary in French.) *Norsk geografisk tidsskrift*, 6 : 6 : 342–358.

DONATH, D. D., 1961. "Experimental study of shear failure in anisotropic rocks." *Bull. Amer. Geol. Soc.*, 72 : 6 : 985–990.

ENGINEERING NEWS-RECORD, 1953. "Landslide destruction of power plant blamed on leaks in penstock tunnel." *Engineering News-Record*, 12 November, 1953.

KIESLINGER, A., 1960. "Residual stress and relaxation in rocks." *Int. Geological Congr., Copenhagen*, Session 21, pp. 270–276.

MACDONALD, D. F., 1913. "Some engineering problems of the Panama Canal in their relation to geology and topography." *U.S. Dept. of Interior, Bureau of Mines*, Bull. 86, pp. 1–88.

ROŠ, M. *and* A. EICHINGER, 1928. "Versuche zur Klärung der Frage der Bruchgefahr. II. Nichtmetallische Stoffe" ("Experimental study of theories of rupture. Non-metallic materials"). Zürich, Eidgenössische Materialprüfungsanstalt der Eidgen. *Techn. Hochschule*, 27 June, 57 pp.

SILVESTRI, T., 1961. "Determinazione sperimentale de resistenza meccanica del materiale constituente il corpo di una diga del tipo 'Rockfill'") ("Experimental determination of the shearing resistance of rockfill"). *Geotechnica*, 8 : 186–191.

STINI, J., 1941. "Unsere Täler wachsen zu" ("Our valleys close up"). *Geologie u. Bauwesen*, 13 : 71–79.

STINI, J., 1942*a*. "Nochmal der Talzuschub" ("Some more about the closing up of our valleys"). *Geologie u. Bauwesen*, 14 : 10–14.

STINI, J., 1942*b*. "Talzuschub u. Bauwesen" ("Engineering consequences of the closing-up of our valleys"). *Die Bautechnik*, 20 : 80.

STINI, J., 1952. "Neuere Ansichten über Bodenbewegungen und ihre Beherrschung durch den Ingenieur" ("New conceptions concerning ground movement and its control by the engineer"). *Geol. u. Bauwesen*, 19 : 1 : 31–54.

STINI, J., 1956. "Wassersprengung und Sprengwasser" ("Blasting effects of water"). *Geologie u. Bauwesen*, 22 : 141–169.

TERZAGHI, K., 1945. "Stress conditions for the failure of saturated concrete and rock." *Proc. Amer. Soc. Test. Mat.*, 45 : 777–801.

TERZAGHI, K., 1962*a*. *Discussion* of Paper by A. Casagrande, "Control of seepage through foundations and abutments of dams." *Géotechnique*, 12 : 1 : 67–71.

TERZAGHI, K., 1962*b*. "Dam foundation on sheeted granite." *Géotechnique*, 12 : 3 : 199–208.

TERZAGHI, RUTH D., 1963. "Sources of error in joint surveys." *Géotechnique* (as yet unpublished).

WILSON, S. D., 1959. "The application of soil mechanics to the stability of open pit mines." *Colorado School of Mines Quarterly*, 54 : 3 : 93–113. Third Symposium on Rock Mechanics.

28

Factors Affecting Sheet and. Rill Erosion

Dwight D. Smith and Walter H. Wischmeier

Abstract—This paper discusses the two principal processes by which sheet erosion occurs and the six factors which effect the magnitude of the losses. The processes are raindrop impact and transportation of soil particles by flowing water. The factors are length and per cent slope, cropping, soil, management and rainfall. The relative effectiveness of each of the three main conservation practices in control of erosion, contour farming, strip cropping, and terracing is presented. The factors and practices are combined in a rational erosion equation for calculating field soil loss for use in application of conservation practices and in assessing land program benefits.

Sheet erosion has been defined [*Baur*, 1952] as "removal of a fairly uniform layer of soil or material from the land surface by the action of rainfall and runoff." Detachment of soil particles by rainfall impact and splash is probably the only process by which a fairly uniform layer of soil is removed from field surfaces. But soil loss in splash from a specific area occurs only when a directional unbalance occurs as on sloping land, with high winds, etc. When runoff enters the picture transportation of detached particles of soil begins and soon reaches a magnitude resulting in channel flow. This marks the transition from sheet to rill erosion, the type of erosion defined [Baur, 1952] as "removal of soil by running water with formation of shallow channels that can be smoothed out completely by cultivation." This differentiation in erosion types is more easily made on paper than on the land. Plot measurements with which this paper is concerned, is the net result of both sheet and rill erosion.

The erosion processes—The water-erosion processes—detachment by rainfall impact and transportation by flowing water—are very complex. There are a large number of variables which combine to give results in some cases in direct opposition to generally accepted theories or popular belief. There are numerous controversial theories which probably result more from localized viewpoints than from erroneous thinking. The complexity of the problem has necessitated use of empirical methods by conservationists for solution of their field problems.

Cook [1936] listed and discussed the two processes of rainfall erosion: (a) that of falling drops striking the land surface, and (b) surface flow or runoff in which soil is moved from an area of land. *Laws* [1940, 1941] began study of effects of rainfall impact on erosion, which included measuring the fall velocities of water and raindrops and the drop-size distribution in natural rain. *Borst* and *Woodburn* [1942] showed that rainfall impact was the principal cause of erosion from a fallow plot. *Ellison* [1947, seven papers] began his work on raindrop splash in 1936 and subsequently developed theories on detachment and transportation. This approach to the erosion problem was further developed by him and other research workers including *Ekern* [1950, 1953] and *Osborn* [1953, 1954]. This work was in general confined to measurements from small splash cans and from plots 12 × 18 inches in size.

Kinetic energy of rainfall—Rainfall records for the ten-year period 1931–1940 at Bethany, Missouri [*Smith* and others, 1945] have been separated into average annual amounts for different intensity intervals. The amount for each interval decreases rapidly with increases in intensity (Table 1). The kinetic energy was computed for each interval (Table 1) by use of an equation (to be published in a future paper) developed by the authors from published data on drop velocity by *Laws* and *Parsons* [1941] and by *Gunn* and *Kinzer* [1949], and drop-size distribution by *Laws* and *Parsons* [1943]. Annual kinetic energy computed by this method averaged about 12 pct less than by using averages of median drop sizes determined by several investigations [*Ekern*, 1953].

Both the amounts of rain and the kinetic energy of rainfall equal to or greater than specified intensities were plotted against rain intensity in Figure 1. The Bethany control plot study shows average annual runoff for continuous corn to be 8.0 inches and for a rotation of corn-wheat-meadow 4.9 inches. For these amounts, the rainfall curve of Figure 1 shows an average infiltration-rate index of two inches per hour for the rotation in contrast to one inch per hour for continuous corn. The energy in rainfall during runoff was 5200 foot tons per year for the rotation in contrast to 8300 foot

TABLE 1 – *Average annual kinetic energy content of rainfall by intensity groups for the ten-year period 1931–40, Bethany, Missouri*

Intensity-interval group	Amount of rain		Kinetic energy of rainfall	
	At group intensity	At intensities ≥ group	Within group	At intensities ≥ group
inches/hr	inch	inch	ft ton	ft ton
0– 0.24	16.00	29.49	9,770	22,632
0.25– 0.49	2.56	13.49	1,979	12,862
0.50– 0.74	1.63	10.93	1,383	10,883
0.75– 0.99	1.30	9.30	1,168	9,500
1.00– 1.49	1.88	8.00	1,785	8,332
1.50– 1.99	1.31	6.12	1,305	6,547
2.00– 2.99	1.90	4.81	1,990	5,242
3.00– 3.99	1.70	2.91	1,858	3,252
4.00– 4.99	0.73	1.21	831	1,394
5.00– 5.99	0.25	0.48	290	563
6.00– 6.99	0.14	0.23	166	273
7.00– 7.99	0.02	0.09	16	107
8.00– 8.99	0.02	0.07	34	91
9.00–10.99	0.05	0.05	57	57
Total.........	29.49		22,632	

tons for continuous corn. With the rotation, only 23 pct of the total annual energy in the rain was applied when runoff was available for movement of the detached soil from the plots. The rotation provided a good measure of cover for the plot 75 pct of the time. With continuous corn (stalks removed in the fall and no fertility treatment) a fair measure of cover was available only 25 pct of the time. This all helps explain why soil loss under continuous corn averaged $5\frac{1}{2}$ times that with the rotation.

Ellison [1947, pp. 245–248] has shown the equation for relative soil detachment as $E = K\,V^{4.33}\,d^{1.07}\,I^{0.65}$ in which E is relative amounts of soil splashed, during a 30-minute period, K is a constant of the soil, V the drop velocity in ft/sec, d the drop diameter in mm, and I the intensity in in/hr. If drop velocity, equivalent drop size and intensity values from the rainfall energy equation previously mentioned are substituted in Ellison's equation, it is shown that relative soil detachment is 7.2 times as much for an intensity of six inches per hour as for one inch per hour. These data, however, were secured with equal time periods for each intensity. If the assumption is made that total detachment for each rain intensity is directly proportional to time of rain, the increase in relative detachments for equal amounts of rain (1 inch) are much less spectacular as is shown in Table 2.

Potential energy of runoff—Most writers have shown a large contrast between energy of rainfall

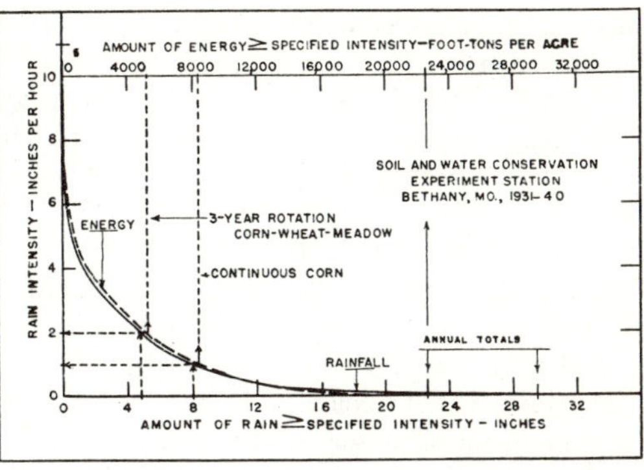

FIG. 1 – Amount and kinetic energy of average annual rainfall equal to or greater than specified rain intensity for ten-year period, 1931–40, Bethany, Missouri, and infiltration rate indices for continuous corn (one inch per hour) and for a three-year rotation (two inches per hour)

and that of runoff. This has resulted from using estimates of kinetic energy of the flowing water as it left the experimental area. The potential energy of runoff is relatively high. For the Shelby soil area with eight per cent average slope 300 ft long, the potential energy of the five-inch average annual runoff is 6800 foot-tons and for the Mexico soil with three per cent average slope 450 ft long, the potential energy of the nine inches average annual runoff is 6900 foot-tons. These figures are about equal to the kinetic energy of the rainfall that occurred during runoff at Bethany although only 30 pct of the total. A large portion of this potential energy of runoff is used in overcoming friction of flow or in work on the surface as runoff passes down the slope. Also, a significant part of the kinetic energy of the rain is transmitted to the runoff water as the raindrops strike the flowing sheet of water and add to its turbulance.

Other factors affecting soil loss—The principal factors in addition to rainfall which are generally considered as affecting the magnitude of soil loss are as follows: (a) per cent slope, (b) length of

TABLE 2 – *Effect of rain intensity on relative detachment, calculated by use of Ellison's equation*

Rain intensity	Relative detachment	
	Per 30-minute period	Per inch of rain
in/hr		
1	1.00	1.00
2	2.18	1.09
4	4.71	1.18
6	7.16	1.20

slope, (c) cover or cropping system, (d) soil, and (e) management.

Per cent slope—The first comprehensive study of the effect of slope on soil loss was published by *Zingg* [1940]. He concluded that soil loss varies as the 1.4 power of the per cent slope. His study was based on data by Duley and Hays working in Kansas, Diseker and Yoder in Alabama, and a series of rainfall simulator studies which he performed at Bethany, Missouri.

The exponential equation, as an empirical description of the slope soil-loss relationship, is quite satisfactory under most field conditions. However, it indicates zero soil loss for a zero per cent slope, a condition which does not exist in the field unless the field is completely diked to prevent runoff. For the extremely steep slopes of soils that are easy to detach and transport, a decrease in rate of increase of soil loss as runoff reaches its capacity to transport soil appears possible. These slopes, however, are seldom in cultivation.

For better description of the relationship on the flatter slopes of Midwest claypan soils, *Smith* and *Whitt* [1947] proposed the following equation, the constants of which were evaluated by use of *Neal's* [1938] rainfall simulator data for Putnam soil

$$R = 0.10 + 0.21\ S^{4/3}$$

in which R is relative soil loss in relation to unity loss for a slope of three per cent. S is per cent of slope.

Seventeen years of data from each of two locations are now available for evaluation of the per cent slope-soil-loss relationship. Data by Hays at the Upper Mississippi Valley Conservation Experiment Station, LaCrosse, Wisconsin, covers four slope groups from 3 to 18 pct on Fayette soil. The plots were cropped to continuous barley for the first five years and to corn-oats-meadow rotation for the succeeding 12 years. Data by Lillard and others working at Blacksburg, Virginia covers 17 years of corn-wheat-meadow rotation on five slope groups ranging from 5 to 25 pct. Similar data comparing only two slopes are available from studies in several other states.

The parabola appears to meet satisfactorily the requirements for an empirical equation to express the relationship between per cent slope and soil loss throughout the full slope range as experienced on cropped fields. It estimates a small soil loss from a zero per cent slope, and the constants in the equation for different soil types can be evaluated readily from data by the method of least squares.

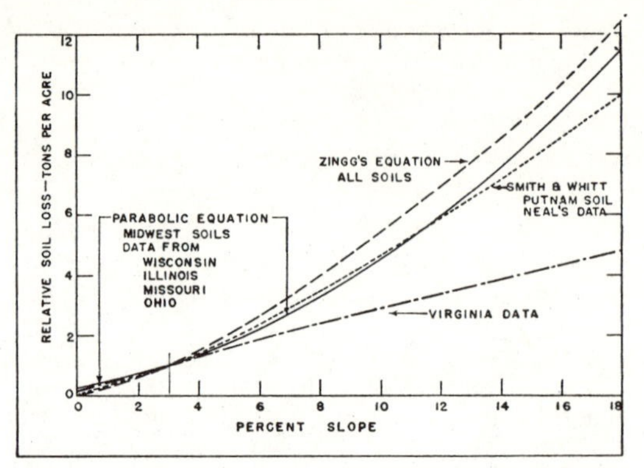

Fig. 2 – The relationship of per cent slope and soil loss as described by parabolic equations derived from presently available data and by other type equations previously published; all equations adjusted to give unity soil loss on a three per cent slope

The constants of the parabolic equation fitted to Hays' data and one fitted to Zingg's rainfall simulator data when adjusted for similar cropping conditions are nearly identical. Data from studies at Dixon Springs, Illinois, and Zanesville, Ohio, comparing two slopes at each location, when adjusted in magnitude to conform with conditions at LaCrosse, also fit the slope of the parabola fitted to the Wisconsin data. The combined data for the four studies give a very good least-squares fit to the equation

$$A = 0.43 + 0.30\ S + 0.043\ S^2$$

in which A is soil loss in tons per acre.

The Virginia data, however, shows a nearly linear relationship. The parabolic equation fitted to these data is

$$A = 0.24 + 0.55\ S - 0.004\ S^2$$

A comparison of the equations for per cent slope-soil-loss relationship is shown in Figure 2 with published equations by Zingg and by Smith and Whitt.

Runoff in Hays' and Zingg's studies increased significantly with increase in per cent slope, although the two soils were quite different. In the Virginia study, it remained nearly constant across the entire range of slopes.

Length of slope—In his study of the effect of slope length on soil loss, Zingg concluded that total soil loss varied as the 1.6 power of the length, and soil loss per unit area as the 0.6 power of the length. He used data from Bethany, Missouri;

Guthrie, Oklahoma; Clarinda, Iowa; LaCrosse, Wisconsin; and Tyler, Texas. A group study in 1945 under the direction of *Musgrave* [1947] proposed 0.35 as the average value for the slope length exponent for soil loss per unit area. In a recent group study (Joint SWC, ARS, and SCS Work Conference on Slope-Practice Values, Purdue University, Lafayette, Indiana, February 27–March 1, 1956) which included the authors, the conclusion was reached that for field use the value of the length exponent should be 0.5 ± 0.1.

Length of slope data are now available from 136 location-years at ten locations where the principal crops were either corn or cotton. The average value of the length exponents by locations weighted for years is 0.46. There is a wide variation in location-year values of the length exponent. In some instances it is negative. Average values for the different locations varied from 0 to 0.9. An explanation of the variations was not readily discernible. Magnitude of the exponent appears to be related both to soil and cover, but more positively to runoff. The larger exponents are, in general, associated with location data which show an increase in runoff with increased slope length and the smaller with locations where the reverse occurs. The largest values are from Bethany, Missouri, and Guthrie, Oklahoma, and the zero or no significant difference from Temple, Texas. It is logical to expect little change in soil loss with increase in slope length when runoff has a limited transport capacity due to a decreasing volume or decreasing turbulance, and is quickly saturated as it passes down a slope of easily detachable soil.

Soil loss is generally expressed in tons per acre. Since the rate of soil loss increases on most soils with an increase in slope length, the loss on the lower segments of slope will be appreciably greater than the average for the entire length. This can be illustrated by computing the relative soil loss for segments of an average slope length assuming the increase in average soil loss per unit area is proportional to $L^{0.5}$. Computed values for four segments of a 300-foot slope length are

Length segment	Relative soil loss per unit area
foot	ton
0–75	1.00
75–150	1.83
150–225	2.37
225–300	2.80
Average for 0–300	2.00

The relationship of length of slope to soil loss has been controversial. This is partly the result of the large variation in experimental results which have not been satisfactorily explained. Another reason has been the different interpretations of what constitutes length of slope. For this reason the following definition of slope length was prepared by the recent slope-practice conference group.

> Slope length is the distance from the point of origin of overland flow to either of the following, whichever is limiting for the major portion of the area under consideration: (1) the point where the slope decreases to the extent that deposition begins, or (2) the point where runoff water enters a well-defined channel. A channel is defined as a part of a drainage network of a size that is not readily obliterated by cultivation and usually suitable for stabilization with grass. It may be a constructed channel such as a terrace or diversion.

Cover—Cover as a variable in erosion control has generally been studied on small runoff plots 70 to 90 ft long and 6 to 14 ft wide. Such plots were established on the ten original soil erosion

TABLE 3 – *Runoff and soil loss from plot studies in Missouri on Shelby loam, 1930–41, and on Mexico silt loam, 1941–50*

Cropping system	Plot data	Annual rainfall	Annual runoff	Annual soil loss
		inch	inch	tons/ac
Continuous Corn[c]	a	29.49	8.20	50.93
Corn-Wheat-Meadow[c]	a	29.49	4.90	9.06
Bluegrass[c]	a	29.49	2.44	0.16
Corn-Oats[c]	b	39.16	11.74	6.08
Corn-Corn-Oats-Wheat & Sw. Cl.	b	39.16	8.83	4.03
Corn-Oats & Sw. Cl. (catch)	b	39.16	8.42	3.07
Corn-Wheat-Sw. Cl. (Hay)	b	39.16	7.89	2.02
Corn-Wheat-Meadow	b	39.16	8.42	1.67
Corn-Wheat-Meadow-Meadow	b	39.16	8.39	1.13
Wheat & Lesp. (Pasture)[d]	b	39.16	9.20	2.20
Tim., Bl. Gr., Sw. C., & Lesp. (Pasture)[d]	b	39.16	6.83	0.26

[a] Plots 6 × 72.6 feet on eight per cent slope; Shelby loam soil.

[b] Plots 10½ × 90 ft., on three per cent slope; Mexico silt loam soil.

[c] No soil treatment; other systems received lime and starter treatments on small grain 1941 through 1946, and full treatment, based on soil tests, 1947 through 1950.

[d] Grazing rate, 1–4 animal units per acre, four to six months per year.

experiment stations by Bennett during the period, 1930–33. They were patterned in general after the earlier work in Missouri by Miller and Duley. An example of the type of data secured from these studies is shown in Table 3, where average annual losses for several rotations are shown. These rotation averages were secured by averaging the annual runoff and soil losses from plots on which all crops of the rotation were grown.

While annual losses under different rotations reflect the effect of different crops or crop sequences they do not show the cause of favorable or unfavorable losses [*Smith*, 1946]. This can be revealed by study of crop-year or preferably crop-period losses. Crop-year losses are those that accumulate during the full calendar year in which a crop is harvested. Crop-period losses, except for meadows with more than one harvest year, are those that accumulate from the beginning of seedbed preparation or harvest of the preceeding crop, to harvest of the crop or beginning of seedbed preparation for the following crop. Crop-period losses obviously represent a more accurate measure of a crop or crop sequence in control of runoff and erosion than do crop year losses. Crop-year losses, however, are more easily tabulated and used for estimating rotation losses for systems not under measurement than are crop period losses. Crop-year soil losses for the different sequences in which crops may be grown have been tabulated by Soil Conservation Service for use by their farm planning technician in limiting field soil loss to a predetermined level. Table 4 shows losses by crops for two rotations of Table 3, tabulated by crop periods and by crop years.

Study of the cover factor on small runoff plots, which represent a small segment of a field area, has been necessary because of the large number of crop and cover combinations that are possible or are in use. Extrapolation of the data has been necessary when applying it to field areas, although relative values for the different cover combinations have been used directly. Erosion measurements from the plots include sheet erosion and some rill erosion, although rill erosion did not develop to the full extent on these short slopes.

Osborn [1954], in studying the effect of density of range cover on erosion, used the rainfall simulator with splash cups and small runoff plots 12 × 18 inches. He showed that a cover of 3500 lb/ac of ordinary crops or grasses and 6000 lb/ac of tall coarse crops and weeds would control 95 pct of rain drop energies. As crop cover decreased below these amounts, rates of soil splash increased very rapidly.

Soils—Soil factors which express the relative erodibility of different soils were developed in 1945 by a group of soil conservation research workers under the direction of *Musgrave* [1947]. Annual losses for continuous corn were adjusted to a common slope of nine per cent and a common 30-minute rainfall amount. Further development was made during the next several years by operations and research workers of the Soil Conservation Service for the mid-western states. Interpolations for soils on which soil losses had not been measured were made by considering the physical properties of the soil after soil technologists had classified the soils by characteristics considered to be related to potential erosion hazards.

TABLE 4 – *Runoff and soil loss tabulated by crop period and by crop year for two rotations of Table 3*

Crop or Cover	Crop Period		Runoff		Soil loss	
	Dates	Rainfall	Crop period	Crop year	Crop period	Crop year
		inch	inch	inch	tons/ac	tons/ac
Corn	Apr. 27–Oct. 8	23.23	5.48	10.89	6.95	7.78
Corn stalks	Oct. 9–Apr. 28	12.48	4.49	...	.95	...
Oats	Apr. 29–July 3	14.15	5.11	12.60	3.18	4.47
Oat stubble	July 4–Apr. 26	28.47	8.40	...	1.08	...
Rotation average		39.16	11.74	11.74	6.08	6.08
Corn	Apr. 27–Oct. 7	23.23	3.58	8.21	3.01	3.65
Wheat	Oct. 8–June 28	26.29	6.45	7.58	2.28	2.05
Sw. Clover (Hay)	June 29–Apr. 26	67.97[a]	13.63	7.87	.78	.37
Rotation average		39.16	7.89	7.89	2.02	2.02

[a] Twenty-two month period.

A more recent attempt at determining soil factors made use of erosion rates under a corn-small grain-meadow rotation from four Soil Conservation Experiment Stations in the midwest. Adjustments were made for differences in slope. Table 5 shows some of these factors. They cannot be considered precise factors, largely because of lack of a satisfactory method of adjusting for rainfall differences.

Osborn [1954] used soil splash cans to determine differences in soil detachability rates. He found a wide range of rates even for the same soil. He reported that detachability rates of medium-textured soils averaged somewhat lower than those of either fine-textured or coarse-textured soils. This technique appears to have excellent possibilities for rating of a large number of soils as to their erosivity.

Management—This factor is a measure of deviation in field application of soil, crop, and conservation practices from standards maintained in operation of the experimental cropping system plots from which runoff and erosion measurements were secured. This includes timeliness of operations, management of crop residues, fertilizer treatments, precision in application of conservation practices, and use of the optimum number of seedbed operations. So far, evaluation has been largely by judgment. It is generally a plus or minus 20 to 30 pct deviation from normal calculated rates.

Practices for reduction of field erosion losses—The three practices for control of sheet and rill erosion other than cover or cropping are contour farming, strip cropping, and terracing. All three are somewhat similar in their actions and operate to control (or limit) both sheet and rill erosion. They are frequently referred to as supporting practices.

Contour farming—This practice relies on a series of contour rows to drain runoff to a waterway. As such, it reduces the steepness of slope over which overland flow passes. But since the capacity of the contour rows to carry runoff is small, the practice breaks down as the storm runoff intensity increases. When this occurs, soil loss approaches, or may even exceed, that for up-and-down-hill farming. The relative value of contouring is expressed as a ratio (decimal) of the up-and-down-hill soil loss. The lowest numerical value of the ratio (highest effectiveness) is for the intermediate slopes. The factor, or ratio, approaches unity for both the flatter and the steeper slopes. As the slope increases, the standard corn row cross-section available for holding runoff decreases and becomes zero at a slope of about 25 pct. Hence, at this point the practice factor value is considered unity. On the extremely flat slopes the up-and-down-hill soil loss approaches that of contouring and becomes equal to it at a zero slope. This too results in a practice factor value of unity. Recommended values for different slope groups are shown in Table 6.

Table 5 – *Soil factors calculated and interpolated for several Midwest soils*

Soil	Location	Relative soil loss
Marshall	Clarinda, Ia.	1.00
Mexico	McCredie, Mo.	1.25
Shelby	Bethany, Mo.	1.25
Fayette	LaCrosse, Wisc.	1.25
Baxter	...	1.00
Union	...	1.25
Ida	...	1.50

Table 6 – *Recommended conservation practice factors for estimating field soil loss*

	Slope group	P_c (contouring)	P_{sc} (strip cropping)[a]	P_{tc} (terracing & contouring)
	pct			
A	1.1- 2	0.60	0.30	...
B	2.1- 4	0.50	0.25	0.10
C	4.1- 7	0.50	0.25	0.10
D	7.1-12	0.60	0.30	0.12
E	12.1-18	0.80	0.40	0.16
F	18.1+	1.00	0.50	...

[a] A system using a four-year rotation of corn-small grain-meadow-meadow.

Strip cropping—Strip cropping functions by filtering suspended soil from runoff as it passes from a cultivated strip to a sod strip. Plot measurements at LaCrosse, Wisconsin, and Bethany, Missouri, indicate that the soil loss from the strip-cropped field is about 50 pct of that with contouring. Since the sod strip acts as a filter, the measured loss from the field is similar to that from the end of a terrace.

Terracing—This practice breaks a long slope into segments of shorter length. Runoff is intercepted by the channel and conducted across the field at a non-eroding velocity. This results in deposition in the terrace channel of part of the soil transported to the channel. The average ratio of soil loss from contour farmed terraced fields to that from unterraced fields not farmed on the contour is about 0.04. A somewhat larger factor is frequently used. Terrace factors (including contouring) were developed by assuming that (1) terraces would divide the average slope length into six segments, (2) average soil loss per unit area is proportional to $L^{0.5}$, and (3) contour farming would make the same reduction as when used alone. Factors thus determined are shown in Table 6.

Estimating field soil loss—Smith [1941] published a graphical method for applying the results of plot measurements to field areas by use of Zingg's length and per cent of slope equations. The method was further developed for use in Iowa by *Browning* and others [1947]. Other articles on the subject were published by *Musgrave* [1947] and by *Smith* and *Whitt* [1947, 1948]. They used the method to prepare crop-practice recommendations for Missouri [*Smith, Whitt* and *Miller*, 1948]. All this work was based on use of the following empirical equation for estimating field soil loss.

$$A = C\,S\,L\,K\,P\,M$$

in which

A is average annual field soil loss in tons per acre.

C is average annual plot soil loss in tons per acre for a selected rotation with farming up and down slope (Table 3, Mexico silt loam plot data).

S and L are relative factors for per cent (S) and length (L) of slope adjusted to give unity loss on a three per cent slope 90 ft long, the same as the Mexico silt loam plot data shown in Table 3; see Tables 7 and 8 for factor values.

K is a soil factor taken directly from Table 5 since the soil factor for Mexico silt loam is 1.0. Soil factor values must be relative to a unity value for the soil of the plots from which C values are secured.

P is the factor for conservation practices in relation to a unity value for up-and-down-hill farming; see Table 6 for factor values.

TABLE 7 – *Percent slope-soil loss factors based on the parabolic equation for four Midwest soils and adjusted to a unity value on a three per cent slope*

Slope	Factor	Slope	Factor
pct		pct	
0	0.25	9	3.85
1	0.45	10	4.49
2	0.70	12	5.94
3	1.00	14	7.59
5	1.75	16	9.44
6	2.19	18	11.48
8	3.04	20	13.73

TABLE 8 – *Length of slope-soil loss factors*

Length	Factors			Length	Factors		
	Exponents of L				Exponents of L		
	0.4	0.5	0.6		0.4	0.5	0.6
ft				ft			
72.6	0.9	0.9	0.9	500	2.0	2.4	2.8
90	1.0	1.0	1.0	600	2.1	2.6	3.1
100	1.0	1.1	1.1	700	2.3	2.8	3.4
200	1.4	1.5	1.6	800	2.4	3.0	3.7
300	1.6	1.8	2.1	900	2.5	3.2	4.0
400	1.8	2.1	2.5	1000	2.6	3.3	4.2

M is the management factor, with management of the crop rotation plots considered unity.

With this equation and a limited amount of field data, the conservationist can select the cropping system and conservation practice necessary to limit field loss to a specific figure. The equation can also be used to estimate rates of erosion before and after application of a land program to determine effectiveness of control and to place a monetary value on benefits. It must be recognized that the method is empirical, and as such it is subject to the limitations of empirical relationships. It is the best approach now available for the uses listed above. With added research and use, its accuracy can be improved.

REFERENCES

BAUR, A. J. (chm.), Soil and water conservation glossary, *J. Soil Water Cons.*, **7**, pp. 41–52, 93–104, 144–156, 1952.

BORST, H. L., AND RUSSEL WOODBURN, The effect of mulching and methods of cultivation on runoff and erosion from Muskingum silt loam, *Agr. Eng.*, **23**, 19–22, 1942.

BROWNING, G. M., C. L. PARISH, AND J. A. GLASS, A method for determining the use and limitation of rotation and conservation practices in control of soil erosion in Iowa, *J. Amer. Soc. Agron.*, **39**, 65–73, 1947.

COOK, H. L., The nature and controlling variables of the water erosion process, *Soil Sci. Soc. Amer. Proc.*, **1**, 60–64, 1936.

EKERN, P. C., Raindrop impact as the force initiating soil erosion, *Soil Sci. Soc. Amer. Proc.*, **15**, 7–10, 1950.

EKERN, P. C., Problems of raindrop impact erosion, *Agr. Eng.*, **34**, 23–25, 1953.

ELLISON, W. D., Soil erosion studies, *Agr. Eng.*, **28**, 145–146, 197–201, 245–248, 297–300, 349–351, 402–405, 442–444, 1947.

GUNN, ROSS, AND G. D. KINZER, The terminal velocity of fall for water droplets, *J. Met.*, **6**, 243–248, 1949.

LAWS, J. O., Recent studies in rain drops and erosion, *Agr. Eng.*, **2**, 431–434, 1940.

LAWS, J. O., Measurement of fall velocity of water drops

and rain drops, *Trans. Amer. Geophys. Union*, **22,** 709–721, 1941.

Laws, J. O., and D. A. Parsons, Measurements of fall velocities of water drops and rain drops, *SCS T. P. 45*, U. S. Dept. Agr., November 1941.

Laws, J. O., and D. A. Parsons, The relation of rain drop size to intensity, *Trans. Amer. Geophys. Union*, **24,** 452–459, 1943.

Musgrave, G. W., The quantitative evaluation of factors in water erosion, a first approximation, *J. Soil Water Cons.*, **2,** 133–138, 1947.

Neal, J. H., The effects of degree of slope and rainfall characteristics on runoff and erosion, *Mo. Agr. Exp. Sta. Res. Bul. 280*, 47 pp., 1938.

Osborn, Ben, Field measurement of soil splash to evaluate ground cover, *J. Soil Water Cons.*, **8,** 255–260, 1953.

Osborn, Ben, Soil splash by raindrop impact on bare soil. *J. Soil Water Con.*, **9,** 33–38, 1954.

Osborn, Ben, Effectiveness of cover in reducing soil splash by raindrop impact, *J. Soil Water Cons.*, **9,** 70–76, 1954.

Osborn, Ben, How rainfall and runoff erode soil, *Water:* Year Book of Agr. U. S. Dept. Agr., pp. 126–135, 1955.

Smith, D. D., The effect of crop sequence on erosion under individual crops, *Soil Soi. Soc. Amer. Proc.*, **11,** 532–538, 1946.

Smith, D.D., Interpretation of soil conservation data for field use, *Agr. Eng.*, **22,** 173–175, 1941.

Smith, D. D., D. M. Whitt, A. W. Zingg, A. G. McCall, and F. G. Bell, Investigations in erosion control and reclamation of eroded Shelby and related soils at the conservation experiment station, Bethany, Missouri, 1930–42, *U. S. Dept. Agr. Tech. Bul. 883*, 175 pp., 1945.

Smith, D. D., and D. M. Whitt, Estimating soil loss from field areas of claypan soils, *Soil Sci. Soc. Amer. Proc.*, **12,** 485–490, 1947.

Smith, D. D., and D. M. Whitt, Evaluating soil losses from field areas, *Agr. Eng.*, **29,** 394–396, 398, 1948.

Smith, D. D., D. M. Whitt, and M. F. Miller, Cropping systems for soil conservation, *Mo. Agr. Exp. Sta. Bul. 518*, 26 pp., Sept. 1948.

Zingg, A. W., Degree and length of land slope as it affects soil loss in runoff, *Agr. Eng.*, **21,** 59–64, 1940.

Department of Soils and Department of Agricultural Engineering, University of Missouri, Columbia, Missouri (*D.D.S.*) *now at Eastern Soil and Water Management Branch, SWC-ARS, Plant Industry Station, Beltsville, Maryland; and, Agricultural Engineering Department, Purdue University, Lafayette, Indiana* (*W.H.W.*)

(Manuscript received April 1, 1957; presented as part of Symposium on Watershed Erosion and Sediment Yields at the Thirty-Seventh Annual Meeting, Washington, D. C., May 1, 1956; open for formal discussion until May 1, 1958.)

This article is reprinted from the TRANSACTIONS of the ASAE (Vol. 12, No. 2, pp. 231, 232, 233 and 239, 1969)
Published by the American Society of Agricultural Engineers, St. Joseph, Michigan

29 Effect of Slope Shape on Erosion and Runoff

R. A. Young and C. K. Mutchler
ASSOC. MEMBER ASAE MEMBER ASAE

EFFECTS of length and degree of slope on soil loss and runoff have been studied extensively throughout the country. From these studies have evolved a number of empirical relationships between length and degree of slope and soil loss. The most recent and commonly accepted percent slope-soil loss relationship is that established by Smith and Wischmeier in 1957 (1)*:

$$A = 0.43 + 0.30\ S + 0.043\ S^2$$

in which A equals loss in tons per acre and S equals percent slope.

In general, soil loss per unit area varies as the slope length to the 0.5 ± 0.1 power (1). Thus the slope-length equation for both length and steepness is

$$LS = \frac{L^{1/2}}{100}(0.76 + 0.53\ S + 0.076\ S^2)$$

where LS is the dimensionless topographic factor used in the universal soil loss equation [2], L is slope length in feet, and S is average percent slope.

Most field data have been collected from erosion plots established on uniform slope segments. Consequently, estimating soil losses and planning soil conservation practices on areas of complex slopes have required much local judgment in adjusting estimate procedures.

The work reported here had two objectives: to extend the use of principles developed for uniform slopes to irregular topography and to investigate the basic concepts of soil movement on land of uniform slope and on land with various slope shapes. Results of the study showed a substantial effect of slope shape on erosion and runoff.

EXPERIMENTAL PROCEDURES

Tests were conducted from 1964 through 1966 on erosion plots located in west central Minnesota on complex topography including concave, uniform, and convex slopes. All plots averaged 8½ percent slope, but the slope on each of the concave and convex plots ranged from 2 to 14 percent. The slope-shape plots were 13⅓ by 75 ft long and situated in groups of three for simultaneous testing. The soil on the convex plots was mainly Buse loam, with some Barnes loam in the lower portions. The remaining plots were primarily Barnes loam, although the concave plots had some Alcester loam near the lower ends. Runoff and soil-loss data were adjusted to a common soil-type base using data from adjacent soil erodibility plots.

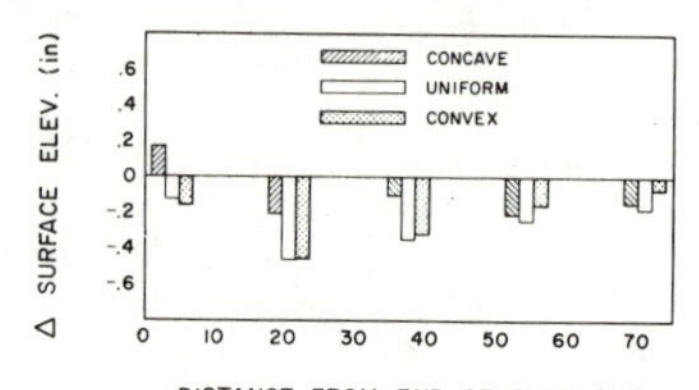

FIG. 1 Change in surface elevation—average for all plots.

The plots were tested under three cover conditions – corn, oats, and cultivated fallow – in order to get a variety of surface conditions for testing the slope-shape variable. All planting and tillage operations were done up-and-down slope. The cultivated fallow treatment received the same tillage as the corn treatment, except that a spring tooth field cultivator was used for cultivation rather than a row cultivator with sweeps. Since erosion tests were planned for only the crop stage period immediately following planting and seeding, the entire sequence of tillage and planting was delayed until summer to facilitate testing with the rainulator.

Standard rainulator procedures (3) were used to obtain runoff and soil loss measurements. Each test consisted of two one-hour applications at an intensity of 2½ in. per hour. The first was conducted at existing moisture conditions and the second about 24 hr later with the soil near saturation. All tests were conducted within 30 days after corn and oats seeding, corresponding to crop-stage period one (4), since this is the period in which a large number of the most erosive rainstorms usually occur in Minnesota.

Just prior to every run, overhead photographs of the surface of each plot were taken to evaluate surface condition and crop canopy, and antecedent soil moisture in the upper 12 in. was determined. Wind velocity and direction, air temperature, water temperature, and relative humidity were also recorded for each run. Particle-size analyses were made of the surface soil and of the subsoil to a depth of 5 ft, and also of the eroded sediment.

Microrelief measurements (5) taken in the fallow plots before and after each run gave indications of the pattern of soil movement within the test plots.

RESULTS AND DISCUSSION

Crop and Slope-Shape Effects

The analysis of variance summary in Table 1 shows the significance of the experimental factors affecting measured soil and water losses. The variation among duplicate runs made over a period of 3 years was used to develop a within-subclass error term.

The type of run greatly influenced both soil and water losses; an average of 2.4 T/A more soil and 0.37 inch more runoff was measured from the wet runs. The use of the dry and wet runs had two purposes: (**a**) to furnish a range of moisture regime for observing other experimental effects and (**b**) to generate data for examining the effects of antecedent moisture and antecedent erosion on soil and water losses. The type of run did not interact with crop or slope shape, so the values obtained from dry and wet runs were combined for analysis of crop and slope-shape effects.

Vegetative growth is normally not enough to provide much protective crop cover during crop-stage period one. However, the oat growth was rapid enough in this case that an effect of cover was observed, as evidenced by the low average soil losses from the oat plots (Table 2). Soil losses from the corn plots were slightly greater than from the fallow plots, but not significantly so. Apparently the crop canopy on the corn plots was insufficient to provide effective protection from erosion, and the surface condition caused by planting on these plots tended to increase the soil loss slightly. The confounding of cover and surface effects precludes further inference on this.

Soil losses from the uniform and convex slopes did not differ significantly. The concave plots, however, eroded about 8 tons per acre less than the uniform plots because of the flattened slope

Paper No. 67-706 was presented at the Winter Meeting of the American Society of Agricultural Engineers at Detroit, Mich., December 1967, on a program arranged by the Soil and Water Division.

The authors—R. A. YOUNG and C. K. MUTCHLER—are research agricultural engineers, corn belt branch, soil and water conservation research division, Agricultural Research Service, U.S. Department of Agriculture.

Author's Note: The research study on which this paper is based was made in cooperation with the Minnesota Agricultural Experiment Station.

* Numbers in parentheses refer to the appended references.

TABLE 1. ANALYSIS OF VARIANCE OF FACTORS CAUSING VARIATION OF SOIL AND WATER LOSSES MEASURED AT THE END OF RAINULATOR PLOTS

Source	Df	Significance	
		Soil loss	Runoff
Within subclasses (error)	90		
Between subclasses	(17)		
Type of run (T)*	1	1%	1%
Crop (C)†	2	1%	NS
Corn vs fallow	1	NS	....
Shape of slope (S)	2	1%	5%
Unif vs convex	1	NS	NS
Unif vs concave	1	1%	NS
T × C	2	NS	NS
T × S	2	NS	NS
C × S	4	NS	NS
T × C × S	4	NS	NS

* Wet or dry run.
† Oats, corn, or fallow.

TABLE 2. AVERAGE SOIL LOSSES FROM SLOPE-SHAPE EROSION PLOTS

Crop	Slope shape			Average
	Concave	Uniform	Convex	
	Tons per acre			
Corn	6.9	16.8	18.1	13.9
Oats	2.5	6.5	7.2	5.4
Fallow	4.6	15.1	15.7	11.9
Average	4.8	12.8	13.7	10.4

at the lower end of the concave plots. The convex plots, although having a higher percent slope at the lower end, perhaps did not have a supply of detached material available from the flatter upper portion of the plot. This may explain the small difference in soil loss between the convex and the uniform slopes.

Water losses due to slope shape followed a different pattern than did soil losses. Although not significant, the greatest difference occurred between the convex and the uniform slopes (Table 3).

Differences in surface condition and crop cover evidently had no net effect on infiltration and the resulting runoff. Since the tillage and planting of all the plots was up-and-down hill, surface detention should have been little affected, and none of the surfaces was deeply furrowed to cause a greatly different initial flow regime.

More water was lost from the convex plots than from either the uniform or the concave plots, indicating either that average runoff velocities on the convex plots were higher than on the other shapes, or that recession times were greater. Since measured recession times varied little, we concluded that higher average runoff velocities caused greater water losses on the convex plots. Unfortunately runoff velocities were not measured.

The above-mentioned results raised some interesting questions on the relationship of erosion to runoff. Erosion is a result of soil detachment caused primarily by raindrop action and soil transport caused primarily by runoff. Since the simulated rainfall on all plots was identical and, assuming that soil detachment per unit rainfall was not significantly affected by the range of slopes involved here, variations in soil loss coming off the end of the plots must have been primarily a function of transport. Transport of eroded particles is thought to be a function of runoff velocity

$$G = K v^x$$

where G is a quantity of transported material, v is velocity, K is a proportionality factor, and the exponent, x, is about 4 (6, 7). Thus, soil transport rates are very sensitive to change in runoff velocity.

As the slope of the convex plots increases, the velocity of the runoff also tends to increase, allowing less infiltration and thus increasing the amount of runoff as well as the amount of soil loss. On the concave plots, the runoff velocity decreases as the slope becomes flatter, resulting in a reduction in the capacity of the water to transport soil. As a result, infiltration increases and runoff and soil loss decrease. This reasoning is supported by the soil loss data shown in Table 2.

The brevity of the above discussion does not imply that the relationship of runoff and soil loss is a simple one. A number of other factors need investigation to allow a fuller discussion.

TABLE 3. AVERAGE WATER LOSSES FROM SLOPE-SHAPE EROSION PLOTS FROM AN APPLICATION OF 2.5 INCHES IN ONE HOUR

Crop	Slope shape			Average
	Concave	Uniform	Convex	
	Inches			
Corn	1.21	1.30	1.43	1.31
Oats	1.26	1.17	1.46	1.30
Fallow	1.05	1.27	1.39	1.24
Average	1.17	1.25	1.43	1.28

Other Factors Affecting Erosion

Various factors of topography, cover, atmospheric condition, and antecedent conditions were examined in regression analyses in an attempt to assess their relative importance to soil loss.

Table 4 gives the multiple-regression coefficients (R^2), the partial regression coefficients (**b**), and the standard partial regression coefficients (b') for corn, oats, fallow, and all crops combined, respectively. For each individual cropping condition, a greater portion of the variance was accounted for by the regression than the 58 percent accounted for when all cropping conditions were combined. Soil erodibility under these different cropping conditions is apparently affected differently by the factors studied or is a function of extraneous variables not included in the regression analysis. Similar increases in variance accountability were found when soil losses were divided into smaller classes on the basis of type of run as well as crop.

Percent Slope Steepness of slope is a primary variable in erosion, and its function has been fairly well defined for a uniform slope. However, in the case of a complex slope, the question is that of determining where to measure it. Or, said another way, how much slope irregularity may be ignored?

Fundamentally the amount and size of sediment that can be carried is dependent on the depth and velocity of the sheet flow (8). In the slope equation, the length factor is primarily a reflection of flow depth since depth accumulates with length. Velocity is reflected primarily in the percent-slope expression. Assuming that slope length and percent effects are independent, a relative percent slope can be obtained from the percent slope expression given in the introduction. Using only the fallow plot data of Table 2, the relative percent slope is 3.6 percent for the concave plot and 9.0 percent for the convex plot. From Table 5, the measured slope is 3.3 percent for the lower 15 ft of the concave slope and 3.7 percent for the lower 20 ft. This supports a hypothesis that the appropriate percent slope for use in the erosion equation can be measured from the portion of the slope immediately above the point of measurement or concern (9). However, the data from the convex plots do not support this notion because the lower half of the plots (Table 5) were essentially uniform and the curvature occurred mostly in the upper portion. Total soil loss from the convex plots differed little from that on the uniform plots.

Simple regressions of soil loss on percent slope of various plot increments indicated that the slopes of the lower 10, 15, and 20-ft increments were most closely correlated with total soil loss

TABLE 4. REGRESSION ANALYSIS FOR SOIL LOSS

	b				b'*			
	Corn	Oats	Fallow	All	Corn	Oats	Fallow	All
15-ft slope	1.702	0.724	1.544	1.312	0.769	0.563	0.638	0.568
Year	2.657	1.055	3.296	2.156	0.600	0.417	0.649	0.461
Days after tillage	−11.340	−7.985	−5.482	−6.893	−0.454	−0.375	−0.203	−0.296
Date of run	4.475	−0.137	2.114	2.153	0.367	−0.020	0.150	0.168
Type of run	2.061	0.284	2.038	1.447	0.274	0.066	0.236	0.182
Wind vector	0.079	−0.024	0.190	0.108	0.077	−0.041	0.162	0.100
0 to 12-in. ant. moisture	−0.170	0.130	−0.076	0.045	−0.114	0.162	−0.045	0.030
Percent cover	0.047	−0.002	1.055	−0.066	0.050	−0.011	0.065	−0.215
Constant	−38.047	1.662	−25.470	−21.455				
R^2	0.691	0.644	0.613	0.581				

* b' represents the relative importance of the variable's effect on erosion.

(Table 6). The slope of the lower 15 ft was chosen as the variable to be included in the multiple regression, since, when it was combined with the other variables, it resulted in a slightly higher R^2 than did the slope of either the lower 10 or 20 ft. The determination of the critical length for computing percent slope needs further study and refinement.

Cover The amount of vegetative cover on the corn plots at the time of testing was only 9 percent, not enough to have a noticeable effect on soil loss (Table 2). An average oat cover of 50 percent at the time of testing significantly reduced the amount of soil loss from all oat plots (Table 1) and tended to mask the effects of most of the other variables measured.

Percent vegetative cover was of little significance in the individual regressions because of its high correlation with crop (r = 0.76). The amount of cover variation within the corn, oats or fallow treatments was not enough to show significant effects. Not until the data for all cover conditions were combined did the percent vegetative cover become significant.

Days after Tillage The effect of tillage was primarily reflected in the number of days following the last tillage operation, since no gross differences in tillage were imposed. This factor proved to be the third most important variable in the regression, as indicated by the b' values in Table 4. Tillage operations result in the mechanical breakup of the soil and the turning over and exposure of new soil material to weathering. With the passage of time after tillage, the finer, more erodible particles are removed from the surface by erosion or eluviation, and the soil gradually becomes less erodible. The greater importance of this variable on the corn plots, as indicated by the standard partial regression coefficients, could be due to the type of tillage tool used. The effect of this variable on the oat plots, which had less tillage than the other plots, was highly correlated (r = 0.63) with the effect of plant cover, since planting was the final tillage operation on those plots before testing.

Date of Run Recent unpublished data from other experiments have indicated that the time of year can have a significant effect on erosion. For this reason, date of run was included in the regression analysis. It did not prove to be a significant factor here, however, because of the narrow range of dates involved.

Antecedent Soil Moisture Most of the significance of antecedent moisture in the regression was probably taken up by the type of run, with which it was highly correlated (r = 0.51); by the slope of the bottom 15 ft, with which it was also highly correlated (r = 0.53), and by the year of testing. Initial runs were always made at existing field moisture, which would not have varied too greatly over the testing season in any one year, and the wet runs were always made at nearly saturated conditions.

Since the concave slopes are naturally less well drained than the convex slopes, they were always somewhat higher in moisture content than the convex plots, especially prior to the wet runs. Thus, the effect of antecedent soil moisture would be partially compensated for by the natural drainage characteristics of the plot shapes.

Wind Wind speed and direction, included in the regression in the form of a vector, proved to be insignificant.

Soil Displacement Measurements of losses from the bottom end of a plot do not necessarily reflect the pattern or extent of soil movement within the plot itself. The concave plots, while yielding relatively small losses, still suffered rather serious erosion on the upper parts of the plot. Two things are considered here: movement of soil from a point and particle size of soil transported.

Soil Movement Microrelief measurements indicated that, on the concave slopes, the average net change in soil-surface elevation within a plot was greatest about 20 ft from the top of the plot and decreased downslope, being usually positive at the bottom as a result of deposition (Fig. 1). Deposition on the lower part of the concave plots was also visually evident. On the convex and uniform slopes, net changes in elevation usually reached a maximum about three-fourths of the way down from the top of the plot and then began to decrease. It may be that, for the amount of rainfall excess, this is the point at which the amount of eroded material entering a unit area from above began to approach the amount of material leaving that area, thus causing the net displacement to decrease.

Sediment Particle Size A particle-size analysis of the surface soils and of the sediment washed off the plots, presented in Table 7, shows that while the texture of the surface soil of all three shapes was very similar, eroded soil from the concave slopes from both the initial and the wet runs was very low in sand and quite high in silt and clay particles, most of the larger sand particles apparently having been deposited on the lower portion of the plots. Soil eroding during the initial runs from the uniform and convex plots was somewhat low in sand particles, but the composition from the wet runs was quite similar to that of the original surface soil.

TABLE 6. COEFFICIENTS OF DETERMINATION FROM SIMPLE REGRESSIONS OF SOIL LOSS ON PERCENT SLOPE OF VARIOUS PLOT INCREMENTS

Distance from plot end (feet):	5	10	15	20	25	30	35	75
r^2	0.429	0.434	0.434	0.434	0.407	0.396	0.394	0.065

TABLE 7. AVERAGE PARTICLE-SIZE DISTRIBUTIONS* OF SURFACE SOIL AND OF ERODED SOIL FROM RAINULATOR APPLICATIONS

Shape	Surface soil	Eroded soil	
		Initial	Wet
	Percent		
Concave			
Sand	49	24	27
Silt	29	37	38
Clay	22	39	35
Uniform			
Sand	50	34	46
Silt	26	35	28
Clay	24	31	26
Convex			
Sand	48	39	42
Silt	27	30	30
Clay	25	31	28

* Sand = > 50 μ.
Silt = 2 μ to 50 μ.
Clay = < 2 μ.

TABLE 5. SLOPE OF FALLOW PLOTS CALCULATED UNDER THE ASSUMPTION OF NO CURVATURE FOR THE PORTION OF THE SLOPE USED

Distance from plot end (feet)	5	10	15	20	25	30	35
				Percent			
Concave	2.4	2.8	3.3	3.7	4.3	4.7	5.1
Uniform	10.6	10.2	9.9	9.9	9.7	9.6	9.7
Convex	9.9	10.5	11.1	11.4	11.6	11.5	11.4

Conclusions

Erosion varies significantly from a given slope length with the same average degree of slope but with different slope configurations. Runoff also varies according to slope configuration but to a lesser degree. These variations are essentially independent of the type of surface cover existing on the ground.

There is a characteristic slope segment, the percent slope of which correlates best with soil loss from that entire slope. For the topographic and rainfall conditions described in this experiment, that slope segment was the bottom 15 ft of the slope.

The amount of lapsed time since the last tillage operation before testing was a significant factor affecting soil loss. This was due to the fact that, with the passage of time, erosion and eluviation tend to remove the more readily eroded particles, leaving the soil in a less erodible condition.

Antecedent soil moisture, while obviously having a significant role in erosion and runoff, appeared insignificant in this experiment because its effects were interrelated with type of run, slope shape, and year of testing.

Maximum soil displacement on the concave slopes took place in the upper one-third of the 75-ft plot, with deposition occurring at the bottom of the plot. On the convex and uniform slopes, the maximum displacement occurred about three-fourths of the way down the slope.

Type of run, initial or wet, and year of testing were the dominant factors affecting runoff from the corn and fallow plots, and amount of vegetative cover was the dominant factor affecting runoff on the oat plots.

Further study is needed to determine the exact pattern of soil movement on these slopes and the change in slope shape resulting from this movement.

References

1 Smith, D. D., and Wischmeier, W. H. Factors affecting sheet and rill erosion. Trans. AGU, 38:(6)889-895, December 1957.

2 Wischmeier, W. H., and Smith, D. D. A universal soil loss equation to guide conservation farm planning. Trans. 7th Internatl. Congr. Soil Sci., Madison, Wis., I:418-425, 1960.

3 Meyer, L. D. Use of the rainulator for runoff plot research. Soil Sci. Soc. Am. Proc. 24:(4)319-322, 1960.

4 Wischmeier, W. H. Cropping management factor evaluations for a universal soil loss equation. Soil Sci. Soc. Am. Proc. 24:(4)322-326, 1960.

5 Allmaras, R. R., Burwell, R. E., Larson, W. E., Holt, R. F., and Nelson, W. W. Total porosity and random roughness of the interrow zone as influenced by tillage. Cons. Res. Rpt. No. 7, USDA, ARS.

6 Laursen, E. M. Sediment-transport mechanics in stable-channel design. ASCE Transactions 123:195-206, 1958.

7 Meyer, L. D., and Monke, E. J. Mechanics of soil erosion by rainfall and overland flow. **Transactions of the ASAE** 8:(4)572-577, 580, 1965.

8 Lutz, J. F., and Hargrove, B. D. Soil movement as affected by slope, discharge, depth and velocity of water. No. Car. State Col. Agr. Exp. Sta. Tech. Bul. 78, 1944.

9 Onstad, C. A., Larson, C. L., Hermsmeier, L. F., and Young, R. A. A method for computing soil movement throughout a field. **Transactions of the ASAE** 10:(6)742-745, 1967.

Applications

VI

Editors' Comments on Papers 30, 31, and 32

Studies of hillslopes by geologists and geomorphologists have usually been highly academic in nature. Meanwhile, as previously noted, agricultural and highway engineers, among others, have produced numerous reports on practical aspects of hillslope form and process. Future hillslope studies may benefit from increased cooperation between these diverse groups. A two-way exchange of information and techniques is necessary for the well-being of any science, and the papers in this part show that this can indeed be the case for slope research.

Geomorphological and slope mapping are of importance in many countries, especially as reconnaissance techniques and as a means of teaching young geomorphologists to view the landscape critically. It is encouraging that soil scientists may be able to employ this geomorphic technique, as the paper by Curtis, Doornkamp, and Gregory demonstrates.

The joint effort by Schumm and Lusby, geomorphologist and hydrologist, respectively, is another example of the way in which slope studies can and should be integrated with other research projects. Hydrologic anomalies detected in data collected over many years in arid watersheds were explained by the results of geomorphic studies of slope erosion processes. In turn a knowledge of the field area's climatic and hydrologic characteristics facilitated geomorphic interpretations.

Finally, the short article by Meyer and Kramer, of the U.S. Department of Agriculture, is an attempt to apply information derived from studies of hillslope form and process to land-use planning. A knowledge of the erosion characteristics of a given slope profile may assist engineers in the planning of construction in order to minimize environmental damage.

We believe that additional applications of geomorphic studies are desirable and necessary if the study of slopes is to progress. The reader who has come this far will realize that much has been accomplished, but far more needs to be done before an understanding of slopes in their diverse geologic and climatic settings can be obtained and these results applied to modern environmental problems.

30

THE DESCRIPTION OF RELIEF IN FIELD STUDIES OF SOILS

L. F. CURTIS, J. C. DOORNKAMP, AND K. J. GREGORY

(*Departments of Geography, Universities of Bristol, Nottingham, and Exeter*)

Summary

Landscape information relevant to site description is discussed. It includes surface form (in plan and profile), micro-relief, aspect, height above sea level, and relative relief of the site in the local setting. Techniques for description of surface form are described which require the recognition and mapping of breaks and changes of slope bounding morphological units of the landscape. These units consist of 'flats' and 'slopes' that make up the facets of the landscape and may possess characteristic angles. The regional setting of the site is described by morphological map, values for drainage density and stream frequency, location with respect to drainage lines of a particular order, and the directional pattern of facets. Methods for the record of surface profiles by observation and measurement of small-scale morphological units are described. A slope classification appropriate to British conditions is put forward.

Introduction

MANY types of field study require accurate descriptions of topography. During a field study of soils, for example, it is usually necessary to describe the topography and an indication of relief is therefore normally given as part of the site description. Recommendations concerning the form which such relief descriptions should take are contained in existing field handbooks (*Soil Survey Staff*, 1951, *Soil Survey of Great Britain*, 1960). However, the development of new geomorphological techniques, including morphological mapping (Waters, 1958; Savigear, 1960; Bridges and Doornkamp, 1963) and methods of slope profile analysis (Savigear, 1956, 1960; Young, 1963), makes it desirable to consider modifications to existing methods of relief description. With this object in mind some new methods are introduced below, together with a slope classification which has been derived in part from the application of these methods.

The principal features of the landscape which characterize a site and are necessary in its description are:

A. The form of the surface, (1) the surface in plan, (2) the surface in profile.
B. The micro-relief features of the surface (e.g. ridge and furrow, earthflows).
C. The aspect of the site.
D. The height of the site above sea-level.
E. The relative relief of the site in the local setting.

Of the above mentioned features, (D) absolute height and (E) relative relief (i.e. the height of the site above the nearest valley floor) can be determined from O.S. maps. The aspect of the site (C) may be measured

Journal of Soil Science, Vol. 16, No. 1, 1965

by taking a compass bearing parallel to the line of steepest slope of the ground on which the soil profile is taken. For the most part the micro-relief features of the ground (B) will be included in the description of the form of the surface. Should these micro-relief features be too small to be included here, their presence and nature can be recorded under the space 'micro-relief features' on the field-data sheet. Particular note should be made of the features which show a repetitive pattern over the slope, e.g. ridge and furrow, terracettes.

The Form of the Surface

1. *The surface in plan*

There are two parts to the description of the surface in plan. First, an account of the area immediately surrounding the soil pit. Secondly, a consideration of the way in which it is related to its surroundings. The description of the immediate surroundings of the soil pit is important for a study of each individual profile but a description of the setting of each profile is of significance when several profiles in one area are to be related.

In most cases a morphological map is the best and most succinct description of the surface in plan. A map of the detailed surface morphology of the surroundings of the site may be made by a technique which requires the recognition of breaks and changes of slope bounding the morphological units of the landscape. Landscape is composed of an intricate network of two kinds of morphological units,[1] namely 'flats' and 'slopes'. A 'flat' is a unit which is less steep than the morphological units immediately above and below. This definition accords with that given by Sparks (1949, p. 167) in respect of the South Downs. These two types of morphological unit may be further subdivided on the basis of either morphology or origin.

The construction of a morphological map is illustrated in Fig. 1. Breaks of slope are angular and well defined in the field and they are represented by a continuous line. Changes of slope in cases where the angle of slope of a morphological unit changes gradually to the next are indicated by a broken line. An arrow-head is then placed against the line such that it lies on the side of steeper slope and pointing downhill. In this way a concave break, or change, of slope has the tip of the arrow-head pointing into the line, while the convex break, or change, of slope has the arrow-head pointing away from the line. By the delimitation of the breaks and changes of slope the area mapped is thereby divided into its morphological units. Rectilinear units are represented by a straight arrow pointing along the direction of maximum (true) slope annotated with the angle of slope in degrees. Convex units have a cross sign (×) on the shaft of the arrow, with the angles of slope near the top and bottom of the unit marked. The concave unit has a single bar across the shaft of the arrow, with slope values similarly recorded. Cliffs and close

[1] The term 'facet', by analogy with the facets of a cut gemstone (Stamp, 1961; Waters, 1958), is probably more appropriate but a facet has been defined in a different way by Savigear (1956).

associations of convex–concave breaks, or changes, of slope may also be recorded as illustrated in Fig. 1.

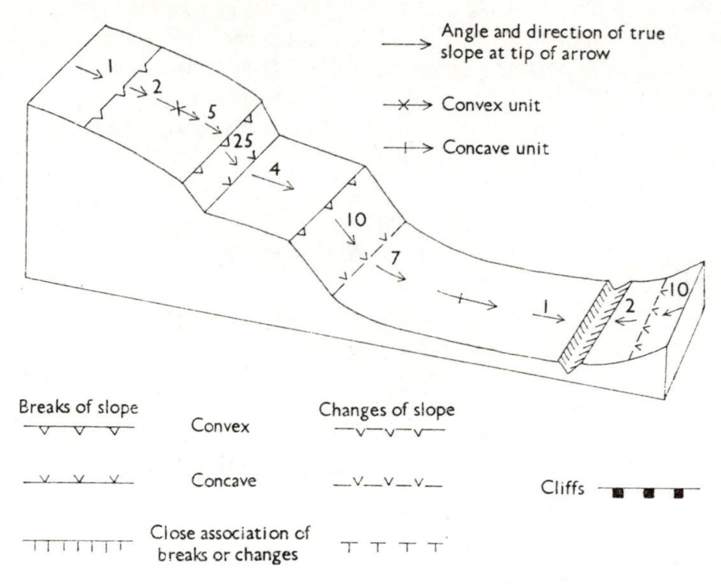

FIG. 1. Construction of morphological map.

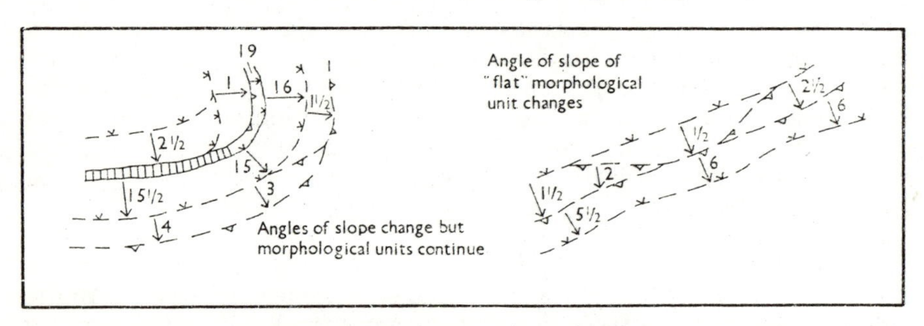

FIG. 2. Interrelationships of breaks and changes of slope.

The breaks and changes of slope which define the area of occurrence of morphological units seldom, if ever, die out. They are all interrelated and form a continuous pattern. The angle of slope of a morphological unit may change and this can be indicated by a change of slope subdividing the unit. The way in which breaks and changes of slope are interrelated is demonstrated in the simplified morphological maps (Figs. 2 and 11).

The general distribution and occurrence of morphological units is dictated by the geomorphological history of the area, while the angle of slope of each unit is a reflection of a series of localized processes and controls which have been exerted upon the facet since its initiation. In the case of a 'flat' unit initiated between two periods of stream incision the slope angle is subsequently adjusted to environmental controls and particularly to geological structure (including lithology) and aspect.

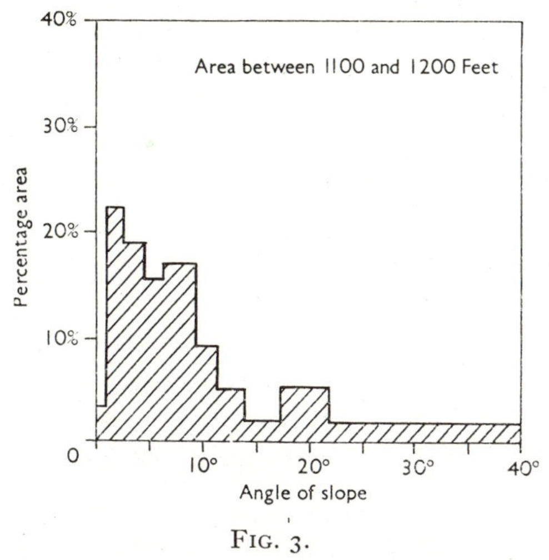

FIG. 3.

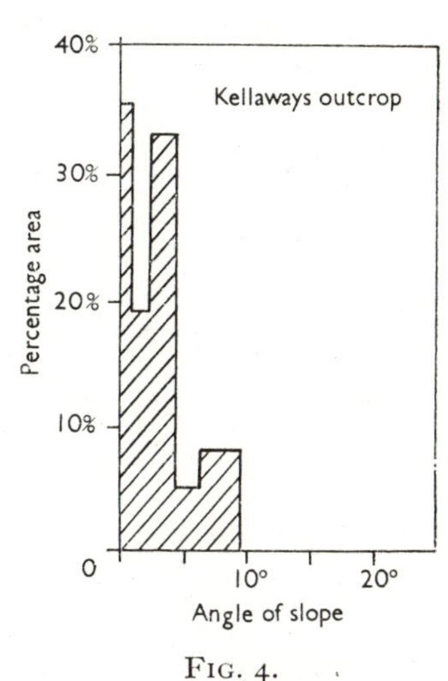

FIG. 4.

FIG. 3. Slope angle and area of slope for a sample region between 1,100 and 1,200 feet in the North York Moors. Percentage area is plotted according to eleven slope categories which are $0-\frac{1}{2}$, 1–2, $2\frac{1}{2}-4$, $4\frac{1}{2}-6$, $6\frac{1}{2}-9$, $9\frac{1}{2}-11$, $11\frac{1}{2}-13\frac{1}{2}$, 14–17, $17\frac{1}{2}-21\frac{1}{2}$, $22-27\frac{1}{2}$, 28 degrees and over.

FIG. 4. Slope angle and area of slope for the Kellaways outcrop in a sample region in the North York Moors. Percentage area is plotted according to the eleven slope categories used in Fig. 3.

Certain types of morphological units will have *characteristic angles* of slope in similar locations or on similar geological outcrops. In a given area certain types of 'flat' and 'slope' units will be found and these will recur throughout the entire area although their characteristic angles will vary according to the controls mentioned above. Fig. 3 shows the relationship between slope angle and area of morphological units mapped for a sample region between 1,100 and 1,200 feet in the North York Moors. The histogram has three maxima, the smallest of which occurs between $17\frac{1}{2}$ and $21\frac{1}{2}$ degrees; this corresponds to morphological units which occur on ground immediately adjacent to streams and rivers. Thus this type of unit has a characteristic angle between $17\frac{1}{2}$ and $21\frac{1}{2}$ degrees. The other two peaks correspond to 'flat' and 'slope' morphological units which occur on the moorland area and have characteristic values of 1–2 degrees and $6\frac{1}{2}-9$ degrees respectively.

Fig. 4 shows the relationship between slope angle and area of slope on the outcrop of Kellaways Rock (Upper Jurassic Sandstones with basal beds of shale in part of the North York Moors). There are three well-marked peaks on the histogram, each of which represents the characteristic angle for a particular type of morphological unit. The Kellaways outcrop is confined exclusively to the moorland areas and two of the maxima represent the 'flats' ($0-\frac{1}{2}$ degree) and the 'slopes' ($2\frac{1}{2}-4$ degrees). The third maximum ($6\frac{1}{2}-9$ degrees) represents strong slopes which occur on the resistant, fine-grained sandstones of the basal beds of the Kellaways outcrop. Seret (1963) reports similar measurements of slope values in Belgium which yielded a statistical diagram on which ten maxima can be traced. Each of the characteristic angles represented was related to the balance between a denudation agent and the resistance of the rock in the corresponding area.

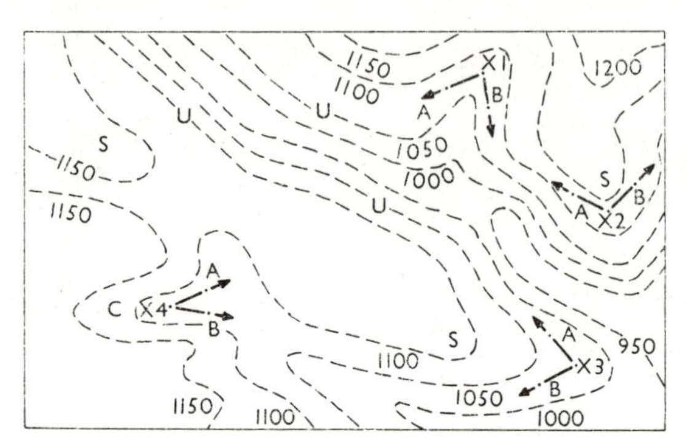

Fig. 5. Contours showing uniform (U), Cove (C), Spur (S) units. Curvatures: ×1 = 74°, ×2 = 102°, ×3 = 78°, ×4 = 33°.

When viewed from above (Fig. 5) the pattern of morphological units may form features which are (i) concave (termed a 'cove' by Aandahl 1948), (ii) convex (termed 'spur'), (iii) straight or uniform. A site description should include a note of the surface pattern (i.e. cove, spur, uniform) and the amount of curvature of the unit. The curvature can be determined by locating two points on a 1: 25,000 map (*A* and *B*, Fig. 5) on either side of the site *X* and at the same level but 200 yards distant. Bearings to *A* and *B* from *X* establish the smaller angle subtended at *X*, thereby giving a measure of the amount of curvature.

The regional setting of the site is largely determined by the type and amount of dissection of the landscape carried out by streams. The pattern of dissection corresponds to texture (Cotton, 1941, p. 216) and to drainage composition (Horton, 1945). Drainage composition is really a combination of drainage density and stream frequency where:

$$\text{Drainage density } (D_d) = \frac{\text{Total stream length}}{\text{Area of drainage basin}}$$

$$\text{Stream frequency } (F_s) = \frac{\text{Number of streams}}{\text{Area of drainage basin}}$$

Thus the regional setting may be stated by reference to drainage density and stream frequency of the drainage basin in which the site occurs and data for these should be included when sufficiently large areas of soil are under consideration (e.g. Moss, 1963).

The position of a site in a drainage basin may be referred to the orders of streams comprising the drainage net. A stream without tributaries is described as a first-order stream, two first-order streams combine to produce a second order, two second orders produce a third, and so on (Horton, 1945; Strahler, 1952, pp. 1120).

Thus the position of the site *S*, Fig. 6, may be described as follows: 'the site is located on the south side of the valley of a second-order stream at a height of *x* feet above the valley floor.' Furthermore, the

FIG. 6. Position of site in relation to stream order.

site can be defined in terms of distances *a* and *b* along the second-order valley.

In the case of sites taken between two valleys the position with regard to the nearest streams may be noted and in the case of watershed sites (called 'length of overland flow' by Horton, 1945) where drainage lines are absent the site may be located with respect to distance from the head of the nearest first-order stream.

The surface in plan may therefore be described by reference to (i) a morphological map, (ii) values for drainage density and stream frequency, (iii) location with respect to drainage lines of a particular order, (iv) surface pattern of morphological units (cove, spur, uniform) and these items are listed on the specimen description sheet (Fig. 10).

2. *The surface in profile*

The description of the profile form of the surface may be carried out in two steps. First, the major divisions of the slope profile are described, and in this case we are dealing with the surface *macro-profile*; secondly, the detailed morphological units contained within the major divisions are distinguished, and in this case we are concerned with the surface *micro-profile*.

(i) *The divisions of the surface macro-profile.* Savigear (1960, p. 158) states that 'any fully developed slope is divisible normally into three sections (i) the crestslope extending from the interfluve to where the slope steepens, (ii) the backslope consisting of the steepest morphological units of the profile, (iii) the footslope, a gently inclined section extending from the backslope to the valley centre. In places the back-slope may be cliffed.'

This division of slopes, Fig. 7*a*, is similar to that proposed by Wood (1942) who distinguished a waxing slope, extending down from the interfluve, a free face, a constant slope, and a waning slope which reaches

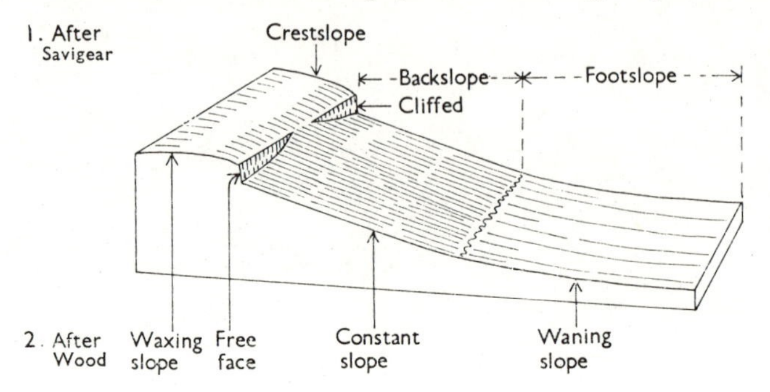

FIG. 7*a*. The subdivisions of a surface, macro-profile.

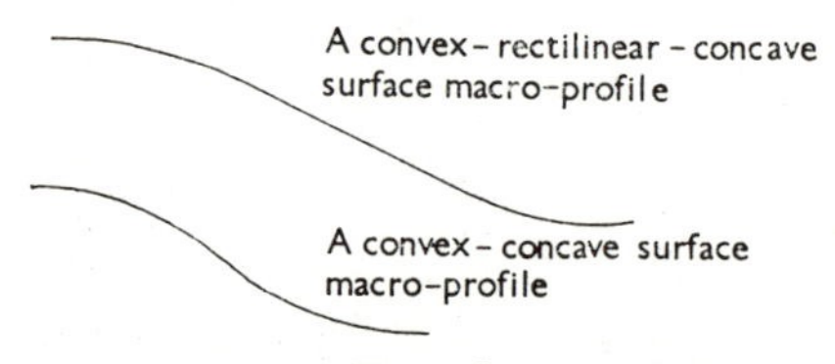

FIG. 7*b*.

to the valley centre. In both cases the general form recognized is that of a convex crest which is separated from a concave lowest portion by a rectilinear unit which may or may not be cliffed. If it is cliffed this usually occurs at its upper extremity.

Variations from this general form often depend on the extent of the rectilinear portion of the profile. If it is very long in comparison with the upper convexity and lower concavity the profile will differ distinctly from the case where the rectilinear portion is barely perceptible. In the latter case the whole of the profile is more accurately described as convex-concave, rather than convex-rectilinear-concave, as would be the case for the former profile (Fig. 7*b*).

Each profile encountered in the field must be described according to its own particular properties, but by making such a description with reference to the general convex-rectilinear-concave form, some standardization of description will result.

Complications arise when a hill-side is polyphase or polycyclic in profile. If there is a marked convex break (or change) of slope below the crestslope–backslope junction, the normal concavity of the lower part of the hill-side is interrupted; for it gives way to a second steepening. For purposes of description, see Fig. 8, such a double-phase slope can be separated into two parts by the marked convex boundary. Two particular sites, *X* and *Y*, Fig. 8, one within each of the two phases, may have the same morphological relationship to a backslope, but the descriptions of these two sites would differ in the following way: for site *X*: 'The hill-side in macro-profile can be divided into two parts, one above and one below a major convex break of slope which separates an

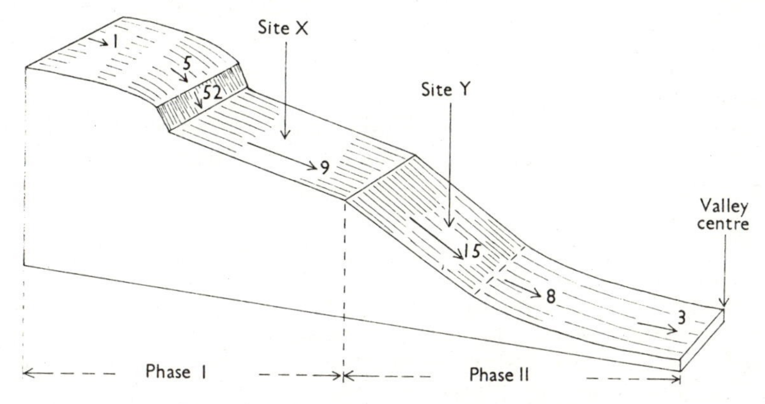

FIG. 8. Polyphase or polycyclic profile.

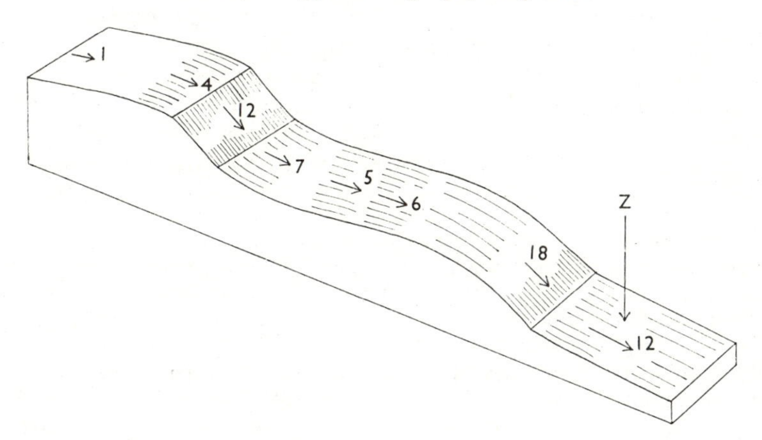

FIG. 9. To illustrate the subdivisions of a possible surface micro-profile for purposes of site description.

upper phase (I) from a lower phase (II) of development. Phase I is composed of a crestslope (1–5 deg., 600 ft) which gives way to a backslope whose upper part is cliffed (52 deg., 50 ft). The site *X* lies half-way (300 ft) between the foot of the cliff and the major convex break of slope, within the rectilinear 9 deg. portion of the backslope.' For site *Y*: 'The marked convex break of slope separating the slopes of phase I from those of phase II forms the upper limit of the backslope within phase II. The backslope (15 deg., 450 ft), with *Y* at its centre, gives way to a lower concave section (8–3 deg., 600 ft) which reaches to the valley centre.'

(ii) *The divisions of the surface micro-profile.* Where the surface form is markedly different from the general convex-rectilinear-concave case the description of the site must rely on a description of the surface profile in terms of the small-scale morphological units of which it is composed, i.e. the surface micro-profile.

The units which will make up the micro-profile can only be of two types, rectilinear or curved. The curved units will either be convex upwards (convex units) or concave upwards (concave units). The succession of such morphological units can be described in much the same way as before. For example, Fig. 9. 'The convex unit (1–4 deg., 25 ft) immediately below the interfluve gives way downslope to a rectilinear slope (12 deg., 8 ft) followed by a concave unit (7–5 deg., 15 ft), a convex unit (6–18 deg., 47 ft), and a rectilinear unit (12 deg., 26 ft) half-way along which occurs the site *Z*.' Such a description could be made for the whole of the profile of the hill-side if so desired.

Each morphological unit, as seen above, can be defined in terms of two quantities, angle(s) of slope and length of slope. In practice a surface profile can be measured in one of two ways, either by measuring the angle of slope between stations spaced at a constant interval, or by measuring the angle of slope only between marked breaks (and/or changes) of slope. In the latter case the length between each station will not be constant, but will depend on the terminal positions of each morphological unit. On an area of smooth topography the constant interval method is preferable. Where individual morphological units are more marked, however, the selected interval method is more informative.

In order to derive the benefits of both methods it is suggested that the surface micro-profile should be recorded in the following manner as illustrated at lower right in the specimen description sheet Fig. 10.

(*a*) Determine the direction of maximum slope from the soil exposure.
(*b*) Measure the angles of slope for successive lengths of 10 yd over a distance of 60 yd upslope from the soil exposure, and similarly six lengths in the downslope direction. Record the readings to the nearest $\frac{1}{2}^{\circ}$. Elevation (E) or Depression (D) as shown on data columns at lower right Fig. 10.
(*c*) Measure slope angles from the soil exposure for three successive lengths of 10 yd to the right and left of the dominant slope direction. The lengths should be at right angles to the dominant slope as indicated by the traverse line Fig. 10, and recorded in the squares at the foot of the vertical column.
(*d*) Make a morphological map of the *micro-profile* features contained within the rectangle bounded by the 120 yd traversed upslope and downslope and the cross traverse of 60 yd. Record the map (using symbols as earlier described) on either side of the vertical data column. The 10-yard intervals indicate the map scale.

In practice the slope angles can be measured satisfactorily with simple levelling instruments, e.g. Abney level, Watkin clinometer. The distances may be paced or measured with a linen tape or chain. The handle of one soil auger may be used as a support for sighting with the level, and the handle of another auger may be used as a sighting mark at the other end of the leg being measured. The augers should be screwed into the ground in such a manner that the handle of the auger on which the level is rested is at the same height above the ground as the handle of the sighting mark auger.

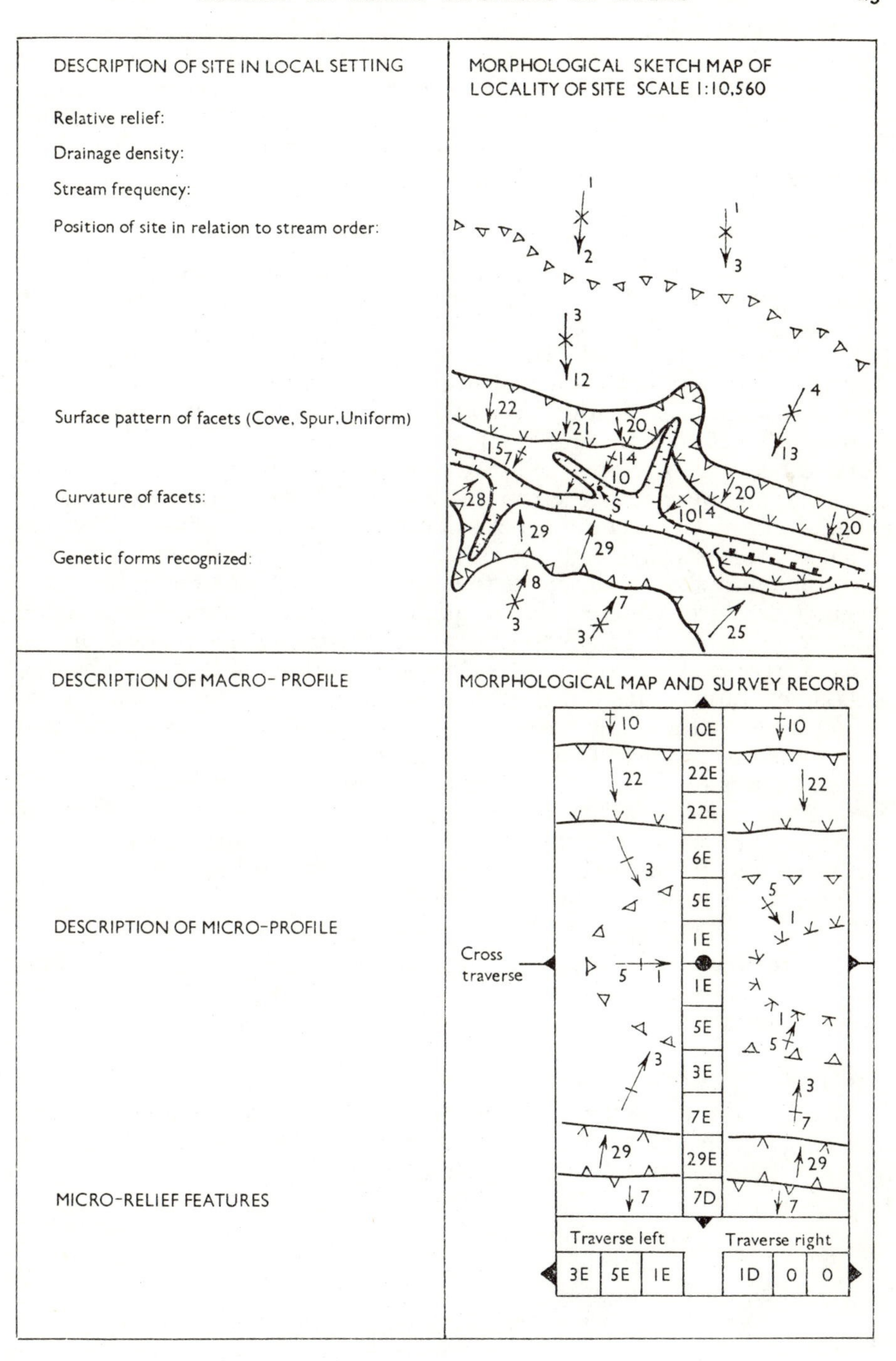
DESCRIPTION OF SITE IN LOCAL SETTING
Relative relief:
Drainage density:
Stream frequency:
Position of site in relation to stream order:
Surface pattern of facets (Cove, Spur, Uniform)
Curvature of facets:
Genetic forms recognized:
MORPHOLOGICAL SKETCH MAP OF LOCALITY OF SITE SCALE 1:10,560
DESCRIPTION OF MACRO-PROFILE
DESCRIPTION OF MICRO-PROFILE
MICRO-RELIEF FEATURES
MORPHOLOGICAL MAP AND SURVEY RECORD
Cross traverse
10E
22E
22E
6E
5E
1E
1E
5E
3E
7E
29E
7D
Traverse left
Traverse right
3E
5E
1E
1D
0
0

FIG. 10.

The principal advantage of carrying out the fixed interval survey of the microprofile is that the quantitative information so gained can then be related, for example, to soil and land-use characteristics. Furthermore, the survey data of the micro-profile provides a satisfactory check in those circumstances where the identification of changes of slope becomes difficult or subjective.

Slope Classifications

The preparation of a soil map normally requires consideration of slope phases. Therefore slope classifications have been developed in various countries and the slope classification of the U.S. Soil Survey has been widely followed. In a study of the subdivision of slopes into slope categories (Curtis, Doornkamp, Gregory, 1962) the authors put forward a slope classification for use in the British Isles based upon three sources of information. First, previous work on the establishment of slope categories was analysed and the results of seventeen previous classifications were combined so that a classification common to most of them was derived. Secondly, a classification was obtained using analyses of slope profiles (Savigear, 1960). Slope profiles measured within the sandstone and shale outcrop of the Lower Coal Measures of the Don Basin, Yorkshire, enabled the frequency of occurrence of each slope value to be determined. Slope categories were then established on the basis of frequency of slope values expressed in units of length. Thirdly, a classification was derived from analysis of the areas of morphological units. This method was based upon analysis of a morphological map constructed in the field (Fig. 11) and gave a direct correlation between angle and area of slope. Seven small drainage basins in Eskdale, North York Moors, on a variety of lithological types were selected for this purpose. The area of each morphological unit of uniform slope was measured and slope categories distinguished on the basis of frequency of slope values expressed in units of area.

The slope categories obtained by the three different means are:

(i) Previous work. 0–½, 1–2½, 3–7, 7½–13, 13½–19, 19½–31, 31½ and over.

(ii) Slope profiles. 0–½, 1–3, 3½–6, 6½–13, 13½–19, 19½–32, 32½ and over, cliffs.

(iii) Slope areas. 0–½, 1–2½, 3–6, 6½–13½, 14–30, 30½ and over, cliffs.

Combining these results with special reference to (ii) and (iii) the following slope categories are recommended for use and they are illustrated on the morphological map (Fig. 11).

Degrees 0–½ Flat
,, 1–2½ Gently sloping.
,, 3–6 Moderately sloping.
,, 6½–13 Strongly sloping.
,, 13½–31 Steep (this category may be subdivided into 13½–19 Moderately steep and 19½–31 Steep).
,, 31½ and over. Very steep.
Cliffs (rock outcrops too steep or irregular to measure).

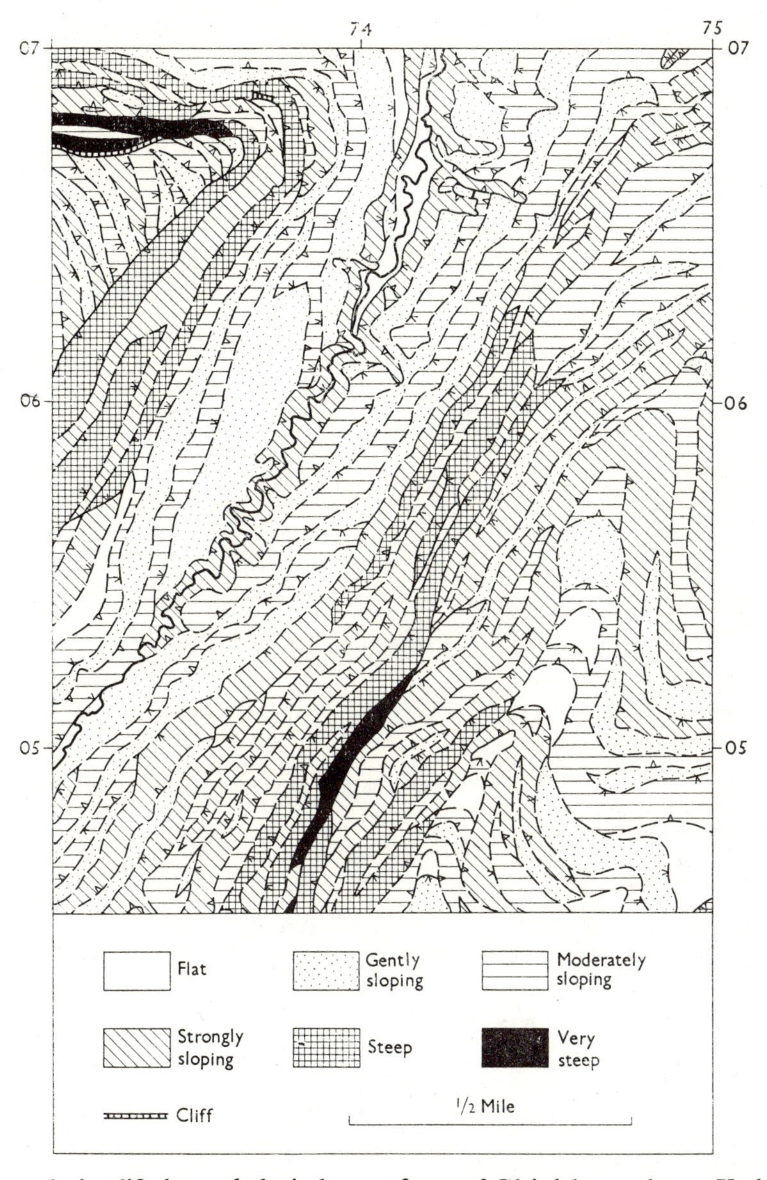

FIG. 11. A simplified morphological map of part of Glaisdale, north-east Yorkshire, shaded to illustrate slope categories.

This slope classification defined slope categories primarily on the basis of frequency of occurrence of particular angles of slope. Thus it is a classification which is likely to be consistent with natural conditions of soil slope but it is not a 'purpose' classification derived from considerations of the land-use characteristics of particular slopes. However, the field scientist is often concerned with agricultural research and so the chosen categories have been examined in relation to data for the limiting ranges of gradient within which agricultural equipment can be used (data supplied by W. J. West, N.I.A.E.).

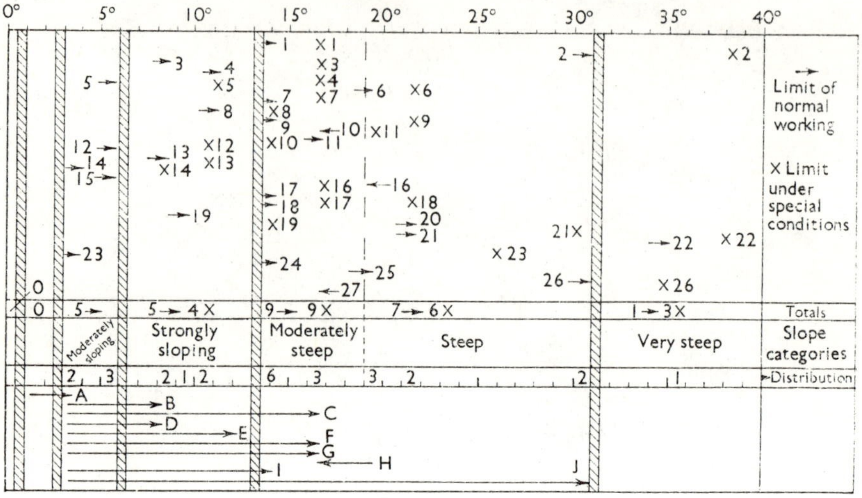

FIG. 12. Relationship between slope categories and use of farm implements. (After W. J. West.)

1. Animal-drawn plough (U) (R).
2. Animal-drawn plough (D) (R).
3. Animal-drawn plough (M).
4. Rotary cultivator (U) (R).
5. Rotary cultivator (D) (R).
6. Track-laying tractor (R).
7. Tractor, standard wheels, multiple-share plough (U) (R).
8. Tractor, standard wheels, multiple-share plough (D) (R).
9. Standard tractor, multiple-share plough mounted (M).
10. Single-acting winch, ploughing.
11. Animal-drawn single row hoeing (M).
12. Standard tractor-towed sowing and hoeing (R).
13. Standard tractor-mounted hoeing (R).
14. Standard tractor-mounted beet hoeing (R).
15. Standard tractor-mounted ridging (R).
16. Sowing and hoeing implements worked by single-acting winch.
17. Four-wheel-drive tractor/track-laying tractor, vineyard work.
18. Standard tractor, mounted reaper (R).
19. Standard tractor, reaper, binder, grubber (R).
20. Animal-drawn, reaper with adjustable seat (R).
21. Rotary cultivator (convertible as self-propelled reaper (R).
22. Self-propelled reaper (R).
23. Combine harvester (R).
24. Rubber-tyred trailer transport (R).
25. Solid-tyred trailer transport (R).
26. Rotary cultivator and powered-axle trailer (M).
27. Winch with hauled trailer (M).

(R) Work along contour.
(M) Work along maximum gradient.
(U) Soil thrown upwards.
(D) Soil thrown downwards.

A. No special difficulties.
B. Level-ground implements can generally be used (but quality of work poorer and difficulty greater).
C. General limit of use of animal traction (C).
D. General limit of use of standard wheeled tractor (C).
E. General limit of use of standard wheeled tractor (M).
F. General limit of use of tractor with four-wheel drive (C).
G. General limit of use of double-acting winch.
H. General limit of use of single-acting winch.
I. General limit of use of rotary cultivator.
J. General limit of use of self-propelled reaper.

The recommended slope categories are shown in Fig. 12 by vertical lines, whereas numbered arrows indicate the approximate limiting angles of slope for normal working of the equipment listed alongside. In some instances the range of use of equipment can be extended by employment of special attachments or, in exceptional circumstances, by an especially skilful operator. Where this is the case a numbered cross indicates the upper limit.

All implements can be used without special difficulty up to the maximum angle of Gentle slopes. Level-ground implements can still be used on Moderate slopes, though with less ease and efficiency; they soon reach their normal working limit on Strong slopes, as does contour work with the standard tractor. However, the standard tractor can work upslope almost to the limit of Strong slopes.

The greatest cluster of normal working limits occurs near the boundary between Strong and Moderately steep slopes. Approximately 74 per cent of the normal working limits and 65 per cent of all limits occur below the maximum angle of Moderately steep slopes.

The data suggests that slope boundaries might be drawn at $2\frac{1}{2}°$, $8\frac{1}{2}°$, 13°, 17°, 31° on the basis of the uses of agricultural implements. Thus three of these coincide with the slope boundaries of the classification based on frequency of natural slopes ($\frac{1}{2}°$, $2\frac{1}{2}°$, 6°, 13°, 19°, 31°).

Acknowledgements

The authors gratefully acknowledge helpful criticism and advice given by D. A. Osmond, B. W. Avery, D. Mackney, J. Catt, and B. C. Clayden.

REFERENCES

AANDAHL, A. R., 1948. The characteristics of slope positions and their influence on the total nitrogen content of a few virgin soils of western Iowa. Soil Sci. Soc. Amer. Proc. **13,** 449.

BRIDGES, E. M., and DOORNKAMP, J. C. 1963. Morphological mapping and the study of soil patterns. Geography. London, **48,** 175–81.

COTTON, C. A., 1941. Landscape. London.

CURTIS, L. F., DOORNKAMP, J. C., and GREGORY, K. J., 1962. The subdivision of slopes into slope categories. Brit. Univ. Geomorph. Res. Group Rep. No. 5, 4–9.

HORTON, R. E., 1945. Erosional development of streams and their drainage basins. Bull. geol. Soc. Amer. **56,** 275–370.

MOSS, R. P., 1963. Soils, slopes and land use in a part of south-western Nigeria. Trans. Inst. brit. Geogr. **32,** 143–68.

SAVIGEAR, R. A., 1956. Technique and terminology in the investigation of slope forms. Premier rapport de la commission pour l'étude des versants. Union Géographique Int. Amsterdam. 66–75.

—— 1960. Slopes and hills in West Africa. Z. Geomorph. Suppl. i, 156–71.

SERET, G., 1963. Essai de classification des pentes en famenne. Z. Geomorph. **7,** 71–85.

SOIL SURVEY OF GREAT BRITAIN, 1960. Soil Survey Handbook.

SOIL SURVEY STAFF, 1951. Soil Survey Manual, U.S. D.A. Handbook No. 18.

SPARKS, B. W., 1949. The denudation chronology of the South Downs. Proc. Geol. Ass. **60,** 165–215.

STAMP, L. D. (Ed.). 1961. A Glossary of Geographical Terms. Longmans, p. 187.

STRAHLER, A. N. 1952. Hypsometric (area-altitude) analysis of erosional topography. Bull. Geol. Soc. Amer. **63,** 1117–41.

WATERS, R. S. 1958. Morphological mapping. Geography, London, **43,** 10–17.

WOOD, A. 1942. The development of hillside slopes. Proc. geol. Ass. **43,** 128–40.

YOUNG, A. 1963. Some field observations of slope form and regolith, and their relation to slope development. Trans. Inst. brit. Geogr. **32,** 1–29.

(*Received 1 April 1964*)

Journal of Geophysical Research Vol. 68, No. 12 June 15, 1963

Seasonal Variation of Infiltration Capacity and Runoff on Hillslopes in Western Colorado

S. A. Schumm and G. C. Lusby

U. S. Geological Survey, Denver, Colorado

31

Abstract. Hillslope erosion was studied during 4 years along 25 slope profiles on Mancos shale hillslopes in western Colorado. Erosion was measured by the movement of markers and exposure of stakes. During winter the soil surface is loosened by frost action, and the stakes show minimum exposure in the spring. Rain-beat during the spring and summer compacts the soil, and the stakes show maximum exposure during the fall. Frost action and compaction cause creep to occur in the upper 2 inches of the lithosols. During spring and summer compaction of the soil by rain-beat decreases infiltration capacity, and runoff increases. Rills form on the slopes, but these are soon destroyed by winter frost action and creep, and the infiltration capacity of the soil is increased. Measurement of sediment yield and runoff from small drainage basins on the Mancos shale shows that mean annual runoff is relatively low, but sediment yields are normal for this type of terrain. An analysis of precipitation and runoff data reveals that average runoff and the ratio of runoff to precipitation are less in the spring than in the fall, reflecting the seasonal changes of soil characteristics. Therefore, seasonal changes in the soil, which cause a seasonal change in infiltration capacity, not only control the rate and process of hillslope erosion but also significantly affect the hydrologic characeristics of these small drainage basins.

Introduction. Large areas of the western United States are underlain by the Mancos shale of Late Cretaceous age and by other formations with similar physical and chemical properties. Distinctive landforms result from the erosion of these formations [*Schumm*, 1956*b*], for under a semiarid or arid climate they exhibit the weathering and erosional characteristics typical of sediments that contain swelling clays. In 1958 a study of hillslope erosion on the Mancos shale of western Colorado was begun at three separate locations to provide information on the rates and mechanics of erosion on these surfaces.

The need for basic hydrologic data applicable to this type of terrain led to the establishment of the Badger Wash research area. In 1953 the Badger Wash drainage basin (Figure 1) was selected as a research area in which the erosional and hydrologic characteristics in several small drainage basins on the Mancos shale could be studied intensively. Five agencies of the federal government are cooperating in this 20-year project [*Lusby et al.*, 1963]. Data are being collected on changes in vegetational cover and type and infiltration rates by the Forest Service, on changes in small mammal populations by the Fish and Wildlife Service, and on precipitation, runoff, and sediment yields by the junior author for the Geological Survey.

The Mancos shale hillslope erosion study was not a part of the Badger Wash project, as originally planned, but because fenced and ungrazed slopes could be found only within Badger Wash, several slopes were selected there for study. Initially it was assumed that the wealth of data to be made available from the investigations in Badger Wash would be a valuable supplement to the data obtained on hillslope erosion and, conversely, that data from the hillslope study might be useful in the interpretation of the Badger Wash data. It soon became apparent that such was the case; precipitation and runoff data collected by Lusby at Badger Wash were necessary for an understanding of the mechanics and rates of slope erosion, and the observations made by Schumm on the Mancos shale explained some apparent anomalies in the hydrologic data.

Description of study areas. The locations of the three areas are shown in Figure 1. All lie within the Canyon Lands section of the Colorado Plateau physiographic province [*Fenneman*, 1931] at elevations from about 5000 ft at Badger Wash to 6000 ft at Montrose. The topography of these areas has been formed by the

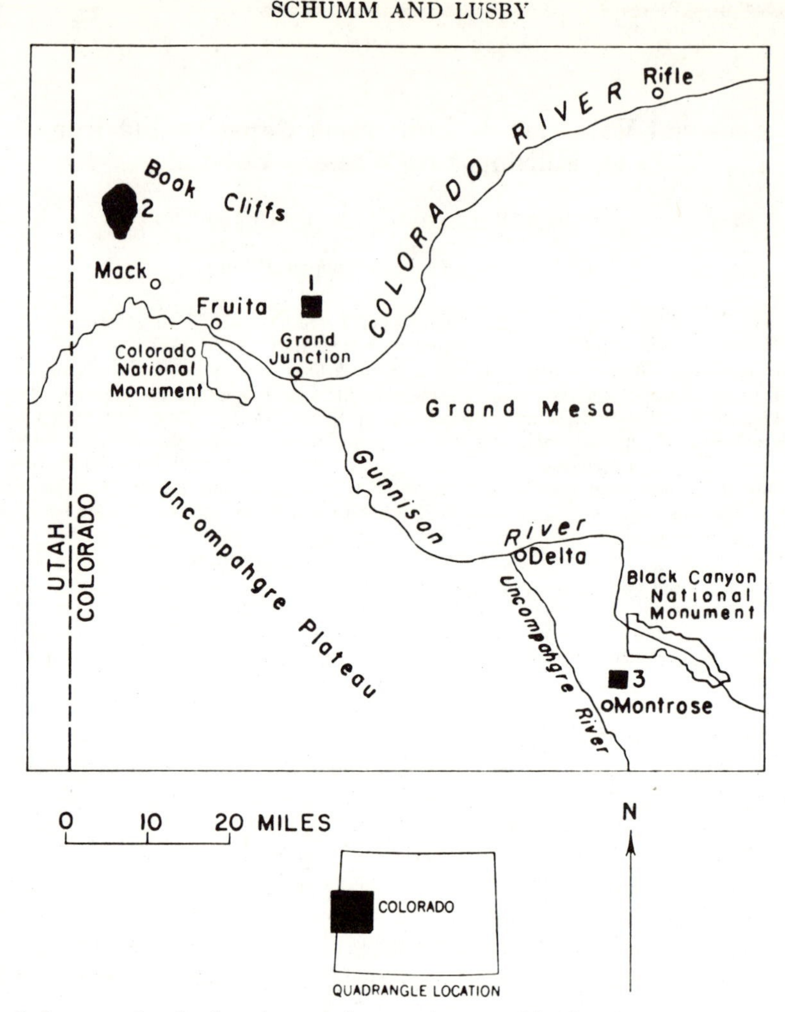

Fig. 1. Index map showing locations of three study areas. Numbers on map designate study areas as follows: (1) Book Cliffs area, (2) Badger Wash area, (3) Montrose area.

dissection of pediments lying near the base of the Book Cliffs and at the base of Black Mesa. The areas are desolate in appearance and poorly vegetated; rounded shale hills are characteristic (Figures 5 and 7). Drainage density for some of the small fifth-order drainage basins at Badger Wash is about 100, indicating a fine-textured topography.

The study areas have a borderline semiarid-arid climate [*Thornthwaite*, 1941]. The average annual precipitation at Fruita, about 16 miles southeast of Badger Wash, is 8.3 inches; at Montrose it is 9.1 inches [*U. S. Weather Bureau*, 1961]. During spring, summer, and fall, precipitation occurs chiefly as thunderstorm rainfall of high intensity; during the winter it occurs mainly as snow or rains of lesser intensity.

The distribution of precipitation throughout the year is almost constant, with an average of between 0.6 and 0.75 inch per month except for two low months, May and June, when the average monthly precipitation is about 0.5 inch, and two months of maximum precipitation, August and September, when the average monthly precipitation is about 0.95 inch.

Summer temperatures at Fruita are generally high during the day and low at night; the average maximum temperature during July is in the middle 90's and the average minimum is in the middle 50's. The yearly average temperature is 51.3°. The frost-free period averages about 130 days, from approximately May 15 to September 20.

Because of the high daytime temperatures and low relative humidity, potential evaporation rates in the area are very high. The average evaporation measured in a Weather Bureau class-A pan at Grand Junction for the months April through October in the period 1958–1960 was about 90 inches. The highest monthly average was 18.8 inches in July.

The small amount of annual precipitation yields only sparse vegetation. Crowns of living perennial plants cover only 8 to 15 per cent of the ground surface at Badger Wash. Vegetation is of the typical desert-shrub type; shadscale, Gardner saltbush, and galleta grass are the dominant plants [*Turner*, 1963].

The study areas are subject to periodic grazing. In Badger Wash both cattle and sheep are allowed to graze from November 15 to May 15. Four small drainage basins have been fenced to exclude grazing, because one purpose of the 20-year study at Badger Wash is the evaluation of changes in type and density of vegetation when grazing is excluded.

All study areas are underlain by the gently dipping marine Mancos shale of Late Cretaceous age. The Mancos shale is about 4000 ft thick in the Book Cliffs and Badger Wash areas [*Fisher et al.*, 1960, p. 8]. It is dark gray, and although thinly bedded, it is not fissile. Veinlets of gypsum and calcite are common, and thin sandstone beds are present.

The Mancos shale weathers to form a lithosol. On the hillslopes a zone (up to about 5 inches thick) of platy shale fragments lies directly on the bedrock. Above this zone of weathered shale fragments is the surface material, which appears either as a mass of loose soil aggregates (Figure 6) or a crust about 2 inches thick (Figure 4). In the Grand Junction area the soil on the slopes is commonly mapped as the Chipeta-Persayo shaly loam [*Knobel et al.*, 1955]. In the Montrose area the Persayo soil series is absent. The Chipeta clay loam, a light grayish-brown soil, and the Chipeta clay, a light to dark gray clay resting on shale, are both present [*Nelson and Kolbe*, 1912].

The bedrock and soils are saline. Calcium and magnesium are present in about the same amounts in the bedrock and soils of both areas, but the sodium content is much greater at Montrose [*Schumm*, 1963]. Illite and chlorite were identified by X-ray analyses in the soils from both Montrose and Badger Wash. In addition, vermiculite is present in the Montrose soil.

Soils containing relatively large amounts of sodium generally swell when wetted. The free swell test showed a 20 per cent increase in volume of the Badger Wash surface material and a 25 per cent increase in volume of the platy material and the bedrock. The surface material sampled at Montrose increased in volume 25 per cent, the platy material 30 per cent, and bedrock 58 per cent. The tests reveal appreciable swelling of soils in both areas. Montrose soils, however, show greater volume changes owing to their higher sodium content and to the presence of vermiculite.

Methods of investigation. In order to measure hillslope erosion, stake profiles were established on the slopes in June 1958. On each slope the profiles extended from the crest of the slope directly down to the adjacent stream channel or, in some cases, across a flat alluvial or erosional basal slope toward the nearest channel. Along each profile iron stakes, 18 inches long and 3/8 inch in diameter, were driven into the slopes at right angles to the surface. The stakes were installed to allow measurement of erosion by progressive exposure of the stakes and to afford stationary reference points to which the position of various types of markers could be related. The markers, placed along the profile lines to indicate by their changing position the effects of creep, were of four kinds: metal washers 1.35 inches in diameter and 0.1 inch thick, wooden disks 1.3 inches in diameter and 0.75 inch thick, wooden blocks 2 inches square and 0.75 inch thick, and rocks of various sizes and shapes, which were marked with aluminum paint. The profiles were visited semiannually, in the spring and fall of the year, when stake exposure and the positions of the markers were measured.

In all, 25 of these profiles were established in June 1958. Measurements were repeated in September 1958, May and July 1959, April and November 1960, March and September 1961. Pin exposure only was measured in April 1962. There are therefore eight periods for which changes in stake exposure are known. Owing to initial measurement problems, measurement of marker movement was not begun until Septem-

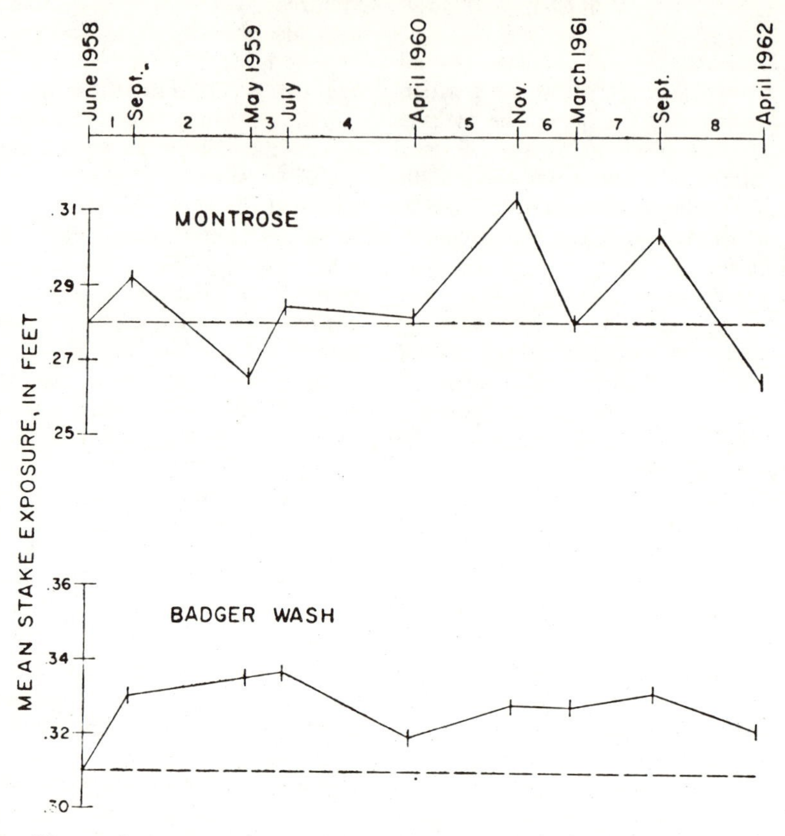

Fig. 2. Mean stake exposure for ten Montrose profiles and ten Badger Wash profiles. Summer periods 1, 3, 5, and 7 coincide with exposure of stakes. Winter periods 2, 4, 6, and 8 coincide with burial of stakes.

ber 1958. There are, therefore, only six periods during which marker movement was measured.

Precipitation at Badger Wash was measured in ten recording rain gages located within the study area in a manner which adequately defined the precipitation pattern over four pairs of grazed and ungrazed drainage basins. Areal precipitation was computed using the Thiessen polygon method.

Each small drainage basin drains directly into a reservoir which is used to measure runoff and sediment yield from that basin. Runoff was computed from a continuous record of water stage in four reservoirs and by manual measurements of water stage in four other reservoirs. Sediment yield was computed from annual reservoir surveys.

All runoff measured at Badger Wash has apparently been surface runoff only. No residual effects of subsurface runoff have been noted on reservoir charts. Runoff is essentially complete within a few minutes after the cessation of rain.

Field observations and measurements. The biannual measurements of stake exposure and marker movement demonstrate conclusively that seasonal changes in the characteristics of the Chipeta-Persayo lithosols are great enough to significantly affect not only the rates of erosion but the erosion process itself [*Schumm*, 1963]. For example, measurements of stake exposure revealed that instead of a progressive exposure with time, the stakes were exposed most in the fall and least in the spring (Figure 2); in addition, the rate of movement of markers was greatest during winter (Figure 3). The differences between the magnitude and in some cases

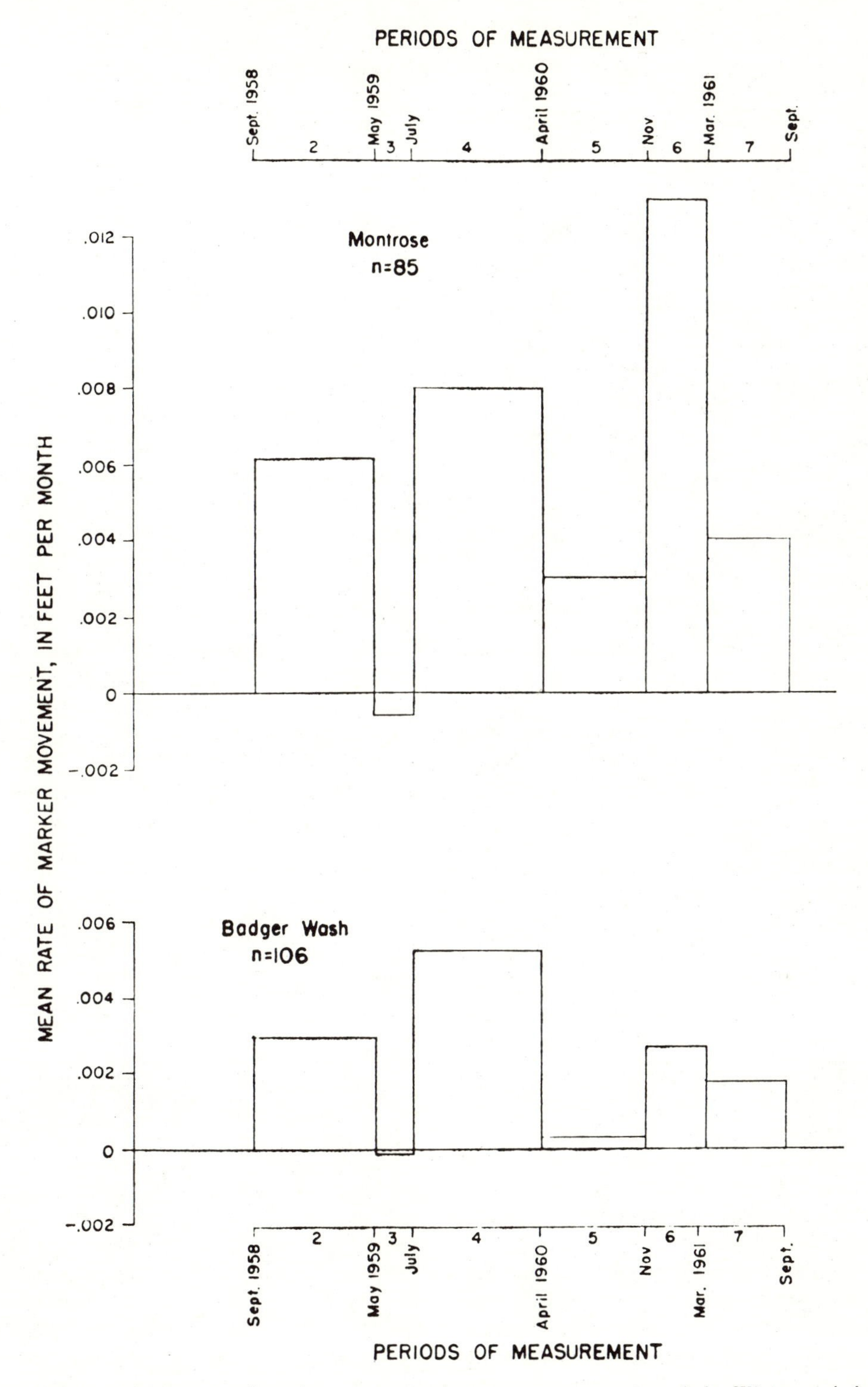

Fig. 3. Average rate of marker movement between measurement periods. Winter periods 2, 4, and 6 show the maximum rates of marker movement. Summer periods 3, 5, and 7 show least marker movement.

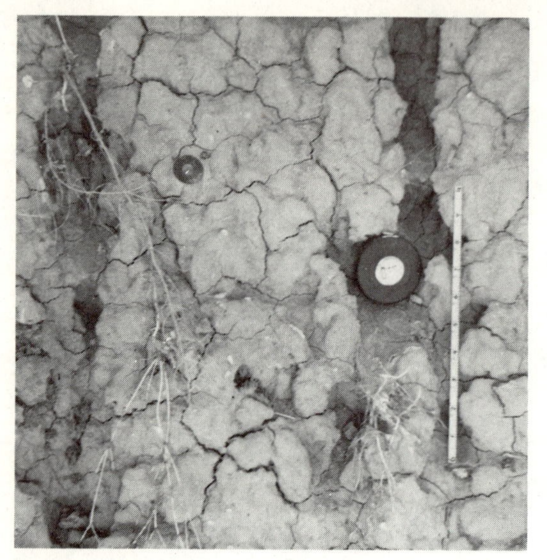

Fig. 4. Soil surface in fall (September 1958) at profile 1, Book Cliffs area. Surface has been sealed and rill channels are present. Infiltration takes place primarily through dessication fractures. Compare with Figure 6.

the direction of the changes from one measurement period to the next and between the Badger Wash and Montrose areas can be explained almost entirely by the different amounts of precipitation which fell during each period and the difference between the soils in the two areas [*Schumm*, 1963].

An explanation of the seasonal differences in rates of creep and stake exposure, illustrated by Figures 2 and 3, can be presented as a summary of the annual cycle of changes in soil characteristics which occur during a typical year on these hillslopes. During the summer, rainfall tends to compact and seal the slope surfaces (Figures 4 and 5); during the winter, frost action in the soil causes heaving and loosening of soil surface (Figures 6 and 7).

Granular ice crystals were observed in the upper soil zone during the winter. The crystals serve as growth centers, and water is drawn to them from surrounding particles. This growth of ice crystals in soil pores and the simultaneous drying of adjacent soil cracks and loosens the dry soil [*Baver*, 1956, p. 153]. The areas between dessication fractures are further reduced in size, and the entire soil surface is raised and loosened by the growth of granular ice crystals in the voids. Several periods of freezing and thawing change the less permeable rilled surface (Figures 4 and 5) to a highly permeable surface composed of soil aggregates without rills (Figures 6 and 7).

In the spring, after the last severe frost, the surface of the slopes is unstable. Tracking by animals at this time of the year will cause important downslope movement of soil (Figure 7). Spring rains soon change this situation by compacting the loose surface by rain-beat (Figures 4 and 5). The edges of the aggregates are destroyed as they sluff into cracks and partly close them. A crust forms over the outer surface of the aggregates, and, although reduced in height, they become more stable. Dessication fractures form as the soil shrinks upon drying, but they close again as the soil swells during rain. The swelling and shrinking of the soils is largely responsible for the creep measured during the summer months. With continued rainfall, runoff again becomes important, and the obliterated rill channels are scoured and become prominent (Figure 5). With the reappearance of the rills the annual cycle on the Mancos shale hillslopes is completed.

An annual cycle of rill formation and obliteration has been observed elsewhere [*Schumm*, 1956*a*], and it is considered a criterion of im-

Fig. 5. Hillslope in Twins drainage basin, Badger Wash in late summer, August 1959. Rill channels are developed by runoff during summer rains. Photograph taken by F. F. Zdenek. Compare with Figure 7.

Fig. 6. Soil surface in spring (March 1961) at profile 1, Book Cliffs area. Surface is very loose and washer is partly buried. Rill channels are almost obliterated. The photograph was taken at a slightly different angle from that of Figure 4, but the washer is the same in both photographs.

portant annual changes in the erosion process. For example, the relatively rapid rates of creep (Figure 3) are associated with winter frost action and destruction of rills. The slower rates of creep are measured when surficial runoff is scouring the rill channels. Thus two different erosion processes are dominant on the Mancos shale slopes at different times of the year.

Hydrology of Mancos shale hillslopes. Data on the hydrologic characteristics of drainage basins on the Mancos shale have been collected for several small areas in Badger Wash. Eight-year records are available for runoff and sediment yield. In addition, two series of infiltration measurements were made with a Rocky Mountain infiltrometer in 1953–1954 and 1958 [*Lusby et al.*, 1963]. These data can be used to indicate how the seasonal changes in soil characteristics affect the hydrology of the slopes and the drainage basins as a whole.

Infiltration rates. The measurements of stake exposure suggest that infiltration capacity of the soil varies greatly during the annual cycle of loosening and compaction. The photographs show a surface capable of a high infiltration capacity in the spring (Figures 6 and 7) and a relatively lower infiltration capacity in the fall (Figures 4 and 5). Infiltration rates were measured by the Forest Service at Badger Wash in the fall of the years 1953, 1954, and 1958. As these measurements were all made in the fall, it is impossible to obtain the annual range of values for rate of infiltration.

It is interesting to note that the infiltration rates measured in 1958 were significantly higher than those measured in 1953 and 1954. On the ungrazed watersheds infiltration rates increased from 0.89 to 1.12 inches/hour for the last 20 minutes of the dry runs and from 0.67 to 0.89 inch/hour for the last 20 minutes of the wet runs [*Thompson*, 1963]. In addition, the water absorbed before the occurrence of runoff almost doubled, increasing from 0.19 to 0.35 inch. Since significant increases in infiltration rates also occurred on the grazed watersheds, the change could not be attributed solely to land use. A check of precipitation data seems to offer an explanation. From April through November 1953, when the infiltration rates were first measured, 6.55 inches of precipitation were recorded at the Fruita weather station. From April through October 1954, when the 1953–1954 infiltration measurements were completed, 4.78 inches of rain fell at Badger Wash. However, only 1.94 inches of rain fell at Badger Wash from April through September 1958, when the second set of infiltration rates was measured. Thus, following the winter periods of soil loosen-

Fig. 7. Hillslope shown in Figure 5 in March 1961, showing unrilled surface with some tracking.

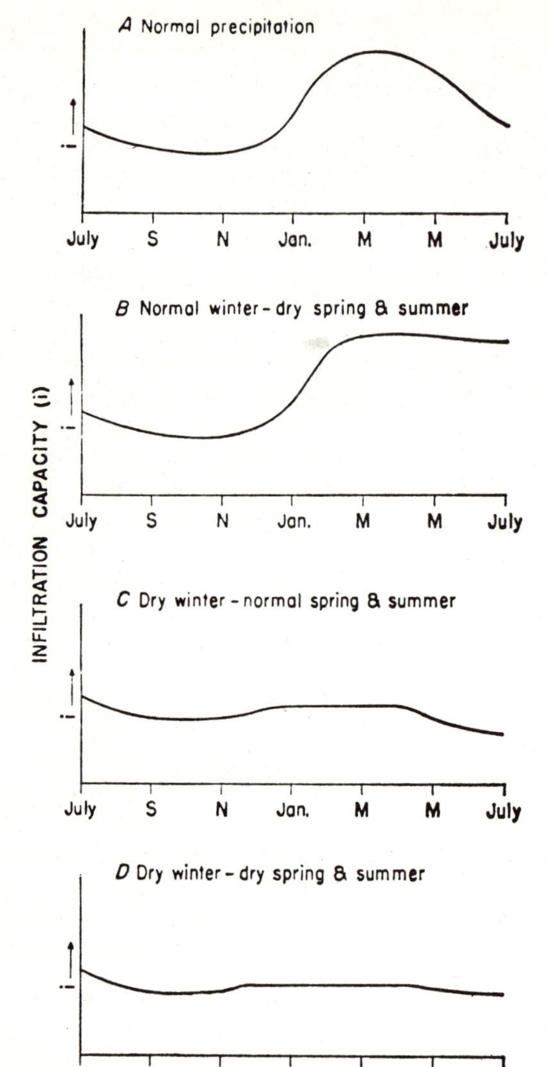

Fig. 8. Diagrams illustrating hypothetical seasonal changes of infiltration capacity on Mancos shale hillslopes. (*A*) Normal precipitation. Infiltration capacity increases during winter and decreases during spring and summer. (*B*) Normal winter precipitation followed by a dry spring and summer. Infiltration capacity increases during winter but does not decrease during spring and summer. (*C*) Dry winter followed by normal spring and summer precipitation. Infiltration capacity remains essentially constant during winter but shows a decrease in the spring and summer owing to normal precipitation. (*D*) Dry winter followed by dry spring and summer. Infiltration capacity remains essentially constant after soil was compacted during previous summer.

ing, only about ⅓ as much rain fell in 1958 as in the summers of 1953 and 1954 [*U. S. Weather Bureau*, 1953–1961]. This small amount of precipitation caused relatively less compaction of the surface, and the increase in infiltration rates between 1953–1954 and 1958 may be explained in this manner. In spite of the compaction and surface sealing after 1.94 inches of rain, the surface absorbed 0.16 inch more water before runoff began in 1958 than in 1953–1954, and the infiltration rate was 0.23 inch/hour greater in the ungrazed watersheds.

Seasonal variations in infiltration rates have been reported by *Horton* [1940] and *Horner and Lloyd* [1940]. Their researches indicated that maximum infiltration rates occurred during the midsummer months as a result of the activity of soil fauna. *Dreibelbis* [1949] reported that in Ohio infiltration is retarded in the spring by frost penetration of the soil. But from our observations at Badger Wash and Montrose we conclude that, on the poorly vegetated Mancos shale hillslopes, infiltration rates are maximum in early spring before the rains begin.

The effects of frost action in loosening the soil can be important on the bare soils of semiarid and arid regions. Infiltration capacity may vary annually as illustrated in Figure 8 for semiarid areas underlain by the Mancos shale and similar formations. The seasonal variation of infiltration capacity illustrated in Figure 8*a* will occur only for a year of normal precipitation. A dry winter will not cause an increase in infiltration capacity (Figures 8*c*, *d*), and a dry summer will not cause a decrease in infiltration capacity (Figures 8*b*, *d*). Some support is given to these diagrams by the variations of stake exposure with varying seasonal precipitation at Badger Wash and Montrose [*Schumm*, 1963].

Sediment yield and runoff. As infiltration capacity is high on these slopes during early spring rains, it can be expected that in comparison with sediment yields and runoff from slopes similar in other respects, but not greatly affected by seasonal changes in infiltration capacity, the Mancos shale slopes should yield less runoff and sediment.

Relationships have been found to exist between average sediment yields and the average slope or relief ratio (relief/length) of a drainage basin and between average annual runoff and drainage density of small drainage basins in

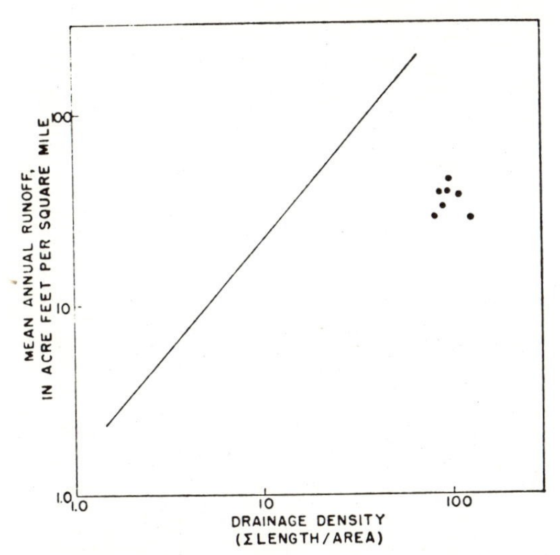

Fig. 9. Relation of Badger Wash runoff data to regression line established between runoff and drainage density for other small drainage basins in a semiarid climate.

Wyoming [*Hadley and Schumm,* 1961; *Schumm and Hadley,* 1961]. When average annual runoff is plotted against drainage density (Figure 9), the Badger Wash drainage basins plot below the regression line which was established for small drainage basins in Wyoming.

Drainage density was obtained for seven Badger Wash drainage basins from very accurate large-scale maps, but the drainage densities from which the regression line of Figure 9 was obtained were measured on aerial photographs of much smaller scale. The two sets of data, therefore, are not comparable, for when drainage density for the Badger Wash basin is measured from aerial photographs, it decreases to about one-half of the value obtained from the special maps. Therefore, the average drainage density should be decreased from about 100 to about 50. In spite of this reduction, runoff is still low, as expected. Even if average runoff is doubled from 35 to 70 acre feet as an adjustment for the higher rainfall of the Wyoming drainage basins, the Badger Wash basins plot below the regression line because a significant proportion of runoff from the early rains will be lost by infiltration into the permeable hillslopes.

When average annual sediment yield for the Badger Wash drainage basins is plotted against relief ratio (Figure 10), the Badger Wash drainage basins plot reasonably close to the regression line which has been established for a number of drainage basins in Wyoming, Utah, and New Mexico. Thus, although runoff is low, sediment yield rates are about normal for this type of topography. It would appear that sediment concentration in the runoff should be high. This may be explained with regard to the rill cycle. The rills are clogged by creep during the winter (Figures 6 and 7), and initial runoff acquires a heavy load of sediment as these channels are cleared.

A further inescapable conclusion is that average runoff and the ratio of runoff to rainfall on the hillslopes should increase from spring to fall. A significant percentage of the total area of the Badger Wash and Montrose study areas is composed of pediments and alluvial valley-fills. The seasonal changes noted on the slopes do not occur to the same extent on these flat surfaces. Therefore, the changes in seasonal infiltration capacity on the slopes can be applied only in a general way to an entire drainage basin. Only the data on runoff measured in the Badger Wash reservoirs are available to demonstrate that the seasonal changes on the slopes may be important enough to influence runoff from the small drainage basins as a whole.

To determine whether the seasonal effects on runoff are measurable, all storms that produced more than 0.1 inch of rain in Badger Wash, during the annual periods of continuous measurement from April through October for the years 1954 through 1961, were examined. In Table 1 the storms are grouped by 0.1-inch increments for a spring period (April through June) and a fall period (August through October). Precipitation and runoff for the storms of each period were averaged in this manner to reduce possible variations in runoff caused by differences in precipitation intensity.

If this reasoning is correct, runoff should be greater in the fall period than in the spring period for comparable storms. Although discrepancies exist between individual storms during some years, average runoff and the ratio of runoff to precipitation are consistently larger during the fall period for all sizes of storms (Table 1).

Two factors which may have an effect on these ratios, intensity of precipitation and antecedent moisture conditions, were investigated. Although the precipitation data indicated that individual storms of slightly greater intensity occurred

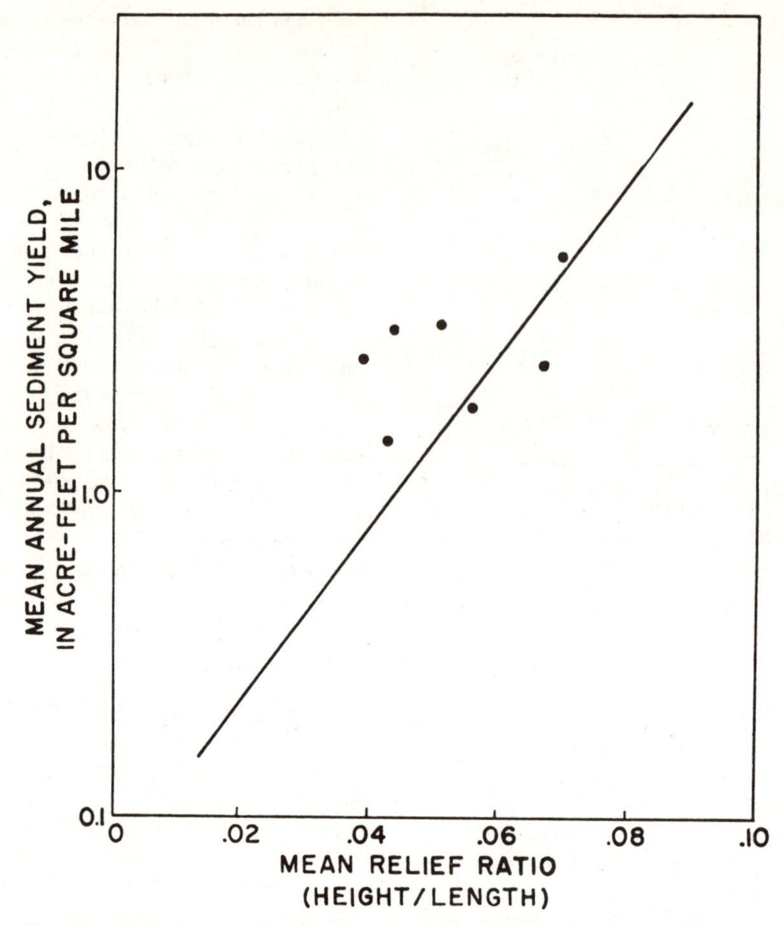

Fig. 10. Relation of Badger Wash sediment yield data to regression line established between sediment yield and relief ratio for other small drainage basins in a semiarid climate.

more frequently during late summer, these storms occurred too infrequently to cause the difference found between average runoff and the runoff-precipitation ratios of Table 1. Individual storms of high intensity occurred during the spring, but, as expected, total unit runoff was less from these high-intensity spring storms than from fall storms of comparable intensity. Differences in rainfall intensity, therefore, are not great enough to cause the disparity between average runoff as measured in the spring and in the fall at Badger Wash.

The effect of antecedent moisture at Badger Wash is more difficult to evaluate, but for the storms listed in Table 1 it is believed to be negligible. During the early part of the spring period there may be some subsurface moisture, which would affect the infiltration rate during extended rains, but most rains in the area are of short duration at all seasons of the year.

The long-term average temperature is 61.1° for the spring period and 63.9° for the fall period. Average maximum temperature for the period under study was about 79.8° for the spring and 79.4° for the fall. Evaporation from a standard Weather Bureau pan at Grand Junction averaged 39.4 inches or 0.43 inch/day for the spring period and 34.1 inches or 0.37 inch/day for the fall period. Average temperatures and rates of evaporation, therefore, are comparable for the two periods.

Although pan evaporation is high, no information is available on evaporation from the soil itself following a storm. However, research on

TABLE 1. Spring and Fall Precipitation and Runoff at Badger Wash, 1954–1961

Range of Storm Precipitation, inches	Period	Number of Storms	Average Precipitation per Storm, inches	Average Runoff per Storm, inches	Runoff-Precipitation Ratio
0.10 to 0.20	Apr.–June	13	0.15	0.001	0.007
	Aug.–Oct.	23	0.15	0.008	0.050
0.21 to 0.30	Apr.–June	4	0.26	0	0
	Aug.–Oct.	11	0.27	0.039	0.145
0.31 to 0.40	Apr.–June	3	0.32	0.007	0.022
	Aug.–Oct.	11	0.34	0.063	0.185
0.41 to 0.50	Apr.–June	3	0.47	0.020	0.043
	Aug.–Oct.	7	0.47	0.093	0.198
0.51 to 0.60	Apr.–June	1	0.58	0.120	0.207
	Aug.–Oct.	2	0.56	0.180	0.322
0.61 to 0.70	Apr.–June	1	0.69	0.004	0.006
	Aug.–Oct.	4	0.66	0.290	0.440
0.91 to 1.0	Apr.–June	0			
	Aug.–Oct.	2	0.94	0.380	0.400
1.31 to 1.40	Apr.–June	0			
	Aug.–Oct.	1	1.34	0.470	0.350
All storms	Apr.–June	25	0.26	0.004	0.015
0.10 to 1.40	Aug.–Oct.	61	0.34	0.079	0.232

evaporation rates from wet soils demonstrates that evaporation from the soil is about equal to that from a free water surface. *Penman* [1948] found that evaporation from a wet, bare soil was 0.9 that from a free water surface, and *Fritschen and van Bavel* [1962] found that evaporation from a wet soil surface was initially 18 per cent higher than from shallow water. The difference between evaporation rates measured during these experiments can be explained by the differences in the texture of soils used in the experiments. Evaporation rates are greatest from fine-grained soils [*Baver,* 1956, p. 282]. Therefore, evaporation rates from the fine-grained Badger Wash soils would probably approach those obtained by Fritschen and van Bavel rather than those obtained by Penman from a sandy loam.

The average length of time between storms at Badger Wash was 8.7 days in the spring period and 6.9 days in the fall period. Average size of storm was 0.26 inch in the spring and 0.34 inch in the fall (Table 1). On the basis of these averages and the knowledge that evaporation from a wet soil is initially the same as that from a water surface, the soils at Badger Wash can unquestionably be dried between most storms.

Of course, the average frequency of occurrence of storms is often exceeded, but this happens about the same number of times in the spring as in the fall. For example, in the spring, rain occurred during the preceding four days for 42 per cent of the storms listed in Table 1, and in the fall rain occurred during the preceding four days for 54 per cent of the storms. Also, during the spring, rain occurred on the preceding day for 25 per cent of the storms and during the fall period for 30 per cent of the storms. Although storms in the fall period of high runoff were more closely spaced, the high rates of evaporation cause rapid drying of the soil during both spring and fall. The large differences in runoff between the spring and fall periods must therefore be attributed predominantly to the large differences in soil characteristics (Figures 4 and 6) rather than to differences in intensity of precipitation or antecedent moisture conditions.

One must be cognizant of the fact that the dominant factor, which causes the seasonal soil changes, is the lack of vegetational cover. The seasonal changes in turn have a critical effect on drainage basin hydrology. Elsewhere at higher altitudes where rainfall is greater, Mancos shale hillslopes are well vegetated, and, although they show evidence of mass movement by slumping, creep induced by frost action in the soil is probably of minor importance. The seasonal variation of soil characteristics is reduced by the presence of vegetation.

Conclusions. Perhaps the most important conclusion to be drawn from this study is that the geomorphic and hydrologic characteristics of Mancos shale hillslopes and drainage basins are determined by seasonal changes in soil characteristics on the slopes. The loosening of the soil surface on the slopes by frost action has a pronounced effect on (1) erosion process (creep is important during winter and spring), (2) infiltration capacity (the soil surface is most permeable during winter and spring), and (3) runoff (average runoff and the ratio of runoff to precipitation are relatively low in the spring).

The compaction of the permeable surface by rain-beat after frost action has a pronounced effect on (1) erosion process (erosion by rills and rainwash are important during summer and fall), (2) infiltration capacity (the soil surface is least permeable during summer and fall), and (3) runoff (average runoff and the ratio of runoff to precipitation are greatest in the late summer and fall).

Similar seasonal changes and their effects may be common on other fine-grained lithosols that are only poorly covered by vegetation. The sparse vegetational cover makes frost action and rain-beat very effective in arid and semiarid regions, whereas frost action and rain-beat are less important under the dense vegetational cover of humid regions.

Acknowledgments. F. F. Zdenek and R. W. Lichty helped with the field work on several occasions. Zdenek made the field measurements in April 1960 and took a number of photographs during 1959–1960. Lichty analyzed the soil in the laboratory. Identification of the clay minerals in samples of the Montrose and Badger Wash soils was made by Karl Ratzlaff. Suggestions for improvement of the manuscript have been made by H. V. Peterson and K. R. Melin of the Geological Survey and by R. J. Chorley of Cambridge University.

Publication authorized by the Director, U. S. Geological Survey.

References

Baver, L. D., *Soil Physics,* 3rd ed., John Wiley & Sons, New York, 1956.

Dreibelbis, F. R. Some influences of frost penetration on the hydrology of small watersheds, *Trans. Am. Geophys. Union, 30,* 279–282, 1949.

Fenneman, N. M., *Physiography of Western United States,* McGraw-Hill Book Co., New York, 1931.

Fisher, D. J., C. E. Erdmann, and J B. Reeside, Jr., Cretaceous and Tertiary formations of the Book Cliffs Carbon, Emery, and Grand Counties, Utah, and Garfield and Mesa Counties, Colorado, *U. S. Geol. Surv. Profess. Paper 332,* 1960

Fritschen, L. J., and C. H. M. van Bavel, Energy balance components of evaporating surfaces in arid lands, *J. Geophys. Res., 67,* 5179–5185, 1962.

Hadley, R. F., and S. A. Schumm, Sediment sources and drainage basin characteristics in upper Cheyenne River basin, *U. S. Geol. Surv. Water-Supply Paper 1531-B,* 137–196, 1961.

Horner, W. W., and C. L. Lloyd, Infiltration-capacity values as determined from a study of an eighteen month record at Edwardsville, Illinois, *Trans. Am. Geophys. Union, 21,* 522–541, 1940.

Horton, R. E., An approach toward a physical interpretation of infiltration capacity, *Soil Sci. Soc. Proc., 5,* 399–417, 1940.

Knobel, E. W., R. K. Dansdill, and M. L. Richardson, Soil survey of the Grand Junction area, Colorado, *U. S. Dept. Agr. Soil Surv. Ser. 1940,* no. 19, 1955.

Lusby, G. C., et al., Hydrologic and biotic characteristics of grazed and ungrazed watersheds of the Badger Wash basin in western Colorado, *U. S. Geol. Surv. Water-Supply Paper 1532-B,* in press, 1963.

Nelson, J. W., and L. A. Kolbe, Soil survey of the Uncompahgre Valley area, Colorado, *U. S. Dept. Agr. Bur. of Soils, Field Operations 1910, 12th Rept.,* 1443–1489, 1912.

Penman, H. L., Natural evaporation from open water, bare soil and grass, *Proc. Royal Soc. London, 193,* 120–145, 1948.

Schumm, S. A., Evolution of drainage systems and slopes in badlands at Perth Amboy, New Jersey, *Bull. Geol. Soc. Am., 67,* 597–646, 1956*a*.

Schumm, S. A., The role of creep and rainwash on the retreat of badland slopes, *Am. J. Sci., 254,* 693–706, 1956*b*.

Schumm, S. A., Seasonal variation of erosion processes and rates on hillslopes in western Colorado, *Z. Geomorphol., 7,* in press, 1963.

Schumm, S. A., and R. F. Hadley, Progress in the application of landform analysis in studies of semiarid erosion, *U. S. Geol. Surv. Circ. 437,* 1961.

Thompson, J. R., Infiltrometer plot records, *U. S. Geol. Surv. Water-Supply Paper 1532-B,* in press, 1963.

Thornthwaite, C. W., Atlas of climatic types in the United States, *1900–1939, U. S. Dept. Agr. Misc. Publ. 421,* 1941.

Turner, G. T., Watershed cover and forage utilization, *U. S. Geol. Surv. Water-Supply Paper 1532-B,* in press, 1963.

U. S. Weather Bureau, *Climatological Data, Colorado, Annual Summary,* vol. 66. pp. 198–207, 1961.

U. S. Weather Bureau, *Climatological Data, Colorado,* vols. 58–66, 1953–1961.

(Manuscript received November 1, 1962; revised February 13, 1963.)

EROSION EQUATIONS PREDICT LAND SLOPE DEVELOPMENT

32

L. Donald Meyer
Member ASAE

Larry A. Kramer
Assoc. Member ASAE

How does slope shape affect sediment load and erosion depth? What changes occur in slope shape as erosion progresses? Which slope shapes are effective in reducing erosion and sediment problems?

THE SHAPE of a hillside may not only affect the rate of erosion at different locations along its slope, but different erosion rates along a slope may appreciably change its shape as erosion progresses. In this research slope shape change, eroded depth and sediment load along slopes of various shapes were computed. This approach may be applied to farm fields and construction sites for predicting locations of critical erosion, rates of sediment movement, and expected slope shape changes.

Present methods for predicting soil loss use the average slope steepness plus other factors to describe expected erosion rates. Thus a relatively uniform slope is assumed, but Nature and man often leave sloping land in shapes that are quite nonuniform. Recent studies of field slope conditions plus research on areas mechanically shaped to more characteristic forms show that slope shape—particularly the steepness of the bottom portion of a slope — is a major factor in determining the relative erosion.

In this project four slope shapes — uniform, concave, convex and complex (upper half convex, lower half concave) — were studied at mean slope steepnesses of 5 and 10 percent. All slopes had a 20-ft elevation difference between top and bottom (Fig. 1).

Three soil-loss prediction equations relating slope steepness and length to total erosion were studied using a digital computer. The coefficient in each was selected so that one erosion period (one iterative solution of the total erosion equation) predicted a soil loss of about 40 tons per acre from a 5 percent slope 400 ft long. This is the average annual soil loss on a Southern Indiana silt loam cropped to continuous corn.

Elevations at 10-ft intervals along each slope were used by the computer program to determine the total erosion or sediment load at the end of each 10-ft increment based on the steepness at and the length from the top of slope to that point. The net erosion or deposition between successive increments was the difference in the sediment load at these points. Erosion depth at each location was determined as the sum of the net erosion for the two adjacent increments divided by their total lengths. New elevations and steepnesses were computed at all points for each successive erosion period. The computer program also produced the input data for an electronic plotter to graph sediment loads and resulting slope profiles.

Both authors are agricultural engineers with SWCRD, ARS, USDA — L. Donald Meyer at Lafayette, Ind., Larry A. Kramer at Columbia, Mo. This is a contribution of the Corn Belt Branch, SWCRD, in cooperation with Purdue Agricultural Experiment Station (Purdue Journal Series Paper No. 3529).

This is a condensation. The full-length report includes a reference list and the analytical methods used in establishing a computer program to describe erosion as a function of slope shape. It also includes samples of program output, including computer-plotted profiles. To order that report, request Paper I-749 from ASAE, St. Joseph, Mich. 49085. Cost is 50¢ per copy — or your ASAE Member Order Form.

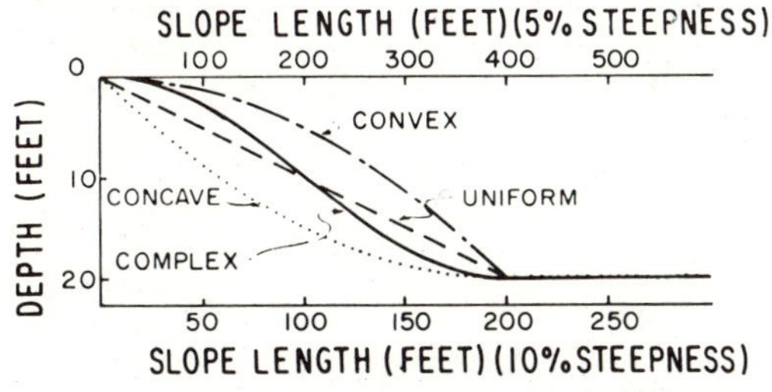

Fig. 1 Four slope shapes studied at 5 and 10 percent steepness, vertical scale is expanded. A flat area was assumed beyond the toe of the slopes

Effect of Slope Shape

Slope profiles of the different shapes develop differently because erosion depth along each varies considerably. The sediment loads during the first erosion period along the four shapes at 5 percent steepness are given in Fig. 2. The relative depth of erosion or deposition is indicated by the difference between the load at successive locations. Where the slope of the curve is positive, erosion is occurring; where negative, deposition is occurring.

Although the sediment load was low at the upper portion of the convex slope where the steepness was small, it increased rapidly to a maximum of 0.92 units as slope steepness and slope length (quantity of runoff) increased. In contrast, the concave slope had its greatest steepness where the least runoff occurred — at the upper part of the slope. The steepness then decreased as the runoff increased so that that total sediment load at any point was low, with a maximum of only 0.14 units. Furthermore, the total sediment movement on the slope (indicated by the area beneath the curves) was much lower for the concave than for the convex slope. Thus, the concave shape not only had a lower sediment load, but also it was changed less rapidly than the convex.

The sediment load of the uniform slope increased steadily as length increased, reaching a maximum of 0.37 units — higher than the maximum for the concave slope but much lower than for the convex one. The sediment load of the complex slope reached a maximum of 0.32 units — nearly equal to the maximum of the uniform shape — about two-thirds the way downslope. Here the slope was still steep and the runoff large. The load then decreased as the slope steepness decreased, even though the runoff increased. For equal elevation changes, the sediment loads for these shapes were similar at 10 percent steepness.

If sediment is a problem, the load off the bottom of the slopes is significant. For the concave and complex slopes,

This article is reprinted from AGRICULTURAL ENGINEERING (vol. 50, no. 9, pp. 522-523, September 1969), the Journal of the American Society of Agricultural Engineers, St. Joseph, Michigan

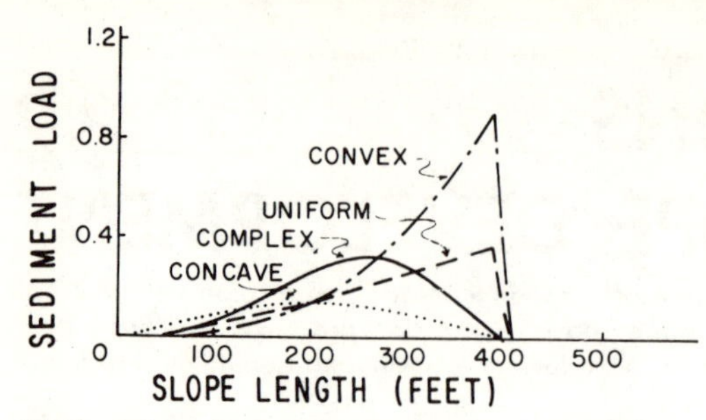

Fig. 2 Sediment load or total erosion (tons per foot of slope width) along the four original slope shapes of 5 percent steepness during their first erosion period. The sediment load was increasing and net erosion taking place where curves have positive slopes. Deposition occurred where curves have negative slopes. The steepness at any point along a curve indicates the depth of erosion or deposition

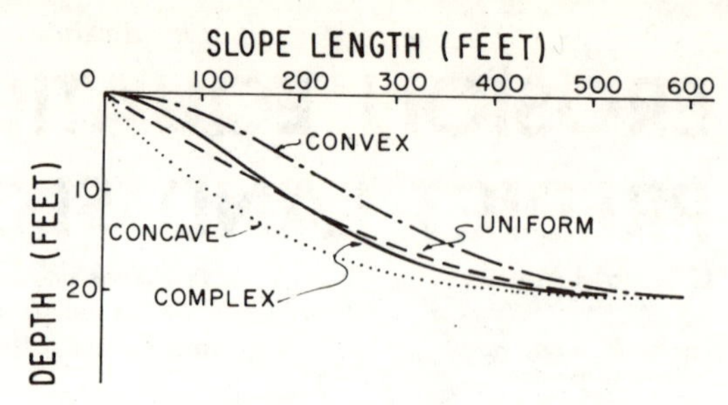

Fig. 3 Slope shapes that developed after 50 periods of erosion from the initial 5 percent slopes of Fig 1

most of the upper-slope sediment load was deposited along the lower one-third of the slope. Little soil movement was indicated beyond the slope base. Sediment load at the base of the uniform and particularly the convex slopes was high. Thus shaping at least the lower portions of long slopes to a concave shape can be quite helpful in reducing sediment losses.

The maximum eroded depths (based on uniform removal) occurred where the sediment load curves of Fig. 2 were of maximum positive slope. This maximum depth was least on the concave slope (0.019 ft) followed by the uniform slope (0.028 ft), the complex slope (0.044 ft) and the convex slope (0.129 ft). At 10 percent steepness, all depths were greater but in the same sequence.

The slope shapes after 50 periods of erosion are shown in Fig. 3. The profile of the convex shape changed the most, the concave shape the least. The uniform slope developed a concave profile; the initially complex slope also developed a concave profile except at the very upper part of the slope. After several hundred periods of erosion, *all* slope shapes tended strongly toward *concave* profiles.

As erosion progressed over succeeding periods, erosion rates decreased because of slope shape changes during the intervening time. Thus erosion intensity tends to diminish with time if other conditions remain the same.

Effect of Slope Steepness

Steeper slopes change shape more rapidly because of their greater depth of erosion per unit of time. At steepnesses averaging 10 percent, shapes changed about twice as fast as the same shapes at 5 percent. However, the sediment loads for the 5 and 10 percent steepnesses were nearly the same at equivalent elevations downslope. The effect of increased steepness on erosion for the 10 percent slope was approximately compensated by the decreased length and consequent reduced runoff. Thus, sediment losses at the toe of slopes with different average steepnesses but the same characteristic shape will be similar for a given elevation change. A moderately steep slope with a flattened portion for deposition beyond the toe may therefore be preferable to a less steep, longer slope with no flattened portion at the end.

To Apply Results

Information on erosion rates along different slope shapes can be applied to various field conditions. Locations of critical erosion for installing terraces or other erosion control measures can be determined for areas with limited good soil. Major sediment sources can be identified so that reservoirs and water courses can be protected from excess sediment. For construction sites, these data indicate those shapes that will minimize sediment movement and slope shape changes.

Suppose a large commercial building is to be constructed on a hillside (Fig. 4). To keep the main floor level, the area will be reshaped so that the slope below the building will be 400 ft long with an average steepness of 5 percent from the building edge to a level residential area beyond the slope base. Runoff from the building area and above will be removed underground. The residential area itself has ditches and storm sewers for the runoff but can handle little eroded sediment.

With a uniform slope from the building to the toe of the hillside, up to 100 tons of sediment per hundred feet of slope width may be expected annually from this site until cover is established. The expected average erosion depth at the slope base will be about 1 in. annually. Rills may be much deeper. For a complex shape on this same area, the sediment loss will be low but the average erosion depth will exceed 1 in. per year near the center of the slope until cover is established. However with a concave slope, the sediment loss will be low *and* the erosion depth shallow. Thus a concave profile would erode less, produce less sediment, and change shape less than the more common uniform or complex slopes. However, any of these would be superior to a pronounced convex slope. ••

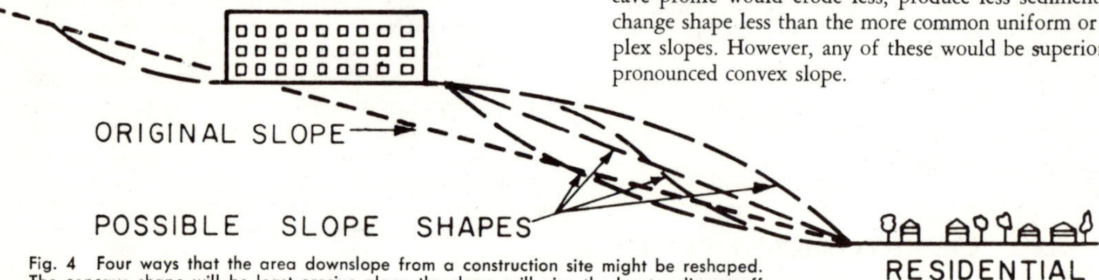

Fig. 4 Four ways that the area downslope from a construction site might be reshaped. The concave shape will be least erosive along the slope, will give the least sediment off the toe of the slope, and will change shape the least

References

Bradley, W. C. (1963). Large-scale exfoliation in massive sandstones of the Colorado Plateau. *Geological Society of America Bulletin,* **75**, 519–528.

Bryan, K. (1940). The retreat of slopes. *Annals of the Association of American Geographers,* **30**, 254–268.

Carson, M. A., and Kirkby, M. J. (1972). "Hillslope form and process." Cambridge University Press, Cambridge. 475 p.

Chorley, R. J., Dunn, H. J., and Beckinsale, R. P. (1964). "The history of the study of landforms," Vol. 1, "Geomorphology Before Davis." Methuen & Co., London. 678 p.

Commission pour l'étude de l'evolution des Versants (1956). *"Premier Rapport de la Commission: International Geographical Congress (Rio de Janeiro)."* 155 p.

——— (1960). International contributions to the morphology of slopes. *Zeit. Geomorphologie,* Supplementband **1**, 240 p.

——— (1963). New international contributions to the morphology of slopes. *Nachrichten der Akademie der Wissenschaften in Göttingen, Mathematisch-Physikalische Klasse,* **13**, 293 p.

——— (1964). International advancement in research on slope morphology. *Zeitschrift für Geomorphologie,* Supplementband **5**, 238 p.

——— (1967). L'Evolution des versants. *Les Congrès et Colloque de l'Université de Liège,* **40**, 384 p.

——— (1970). New contributions to slope evolution, *Zeitschrift für Geomorphologie,* Supplementband **9**, 186 p.

Demek, J. (1972). "Manual of detailed geomorphological mapping." Prague Academia. 344 p.

Dylik, J. (1968). The significance of the slope in geomorphology. *Societé Sciences et Lettres de Lodz, Bull.,* **19**, 1–19.

Hack, J. T. (1960). Interpretation of erosional topography in humid temperate regions. *American Journal of Science,* **258-A**, 80–97.

Hirano, M. (1968). A mathematical model of slope development. *Journal of Geosciences, Osaka University,* **11**, 13–52.

Johnson, D. W. (1954). *"Geographical essays by William Morris Davis."* Dover Publications, New York. 777 p.

Melton, M. A. (1965). Debris-covered hillslopes of the southern Arizona desert—consideration of their stability and sediment contribution. *Journal of Geology,* **73**, 715–729.

Penck, W. (1953). "Morphological analysis of land forms." Macmillan & Co., London. 429 p.

Rapp, A. (1960). Recent development of mountain slopes in Karkevagge and surroundings, northern Scandinavia. *Geografiska Annaler,* **42**, 71–200.
Roth, E. S. (1965). Temperature and water content as factors in desert weathering. *Journal of Geology,* **73**, 454–468.
Scheidegger, A. E. (1970). "Theoretical geomorphology," 2nd ed. Springer-Verlag, Berlin. 435 p.
Schumm, S. A. (1966). The development and evolution of hillslopes. *Journal of Geological Education,* **14**, 98–104.
——— and Chorley, R. J. (1964). The fall of Threatening Rock. *American Journal of Science,* **262**, 1041–1054.
——— and Chorley, R. J. (1966). Talus weathering and scarp recession in the Colorado Plateaus. *Zeitschrift für Geomorphologie,* **10**, 11–36.
Simons, M. (1962). The morphological analysis of landforms: a new review of the work of Walther Penck. *Institute of British Geographers Transactions,* **31**, 1–14.
Stoddart, D. R. (1969). Climatic geomorphology: review and re-assessment, *in* "Progress in geography," Vol. 1. Edward Arnold, London, p. 160–222.
Tylor, A. (1875). Action of denuding agencies. *Geological Magazine,* **2**, 433–473.
Von Engeln, O. D. (1940). Symposium: Walther Penck's contribution to geomorphology. *Annals of the Association of American Geographers,* **30**, 219–289.
Young, A. (1972). "Slopes." Oliver & Boyd, Edinburgh. 288 p.

Author Citation Index

Subject Index